RECOMBINANT MOLECULES:
Impact on Science and Society

MILES INTERNATIONAL SYMPOSIUM SERIES

Number 10: **Recombinant Molecules: Impact on Science and Society**
Roland F. Beers, Jr., and Edward G. Bassett, editors. 556 pp., 1977.

Number 9: **Cell Membrane Receptors for Viruses, Antigens and Antibodies, Polypeptide Hormones, and Small Molecules**
Roland F. Beers, Jr., and Edward G. Bassett, editors. 560 pp., 1976.

Number 8: **The Role of Immunological Factors in Infectious, Allergic, and Autoimmune Processes**
Roland F. Beers, Jr., and Edward G. Bassett, editors. 560 pp., 1976.

Recombinant Molecules:
Impact on Science and Society

Miles International Symposium Series
Number 10

Editors:

Roland F. Beers, Jr., M.D., Ph.D.
Miles Laboratories, Inc.
Elkhart, Indiana

Edward G. Bassett, Ph.D.
Miles Laboratories, Inc.
Elkhart, Indiana

Raven Press ■ New York

Raven Press, 1140 Avenue of the Americas, New York, New York 10036

Made in the United States of America

Main entry under title:

Recombinant molecules.

(Miles international symposium series; no. 10)
Proceedings of the 10th Miles international symposium held at the Massachusetts Institute of Technology and sponsored by Miles Laboratories.
Includes bibliographical references and index.
1. Genetic recombination--Congresses. 2. Genetic engineering--Congresses. 3. Genetic engineering--Social aspects--Congresses. I. Beers, Roland R. II. Bassett, Edward Graham, 1927- III. Miles Laboratories, inc., Elkhart, Ind. IV. Series. [DNLM: 1. Recombination, Genetic--Congresses. W3 MI543 no. 10 1976 / QH443 R311 1976]
QH443.R4 575.2 76-5675
ISBN 0-89004-131-8

Preface

One of the major events in the history of biological research occurred when it was found that excised segments of genetic material (DNA) from two different species could be annealed *in vitro* to form a hybrid DNA molecule which, on reintroduction into the cell, could impose entirely new genetic controls on that cell. The technology that enables the molecular basis of gene expression and heredity to be established, while providing a foundation for the creation of new organisms with desired genetic characteristics, has evoked a serious concern among scientists and laymen. This concern emanates from the theoretic creation of unique forms of agents of infection (or those adversely affecting the environment) whose biological properties cannot be completely predicted.

To provide a vehicle for a discussion of the scientific and societal ramifications of this technology, Miles Laboratories, Inc. sponsored the Tenth Miles International Symposium held at the Massachusetts Institute of Technology. Actively participating and sharing views during this conference were involved scientists from all over the world. These proceedings are the papers delivered at this 3-day conference.

The editors extend grateful thanks to Frank E. Young, M.D., Ph.D., Chairman of the Program Committee, and to members of that Committee who also chaired individual sessions. Acknowledgment is also given to Walter A. Compton, M.D., Chairman and Chief Executive Officer of Miles Laboratories, Inc., whose interest and encouragement have made these symposia a traditional annual event.

We are grateful to the authors for assisting in the prompt publication of this volume. Within the time allowed for editing, it was not possible for each discussant to review his remarks; if error or misinterpretation of the discussions has resulted, the editors take full responsibility.

Roland F. Beers, Jr.
Edward G. Bassett

Contents

Section F: Societal Impact—Issues and Policies

Contributors

Laura Alexander
Department of Microbiology
Institute of Dental Research
University of Alabama, Birmingham
Birmingham, Alabama 35294

David Baltimore
Center for Cancer Research
Massachusetts Institute of Technology
Cambridge, Massachusetts 02139

Roland F. Beers, Jr.
Miles Laboratories, Inc.
Elkhart, Indiana 46514

Paul Berg
Department of Biochemistry
Stanford University School of Medicine
Stanford, California 94305

Mary Betlach
Department of Biochemistry and Biophysics
University of California, San Francisco
San Francisco, California 94143

Frederick R. Blattner
Laboratory of Genetics
University of Wisconsin
Madison, Wisconsin 53706

Ann E. Blechl
Laboratory of Genetics
University of Wisconsin
Madison, Wisconsin 53706

Francisco Bolivar
Department of Biochemistry and Biophysics
University of California, San Francisco
San Francisco, California 94143

Herbert W. Boyer
Department of Biochemistry and Biophysics
University of California, San Francisco
San Francisco, California 94143

Winston J. Brill
Department of Bacteriology
College of Agricultural and Life Sciences
University of Wisconsin
Madison, Wisconsin 53706

H. Bursztyn-Pettegrew
Department of Genetics
Stanford University School of Medicine
Stanford, California 94305

Felipe Cabello
Department of Medicine
Stanford University School of Medicine
Stanford, California 94305

John Carbon
Department of Biological Sciences
University of California
Santa Barbara, California 93106

Peter S. Carlson
Department of Crop and Soil Science
Michigan State University
East Lansing, Michigan 48824

Annie C. Y. Chang
Department of Medicine
Stanford University School of Medicine
Stanford, California 94305

M.-D. Chilton
Department of Microbiology
University of Washington
Seattle, Washington 98195

Josephine E. Clark
Department of Microbiology
Institute of Dental Research
University of Alabama, Birmingham
Birmingham, Alabama 35294

Louise Clarke
Department of Biological Sciences
University of California
Santa Barbara, California 93106

Edward C. Cocking
Department of Botany
University of Nottingham
Nottingham, NG7 2RD England

Stanley N. Cohen
Department of Medicine
Stanford University School of Medicine
Stanford, California 94305

Roy Curtiss, III
Department of Microbiology
Institute of Dental Research
University of Alabama, Birmingham
Birmingham, Alabama 35294

Edward C. Dart
Bioscience Group, Corporate Laboratory
Imperial Chemical Industries, Ltd.
Runcorn, Cheshire, England

Dana Davoli
Department of Biological Chemistry
Harvard Medical School
Boston, Massachusetts 02115

Igor B. Dawid
Department of Embryology
Carnegie Institute of Washington
Baltimore, Maryland 21210

M. Drummond
Department of Microbiology
University of Washington
Seattle, Washington 98195

Craig Duncan
Department of Microbiology
University of Rochester
Rochester, New York 14642

Argiris Efstratiadis
Biological Laboratories
Harvard University
Cambridge, Massachusetts 02138

S. D. Ehrlich
Institut de Biologie Moleculaire
Faculté de Science
Paris 75005, France

Stanley Falkow
Department of Microbiology and Immunology
University of Washington School of Medicine
Seattle, Washington 98195

George C. Fareed
Department of Microbiology and Immunology and
Molecular Biology Institute
University of California, Los Angeles
Los Angeles, California 90024

David Figurski
Department of Biology
University of California, San Diego
La Jolla, California 92093

Stephen P. Goff
Department of Biochemistry
Stanford University School of Medicine
Stanford, California 94305

Raúl Goldschmidt
Department of Microbiology
Institute of Dental Research
University of Alabama, Birmingham
Birmingham, Alabama 35294

Howard M. Goodman
Department of Biochemistry and Biophysics
University of California, San Francisco
San Francisco, California 94143

M. P. Gordon
Department of Biochemistry
University of Washington
Seattle, Washington 98195

David J. Grunwald
Laboratory of Genetics
University of Wisconsin
Madison, Wisconsin 53706

Dean H. Hamer
Laboratory of Molecular Genetics
National Institute of Child Health and Human Development
National Institutes of Health
Bethesda, Maryland 20014

Donald R. Helinski
Department of Biology
University of California, San Diego
La Jolla, California 92093

Vickers Hershfield
Department of Microbiology
Duke University Medical Center
Durham, North Carolina 27710

Herbert L. Heyneker
Department of Biochemistry and Biophysics
University of California, San Francisco
San Francisco, California 94143

J. Charles Hsu
Department of Microbiology
Institute of Dental Research
University of Alabama, Birmingham
Birmingham, Alabama 35294

Sheila C. Hull
Department of Microbiology
Institute of Dental Research
University of Alabama, Birmingham
Birmingham, Alabama 35294

Christine Ilgen
Department of Biological Sciences
University of California
Santa Barbara, California 93106

Matsuhisa Inoue
Department of Microbiology
Institute of Dental Research
University of Alabama, Birmingham
Birmingham, Alabama 35294

Leon Jacobs
Office of Collaborative Research
National Institutes of Health
Bethesda, Maryland 20014

Laurence H. Kedes
Department of Medicine
Stanford University School of Medicine
Palo Alto, California 94305

J. Lederberg
Department of Genetics
Stanford University School of Medicine
Stanford, California 94305

Seymour Lederberg
Division of Biology and Medicine
Brown University
Providence, Rhode Island 02912

Larry J. Maturin, Sr.
Department of Microbiology
Institute of Dental Research
University of Alabama, Birmingham
Birmingham, Alabama 35294

Georg Melchers
Max-Planck-Institut für Biologie
74 Tübingen, West Germany

D. Merlo
Department of Microbiology
University of Washington
Seattle, Washington 98195

Richard J. Meyer
Department of Biology
University of California, San Diego
La Jolla, California 92093

A. Montoya
Department of Microbiology
University of Washington
Seattle, Washington 98195

Robert Moody
Department of Microbiology
Institute of Dental Research
University of Alabama, Birmingham
Birmingham, Alabama 35294

David D. Moore
Laboratory of Genetics
University of Wisconsin
Madison, Wisconsin 53706

John F. Morrow
Department of Biological Chemistry
Harvard Medical School
Boston, Massachusetts 02115

Kenneth Murray
Department of Molecular Biology
University of Edinburgh
Edinburgh EH9 3JR, Scotland

Noreen E. Murray
Department of Molecular Biology
University of Edinburgh
Edinburgh EH9 3JR, Scotland

Daniel Nathans
Department of Microbiology
Johns Hopkins University School of Medicine
Baltimore, Maryland 21205

E. W. Nester
Department of Microbiology
University of Washington
Seattle, Washington 98195

Alexander L. Nussbaum
Boston Biomedical Research Institute
Boston, Massachusetts 02114

Dennis A. Pereira
Department of Microbiology
Institute of Dental Research
University of Alabama, Birmingham
Birmingham, Alabama 35294

Michel Perricaudet
Department of Molecular Biology
Institut Pasteur
75015 Paris, France

Ulf Pettersson
Department of Microbiology
The Wallenberg Laboratory
Uppsala University
75277 Uppsala, Sweden

Lennart Philipson
Department of Microbiology
The Wallenberg Laboratory
Uppsala University
75277 Uppsala, Sweden

Barry Ratzkin
Department of Biological Sciences
University of California
Santa Barbara, California 93106

Thomas B. Rice
Plant Genetics Group
Pfizer Central Research
Groton, Connecticut 06340

Brian M. Richards
Molecular Biological Research Laboratories
G. D. Searle & Co.
High Wycombe, Buckinghamshire, England

Mark H. Richmond
Department of Bacteriology
University of Bristol
Bristol BS8 1TD, England

Richard J. Roberts
Cold Spring Harbor Laboratory
Cold Spring Harbor, New York 11724

Raymond L. Rodriguez
Department of Biochemistry and Biophysics
University of California, San Francisco
San Francisco, California 94143

James W. Schumm
Laboratory of Genetics
University of Wisconsin
Madison, Wisconsin 53706

D. Sciaky
Department of Microbiology
University of Washington
Seattle, Washington 98195

V. Sgaramella
Laboratorio di Genetica Biochemica ed Evoluzionistica, C.N.R.
27100 Pavia, Italy

John Shine
Department of Biochemistry and Biophysics
University of California, San Francisco
San Francisco, California 94143

Magdalene So
Department of Microbiology and Immunology
University of Washington School of Medicine
Seattle, Washington 98195

I. Stroynowski
Department of Genetics
Stanford University School of Medicine
Stanford, California 94305

Waclaw Szybalski
McArdle Laboratory for Cancer Research
University of Wisconsin
Madison, Wisconsin 53706

Natalie M. Teich
Imperial Cancer Research Fund Laboratories
Lincoln's Inn Fields
London WC2A 3PX, England

Charles A. Thomas, Jr.
Department of Biological Chemistry
Harvard Medical School
Boston, Massachusetts 02115

Kenneth Timmis
Max-Planck-Institut für Molekulare Genetik
1 Berlin 33 (Dahlem), West Germany

Pierre Tiollais
Department of Molecular Biology
Institut Pasteur
75015 Paris, France

John Tooze
European Molecular Biology Organization
6900 Heidelberg 1, West Germany

Frances R. Warshaw
The Group on Genetics and Social Policy
Boston Area Science for the People
Cambridge, Massachusetts 02139

Robin A. Weiss
Imperial Cancer Research Fund Laboratories
Lincoln's Inn Fields
London WC2A 3PX, England

Peter K. Wellauer
Swiss Cancer Research Institute
1011 Lausanne, Switzerland

Bill G. Williams
Laboratory of Genetics
University of Wisconsin
Madison, Wisconsin 53706

Gary A. Wilson
Department of Microbiology
University of Rochester
Rochester, New York 14642

John M. Wozney
Department of Biological Chemistry
Harvard Medical School
Boston, Massachusetts 02115

Frank E. Young
Department of Microbiology
University of Rochester School of Medicine and Dentistry
Rochester, New York 14642

Recombinant Molecules: Impact on Science and Society, edited by R. F. Beers, Jr. and E. G. Bassett. Raven Press, New York © 1977.

1. Introduction

Roland F. Beers, Jr.

Miles Laboratories, Inc., Elkhart, Indiana 46514

One of the original purposes for establishing the annual Miles symposium series was to follow and record the evolution of the science of molecular biology into a technology of molecular biology. The intellectual tour de force of twentieth century biology had taken place: the discovery of the structural basis for genetic transfer from one generation to the next and the beginning of an understanding of possible mechanisms for translating genetic information to somatic structure and function. The double helix of DNA and the genetic code for amino acids formed the basis for the central dogma of molecular biology in 1967, the year of the first Miles symposium on the subject of messenger RNA. Today's symposium is a clear recognition of the fact that molecular biology is on the threshold of becoming a technology.

Nevertheless, even though the current status of molecular biology has been the expectation of both the scientist and the public supporting him, there is today a growing feeling of uneasiness and outright fear of this new technology, a concern that received its first major public recognition at the now famous Asilomar conference in February, 1975. The immediate issue, like that of the first nuclear chain reaction, is the uncertainty of the potential for what is popularly referred to as genetic engineering. Four technologies from molecular recombinant research with decreasing probability of success and increasing lag time until success is achieved appear to me to be:

1. genetic modification of microbial organisms for the purpose of increasing the quality and quantity of a desired microbial product such as an enzyme;
2. genetic modification of higher plant organisms to increase their productivity with respect to yield, caloric and nutritional content, and the special challenge of developing cereal having symbiotic relationships with nitrogen-fixing microorganisms;
3. transfer of animal (mammalian) genomes to microorganisms for the synthesis of specific proteins or hormones such as insulin; and
4. genetic transformation of somatic cells to correct genetic defects such as sickle cell anemia or phenylketonuria through transformation

of the cells *in vitro* or *in vivo* with virus vectors. The risks of harmful consequences to man and his environment appear to increase in the same order.

The analogy with nuclear physics is not inappropriate, but the uncertainties of potential for risks appear to be greater than those predicted or encountered after the first successful nuclear chain reaction. The threat of nuclear technology was initially identified as a human threat, that is, deliberate use of the technology for destructive purposes. Later, with the advent of nuclear power generation, the risks of errors in judgment or accident received the major emphasis. The primary emphasis of risks in molecular recombinant research is addressed to errors of judgment and accident. This places the issue of risks in a slightly different ethical framework.

Most of the discussions and deliberations held in the past and continued in this volume have been concerned with the mechanisms by which society reaches decisions for minimizing the risks and then establishing standards of conduct to implement those decisions. Some thought has been given to the appropriateness of the goals and benefits to be accrued from molecular recombinant research, although the motivations behind these goals are highly diversified. It is appropriate at this time to reflect on the philosophical ramifications of this new technology with special attention given to proper historical perspective. So revolutionary is this new technology that there is a tendency to consider it a unique event in history without precedent. In fact, there is a precedent that has run throughout the entire history of mankind and should be examined in terms of society's attitude toward and response to the uncertainties of any new technology that threatens to alter traditional beliefs and status quo or promises to bring forth a new utopia. This social environment in which the molecular recombinant research is carried out should be recognized and understood by the practitioners of research and by those who control and support this research. It should also be recognized by those groups in society who assume an adversary position with respect to decisions for goals and their implementation by society.

Two elements of society's attitude and response that strike me as significant are (a) an anti-intellectual attitude and (b) an unjustified expectation that the new technology can be used to solve major problems of society without concurrent institutional and behavioral reforms. I use the term intellectual to identify the rational activities of the human mind as distinguished from any philosophical interpretation of what is intellectual or nonintellectual.

Man's response to uncertainty generated by knowledge is recorded in biblical times in the first few chapters of Genesis. Knowledge provides the basis for control over the present and the future, that is, power. Yet, because knowledge is often incomplete, so is the power it provides, hence the source of the uncertainty. Indeed, the incompleteness of that power has led to the creation of the major religions of the world. Depending on one's religious convictions, man either created or discovered a transcendental Being to compensate or make allowances for the incompleteness of his own knowledge and power. Intercession in man's behalf has been sought through supplication and ritualistic sacrifices, a form of power bargaining. Inevitably, this struggle for certainty through power developed a strong ethical character that eventually became highly legalistic in its interpretation and enforcement. The key ethical component that is today as important as at the time of Genesis is the assignment and acceptance of the moral responsibility for the possession of that power.

Two major institutions of civilization evolved in parallel and inevitable conflict. In broad terms, one is religious or transcendental, the other is intellectual or scientific. Each proclaimed itself as the authority for the ultimate source of knowledge, and each asserted its right to use that knowledge in its quest for power and certainty. The boundary of these two areas has, of course, shifted dramatically during the last 200 years in favor of the intellectually based institutions: science and technology.

However, the struggle over the authority for power and the responsibility for the use of that power still remains a major struggle today. Indeed, the dilemma facing mankind is an imponderable paradox: absolute power controlled by either of these institutions contains the seeds of destruction not only of civilization but of man, the species. Authoritarian institutions are on the increase worldwide. The transcendental Being may not be recognized as such, but any ideology whose power resides in its position of authority provides the basis for governing a society as if a transcendental Being existed and was not accountable to the critical intellectual processes of man. On the other hand, the imperfect state of man's knowledge and, hence, his capacity to predict and control his future is equally dangerous if this limitation is not clearly recognized by society. Scientism is the ideology that does not recognize those limitations.

The struggle today is not over the ownership of the source of knowledge but rather over the assignment of the responsibility for the use of that knowledge. Perhaps the clearest example of this is seen in the current posture of the Roman Catholic Church toward the goals and methods of controlling the size of the human population. Underlying its refusal to sanction a technological solution is the explicit premise that the responsibility for meeting this problem cannot be given to man. In other words, whatever conclusions man may reach regarding the need for controlling populations and the immediate as well as long-term means for meeting that need, the authority to assume this responsibility and, therefore, the power to implement the means are *de facto* denied to man.

This is an extreme case of what has become during the last half of the twentieth century a growing threat to the assertion that the human mind through its intellectual processes can indeed be the basis for man's assuming responsibility for his own fate. The proponents of moratoria on research for the purpose of restricting knowledge about man and his universe and the direct attacks on technological change are to be recognized as human responses to the fear of the unknown, but implicit in these responses are a distrust of the human intellect to perform adequately toward the threatened risks.

Compounding and confusing society's response to the new technology is its oversimplified belief that a technology can solve a major social problem without regard to the institutional and behavioral changes that must ultimately occur. The recognized inadequacy of food, clothing, and housing in two-thirds of the world population is identified as a problem in inadequate availability, distribution, and utilization of the resources of the earth relative to the population size and rate of growth. The solution to this problem is identified as an increase of these resources. I need not remind you of the basic fallacy underlying the assumptions of the "Green Revolution" except to emphasize that the complexity and cost of the technology were immensely greater than its proponents had considered. However, any technology, including genetic engineering, designed to increase the productivity of the earth can do no more than buy the time necessary for man to discover and design the proper solution of institutional and behavioral reform to obliterate the underlying cause, namely, unlimited growth.

If genetic engineering is to be used to continue this treadmill of mankind, where its objective is directed toward quantity rather than quality of life, then the tragedy of man lies in the contrast between the genius of his intellect to develop a technology as brilliant and beautiful as genetic engineering and the goals to which that technology is applied. It is analogous to the contrast between marvels of a color television system and the usual programs displayed on a TV screen. Yet, the analogy is not entirely appropriate. For in buying time, this new technology introduces an additional cost. If during this interim period a solution is not found and implemented, the magnitude of the ultimate catastrophe by virtue of the numbers of mankind involved will be even greater than if the technology had not been developed.

Thus, I must conclude these remarks by reminding you that as real and serious as the risks of molecular recombinant research are from a technical standpoint, a far more serious threat is the role such a technology will play in not affording man the opportunity to set his own house straight but in providing him a Cyrenaic sense of complacency about his future on this earth.

Recombinant Molecules: Impact on Science and Society, edited by R. F. Beers, Jr. and E. G. Bassett. Raven Press, New York © 1977.

2. Introduction to Section A: Technological Advances

Frank E. Young

Department of Microbiology, University of Rochester School of Medicine and Dentistry, Rochester, New York 14642

In this section and others in this volume we discuss the remarkable explosion of technological advances that now enable scientists to manipulate genomes and cross major species barriers. It is therefore important to realize that these elegant techniques have emanated from the initial discovery of transformation by Griffith in 1928. As with many unexpected observations, the implications of these early experiments were not appreciated by either the investigator or most of his contemporaries. Essentially Griffith demonstrated that a factor in a culture of heat-killed smooth *Pneumococcus* could transform a nonvirulent rough strain into a virulent smooth strain when both were injected subcutaneously into mice (7). Although he erroneously attributed this phenomenon to a phenotypic modification of the rough strain by a pabulum of capsular remnants, he described his work in sufficient detail to stimulate Avery to explore this unusual result. Subsequent work in Avery's laboratory by Dawson, Sia, and Alloway (1,6,19) demonstrated that this remarkable event could be produced *in vitro* by purified "fibers" obtained from crude extracts of smooth microorganisms. In a classic study published in 1944 (3), Avery and co-workers unequivocally established that DNA was the transforming principle. This capacity to isolate DNA, manipulate it *in vitro,* and introduce it into other cells was the essential foundation for the technological advances described in this volume.

Four major discoveries were required before investigators could readily manipulate DNA rather than merely adding purified DNA fragments to competent bacteria and subsequently selecting random recombinant events. The most significant advance resulted from the identification of enzymes responsible for the variation of bacteriophage-cell interactions that were discovered almost 25 years ago by Luria and Human (12). The class I enzymes identified by Meselson and Yuan (15) and Linn and Arber (11) have not been as important in recombinant molecule technology as the class II enzymes first observed in *Haemophilus influenzae* serotype d by Smith and Wilcox (20) and Kelly and Smith (10). As will be discussed by Dr. Roberts in this section, class II enzymes with different substrate specificities have been isolated from many organisms. Interestingly, a number of organisms have enzymes that recognize the same sequence of nucleotides. Roberts has suggested the term *isoschizomers* for enzymes from different organisms that recognize the same nucleotide sequence.

The development of both plasmid and bacteriophage vectors required the capacity to join the fragments from different species. Three major procedures have been employed. Dr. Sgaramella will describe the use of T4 ligase to join flush-ended DNA segments. This technique was a product of the systematic attempt by Khorana and Sgaramella (17) to synthesize DNA *in vitro*. A second method was developed to take advantage of the joining of cohesive ends generated by site-specific endonucleases (14,16). Still other workers have added a complementary stretch of nucleotides to each strand to provide specificity for the joining reactions as well as complementary sequences to promote ligation (8).

Another essential requirement for the development of genetic engineering was the transformation of the most extensively characterized genetic model system *Escherichia coli*. Although transformation of "colon bacilli" was described as early as 1947 by Boivin (4), genetic manipulation of *E. coli* was not accomplished on a routine basis until 1960. The primary work of Kaiser and Hogness (9) led to the transformation of *E. coli* with DNA from λdg in the presence of helper phage. Subsequent studies by Mandel and Higa (13) and Cohen and co-workers (5) resulted in the development of procedures to transfect and transform *E. coli* in the absence of helper phage. The investigations of Boyer and co-workers and others presented in this volume support the contention that a basic understanding of a microbial model system is essential for the application of genetics to medical research. In addition, these experiments will serve to illustrate the application of recombinant molecule technology to gene cloning through transformation of *E. coli*.

A fourth significant technologic advance was the development of rapid procedures to visualize DNA fragments and to subsequently isolate fragments in preparative amounts. Prior to the introduction of agarose-ethidium bromide electrophoresis by Sharp and co-workers (18), fragmentation of genomes could be measured only by changes in viscosity, differences observed in velocity sedimentation through sucrose gradients or polyacrylamide electrophoresis. Sharp's simple, rapid method has permitted direct visualization of DNA, calculation of molecular weights of the fragments, and development of recognition patterns for different enzymes reacting with reference viral DNA preparations such as λ, adenovirus-2, and SV40.

The final topic of this section focuses on the choice of the cell used in cloning experiments. Currently, two genetic systems are being exploited for recombinant molecule technology: *E. coli* and *B. subtilis*. Because *E. coli*, unlike *B. subtilis*, is a pathogen for humans, efforts were initiated following the Asilomar meeting to "enfeeble this organism and thereby minimize its survival outside the laboratory." Dr. Roy Curtiss will summarize the progress that he has made in "disarming" the laboratory strain *E. coli* K-12 to render it less capable of existing in nature. If the host can be enfeebled but still retain transformability, it may be possible to significantly reduce its biohazard. *B. subtilis,* a soil microorganism, does not usually produce infection in the uncompromised host. Because of its use in microbial fermentation, it may represent an alternate system. Young and co-workers have summarized the current status of the *B. subtilis* transformation system with particular emphasis on heterospecific transformation. They have established that genes from a bacteriophage (ϕ3T) that infects *B. subtilis* can be introduced into pMB9 to form a chimeric plasmid, pCD1. The bacteriophage gene, *thy*P3, can function in the cytoplasm of *E. coli,* can be reextracted as the chimeric plasmid, and can be used to transform *B. subtilis* at an extremely high frequency. The simultaneous and independent work of Ehrlich presented in this volume shows that the *thy*P3 gene can also be incorporated into pSC101.

Even prior to the discovery of the technology required for these manipulations, Aposhian (2) presented a provocative address in 1969 on "The Use of DNA for Gene Therapy—the Need, Experimental Approach and Implications." Although the imagined technology was subsequently proven to be far less effective than the contemporary approach, he appropriately described some of the imagined benefits and enunciated the concerns that have been raised about the application of recombinant molecule technology to genetic engineering and gene therapy. He also questioned "should the health science community begin to prepare the political and lay community for the moral and other implications of gene therapy?" This volume is designed to explore the many facets of recombinant molecule technology that have even broader implications than the fantasy of gene therapy. The basic technological advances and the potential opportunities for genetic analyses are legion. In addition, risks accompany the opening of Pandora's box. The exploration of benefits and risks of a new technology illustrates an increased awareness by scientists of their moral and ethical responsibilities to society. The studies to be presented in this section serve merely to illustrate our primitive attempts to manipulate genomes and to outline some of the major lines of research that will continue for many years.

REFERENCES

1. Alloway, J. L. (1932): The transformation in vitro of R pneumococci into S forms of different types by the use of filtered pneumococcus extracts. *J. Exp. Med.,* 55:91–99.
2. Aposhian, H. V. (1970): The use of DNA for gene therapy—the need, experimental approach and implications. *Perspect. Biol. Med.,* 14:98–108.
3. Avery, O. T., MacLeod, C. M., and McCarty, M. (1944): Studies on the chemical nature of the substance inducing transformation of pneumococcal types. I. Induction of trans-

formation by a deoxyribonucleic acid fraction isolated from pneumococcus types III. *J. Exp. Med.*, 79:137–158.
4. Boivin, A. (1947): Directed mutations in colon bacilli, by an inducing principle of deoxyribonucleic nature: its meaning for the general biochemistry of heredity. *Cold Spring Harbor Symp. Quant. Biol.*, 12:7–17.
5. Cohen, S. N., Chang, A. C. Y., and Hsu, L. (1972): Nonchromosomal antibiotic resistance in bacteria: genetic transformation of Escherichia coli by R-factor DNA. *Proc. Natl. Acad. Sci. U.S.A.*, 69:2110-2114.
6. Dawson, M. H., and Sia, R. H. P. (1931): In vitro transformation of pneumococcal types. I. A technique for inducing transformation of pneumococcal types in vitro. *J. Exp. Med.*, 54:681–699.
7. Griffith, F. (1928): The significance of pneumococcal types. *J. Hygiene*, 27:113–159.
8. Jackson, D. A., Symons, R. H., and Berg, P. (1972): Biochemical method for inserting new genetic information into DNA of simian virus 40: circular SV40 DNA containing lambda phage genes and the galactose operon of Escherichia coli. *Proc. Natl. Acad. Sci. U.S.A.*, 69:2904–2909.
9. Kaiser, A. D., and Hogness, D. S. (1960): The transformation of Escherichia coli with deoxyribonucleic acid isolated from bacteriophage λdg. *J. Mol. Biol.*, 2:392–415.
10. Kelly, T. J., and Smith, H. O. (1970): A restriction enzyme from Haemophilus influenzae. II. Base sequence of the recognition site. *J. Mol. Biol.*, 51:393–409.
11. Linn, S., and Arber, W. (1968): Host specificity of DNA produced by Escherichia coli. In vitro restriction of phage fd replicative form. *Proc. Natl. Acad. Sci. U.S.A.*, 59:1300–1306.
12. Luria, S. E., and Human, M. L. (1952): A non-hereditary, host-induced variation of bacterial viruses. *J. Bacteriol.*, 64:557–569.
13. Mandel, M., and Higa, A. (1970): Calcium dependent bacteriophage DNA infection. *J. Mol. Biol.*, 53:159–162.
14. Mertz, J. E., and Davis, R. W. (1972): Cleavage of DNA by RI restriction endonuclease generates cohesive ends. *Proc. Natl. Acad. Sci. U.S.A.*, 69:3370–3374.
15. Meselson, M., and Yuan, R. (1968): DNA restriction enzyme from E. coli. *Nature*, 217:1110–1114.
16. Sgaramella, V. (1972): Enzymatic oligomerization of bacteriophage P22 DNA and of linear simian virus 40 DNA. *Proc. Natl. Acad. Sci. U.S.A.*, 69:3389–3393.
17. Sgaramella, V., and Khorana, H. G. (1972): Studies on polynucleotides. CXII. Total synthesis of the structural gene for an alanine transfer RNA from yeast: enzymatic joining of the chemically synthesized polynucleotides to form the DNA duplex representing the nucleotide sequence 1 to 20. *J. Mol. Biol.*, 72:427–444.
18. Sharp, P. A., Sugden, B., and Sambrook, J. (1973): Detection of two restriction endonuclease activities in Haemophilus parainfluenzae using analytical agarose-ethidium bromide electrophoresis. *Biochemistry*, 12:3055–3063.
19. Sia, R. H. P., and Dawson, M. H. (1931): In vitro transformation of pneumococcal types II. The nature of the factor responsible for transformation of pneumococcus types. *J. Exp. Med.*, 54:701–710.
20. Smith, H. O., and Wilcox, K. W. (1970): A restriction enzyme from Haemophilus influenzae. I. Purification and general properties. *J. Mol. Biol.*, 51:379–391.

Recombinant Molecules: Impact on Science and Society, edited by R. F. Beers, Jr. and E. G. Bassett. Raven Press, New York © 1977.

3. The Construction of Molecular Cloning Vehicles

Herbert W. Boyer, Mary Betlach, Francisco Bolivar, Raymond L. Rodriguez, Herbert L. Heyneker, John Shine, and Howard M. Goodman

Department of Biochemistry and Biophysics, University of California, San Francisco, California 94143

INTRODUCTION

The technology for the molecular cloning of DNA fragments has its conceptual origin in the replicon model of Jacob et al. (9). This model delineated two elements necessary for the replication of a DNA molecule *in vivo* other than the battery of enzymes and proteins required for extension of the replication fork. These elements are (a) a region on the DNA molecule, equivalent to the origin of DNA replication, which can be recognized by another protein, and (b) the gene for this protein, which is found on the same molecule. Any DNA fragment inserted or recombined into such a molecule, without disturbing either function, can replicate as part of and under the inherent replication control of that molecule. For years geneticists have used *in vivo* recombination mechanisms to manipulate DNA fragments from one replicon to another, which often serves as a purification of various regions of a more complex genome. In recent years techniques for *in vitro* recombination of DNA fragments have expanded this procedure to extend beyond ordinary biological barriers (3,15).

There are several important technical components to *in vitro* recombinant technology which ultimately result in the insertion of DNA fragments from any source into replicons (viral or plasmid DNA molecules) and their recovery as replicating elements in bacteria. These components are:

1. the systematic dissection of the DNA molecules of interest with restriction endonucleases (see Roberts, *this volume,* for a discussion of these enzymes);
2. the rejoining of DNA fragments to an appropriate cloning vehicle (or replicon);
3. the transformation of a cell with the recombinant DNA and selection of cells containing the recombinant plasmid;

4. the identification and characterization of the "cloned" fragment of DNA. We would like to discuss here various aspects and developments in several of these components of recombinant DNA technology.

IN VITRO LIGATION OF DNA FRAGMENTS

Three methods can be used to ligate duplex DNA molecules *in vitro*. The short complementary termini generated by restriction endonucleases, such as the *Eco*RI endonuclease, can be stabilized at low temperatures (10 to 20°C), and the phosphodiester bonds can be reesterified by either *E. coli* or phage T4 DNA ligase (4,14). The concentration and size of the DNA fragments, of course, influence the relative intermolecular and intramolecular ligation events (4). As a result of the relative instability of the short complementary termini generated by restriction endonucleases (2 to 5 nucleotides), the circularization of molecules of length greater than 25 to 30 kilobases through these termini occurs infrequently. Since the establishment of a stable replicative plasmid after transformation of *E. coli* by linear molecules is less efficient than with circular molecules, the recovery of large recombinant plasmids is minimized by this *in vitro* technique.

In contrast, the construction of longer homopolymeric tails on DNA fragments afforded by the terminal transferase enzyme can result in the recovery of larger plasmids (8,10). The disadvantages of this approach in the past have been the additional enzymes required (that is, λ-exonuclease and terminal transferase) and the inability to excise the inserted fragment as an intact fragment by restriction endonuclease treatment of the recombinant plasmid.

The third approach takes advantage of the unique property of T4 DNA ligase, which can covalently join blunt-ended DNA molecules (19,20). This reaction, which can be stimulated about 10-fold by T4 RNA ligase, requires a rather high concentration of termini ($K_m = 5 \times 10^{-5}$ moles) in order to proceed with a reasonable rate (21). Blunt-ended DNA molecules can be generated by some restriction endonucleases (e.g., *Hin*dII) and randomly cleaved; blunt-ended DNA molecules can be generated by bovine pancreatic endonuclease I in the presence of Mn^{++} (11). DNA molecules with cohesive termini can be made blunt-ended by controlled S1 nuclease digestion or repaired with DNA polymerase (2). In some cases the covalent joining of combinations of blunt-ended DNA fragments creates new substrate sites for restriction endonucleases. As a caveat, we have found that unligated linear DNA can be circularized *in vivo* at a low frequency via recombination events not necessarily near the termini of the molecule. The recovery of plasmid molecules of this type can be minimized by a molecular selection, that is, destruction of linear DNA molecules by exonuclease V digestion prior to transformation.

BLUNT-END LIGATION WITH RESTRICTION LINKERS

A short (eight residues), synthetic, self-complementary DNA molecule which includes the *Eco*RI substrate site (5) can be covalently joined as a duplex DNA structure *in vitro* to other blunt-ended DNA molecules (2,7, 18). After treatment with the *Eco*RI restriction endonuclease, these covalently joined molecules can be received into the *Eco*RI site of a plasmid cloning vehicle. We have used this scheme (Fig. 1) to clone a chemically synthesized *lac* operator DNA segment (7). On the basis of this demonstration, three restriction site "linkers" containing either the *Eco*RI (CCGAATTCGG), *Bam*HI (CCGGATCCGG), or *Hin*dIII (CCAAGCTTGG) restriction site were chemically synthesized for cloning experiments (18). These molecules are self-complementary and form a high percentage of stable duplex structures at temperatures up to 20°C. They have been successfully ligated *in vitro* into multimers and have been ligated to blunt-ended plasmid DNA and cDNA molecules.

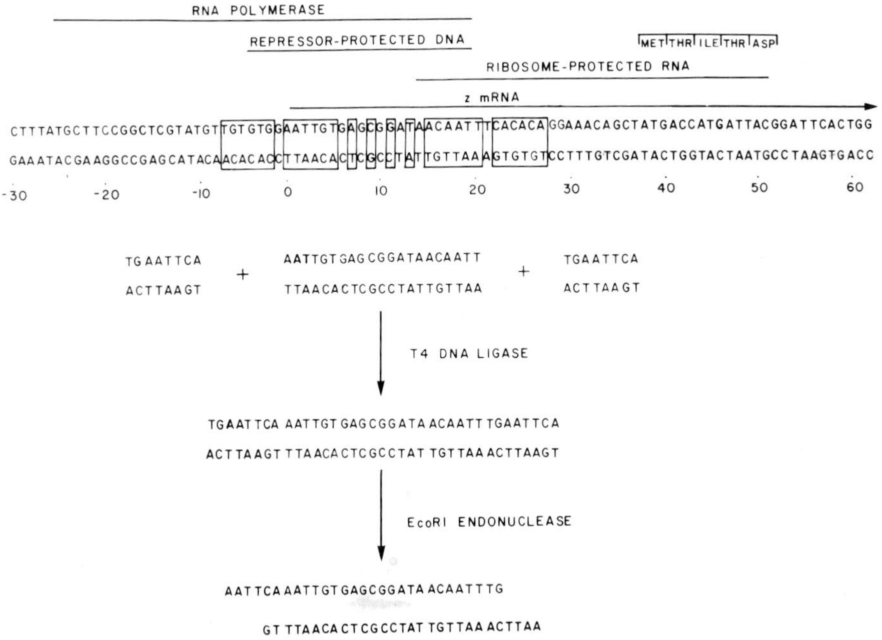

FIG. 1. Cloning of synthetic *lac* operator with *Eco*RI "linkers." The nucleotide sequence at the top of the figure represents the controlling elements of the *lac* operon. A synthetic, self-complementary octadeoxyribonucleotide containing the *Eco*RI substrate site (5) was blunt-end joined (7,19–21) to the synthetic *lac* operator sequence. Treatment with *Eco*RI endonuclease of the enzymatically joined fragments yields a fragment which can in turn be joined to a plasmid cleaved with *Eco*RI termini. (From ref. 7.)

This cloning innovation should be quite helpful for DNA sequence technology. Adequate DNA sequence determination procedures exist for DNA molecules of 100 to 150 nucleotides (12). These procedures rely on labeling one of the 5′ termini of a DNA molecule with ^{32}P, which can be accomplished by cleaving the DNA of interest with a restriction endonuclease and labeling the 5′ termini with ^{32}P-γ-ATP and then fractionating the labeled strands by a second restriction endonuclease treatment and/or gel electrophoresis. Of course, fragments of DNA larger than 200 to 300 nucleotide base pairs without restriction sites become impermeable to this approach. However, fragments of DNA of greater than two kilobases, for example, without a restriction site can be randomly cleaved with bovine pancreatic endonuclease I and ligated to restriction substrate linkers. These molecules can be cloned and the nucleotide sequence for a set of overlapping fragments can be determined.

PLASMID CLONING VEHICLES

The most critical component of the cloning technique is the cloning vehicle (17). One can itemize some of the essential features required for a useful cloning vehicle as follows. First of all, the plasmid must be purified easily and in sufficient quantities for experimentation. Therefore it would be desirable to delete as much of the nonessential plasmid genetic material as possible from the cloning vehicle. Secondly, one must have genetic markers associated with the plasmid which provide for the screening or selection of transformants with recombinant plasmids. Thirdly, the presence of a number of different restriction endonuclease substrate sites on the molecule would provide the necessary versatility for cloning DNA molecules. Toward this end, that is, the construction of a versatile plasmid cloning vehicle, we have been engaged in the construction and characterization of a series (Table 1) of plasmids with the above properties (1,2,17). In addition, we feel that the genetic characterization of these plasmids might provide information for the design and construction of a biologically containable plasmid for experiments deemed biohazardous. We will summarize the essential features of several plasmids we have constructed and present our analysis of two genetic features of these plasmids, namely, the nature of tetracycline resistance used as one of the genetic markers and the genetic control of the autonomous replication of these molecules. Our approach has been to extract, via recombinant DNA techniques, the minimal genetic component of the ColEl replication mechanism (6) and combine this with the minimal components for two selective markers (17), ampicillin resistance (A_p^r) and tetracycline resistance (T_c^r). Our current estimates for these components are approximately 900, 300, and 1,500 nucleotide base pairs.

The smallest plasmid we have constructed (pBR345) represents about one-sixth (~1,000 nanobase pairs) the mass of the ColE1 plasmid, from

TABLE 1. *Compilation of plasmid data*

Plasmid	MW (md)	MW (kb)	Phenotype conferred by plasmid[a]	Restriction sites: *Bam*HI	*Bgl*I	*Eco*RI	*Eco*RII	*Hae*II	*Hinc*II	*Hind*III	*Hpa*I	*Pst*I	*Sal*I	*Sma*I	Enzyme	Substrate
pSC101	5.8	8.9	Tc^r	1		1			4	1	1	0	1	1		
ColE1	4.2	6.5	Col^{imm}	0		1				0		2	0		*Bam*HI	G↓GATCC
pSF2124	7.4	11.4	Ap^r Col^{imm}	1		1				0		5	0		*Bgl*I	–
pMB8	1.7	2.6	Col^{imm}	0		1	5	2	0	0	0	2	0	1	*Eco*RI	G↓AATTC
pMB9	3.5	5.4	Tc^r Col^{imm}	1		1	9		2	1	1	0	1	1	*Eco*RI*	↓AATT
pBR312	6.7	10.3	Tc^r Ap^r Col^{imm}	2		1	11		4	1	1	3	1	1	*Eco*RII	↓CCTGG ↓CCAGG
pBR313	5.8	8.9	Tc^r Ap^r Col^{imm}	1	5	1			4	1	1	3	1	1	*Hae*II	PuGcGc↓Py
pBR317	5.4	8.3	Tc^r Ap^r Col^{imm}	1	5	1			3	1	1	2	1	1	*Hae*III	GG↓CC
pBR318	3.8	5.9	Tc^r Col^{imm}	1	3	1			3	1	1	1	1	1	*Hinc*II	GTPy↓PuAC
pBR320	1.9	2.9	Ap^r	1	1	1	3	5	2	1	0	1	1	0	*Hind*III	A↓AGCTT
pBR321	3.2	4.9	Tc^r Ap^r	1	4	1			2	1	0	1	1	0	*Hpa*I	GTT↓AAC
pBR322	2.6	4.0	Tc^r Ap^r	1	3	1	5	9	2	1	0	1	1	0	*Pst*I	CTGCA↓G
pBR333	1.2	1.8	Ap^r	0	1	1	1	1	1	1	0	1	0	0	*Sal*I	G↓TCGAC
pBR341	5.1	7.8	Kan^r (op)	0		4			2	1	0	0	1	0	*Sma*I	CCC↓GGG
pBR345	0.7	0.9	(op)	0	0	2	1	1	0	0	0	0	0	0		
pBR350	2.2	3.4	Tc^r	1	3	2		5	1	1	0	0	1	0		
pBR351	1.7	2.6	Ap^r	0	1	2	2	1	1	0	0	1	0	0		

[a] Tc^r, tetracycline resistance; Ap^r, ampicillin resistance; Kan^r, kanamycin resistance; Col^{imm}, colicin E1 immunity; op, lac operator.

which it was derived. The plasmid exists as a head-to-tail dimer of the 1-kilobase (kb) unit. pBR345 contains a synthetic *lac* operator (7) which imposes a β-galactosidase constitutive phenotype to cells which contain it. A current restriction map of pBR345 is presented in Fig. 2. The nucleotide sequence of this plasmid is being determined and two features of this plasmid are of particular interest: (a) the nature of the origin of replication (23,24) and (b) whether a structural gene is present on the molecule which contributes to the replication or maintenance of this molecule. The latter area of interest is the most straightforward in terms of experimental investigation since one can determine from the nucleotide base sequence if a structural gene is present and can also look for plasmid-controlled expression of such a protein (13).

There are several ways to probe this structural gene, once identified, for its essential role in DNA replication. With other miniplasmids of ColEl (pMB8 and pVH51) we have not detected a protein of molecular weight above 15,000 daltons (13), which suggests that perhaps one-third to one-

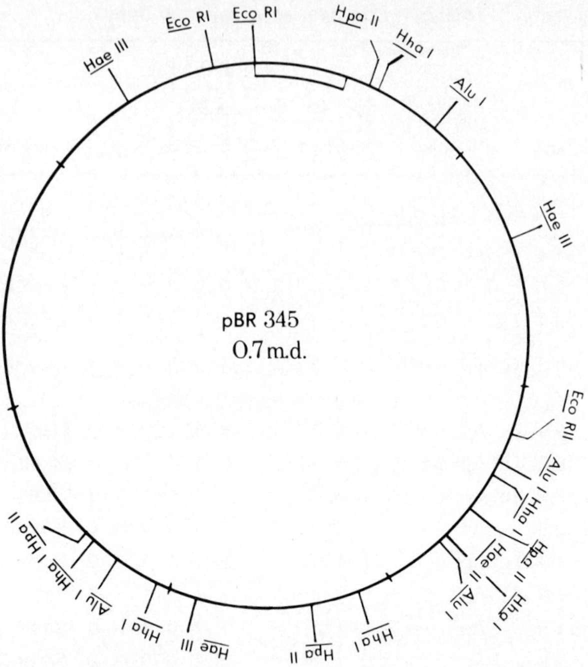

FIG. 2. Restriction map of pBR345. A monomeric representation of pBR345 with the relative positions of 20 restriction sites. Positions were determined by comparing the molecular weights of DNA fragments resulting from various combinations of restriction endonuclease digests after electrophoresis on polyacrylamide gels.

half of the pBR345 plasmid molecule might be dispensable as far as genetic information required for replication is concerned. Perhaps, at most, one modest-sized protein (10,000 to 15,000 daltons) must be provided for replication of the plasmid pBR345. If this putative structural gene can be physically separated from the origin of replication, then it should be possible to construct a plasmid cloning vehicle which is biologically containable. The identification of the origin of replication for pBR345 can be approached from several directions, namely, precise localization of the unique RNA primer for DNA synthesis associated with the ColE1 molecule (23,24). One can imagine that the origin of DNA replication for this molecule in its simplest form would consist of the 38 oligoribonucleotide primer (24) plus the promoter for the initiation of its transcription and a transcriptional stop sequence for its termination, or approximately 100 to 150 nucleotide base pairs.

Thus, the minimal size for an autonomously replicating plasmid vehicle with two selective markers such as A_p^r and T_c^r would be about 2 megadaltons or 3 kilobase pairs. If the putative structural gene required for the specific replication of this molecule can be removed and relocated in the host chromosome, for example, the size of the plasmid could be reduced further to about 1.5 megadaltons. The utility of having the smallest and therefore most defined replicating plasmid is severalfold. First of all, the relative yield of the cloned fragment is maximized and the separation of the cloned fragment from the vehicle becomes more manageable. Other advantages are offered through the decreased number of inherent restriction sites in the cloning vehicle. For example, we have removed from some plasmids (*see below*) several *Bam*HI or *Pst*I restriction sites in nonessential regions of the plasmid so that only one *Bam*HI or *Pst*I restriction site remains in the plasmid for cloning purposes (1,2). In other cases we have been able to insert chemically synthesized restriction endonuclease substrate sites into defined regions of the plasmid for cloning purposes (2).

The smallest and most versatile plasmid cloning vehicle we have constructed to date is pBR322 (2). This plasmid is 2.6 megadaltons in mass and contains two selective markers, A_p^r and T_c^r. Neither of these genes is on transposable elements and neither can be transposed from the cloning

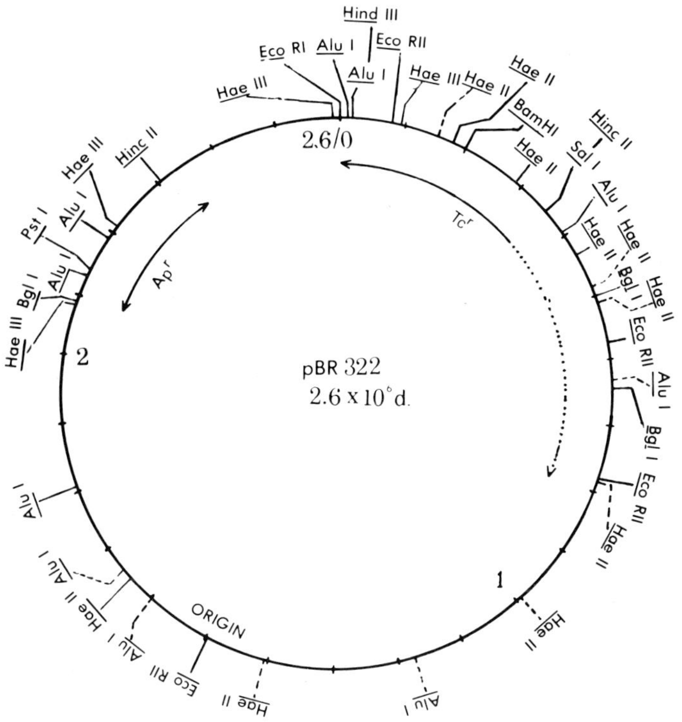

FIG. 3. Restriction map of pBR322. Positions of restriction sites were determined by estimating fragment sizes of combined restriction endonuclease digests. The sizes of the A_p^r and T_c^r genes were estimated from the measured values of the proteins thought to be associated with these antibiotic resistance mechanisms (1).

vehicle to another plasmid. pBR322 has one *Pst*I site located in the A_p^r gene, one *Bam*HI, one *Hin*dIII, and one *Sal*I site all located in the T_c^r gene. It has one *Eco*RI site which lies between the A_p^r and T_c^r genes. The relative positions of 36 restriction endonuclease substrate sites cleaved by 11 different endonucleases have been determined (Fig. 3).

CLONING DNA IN pBR322

The advantage of having two selective markers becomes obvious in view of the locations of the *Pst*I, *Sal*I, *Bam*HI, and *Hin*dIII restriction sites. Fragments of DNA generated by the *Pst*I endonuclease, for example, when cloned in the pBR322 plasmid result in the loss of A_p^r (2). T_c^r transformants derived from such an experiment can be identified as A_p^s. In addition, the positions of the cleavages made by the *Pst*I endonuclease leave protruding 3′-OH single strands of DNA which are excellent primers for terminal transferase addition of deoxyribonucleotides. If deoxyguanosine triphosphate is used as a substrate for tailing of the plasmid, and deoxycytosine triphosphate is used for tailing DNA fragments, then the *Pst*I restriction site can be restored upon reannealing of the two DNA components (16). For example, this approach has been successful for cloning cDNA (A. Dugaiczyk, *personal communication*). Thus, the cloned DNA can be precisely removed with *Pst*I endonuclease treatment of the recombinant plasmid.

Cloning of DNA fragments generated by *Hin*dIII, *Sal*I, or *Bam*HI restriction endonucleases can likewise be detected by screening A_p^r transformants for T_c^s (1,17). More importantly, T_c^s transformants can be quantitatively selected by inhibiting their growth with tetracycline and killing the actively growing T_c^r (nonrecombinant plasmid) transformants with D-cycloserine. In these cases, one of the antibiotic-resistant genes is inactivated by the cloning of a DNA fragment which is an important factor to be considered, in view of the biohazard imagined to exist by employing several antibiotic-resistant markers in a cloning vehicle. Other plasmids can be used for cloning *Eco*RI endonuclease–generated fragments of DNA (e.g., pMB9 or pBRH3; *see below*). The other advantage to this cloning vehicle is the possibility of using combinations of two different restriction sites for cloning fragments of DNA. In particular, this can be used for subcloning large fragments of DNA for sequence analysis.

GENETIC CONTROL OF T_c^r IN pSC101 AND THE pMB9 SERIES OF PLASMIDS

The tetracycline-resistant genotype of pMB9, pBR312, and its derivatives (pBR313, pBR322, etc.) was derived from the plasmid pSC101 (17). We found initially that the cloning of DNA fragments in the *Hin*dIII site of pBR313 and pBR322 resulted in a T_c^s phenotype. The report (M. Ptashne, *personal communication*) of a cloned fragment in the *Hin*dIII site of pMB9 which did not inactivate tetracycline resistance prompted us to initiate a more detailed analysis of the tetracycline resistance conferred on cells by the plasmids pSC101 and pMB9. We would like to summarize our current understanding of this antibiotic resistance mechanism (most of which is derived from the application of cloning technology and serves to illustrate the fourth component of this technology) and the effects of cloning fragments of DNA in several restriction sites associated with the T_c^r genes.

E. coli cells harboring the plasmid pSC101 (3 to 5 copies per cell) exhibit a basal level of resistance to tetracycline which can be further induced by exposure to tetracycline (22). Uninduced cells plate with 50% efficiency on enriched medium containing about 18 μg/ml of tetracycline, whereas induced cells plate with 50% efficiency on medium containing 36 μg/ml of tetracycline. Cells without the pSC101 plasmid do not survive on medium with as little as 0.1 μg/ml of tetracycline. We propose that the wild-type phenotype of tetracycline resistance is inducible or T_c^I.

Examination of the level of tetracycline resistance determined by eight different pSC101 recombinant plasmids (with DNA fragments from different sources) at the *Eco*RI site revealed that, in all cases, the level of resistance to tetracycline and/or the T_c^I phenotype was altered. In some cases the recombinants were sensitive to 1 to 2 μg/ml of tetracycline. We do not know if some DNA fragments cloned at this site completely inactivate the T_c^r mechanism. The explanation for these observations is obscure but can be correlated with an alteration in one of the pSC101 plasmid-directed polypeptides. In minicells containing pSC101, six polypeptides are detected by radioisotopic label and separation on SDS-polyacrylamide gels (13). One of these polypeptides (14,000 daltons) is missing when the products of different *Eco*RI recombinant pSC101 plasmids are analyzed in a similar way and a new polypeptide of slightly larger molecular weight (depending on the recombinant plasmid) is apparent. The relative amount of a second polypeptide (26,000 daltons) varies from one recombinant plasmid to another. It has been speculated that the *Eco*RI site of pSC101 is at the terminus of the 14,000-dalton structural gene in order to partly account for these observations. Insertion of a fragment of DNA at such a site would, on the average, modestly increase the mass of the 14,000-dalton polypeptide. The specific alteration of the 14,000-dalton polypeptide may in turn influence the relative amount of 26,000-dalton polypeptide observed.

Three pSC101 plasmid-directed polypeptides (34,000, 26,000, and 14,000 daltons) are measurably induced two- to tenfold when minicells are labeled in the presence of tetracycline. Only a part of the pSC101 genome has been incorporated into the pMB9 series of plasmids. Cells containing these plasmids are constitutive in their resistance to tetracycline but at a level of about 100 μg/ml. There are an estimated 20 to 30 copies of this type of plasmid per cell, which probably accounts for the high level of resistance. In minicells containing the pMB9 plasmid, only the 34,000- and 17,000-dalton polypeptides of the pSC101 set of products are present. Cloning at the *Bam*HI and *Sal*I sites of the pBR313 plasmids results in an altered 34,000-dalton protein, and a deletion located clockwise to the *Sal*I site missing the 17,000-dalton protein is T_c^S.

Our current working hypothesis is that the 34,000- and 17,000-dalton polypeptides account for the high level of resistance to T_c^r which is observed in cells containing the pSC101 plasmid and the pMB9 series of plasmids. In the former case, this requires some induction mechanism to produce large amounts of these two polypeptides and which perhaps depends on the 26,000- and 14,000-dalton polypeptides in some way as yet to be determined. In the multicopy plasmid, the increased gene dosage yields a high constitutive level of these polypeptides and bypasses the requirement for any induction mechanism.

In order to examine the effect of cloning DNA fragments at the *Hin*dIII site of the pBR313 and pBR322 plasmids, we have determined the nucleotide sequence of this region of the plasmid (Fig. 4). Of interest is the region around the *Hin*dIII site, which falls at the end of a 16-nucleotide symmetrical sequence. We have found that this *Hin*dIII site is protected from

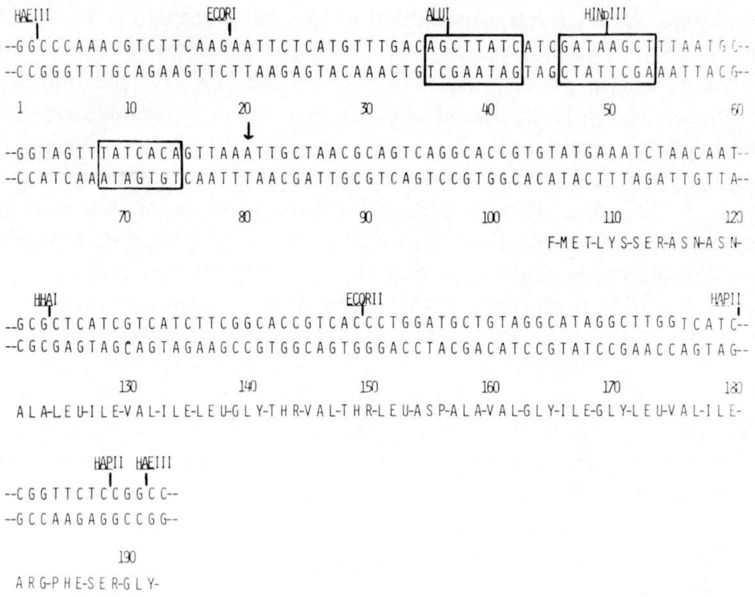

FIG. 4. Nucleotide sequence in the region of the *Eco*RI and *Hin*dIII endonuclease restriction sites. The new deoxyribonucleotide sequence method of Maxam and Gilbert (12) was used to determine the sequence depicted here. A 16-nucleotide region of symmetry is noted as well as a possible promoter initiation site and the potential N-terminal region of the gene coding for an 18,000-dalton protein involved with T_c^r.

endonucleolytic cleavage by *Hin*dIII endonuclease but not the *Eco*RI endonuclease in the presence of *E. coli* DNA-dependent RNA polymerase. Also of interest is the possibility that the sequence at position 106 might delineate the N-terminus of the 34,000-dalton polypeptide, since the sequence reads from the start codon at that position through 29 amino acid codons without a nonsense codon. The sequence is being extended in order to make this a more statistically significant observation. One can then see that insertions of DNA at the *Hin*dIII site could have drastic effects on the possible recognition and/or initiation of a mRNA molecule for the 34,000-dalton polypeptide. A systematic analysis of such alterations is underway. In many cases we find that fragments of DNA up to several megadaltons in size lead to 90 to 99% loss of tetracycline resistance. Thus, in order to score or select for recombinants at the *Hin*dIII site, it is important to use tetracycline concentrations of 50 to 100 μg/ml in the medium.

It is impossible to resist pointing out that the *Eco*RI site of the sequence depicted in Fig. 4 (which is identical to that of the pSC101 plasmid from that point and reading to the right) is followed by a nonsense codon (position 31), which might represent the C-terminus of the 14,000-dalton polypeptide. The sequence counterclockwise to the *Eco*RI site in pSC101 is being determined in order to obtain more support for this hypothesis.

SUMMARY

The plasmid pBR322 is the most versatile plasmid we have constructed. It approaches the minimal size for a plasmid carrying the genetic information required for a versatile cloning vehicle. A characterization of this and other plasmids is underway with particular emphasis on the genetic aspects of replication and tetracycline resistance.

ACKNOWLEDGMENTS

This work was supported by grants from the NSF and the NIH. H.W.B. is an investigator for the Howard Hughes Medical Institute. H.L.H. is the recipient of a fellowship grant from the Netherlands Organization for the Advancement of Pure Research (Z.W.O.). J.S. is the recipient of a CSIRO postdoctoral fellowship.

REFERENCES

1. Bolivar, F., Rodriguez, R. L., Betlach, M. C., and Boyer, H. W. (1977): Construction and characterization of new cloning vehicles. (*submitted for publication*).
2. Bolivar, F., Rodriguez, R. L., Greene, P. J., Betlach, M. C., Heyneker, H. L., and Boyer, H. W. (1977): A new multipurpose plasmid cloning vehicle, pBR322. (*submitted for publication*).
3. Chang, A. C. Y., and Cohen, S. N. (1974): Genome construction between bacterial species

in vitro: replication and expression of staphylococcus plasmid genes in Escherichia coli. *Proc. Natl. Acad. Sci. U.S.A.,* 71:1030–1034.
4. Dugaiczyk, A., Boyer, H. W., and Goodman, H. M. (1975): Ligation of EcoRI endonuclease-generated fragments into linear and circular structures. *J. Mol. Biol.,* 96:171–184.
5. Greene, P. J., Poonian, M. S., Nussbaum, A. L., Tobias, L., Garfin, D. E., Boyer, H. W., and Goodman, H. M. (1975): Restriction and modification of a self-complementary octanucleotide containing the EcoRI substrate. *J. Mol. Biol.,* 99:237–261.
6. Helinski, D., Lovett, M., Williams, P., Katz, L., Collins, J., Kuperstooch-Portnoy, K., Sato, S., Levitt, R., Sparks, R., Hershfield, V., Guiney, D., and Blair, D. (1975): Modes of plasmid DNA replication in Escherichia coli. In: *DNA Synthesis and its Regulation, Vol. 3,* edited by M. Goulian, P. Hanawalt, and C. F. Fox, pp. 514–536. W. A. Benjamin, Menlo Park, Calif.
7. Heyneker, H. L., Shine, J., Goodman, H. M., Boyer, H. W., Rosenberg, J., Dickerson, R. E., Narang, S. A., Itakura, K., Lin, S., and Riggs, A. D. (1976): Synthetic lac operator DNA is functional in vivo. *Nature,* 263:748–752.
8. Jackson, D. A., Symons, R. H., and Berg, P. (1972): Biochemical method for inserting new genetic information into DNA of simian virus 40: circular SV40 DNA molecules containing lambda phage genes and the galactose operon of Escherichia coli. *Proc. Natl. Acad. Sci. U.S.A.,* 69:2904–2909.
9. Jacob, F., Brenner, S., and Cuzin, F. (1963): On the regulation of DNA replication in bacteria. *Cold Spring Harbor Symp. Quant. Biol.,* 28:329–348.
10. Lobban, P., and Kaiser, A. D. (1973): Enzymatic end-to-end joining of DNA molecules. *J. Mol. Biol.,* 78:453–471.
11. Matsuda, M., and Ogoshi, H. (1966): Specificity of DNase I. Estimation of nucleosides present at the 5′-phosphate terminus of a limit digest of DNA by DNase I. *J. Biochem. (Tokyo),* 59:230–235.
12. Maxam, A. M., and Gilbert, W. (1977): A new method for sequencing DNA. *Proc. Natl. Acad. Sci. U.S.A. (in press).*
13. Meagher, R. B., Tait, R. C., Betlach, M. C., and Boyer, H. W. (1977): Protein expression in E. coli minicells by recombinant plasmids containing eucaryotic DNA. *Cell (in press).*
14. Mertz, J. E., and Davis, R. W. (1972): Cleavage of DNA by RI restriction endonuclease generates cohesive ends. *Proc. Natl. Acad. Sci. U.S.A.,* 69:3370–3374.
15. Morrow, J. F., Cohen, S. N., Chang, A. C. Y., Boyer, H. W., Goodman, H. M., and Helling, R. B. (1974): Replication and transcription of eucaryotic DNA in Escherichia coli. *Proc. Natl. Acad. Sci. U.S.A.,* 71:1743–1747.
16. Otsuka, A. (1977): Regeneration of PstI restriction sites during addition of deoxynucleotide tails. *Proc. Natl. Acad. Sci. U.S.A. (in press).*
17. Rodriguez, R. L., Bolivar, B., Goodman, H. M., Boyer, H. W., and Betlach, M. C. (1976): Construction and characterization of cloning vehicles. In: *Molecular Mechanisms in the Control of Gene Expression,* edited by D. P. Nierlich, W. J. Rutter, and C. F. Fox, pp. 471–477. Academic Press, New York.
18. Scheller, R. H., Dickerson, R. E., Boyer, H. W., Riggs, A. D., and Itakura, I. (1977): The chemical synthesis of "linkers": restriction enzyme recognition sites useful for cloning. *Proc. Natl. Acad. Sci. U.S.A. (in press).*
19. Sgaramella, V. (1972): Enzymatic oligomerization of bacteriophage P22 DNA and of linear simian virus 40 DNA. *Proc. Natl. Acad. Sci. U.S.A.,* 69:3389–3393.
20. Sgaramella, V., van de Sande, J. H., and Khorana, H. G. (1970): Studies on polynucleotides, C. A novel joining reaction catalyzed by the T4-polynucleotide ligase. *Proc. Natl. Acad. Sci. U.S.A.,* 67:1468–1475.
21. Sugino, A., Cozzarelli, N. R., Shine, J., Heyneker, H. L., Boyer, H. W., and Goodman, H. M. (1977): Interaction of bacteriophage T4 RNA and DNA ligases in the joining of duplex DNA at base-paired ends. *J. Biol. Chem. (in press).*
22. Tait, R. C., Rodriguez, R. L., and Boyer, H. W. (1977): Altered tetracycline resistance in pSC101 recombinant plasmids. *Mol. Gen. Genet. (in press).*
23. Williams, P. H., Boyer, H. W., and Helinski, D. R. (1973): Size and base composition of RNA in supercoiled plasmid DNA. *Proc. Natl. Acad. Sci. U.S.A.,* 70:3744–3748.
24. Williams, P. H., Boyer, H. W., Leavitt, R. W., and Helinski, D. R. (1977): Unique base composition of RNA segments in complementary strands of supercoiled plasmid ColE1 DNA. *(submitted for publication).*

Recombinant Molecules: Impact on Science and Society, edited by R. F. Beers, Jr. and E. G. Bassett. Raven Press, New York © 1977.

4. The Role of Restriction Endonucleases in Genetic Engineering

Richard J. Roberts

Cold Spring Harbor Laboratory, Cold Spring Harbor, New York 11724

INTRODUCTION

The class II restriction endonucleases[1] have played a key role in the development of recombinant DNA technology although, of the many enzymes now available, only *Eco*RI and *Hin*dIII have been used extensively. In part this stems from their being among the first restriction endonucleases to be discovered, but also because they give fragments containing cohesive termini and they cleave several small plasmid DNAs and certain derivatives of bacteriophage lambda DNA within nonessential regions of the genome. Thus, they can be used to produce vectors into which foreign pieces of DNA can be inserted. The formation of these recombinant DNAs may be achieved using either the restriction endonuclease–DNA ligase method (22,37) or the terminal deoxynucleotidyl transferase method (11,16). In the first method, both the fragment to be cloned and the vector have to contain identical cohesive termini, so the size of the donor fragment is dependent on the particular restriction enzyme employed. If this enzyme cleaves within the region of interest, then partial digests of the donor DNA can be used but this greatly reduces the efficiency of cloning. Rejoining of the original vector means that this method has a high background of nonrecombinant molecules. These limitations are offset by the technical simplicity of the method, together with the capability of recovering the inserted fragment from the recombinant by digestion with the restriction enzyme used to prepare it. With the terminal deoxynucleotidyl transferase method, fragments generated by both restriction endonucleases or by shear can be cloned with little background due to rejoining of the vector; however, in neither case is it currently possible to recover the cloned segment of DNA free of the homopolymer blocks used to insert it plus some neighboring vector DNA. The present abundance of restriction endonucleases with different specificities

[1] Restriction endonucleases are named in accordance with the proposal of Smith and Nathans (33).

will allow manipulation of specific fragments of DNA in a manner that has not hitherto been possible.

This chapter describes some of the newly discovered restriction endonucleases which seem to provide alternative possibilities for genetic engineering and suggests schemes whereby the specificity of the nucleases can be exploited in the creation of new recombinant genomes.

SPECIFIC ENDONUCLEASES

At the present time more than 40 different specific endonucleases have been discovered, and the recognition sequences for 23 of these have been deduced, as depicted in Table 1 (26). The enzymes fall into three groups differing in the length of the recognition sequence and hence in the frequency

TABLE 1. *Specific endonucleases and their recognition sequences*

Tetranucleotide		Pentanucleotide		Hexanucleotide	
*Alu*I[a]	AG↓CT	*Eco*RII	↓CC(A/T)GG	*Ava*I	C↓PyCGPuG
*Hae*III	GG↓CC			*Bam*I	G↓GATCC
*Hha*I	GCG↓C			*Bgl*II	A↓GATCT
*Hpa*II	C↓CGG			*Bal*I	TGG↓CCA
*Mbo*I	↓GATC			*Eco*RI	G↓AATTC
*Taq*I	T↓CGA	*Hph*I	GGTGA→8 bp	*Hin*dIII	A↓AGCTT
		*Mbo*II	GAAGA→8 bp	*Hpa*I	GTT↓AAC
				*Pst*I	CTGCA↓G
*Hin*fI	G↓ANTC			*Xma*I	C↓CCGGG
*Dpn*I	↓GATC when modified			*Hae*I	(A/T)GG↓CC(T/A)
				*Hae*II	PuGCGC↓Py
				*Hin*dII	GTPy↓PuAC

[a] *Alu*I is from *Arthrobacter luteus* (28); *Hae*I, *Hae*II, and *Hae*III from *Haemophilus aegyptius* (4,18,27; B. G. Barrell and P. Slocombe; K. Murray, A. Morrison, H. W. Cooke, and R. J. Roberts, *unpublished observations*); *Hha*I from *Haemophilus haemolyticus* (29); *Hpa*I and *Hpa*II from *Haemophilus parainfluenzae* (7,31); *Mbo*I and *Mbo*II from *Moraxella bovis* (N. L. Brown, C. A. Hutchison, III, and M. Smith; R. E. Gelinas, P. A. Myers, R. J. Roberts, K. Murray, and S. A. Bruce, *unpublished results*); *Taq*I from *Thermus aquaticus* (S. Sato, C. A. Hutchison, III, and J. I. Harris, *unpublished observations*); *Hin*fI from *Haemophilus influenzae* serotype f (17; C. A. Hutchison, III, and B. G. Barrell, *unpublished observations*); *Dpn*I from *Diplococcus pneumoniae* (15; S. Lacks, *unpublished observations*); *Eco*RII from *Escherichia coli* (2,3); *Hph*I from *Haemophilus parahaemolyticus* (14,17); *Ava*I from *Anabaena variabilis* (21); *Bam*I from *Bacillus amyloliquefaciens* H (38; R. J. Roberts, G. A. Wilson, and F. E. Young, *unpublished observations*); *Bgl*II from *Bacillus globigii* (G. A. Wilson and F. E. Young; B. S. Zain and R. J. Roberts, *unpublished results*); *Bal*I from *Brevibacterium albidum* (R. E. Gelinas, G. A. Weiss, R. J. Roberts, A. Morrison, and K. Murray, *unpublished results*); *Eco*RI from *Escherichia coli* (8,10); *Hin*dII and *Hin*dIII from *Haemophilus influenzae* serotype d (13,24,34); *Pst*I from *Providencia stuartii* (5,32); and *Xma*I from *Xanthomonas malvacearum* (S. A. Endow and R. J. Roberts, *unpublished observations*).

of cleavage of DNA. Within each group there are considerable variations in the type of cleavage produced. Thus, in the group recognizing tetranucleotides, examples may be found of endonucleases cleaving to produce flush ends: (*Hae*III, GG↓CC); a 3′-terminal dinucleotide extension (*Hha*I, GCG↓C); a 5′-dinucleotide extension (*Hpa*II, C↓CGG); and a 5′-tetranucleotide extension (*Mbo*I, ↓GATC). In addition, the enzyme *Hinf*I recognizes a staggered tetranucleotide palindrome (G↓ANTC), where N may be any nucleotide and cleavage products have a 5′-trinucleotide extension. Because any tetranucleotide palindrome occurs once in 256 base pairs by chance, rather small fragments frequently appear as products of cleavage. Exceptions to this are the enzymes *Hha*I and *Hpa*II, which cleave eukaryotic DNA much less frequently than expected on the basis of statistical probability. Presumably the explanation for this is that these restriction endonucleases contain the dinucleotide CG within their recognition sequences, and this dinucleotide occurs rarely in eukaryotic DNA (12,35). One interesting member of the group recognizing tetranucleotides is the enzyme *Dpn*I from *Diplococcus pneumoniae* (15), which recognizes the sequence GATC only when that sequence is modified (S. Lacks, *unpublished observations*). This is the only known example of a restriction endonuclease with an absolute requirement for modified DNA. Another strain of *Diplococcus pneumoniae* contains a related enzyme, *Dpn*II, which recognizes and cleaves the same sequence in its unmodified form (15).

Only three members of the second class of enzymes, which recognize pentanucleotide sequences, have been reported so far. Of these, the enzymes *Hph*I and *Mbo*II are unusual in that they recognize a pentanucleotide sequence which lacks any element of symmetry, and furthermore, they do not cleave within that sequence but rather 8 base pairs away from it (14; N. L. Brown, C. A. Hutchison, III, and M. Smith, *unpublished observations*). The third set of specific endonucleases are those recognizing hexanucleotide palindromes, and again a wide variety is found both in the type of sequences cleaved and in the site of cleavage within the recognition sequence. Thus, *Hpa*I cleaves GTT AAC to leave flush-ended fragments, *Bam*I cleaves G GATCC to leave a 5′-tetranucleotide extension, whereas *Pst*I cleaves CTGCA G to leave a 3′-tetranucleotide extension. In addition, the enzymes *Hae*I, *Hae*II, and *Hind*II do not require an absolute specificity at all bases within the recognition sequence and thus cleave several different hexanucleotide sequences. One feature of all hexanucleotide recognition sequences is that each must be a subset of sequences recognized by one of the tetranucleotide-recognizing enzymes, and several specific examples are now known. *Mbo*I recognizes the sequence of GATC, whereas *Bam*I and *Bgl*II recognize two specific subsets of this sequence because they contain the tetranucleotide GATC within their own recognition sites. This fact aids the determination of the recognition sequences and also has some implications for genetic engineering, as will be discussed below. Similarly, *Hae*III and *Bal*I share the recognition sequence GGCC, and *Alu*I and *Hin*dIII share the recognition sequence AGCT.

Some of the other specific endonucleases which have been characterized (although their recognition sequences are not yet known) are listed in Table 2. These enzymes have been chosen because they are present in fairly large amounts in the bacterial cells and are easily purified. The enzyme *Mnl*I from *Moraxella nonliquefaciens* produces many fragments in SV40 and ϕX174 DNA and probably recognizes a tetranucleotide whereas the other enzymes show greater specificity. In the case of *Hga*I from *Haemophilus gallinarum* (36), the large number of cleavages on bacteriophage lambda DNA and adenovirus-2 DNA suggests either a tetra- or pentanucleotide recognition sequence or possibly some set of hexanucleotides. This enzyme fails to cleave SV40, perhaps because it contains the dinucleotide CG within its recognition sequence. Enzymes containing this dinucleotide within their recognition sequence cut eukaryotic DNA infrequently as mentioned above and, in this regard, SV40 DNA behaves like eukaryotic DNA (20). In contrast, adenovirus-2 DNA and bacteriophage lambda DNA are cut frequently by *Hga*I and also by *Hpa*II, *Hha*I, and *Taq*I (1, 29; R. J. Roberts and P. A. Myers, *unpublished results*). Polyoma DNA, like adenovirus-2 DNA, is also cleaved quite frequently by enzymes containing CG in the recognition sequence (6), perhaps pointing to an unusual evolutionary origin for both viruses. The other members of this group are still poorly characterized; however, it is known that *Sal*I produces cohesive termini (9). These enzymes offer the best practical means of extending the range of sites presently accessible by the enzymes listed in Table 1. Other enzymes have been isolated (26); however, they are present in rather small quantities or else have not been purified to a point where they are yet useful.

One of the most widely used methods for joining DNA fragments to prepare recombinant molecules involves the use of DNA ligase. So far, this has been used exclusively for enzymes that produce fragments with cohesive termini, and a list of such enzymes is found in Table 3. The enzymes *Bam*I, *Bgl*II, and *Mbo*I all produce fragments with the same 5′-tetranucleotide extension, GATC, which means that a common vector could be used for fragments generated by any of these three enzymes. Similarly, both *Eco*RI and *Eco*RI* fragments can be cloned in an *Eco*RI vector. *Eco*RII leaves a 5′-terminal pentanucleotide extension, and because this contains four GC base pairs, such fragments should be ligated with high efficiency. Unfortunately, no suitable vector has been described. Similarly, fragments produced by *Hae*II contain cohesive termini with four GC base pairs, and ligation should proceed with high efficiency but no vector has yet been described. Among other enzymes in this list, vectors have been prepared for *Hin*dIII (23) and *Sal*I (9), and ligation of *Hha*I fragments has been reported (19). Because *Hha*I fragments contain only a dinucleotide 3′-extension, it might have been expected that ligation would be an inefficient process. In-

TABLE 2. *Partially characterized specific endonucleases*

Enzyme	Source	Frequency of cleavage			Reference
		λ	Ad-2	SV40	
*Bgl*I	*Bacillus globigii*	22	12	1	G. A. Wilson and F. E. Young, *unpublished observations*
*Blu*I	*Brevibacterium luteum*	1	7	0	R. J. Roberts, *unpublished results*
*Hga*I	*Haemophilus gallinarum*	>30	>30	0	(36)
*Kpn*I	*Klebsiella pneumoniae*	2	8	1	(32)
*Mnl*I	*Moraxella nonliquefaciens*	>50	>50	>10	R. Greene, *unpublished results*
*Sac*I	*Streptomyces achromogenes*	2	>7	0	J. R. Arrand, P. E. Myers, and R. J. Roberts, *unpublished observations*
*Sac*II	*Streptomyces achromogenes*	3	>20	0	"
*Sal*I	*Streptomyces albus* G	2	3	0	"
*Xba*I	*Xanthomonas badrii*	1	4	0	B. S. Zain and R. J. Roberts, *unpublished observations*

TABLE 3. *Fragments with cohesive termini*

*Bam*I	G↓GATCC	*Mbo*I	↓GATC
*Bgl*II	A↓GATCT		
*Eco*RI	G↓AATTC	*Eco*RI*	↓AATT
*Eco*RII	↓CC(A/T)GG		
*Hae*II	PuGCGC↓Py		
*Hha*I	GCG↓C		
*Hind*III	A↓AGCTT		
*Hinf*I	G↓ANTC		
*Hpa*II	C↓CGG		
*Pst*I	CTGCA↓G		
*Taq*I	T↓CGA		
*Xma*I	C↓CCGGG		
*Sal*I	?		
*Sst*I	?		

References to the information given in this table may be found in the legend to Table 1. *Eco*RI* refers to the activity detected when *Eco*RI is used under abnormal conditions of pH and ionic strength (25).

deed, in the case of *Hpa*II fragments (which also contain a dinucleotide extension), *in vitro* ligation proceeds very inefficiently (R. J. Roberts, *unpublished results*). One possible explanation for this apparent anomaly could lie in base-stacking phenomena associated with the bases adjacent to the cohesive termini. Finally, the enzyme *Sst*I, which is an isoschizomer of *Sac*I (Table 2), also produces cohesive termini (S. Goff and A. Rambach, *unpublished observations*). It is not yet known if the site of cleavage within the recognition sequence is identical for both enzymes.

SELECTION OF RECOMBINANT DNAs

There are several ways in which the known specificity of certain restriction endonucleases might be used to advantage in the construction of recombinant DNAs. As mentioned previously, the enzymes *Bam*I, *Bgl*II, and *Mbo*I all produce fragments containing the same 5′-tetranucleotide extension GATC, and a vector for any one of these fragments would be suitable for the other members of the set. However, if a *Bam*I vector were actually used to clone a *Bgl*II fragment, the resulting vector-fragment recombinant would be resistant to the action of both *Bam*I and *Bgl*II. This arises because the recombinant site is a hybrid of the *Bam*I-*Bgl*II sites with only a common central tetranucleotide. This fact could be used to select hybrid recombinant DNAs from a mixture of all possible recombinants as illustrated in Fig. 1.

If a mixture of *Bam*I fragments and *Bgl*II fragments is incubated with DNA ligase, three possible kinds of recombinants result: homodimers (composed of two *Bam*I fragments or two *Bgl*II fragments), plus hetero-

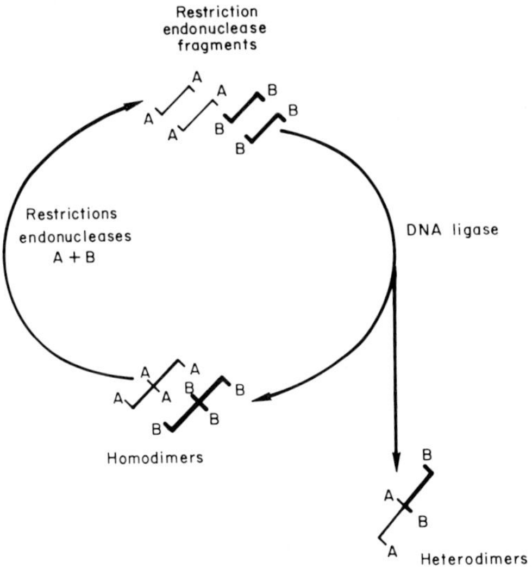

FIG. 1. A scheme for the *in vitro* selection of recombinant DNA molecules. Restriction endonucleases A and B would be chosen so as to produce either fragments with identical cohesive termini or flush-ended fragments.

dimers (in which a *Bgl*II fragment has become joined to a *Bam*I fragment). When this mixture is treated with the two restriction endonucleases, the homodimers would be cleaved to generate the original fragments, whereas the heterodimers would be resistant. If two DNAs are incubated with DNA ligase in the presence of the restriction endonucleases *Bam*I and *Bgl*II, a reaction pathway can be envisaged as shown in Fig. 1. The homodimers are able to cycle between cleavage and ligation whereas the heterodimers, once formed, are effectively removed from the reaction pathway, being a substrate for none of the enzymes present. Under appropriate conditions, it should be possible to drive the reaction so that only heterodimers remain. Such heterodimers, although resistant to the reaction of *Bam*I and *Bgl*II, would still be susceptible to the action of *Mbo*I which recognizes only the central tetranucleotide palindrome, GATC. Consequently, *Mbo*I could be used to reduce the size of the cloned fragment if it contains more genetic information than is desired. If the *Bam*I fragment were to contain a gene of interest, surrounded by a large amount of sequence of less interest, then partial digestion of the recombinant DNA with *Mbo*I offers the chance of removing small *Mbo*I fragments which contain the undesired segments of DNA while allowing the desired region to be cloned in the original vector. Similar considerations apply to the *Eco*RI* activity, which may be used to reduce the size of an *Eco*RI fragment.

Of course, another set contains all enzymes producing flush-ended fragments since it has been shown that such fragments can also be joined using

DNA ligase (30). Thus, *Hpa*I fragments might be joined to *Ava*I fragments and the resulting heterodimers selected by the scheme illustrated in Fig. 1. Also, *Bal*I fragments could be reduced in size by the action of *Hae*III. In view of the prolific occurrence of restriction endonucleases, it seems likely that many similar sets of enzymes will occur.

Thus, it is rapidly becoming possible to choose exactly which fragments are to be joined and, having chosen, to use a vector system from which the recombinant DNA can be selected *in vitro* prior to cloning. These considerations will be of great importance when undertaking the construction of recombinant DNA molecules that contain promoters, ribosome binding sites, genes, and terminators, constructed in such a way that the gene is under some known control system. For example, it may soon be possible to construct a recombinant DNA containing a eukaryotic gene under known prokaryotic control and guarantee its expression by the correct placement of the necessary prokaryotic control signals.

ADAPTORS

For plasmid vectors, many attempts have been made to reduce the size of these vectors so that they contain no unwanted genetic information. An ideal vector should have the following properties:

1. It should be capable of autonomous replication.
2. It should contain the genes for some selectable marker.
3. It should contain a restriction site which allows it to be opened without affecting 1. or 2.

Clearly, as the amount of nonessential information is reduced, the possibility of removing the desired restriction enzyme site(s) increases. One possible way to overcome this difficulty would be to chemically synthesize an adaptor containing the particular site that was desired and to insert it into the plasmid genome. For instance, if a plasmid had one site for *Eco*RI but no site for *Bam*I, then an adaptor could be constructed such that it contained a *Bam*I site flanked by two *Eco*RI cohesive ends. Upon opening the plasmid with *Eco*RI, followed by ligation in the presence of the adaptor, some recombinant molecules would be formed which now contained two *Eco*RI sites flanking a *Bam*I site. This scheme is illustrated in Fig. 2. The newly introduced site would be available for insertion using either DNA ligase to insert *Bam*I fragments or terminal deoxynucleotidyl transferase to introduce other fragments. In the latter case, such an adaptor would have one other advantage. One of the limitations imposed by the terminal deoxynucleotidyl transferase method is that it is often difficult, if not impossible, to recover the newly inserted DNA from the plasmid. This arises because the addition of the homopolymer tails destroys the restriction enzyme recognition site to which it is added (however, *see below*).

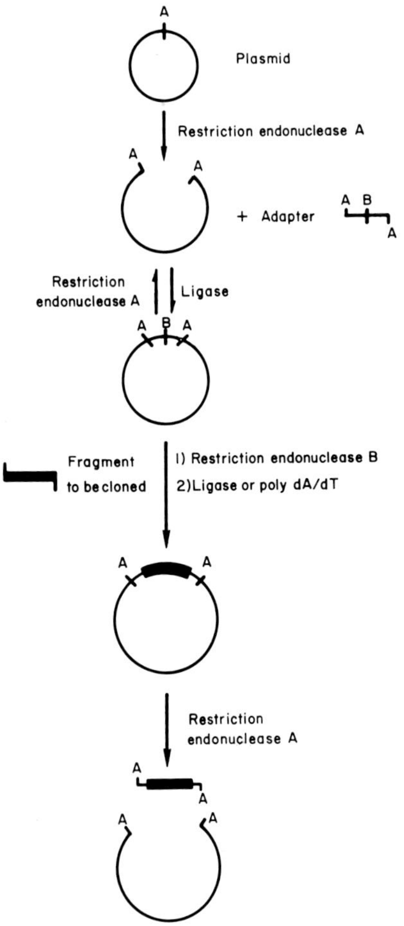

FIG. 2. A scheme showing the use of adaptors.

In the example shown in Fig. 2, although the restriction enzyme site B has been destroyed during the formation of the recombinant, the two flanking restriction enzyme sites A are still intact and could be used to remove the newly inserted fragment together with its homopolymer tails. Although some DNA other than the cloned fragment is present, it would be minimal, being composed mainly of the homopolymers. Even this difficulty could be overcome if a suitable enzyme and an appropriate homopolymer were used to generate the recombinant DNA. Already three such enzymes are known: *Pst*I, *Hae*II, and *Hha*I all generate fragments with 3′-terminal nucleotide extensions. In the case of *Hha*I, the addition of deoxycytidylate residues would actually regenerate the *Hha*I recognition site during its introduction into the vector. Now, after cleavage with *Hha*I, the homopolymer tail would remain associated with the plasmid. Similarly, the addition of a homopolymer tail of deoxyguanylate residues to *Pst*I fragments and either deoxycytidylate or thymidylate residues to *Hae*II fragments would again result in the regeneration of the restriction enzyme recognition site. Furthermore, the presence of these 3′-terminal nucleotide extensions allows the efficient addition of homopolymers by terminal deoxynucleotidyl transferase without the need for prior resection with lambda exonuclease. The combination of this approach with suitable adaptors would thus allow almost any plasmid to be used as a vector. In addition, both the DNA ligase and the terminal deoxynucleotidyl transferase methods would allow recovery of just the cloned fragment from the recombinant DNA.

From the foregoing discussion, it is clear that the availability of many different restriction endonucleases will have important implications for the field of genetic engineering. In addition to being able to degrade DNA into precisely defined fragments, it will also be possible to construct new DNA molecules in a precise manner. As an alternative to chemical synthesis of the $tRNA^{Tyr}$ gene, it would now be possible to excise that DNA from the appropriate transducing phage either with its promoter intact or in a form in which it could be attached to a new promoter that might be amenable to external control. Obvious candidates would be the control systems that regulate the lactose or tryptophan operons.

ACKNOWLEDGMENTS

The author thanks D. Botstein and T. Otsuka for many stimulating discussions and his many colleagues who have generously provided access to their unpublished work. Thanks must also be extended to John Arrand, Sharyn Endow, Richard Gelinas, Phyllis Myers, and Sayeeda Zain, without whose hard work Tables 1 and 2 would be much shorter. This work was supported by grants from the National Cancer Institute, CA13106, and the National Science Foundation, GB43912.

REFERENCES

1. Allet, B. (1973): Fragments produced by cleavage of lambda deoxyribonucleic acid with Haemophilus parainfluenzae restriction enzyme HpaII. *Biochemistry,* 12:3972–3977.
2. Bigger, C. H., Murray, K., and Murray, N. E. (1973): Recognition sequence of a restriction enzyme. *Nature* [*New Biol.*], 244:7–10.
3. Boyer, H. W., Chow, L. T., Dugaiczyk, A., Hedgpeth, J., and Goodman, H. M. (1973): DNA substrate site for the EcoRII restriction endonuclease and modification. *Nature* [*New Biol.*], 244:40–43.
4. Bron, S., and Murray, K. (1975): Restriction and modification in B. subtilis. Nucleotide sequence recognized by restriction endonuclease R. BsuR from strain R. *Mol. Gen. Genet.,* 143:25–33.
5. Brown, N. L., and Smith, M. (1976): The mapping and sequence determination of the single site in ϕX174 amber 3 replicative form DNA cleaved by restriction endonuclease PstI. *FEBS Lett.,* 65:284–287.
6. Fried, M., and Griffin, B. E. (1977): Organization of the genomes of polyoma and SV40. *Adv. Cancer Res.,* 24:67–113.
7. Garfin, D. E., and Goodman, H. M. (1974): Nucleotide sequences at the cleavage sites of

two restriction endonucleases from Haemophilus parainfluenzae. *Biochem. Biophys. Res. Commun.*, 59:108–116.
8. Greene, P. J., Betlach, M. C., Boyer, H. W., and Goodman, H. M. (1976): The EcoRI restriction endonuclease. In: *Methods in Molecular Biology, Vol. 7: DNA Replication,* edited by R. B. Wickner, pp. 87–111. Marcel Dekker, New York.
9. Hamer, D. H., and Thomas, C. A., Jr. (1976): Molecular cloning of DNA fragments produced by restriction endonucleases SalI and BamI. *Proc. Natl. Acad. Sci. U.S.A.*, 73: 1537–1541.
10. Hedgpeth, J., Goodman, H. M., and Boyer, H. W. (1972): DNA nucleotide sequence restricted by the RI endonuclease. *Proc. Natl. Acad. Sci. U.S.A.*, 69:3448–3452.
11. Jackson, D. A., Symons, R. H., and Berg, P. (1972): Biochemical method for inserting new genetic information into DNA of simian virus 40: circular SV40 DNA molecules containing lambda phage genes and the galactose operon of Escherichia coli. *Proc. Natl. Acad. Sci. U.S.A.*, 69:2904–2909.
12. Josse, J., Kaiser, A. D., and Kornberg, A. (1961): Enzymatic synthesis of deoxyribonucleic acid. VIII. Frequencies of nearest neighbor base sequences in deoxyribonucleic acid. *J. Biol. Chem.*, 236:864–875.
13. Kelly, T. J., and Smith, H. O. (1970): A restriction enzyme from Haemophilus influenzae. II. Base sequence of the recognition site. *J. Mol. Biol.*, 51:393–409.
14. Kleid, D., Humayun, Z., Jeffrey, A., and Ptashne, M. (1976): Novel properties of a restriction endonuclease isolated from Haemophilus parahaemolyticus. *Proc. Natl. Acad. Sci. U.S.A.*, 73:293–297.
15. Lacks, S., and Greenberg, B. (1973): A deoxyribonuclease of Diplococcus pneumoniae specific for methylated DNA. *J. Biol. Chem.*, 250:4060–4066.
16. Lobban, P. E., and Kaiser, A. D. (1973): Enzymatic end-to-end joining of DNA molecules. *J. Mol. Biol.*, 78:453–471.
17. Middleton, J. H. (1973): Restriction endonucleases from Haemophilus species: Enzymes for specific fragmentation of DNA. Ph.D. Thesis, University of North Carolina, Chapel Hill.
18. Middleton, J. H., Edgell, M. H., and Hutchison, C. A., III (1972): Specific fragments of ϕX174 deoxyribonucleic acid produced by a restriction enzyme from Haemophilus aegyptius, endonuclease Z. *J. Virol.*, 10:42–50.
19. Miller, L. K., and Fried, M.: Construction of infectious polyoma hybrid genomes in vitro. *Nature,* 259:598–601.
20. Morrison, J. M., Keir, H. M., Subak-Sharpe, H., and Crawford, L. V. (1967): Nearest neighbor base sequence analysis of the deoxyribonucleic acids of a further three mammalian viruses: simian virus 40, human papilloma virus and adenovirus type 2. *J. Gen. Virol.*, 1:101–108.
21. Murray, K., Hughes, S. G., Brown, J. S., and Bruce, S. A. (1976): Isolation and characterization of two specific endonucleases from Anebaena variabilis. *Biochem. J.*, 159:317–322.
22. Murray, K., and Murray, N. E. (1974): Manipulation of restriction targets on phage lambda to form receptor chromosomes for DNA fragments. *Nature,* 251:476–481.
23. Murray, K., and Murray, N. E. (1975): Phage lambda receptor chromosomes for DNA fragments made with restriction endonuclease III of Haemophilus influenzae and restriction endonuclease I of Escherichia coli. *J. Mol. Biol.*, 98:551–564.
24. Old, R., Murray, K., and Roizes, G. (1975): Recognition sequence of restriction endonuclease III from Haemophilus influenzae. *J. Mol. Biol.*, 92:331–339.
25. Polisky, B., Greene, P., Garfin, D. E., McCarthy, B. J., Goodman, H. M., and Boyer, H. W. (1975): Specificity of substrate recognition by the EcoRI restriction endonuclease. *Proc. Natl. Acad. Sci. U.S.A.*, 72:3310–3314.
26. Roberts, R. J. (1976): Restriction endonucleases. *CRC Crit. Rev. Biochem.*, 4:123–164.
27. Roberts, R. J., Breitmeyer, J. B., Tabachnik, N. F., and Myers, P. A. (1975): A second specific endonuclease from Haemophilus aegyptius. *J. Mol. Biol.*, 91:121–123.
28. Roberts, R. J., Myers, P. A., Morrison, A., and Murray, K. (1976*a*): A specific endonuclease from Arthrobacter luteus. *J. Mol. Biol.*, 102:157–166.
29. Roberts, R. J., Myers, P. A., Morrison, A., and Murray, K. (1976*b*): A specific endonuclease from Haemophilus haemolyticus. *J. Mol. Biol.*, 103:199–208.
30. Sgaramella, V., van de Sande, J. H., and Khorana, H. G. (1976): Studies on polynucleo-

tides. C. A novel joining reaction catalyzed by the T4-polynucleotide ligase. *Proc. Natl. Acad. Sci. U.S.A.,* 67:1468–1475.
31. Sharp, P. A., Sugden, B., and Sambrook, J. (1973): Detection of two restriction endonuclease activities in Haemophilus parainfluenzae using analytical agarose-ethidium bromide electrophoresis. *Biochemistry,* 12:3055–3063.
32. Smith, D. I., Blattner, F. R., and Davies, J. (1976): The isolation and partial characterization of a new restriction endonuclease from Providencia stuartii. *Nucleic Acids Res.,* 3:343–353.
33. Smith, H. O., and Nathans, D. (1973): A suggested nomenclature for bacterial host modification and restriction systems and their enzymes. *J. Mol. Biol.,* 81:419–423.
34. Smith, H. O., and Wilcox, K. W. (1970): A restriction enzyme from Haemophilus influenzae. I. Purification and general properties. *J. Mol. Biol.,* 51:379–391.
35. Swartz, M. N., Trautner, T. A., and Kornberg, A. (1962): Enzymatic synthesis of deoxyribonucleic acid. XI. Further studies on nearest neighbor base sequences in deoxyribonucleic acids. *J. Biol. Chem.,* 237:1961–1967.
36. Takanami, M. (1974): Restriction endonucleases AP, GA, and H-I from three Haemophilus strains. In: *Methods in Molecular Biology, Vol. 7: DNA Replication,* edited by R. B. Wickner, pp. 113–133. Marcel Dekker, New York.
37. Thomas, M., Cameron, J. R., and Davis, R. W. (1974): Viable molecular hybrids of bacteriophage lambda and eukaryotic DNA. *Proc. Natl. Acad. Sci. U.S.A.,* 71:4579–4583.
38. Wilson, G. A., and Young, F. E. (1975): Isolation of a sequence-specific endonuclease (BamI) from Bacillus amyloliquefaciens H. *J. Mol. Biol.,* 97:123–125.

Recombinant Molecules: Impact on Science and Society, edited by R. F. Beers, Jr. and E. G. Bassett. Raven Press, New York © 1977.

5. Development of the *Bacillus subtilis* Model System for Recombinant Molecule Technology

Frank E. Young, Craig Duncan, and Gary A. Wilson

Department of Microbiology, University of Rochester School of Medicine and Dentistry, Rochester, New York 14642

INTRODUCTION

The discovery of transformation of *Bacillus subtilis* by Spizizen in 1958 (23) stimulated the development of this model system for investigations in biochemical genetics and developmental biology. As early as 1963, studies (17) were initiated to determine the frequency of heterologous transformation among closely related species. Subsequent work demonstrated that heterologous transformation was influenced by two major factors: (a) the homology of the neighborhood surrounding the site of integration of the fragment of heterologous DNA (31,32); and (b) the presence of unmethylated, unprotected restriction sites (34). The recent advent of recombinant molecule technology has provided an additional method for the introduction of heterologous DNA. Therefore, this discussion will focus on:

1. The attributes and liabilities of the *B. subtilis* model system in relation to its use as a cloning system.
2. The development of bacteriophage ϕ3T that carries a gene encoding thymidylate synthetase (*thy*P3) as a cloning vector.
3. Recent studies with a chimeric plasmid, pCDI, formed from the *Escherichia coli* plasmid pMB9 and the *thy*P3 gene from ϕ3T. Surprisingly, thymidylate synthetase encoded by *thy*P3 in this plasmid (pCDI) functions in both *E. coli* and *B. subtilis* 168.

ATTRIBUTES OF THE *B. subtilis* MODEL SYSTEM

The major attributes of the *B. subtilis* model system are shown in Table 1. Four points merit emphasis. First, *B. subtilis* is a nonpathogenic organism that rarely causes disease in hosts that are not compromised by underlying disease processes (11,13,19). In an analysis of one series of cases, Farrar (11) divided infections with nonpathogenic bacilli into three types: local infections of damaged organs such as the eye, mixed infection with organisms of recognized pathogenicity, and disseminated infections. In the dis-

TABLE 1. *Major attributes of the* Bacillis subtilis *model system*

1. Nonpathogenic soil bacillus
2. High-frequency transformation (4%)
3. Circular map (220 defined loci)
4. Bacteriophage vectors (ϕ3T)
5. Cryptic plasmids
6. Incorporation of ColE1
7. Site-specific endonucleases

seminated cases, an underlying cause of the infection could be identified that was usually related to trauma.

In a more recent series, Ihde and Armstrong (13) pointed out that these types of infections were also seen in immunologically compromised hosts. Therefore, unlike *E. coli,* bacillus species other than *Bacillus anthracis* rarely cause disease unless there is an underlying disease or traumatic process. Furthermore, infections with bacilli are rarely classified, and many organisms called *B. subtilis* are actually other bacilli. Second, unlike the gram-negative organisms, *B. subtilis* has a simple cell surface composed only of peptidoglycan and teichoic acid (37). Therefore, substances produced by *B. subtilis,* unlike those produced by *E. coli,* would not be contaminated by endotoxin. Because even trace amounts of cell wall products from the gram-negative organisms can produce a febrile reaction, this is an important consideration. Third, the frequencies of DNA-mediated transformation are very high. Although one can readily achieve frequencies of 1 to 4%, it is possible to obtain even greater levels of transformation (approximately 10%) with DNA prepared from gently lysed L-forms or protoplasts (2). Fourth, a number of site-specific endonucleases have been isolated in the *B. subtilis* genospecies.[1] *Bsu* was isolated by Trautner and co-workers (26) from a strain of *B. subtilis* designated strain R, and it recognizes the sequence $\begin{matrix}5'\text{GGCC}3'\\3'\text{CCGG}5'\end{matrix}$ (3). Studies in our laboratory led to the isolation of three more site-specific endonucleases, *Bam*HI, *Bgl*I, and *Bgl*II (33,34) (Table 2). *Bam*HI recognizes the sequence $\begin{matrix}5'\text{GGATCC}3'\\3'\text{CCTAGG}5'\end{matrix}$ (34). The *Bgl*II recognition site has been established as $\begin{matrix}5'\text{AGATCT}3'\\3'\text{TCTAGA}5'\end{matrix}$ (R. J. Roberts, *personal communication;* V. Pirrotta, *personal communication*). The *Bgl*I site has not been defined. These site-specific nucleases rapidly hydrolyze heterologous and intergenotic DNA in transformation or transfection studies (34). Bacteriophages that can be propagated on various members

[1] A genospecies was defined by Ravin (20) as those organisms that can exchange genes, in contrast to a taxospecies, which encompasses organisms that share a high proportion of similar traits.

TABLE 2. *Site-specific nucleases in the* Bacillus *genospecies*

Strain	Enzyme	Site	No. of sites λ	No. of sites ϕ3T
B. subtilis R	*Bsu*	GG↓CC	>50	>50
B. amyloliquefaciens	*Bam*HI	G↓GATCC	5	4
B. globigii	*Bgl*I	Unknown	22	>50
B. globigii	*Bgl*II	A↓GATCT	4	21

TABLE 3. *Negative features of the* Bacillus subtilis *model system*

1. Maps of potential vectors not defined
2. Bacillus plasmids predominantly cryptic
3. Disabled strains still under development

of the genospecies are the most efficient probes for biologically detecting the presence of site-specific endonucleases and are useful in determining subtle differences among endonucleases and methylases in closely related organisms.

In comparison with *E. coli,* however, there are a number of negative features (Table 3). First, the maps of the potential bacteriophage vectors have not been defined to the same extent as have the elegant studies with lambda. The bacteriophages ϕ29, SPP1, and SP01 have been more extensively characterized (see ref. 12 for a review of *B. subtilis* bacteriophages). None of these bacteriophages, however, has been used as vectors to date. Furthermore, only one report of specialized transduction has been presented (21); unfortunately, the level of transduction with this virus, bacteriophage ϕ105, was very low. Second, the plasmids studied thus far are cryptic in both *B. subtilis* and *B. pumilus* (15,16). Finally, no attempts have been made to develop disabled strains that would be incapable of surviving in soil. However, most of these objections should yield to sustained and concentrated investigations of this model system.

HETEROLOGOUS TRANSFORMATION

Transformation of bacteria by DNA from other species has been termed heterologous or interspecific transformation. The frequency of heterologous transformation in bacilli has been examined by studying the capacity of *B. subtilis* 168 to be transformed by DNA from various strains such as *Bacillus amyloliquefaciens, B. natto, B. licheniformis, B. pumilus,* and *B. globigii.* Heterologous transformation (genetic modification produced by uptake of DNA from nonhomologous strains) usually occurs at a low frequency (31). The transformant obtained by integration of the heterologous DNA is defined as an intergenote (31). When DNA is extracted from and used as a donor in transformation experiments, the process is termed intergenotic transformation. Heterologous transformation can be accomplished by direct selection of the desired trait (31) or by analysis of a nonselected trait introduced by congression (37). In addition, prophage DNA can be used as a source of heterologous DNA when the virus can be integrated into the chromosome of closely related bacilli (32).

In their initial experiments, Marmur and co-workers (17) demonstrated that DNA-DNA homology was one of the more significant factors that influence the capacity of heterologous DNA to transform *B. subtilis.* Although the competent cell readily and irreversibly binds DNA from a variety of microorganisms, only DNA with significant base homology is integrated into the chromosome (14). In one of the most probing studies, Chilton and McCarthy (6) established that certain regions of the chromosome were conserved, particularly the regions extending from *pur*A to *cys*A. Another region of relative homology between *B. subtilis* and *B. globigii* occurred near *leu*A. In our own studies it was established that the frequency of transformation of the recipient strain by heterologous DNA was much lower than by homologous DNA (Table 4). Once the intergenote was formed, however, there was a marked increase in the efficiency of transformation. Surprisingly, the efficiency of heterologous and intergenotic transformation varies with different markers (Table 4). In other experiments using DNA from heterologous strains lysogenic for bacteriophage ϕ105, it was possible to establish conclusively that the homology of the neighborhood surrounding the marker was the dominant factor that influenced the outcome of heterologous transformation. Therefore, in DNA-mediated transformation of *B. subtilis* by heterologous DNA, once the initial recombinant was formed, subsequent transformation events occurred at high frequency. However, unless the marker was in a highly conserved region it was difficult to obtain the intergenote, even when the DNA was isolated from closely related strains (31). DNA from distantly related organisms such as *E. coli* was not usually expressed, even though it was irreversibly bound to and transported into the cell.

TABLE 4. *Interspecific transformation of* Bacillus subtilis *168*

	Source of DNA		
Marker	Heterologous	Intergenotic	Homologous
His^+	22	77,400	111,000
Rif^r	76,000	1,670,000	650,000
ϕ105	409	2,267	5,060

From refs. 31 and 32.

DEVELOPMENT OF THE *B. subtilis* MODEL SYSTEM FOR RECOMBINANT MOLECULE TECHNOLOGY

B. subtilis is a spore-forming rod that naturally inhabits the soil. This organism is nonpathogenic and rarely produces disease unless the host is severely compromised (11,13,19). Therefore, it is not likely to be a hazard to humans. *B. subtilis,* however, exists for prolonged periods in nature because of the formation of spores. It is important to realize that even organisms not forming spores can persist, as in the case of *Mycobacterium tuberculosis* in dust or dried sputum or *E. coli* in water samples. The problem of persistence of *B. subtilis* can be overcome by the isolation of asporogenic variants that cannot survive well in nature. Because autolysins are inactivated by protease produced during sporulation, these asporogenic mutants rapidly autolyze during the stationary phase of growth (4). Many of the asporogenic mutants can still be transformed (37), thus permitting genetic manipulation. The introduction of additional auxotrophic requirements such as D-alanine and purine can further reduce survival of these strains (F. E. Young and V. Clark, *unpublished observations*).

The construction of vectors is a more difficult problem. Although a number of plasmids have been identified in *B. pumilus* and *B. subtilis,* these do not carry any selective traits (15,16). Therefore, it was necessary to develop bacteriophage and plasmid vectors. To this end we have initially concentrated on the bacteriophage ϕ3T (24,29,30,34,36). This virus, originally isolated by Tucker (27), carries a gene encoding thymidylate synthetase that we have designated *thy*P3 (28,29). Bacteria that require thymine for growth because of mutations in *thy*A and *thy*B can be converted to Thy$^+$ by infection (28–30) or transfection (29,36) with this virus. The frequency of transformation of *B. subtilis* to Thy$^+$ by bacteriophage DNA or prophage DNA is only 10- to 20-fold lower than the transformation of other auxotrophic markers, even though the *thy*P gene is from a bacteriophage (Fig. 1). Transformation to Thy$^+$ does not require the complete bacteriophage genome. Treatment of bacteriophage ϕ3T with the site-specific endonucleases *Eco*RI, *Bam*HI, *Bgl*II, *Bsu, Hin*dII, and *Hin*dIII does not result in a complete inactivation of the *thy*P3 gene, although a marked reduction in Thy$^+$ transforming activity occurs. The endonuclease *Hpa*II and pancreatic deoxyribonucleases, however, do abolish the Thy$^+$ transforming activity.

Mapping studies with PBS1 establish that this unusual bacteriophage integrates in the chromosome of *B. subtilis* in the terminus between the two bacterial genes encoding the biosynthesis of thymine (*thy*A and *thy*B) as shown in Fig. 2. Dr. J. Hoch isolated a virus from soil that also "converts" Thy$^-$ to Thy$^+$ strains. Dean et al. (9) have shown that this bacteriophage, designated ρ11, is quite similar to ϕ3T as evidenced by agarose gel patterns of *Eco*RI digests of DNA of ρ11 and ϕ3T and neutralization with antibody. Preliminary studies indicate that the bacteriophage ρ11 integrates at the same site as ϕ3T (M. Williams, *unpublished observations*).

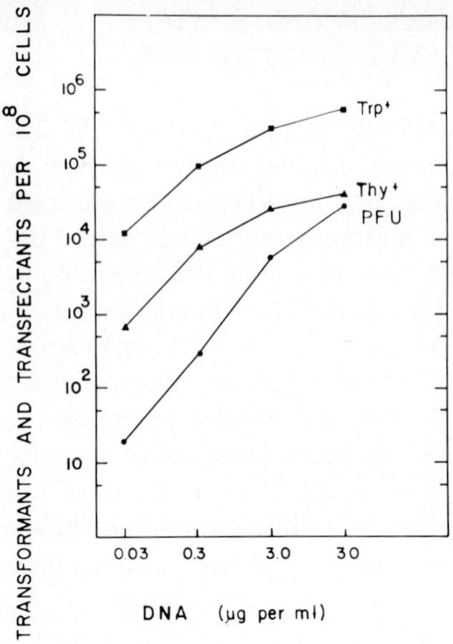

FIG. 1. Dose-response curve for transformation with ϕ3T prophage DNA. DNA from strain CU809 (ϕ3T) was used to transform competent RUB830. Thy^+ and Trp^+ CFU and ϕ3T PFU were assayed. (Reprinted from ref. 29.)

In order to develop novel chimeric plasmids for cloning experiments, we used the plasmid pMB9 that has a single *Eco*RI site and carries the gene for tetracycline resistance (Tet^r). The plasmid was propagated in *E. coli* strain C600 and purified by density gradient centrifugation in the presence

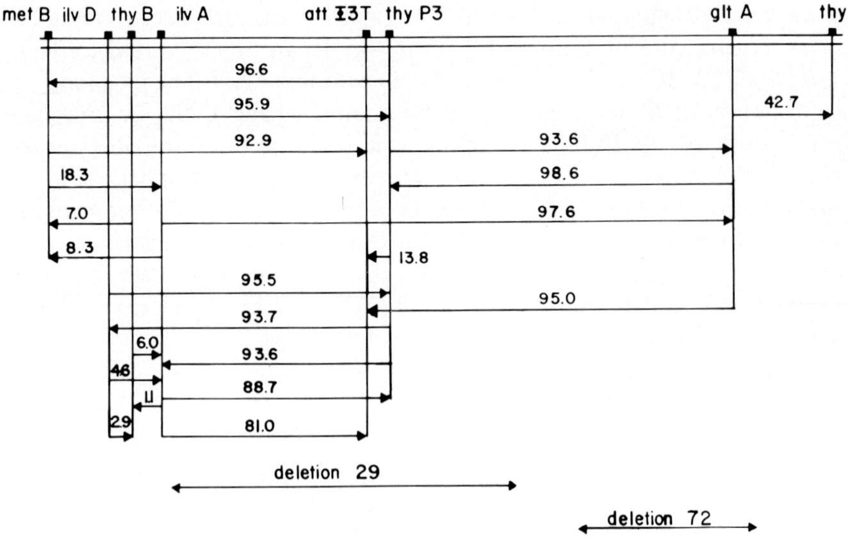

FIG. 2. Genetic map of the terminal region of the *B. subtilis* chromosome showing the attachment site of the ϕ3T genome and *thy*P3 locus. Map distances are based on the PBS1 transduction analysis. (Reprinted from ref. 29.)

of ethidium bromide (7). Plasmid pMB9 and bacteriophage ϕ3T were treated with *Eco*RI, the endonuclease inactivated by heat, and the mixture incubated with ligase (T. Maniatis, *personal communication*). Samples of the reaction mixture were added to a Thy$^-$ strain of *E. coli* C600 (18) under conditions for optimal transformation (5,8), and Thy$^+$ and Tet^R recombinants were selected on minimal medium containing tetracycline. The plasmid was recovered from lysates of two of these Thy$^+$ Tet^r *E. coli* transformants by density gradient centrifugation in the presence of ethidium bromide (7) and analyzed by agarose gel electrophoresis using a variety of site-specific endonucleases. The molecular weights of the fragments were calculated from the mobility of a standard digest of adenovirus-2 viral DNA as described previously (24). Based on the preliminary analysis of the molecular weights of the fragments shown in Fig. 3, it was concluded that the chimeric plasmid pCDl is 5.54 megadaltons. pCDl has two *Eco*RI sites, two *Bgl*II sites, and one *Bam*HI site. The fragment encoding *thy*P3 derived from bacteriophage ϕ3T is approximately 2.1 megadaltons.

To determine the biologic activity of the chimeric plasmid pCDl, we performed two additional transformation experiments. First, the plasmid was used to transform *E. coli* to Tet^r. Of 1,200 Tet^r clones tested, all were Thy$^+$. Furthermore, all of the Thy$^+$ recombinants were also Tet^r. Second, we examined the capacity of DNA from pCDl to transform a Thy$^-$ strain of *B. subtilis*. In these experiments the efficiency of the *thy*P3 gene in pCDl was compared to the efficiency of transformation with mature ϕ3T DNA and prophage ϕ3T DNA. As shown in Fig. 4, the frequency of transformation of Thy$^+$ was similar with the three preparations of DNA. Transformants could be detected with 10 pg of DNA. Considering that the bacteriophage DNA is approximately 78 megadaltons (24), the maximum enrichment of the *thy*P3 gene is 34-fold in pCDl as compared with ϕ3T. Thus, the frequency of transformation of *B. subtilis* with plasmid DNA obtained from *E. coli* is extraordinarily high.

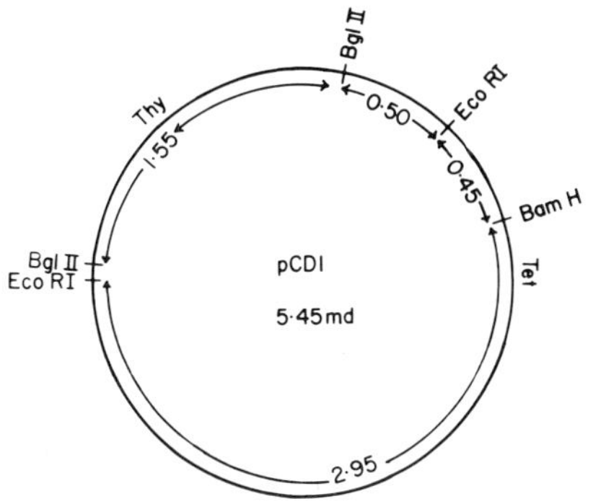

FIG. 3. Physical map of pCD1. The chimeric plasmid pCD1 was treated with *Bam*HI, *Bgl*II, and *Eco*RI. Fragments were separated by electrophoresis in agarose gels and the mobility compared with that of digests of adenovirus-2 DNA.

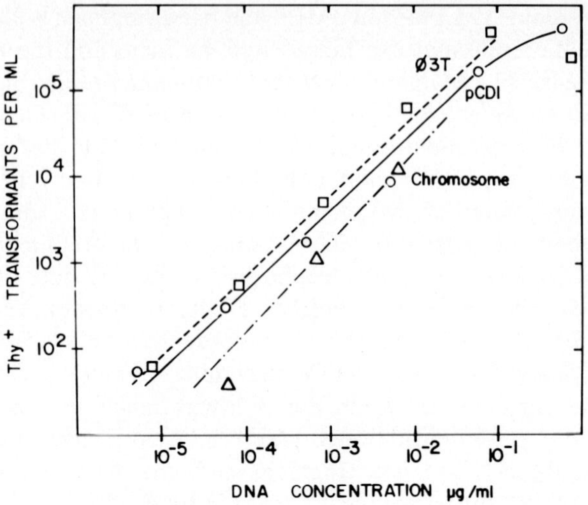

FIG. 4. DNA-mediated transformation of *B. subtilis* with *thy*P3 from various sources. Strain RUB830 carrying *phe*A, *trp*C2, *thy*A, and *thy*B was transformed with varying concentrations of pCD1, mature and prophage DNA. Transformants were selected on minimal glucose agar supplemented with casein hydrolysate.

DISCUSSION

Transformable strains of bacteria usually do not discriminate between DNA from heterologous and homologous DNA during uptake. For instance, *B. subtilis* and *Streptococcus pneumoniae* will irreversibly bind DNA from calf thymus as well as the homologous transforming DNA. Only extensive modification, such as glucosylation as in the case of bacteriophage T6 DNA, significantly influences DNA uptake and the efficiency of DNA as a competitor of homologous DNA in transformation experiments (22). Unlike the rapid incorporation of the homologous DNA with the recipient chromosome, the heterologous DNA in *S. pneumoniae* persists for longer periods of time prior to its eventual degradation and reincorporation with the chromosome of the host (14). Despite its intracellular persistence, heterologous DNA from *E. coli* is rarely biologically active. Therefore, the exchange of chromosomal genetic traits from foreign strains has, in large measure, been influenced both by the nucleotide homology between the donor and recipient chromosome and by the capacity of the competent cell to restrict the incoming fragment of DNA. Alternatively, a foreign segment of DNA could be integrated in the form of a covalently closed, circular DNA molecule as either a plasmid or a virus. Although isolated reports of transfection of *B. subtilis* by polyoma virus (1) and transformation by R factors (10) exist, there have been no substantiated reports of transformation of *B. subtilis* by donor DNA from widely divergent organisms. Nevertheless, the studies of Chang and Cohen (5) clearly demonstrate that a plasmid gene encoding ampicillin resistance from *Staphylococcus aureus* can function in the cytoplasm of *E. coli* when a chimeric plasmid is formed between the *E. coli* plasmid pSC101 and the *S. aureus* plasmid pI258. Furthermore, such interspecies chimeras may have some advantages in cloning experiments owing to their ease of separation from resident plasmids (25). Because of the low frequency of transformation in *S. aureus*, it is unlikely that reciprocal transformation of *S. aureus* would occur readily. The successful expression of the *E. coli–S. aureus* chimeric plasmid in *E. coli* indicates that it may be possible to overcome species barriers or detect previously unobserved natural transformation by developing a chimeric plasmid using pMB9 and the gene encoding thymidylate synthetase, *thy*P3. Nevertheless, we did not expect such a high level of function of *thy*P3 in *E. coli* nor the high efficiency transformation of *B. subtilis* with DNA from pCDl. Apparently, *B. subtilis* 168 does not distinguish between the *thy*P3 gene in the mature ϕ3T DNA, prophage DNA, or the chimeric plasmid based on the efficiency of transformation of the recipient strain RUB830 by DNA from these sources (Fig. 4). Furthermore, as little as 10 pg of chimeric plasmid DNA would transform *B. subtilis*. Therefore, genetic exchange between *B. subtilis* and *E. coli* can now be viewed as a reality.

At present a number of questions remain. Why does the chimeric plasmid escape restriction and function in *B. subtilis?* Does the *thy*P3 encoding fragment carry a promoter that enables it to function in *B. subtilis?* Does the chimeric plasmid exist as a plasmid in *B. subtilis* or is it integrated into the chromosome near the attachment site of ϕ3T? Unfortunately, these questions are at present unanswered. The high frequency of transformation of *B. subtilis* by the chimeric plasmid and the low probability of transformation of *B. subtilis* by *E. coli* genes indicate that *thy*P3 and not a contaminating gene obtained from *E. coli* is most probably responsible for the Thy^+ transforming activity. At present, the most plausible model is that the chimeric plasmid contains *E. coli* promoters obtained from pMB9 and at least one *B. subtilis* promoter derived from ϕ3T. Such molecular anatomy would enable the chimeric plasmid to function in two cytoplasms.

Independent studies by Ehrlich and co-workers (*this volume*) have also documented that the *thy*P3 gene studied in our laboratory can be incorporated into the plasmid pSC101. Based on these investigations from Lederberg's laboratory and our own studies, it is clear that genes can be transferred readily from *B. subtilis* to *E. coli*. This discovery should greatly facilitate the transfer of genes between these species on the chimeric plasmids. Furthermore, it is our opinion that transformable, enfeebled vehicles in *B. subtilis* should be developed to facilitate the use of this nonpathogenic host for cloning experiments. Studies are in progress to develop cloning vehicles carrying *thy*P3 as the selective trait. Through these methods the classic barriers to heterologous transformation should be breached.

ACKNOWLEDGMENT

These studies were aided in part by Grant VC 27 from the American Cancer Society.

REFERENCES

1. Bayreuther, K. E., and Romig, W. R. (1964): Polyoma virus: Production in Bacillus subtilis. *Science,* 146:778–779.
2. Bettinger, G. E., and Young, F. E. (1975): Transformation of Bacillus subtilis: Transforming ability of deoxyribonucleic acid in lysates of L-forms or protoplasts. *J. Bacteriol.,* 122:987–993.
3. Bron, S., and Murray, K. (1975): Restriction and modification in B. subtilis. Nucleotide sequence recognized by restriction endonuclease R. Bsu R from strain R. *Mol. Gen. Genet.,* 143:25–33.
4. Brown, W. C., and Young, F. E. (1970): Dynamic interactions between cell wall polymers, extracellular proteases, and autolytic enzymes. *Biochem. Biophys. Res. Commun.,* 38:564–568.
5. Chang, A. C. Y., and Cohen, S. N. (1974): Genome construction between bacterial species in vitro: Replication and expression of Staphylococcus plasmid genes in Escherichia coli. *Proc. Natl. Acad. Sci. U.S.A.,* 71:1030–1034.
6. Chilton, M.-D., and McCarthy, B. J. (1969): Genetic and base sequence homologies in Bacilli. *Genetics,* 62:697–710.
7. Clewell, D. B., and Helinski, D. R. (1969): Supercoiled circular DNA-protein complex in Escherichia coli: Purification and induced conversion to an open circular DNA form. *Proc. Natl. Acad. Sci. U.S.A.,* 62:1159–1166.
8. Cohen, S. N., Chang, A. C. Y., and Hsu, L. (1972): Nonchromosomal antibiotic resistance in bacteria: Genetic transformation of Escherichia coli by R-factor DNA. *Proc. Natl. Acad. Sci. U.S.A.,* 69:2110–2114.
9. Dean, D. H., Orrego, J. C., Hutchinson, K. W., and Halvorson, H. O. (1976): A new temperate phage for Bacillus subtilis, ρ11. *J. Virol.,* 20:509–519.
10. Domaradskii, I. V., Levadnaia, T. B., Sitrikov, B. S., Rassadin, A. S., and Denisova, T. S. (1976): Transformation of Bacillus subtilis by isolated DNA of R plasmids. *Dokl. Akad. Nauk SSSR,* 226:1443–1445.
11. Farrar, W. E. (1963): Serious infections due to "non pathogenic" organisms of the genus Bacillus. *Am. J. Med.,* 34:134–141.
12. Hemphill, H. E., and Whiteley, H. R. (1975): Bacteriophages of Bacillus subtilis. *Bacteriol. Rev.,* 39:257–315.
13. Ihde, D. C., and Armstrong, D. (1973): Clinical spectrum of infection due to Bacillus species. *Am. J. Med.,* 55:839–845.
14. Lacks, S., Greenberg, B., and Carlson, K. (1967): Fate of donor DNA in pneumococcal transformation. *J. Mol. Biol.,* 29:327–347.
15. Lovett, P. S., and Bramucci, M. G. (1974): Biochemical studies of two Bacillus pumilus plasmids. *J. Bacteriol.,* 120:488–494.
16. Lovett, P. S., and Bramucci, M. G. (1975): Plasmid deoxyribonucleic acid in Bacillus subtilis and Bacillus pumilus. *J. Bacteriol.,* 124:484–490.
17. Marmur, J., Seaman, E., and Levine, J. (1963): Interspecific transformation in Bacillus. *J. Bacteriol.,* 85:461–467.
18. Miller, J. H. (Ed.) (1972): *Experiments in Molecular Genetics.* Cold Spring Harbor Laboratory, Cold Spring Harbor, New York.
19. Pearson, H. E. (1970): Human infections caused by organisms of the Bacillus species. *Am. J. Clin. Pathol.,* 53:506–515.
20. Ravin, A. W. (1963): Experimental approaches to the study of bacterial phylogeny. *Am. Natural.,* 97:307–318.
21. Shapiro, J. A., Dean, D. H., and Halvorson, H. O. (1974): Low-frequency specialized transduction with Bacillus subtilis bacteriophage ϕ105. *Virology,* 62:393–403.

22. Sołtyk, A., Shugar, D., and Piechowska, M. (1975): Heterologous deoxyribonucleic acid uptake and complexing with cellular constituents in competent Bacillus subtilis. *J. Bacteriol.,* 124:1429–1438.
23. Spizizen, J. (1958): Transformation of biochemically deficient strains of Bacillus subtilis by deoxyribonucleate. *Proc. Natl. Acad. Sci. U.S.A.,* 44:1072–1078.
24. Thomas, E. K., Hailparn, E. M., Wilson, G. A., and Young, F. E. (1976): Physical mapping of bacteriophage DNA by exonuclease III and endodeoxyribonuclease *Bam*HI. In: *Proceedings of the 1976 ICN–UCLA Conference on Molecular Mechanisms in the Control of Gene Expression,* edited by D. P. Nierlich, W. J. Rutter, and C. F. Fox, pp. 605–610. Academic Press, New York.
25. Timmis, K., Cabello, F., and Cohen, S. N. (1975): Cloning, isolation and characterization of replicating regions of complex plasmid genomes. *Proc. Natl. Acad. Sci. U.S.A.,* 72: 2242–2246.
26. Trautner, T. A., Pawlek, B., Bron, S., and Anagnostopoulos, C. (1974): Restriction and modification in B. subtilis: Biologic aspects. *Mol. Gen. Genet.,* 131:181–191.
27. Tucker, R. G. (1964): The desoxyribonucleic acid composition of a temperate Bacillus subtilis bacteriophage. *Biochem. J.,* 92:58P–59P.
28. Tucker, R. G. (1969): Acquisition of thymidylate synthetase activity by a thymine-requiring mutant of Bacillus subtilis following infection by the temperate phage ϕ3. *J. Gen. Virol.,* 4:489–504.
29. Williams, M. T., and Young, F. E. (1976): The temperate Bacillus subtilis bacteriophage ϕ3T: Chromosomal attachment site and comparison to temperate phages ϕ105 and SP02. *J. Virol.,* 21:522–529.
30. Wilson, G. A., Williams, M. T., Baney, H. W., and Young, F. E. (1974): Characterization of temperate bacteriophages of Bacillus subtilis by the restriction endonuclease EcoRI: Evidence for three different temperate bacteriophages. *J. Virol.,* 14:1013–1016.
31. Wilson, G. A., and Young, F. E. (1972): Intergenotic transformation of the Bacillus subtilis genospecies. *J. Bacteriol.,* 111:705–716.
32. Wilson, G. A., and Young, F. E. (1973): Intergenotic and heterospecific transformation: The mechanism of restriction of genetic exchange in Bacillus subtilis. In: *Bacterial Transformation,* edited by L. J. Archer, pp. 269–292. Academic Press, New York.
33. Wilson, G. A., and Young, F. E. (1975): Isolation of a sequence-specific endonuclease (BamI) from Bacillus amyloliquefaciens H. *J. Mol. Biol.,* 97:123–125.
34. Wilson, G. A., and Young, F. E. (1976): Restriction and modification in the Bacillus subtilis genospecies. In: *Microbiology 1976,* edited by D. Schlessinger, pp. 350–357. American Society for Microbiology, Washington, D. C.
35. Yamaguchi, K., Nagata, Y., and Maruo, B. (1974): Genetic control of the rate of α-amylase synthesis in Bacillus subtilis. *J. Bacteriol.,* 119:410–415.
36. Young, F. E., Williams, M. T., and Wilson, G. A. (1976): Development of biochemical genetics in Bacillus subtilis. In: *Microbiology 1976,* edited by D. Schlessinger, pp. 5–13. American Society for Microbiology, Washington, D. C.
37. Young, F. E., and Wilson, G. A. (1972): Genetics of Bacillus subtilis and other gram-positive sporulating bacilli. In: *Spores V,* edited by H. O. Halvorson, R. Hanson, and L. L. Campbell, pp. 77–106. American Society for Microbiology, Washington, D.C.

Recombinant Molecules: Impact on Science and Society, edited by R. F. Beers, Jr. and E. G. Bassett. Raven Press, New York © 1977.

6. Biological Containment: The Subordination of *Escherichia coli* K-12

Roy Curtiss, III, Dennis A. Pereira, J. Charles Hsu, Sheila C. Hull, Josephine E. Clark, Larry J. Maturin, Sr., Raúl Goldschmidt, Robert Moody, Matsuhisa Inoue, and Laura Alexander

Department of Microbiology, Cancer Research and Training Center, Institute of Dental Research, University of Alabama in Birmingham, Birmingham, Alabama 35294

INTRODUCTION

The idea to manipulate genetically host strains and vectors to make them safer for recombinant DNA molecule research was strongly endorsed at the International Conference on Recombinant DNA Molecules held at the Asilomar Conference Center in Pacific Grove, California, in February 1975. The birth of this concept of biological containment as a means for augmenting the safety afforded by physical containment facilities and procedures was relatively painless as was the initial formulation of the means to manipulate hosts and vectors genetically to preclude the survival and/or transmission of cloned recombinant DNA to other hosts and/or vectors in the biosphere. The disarming of *Escherichia coli* K-12 and its plasmid and phage cloning vectors has, however, been considerably more difficult, frustrating, and time-consuming than originally anticipated.

Immediately after the completion of the Asilomar meeting, our laboratory group commenced to construct a number of *E. coli* K-12 strains that would have numerous built-in safety features for recombinant DNA research. Our specific goal was to construct strains possessing mutations that would

1. Increase usefulness for recombinant DNA molecule research.
2. Preclude colonization of and survival in the intestinal tract.
3. Preclude biosynthesis of the rigid layer of the cell wall in nonlaboratory controlled environments.
4. Lead to degradation of DNA in nonlaboratory controlled environments.
5. Permit cloning vector replication to be dependent on the host strain.

6. Preclude or minimize transmission of recombinant DNA to other bacteria that might be encountered in nature should the host strain escape from the confines of the laboratory.
7. Permit monitoring of the strains.

Our initial expectation was that the task could be easily attained. To the contrary, we have learned that *E. coli* has enormous resiliency and a great capacity to do the illogical and unexpected. With each passing month, our respect for the sophisticated biological mechanisms that have evolved in *E. coli* has increased. Moreover, we have learned a great deal about which mutations do and which mutations do not contribute to either the safety, utility, or both of *E. coli* strains for recombinant DNA molecule research. Toward the end of January 1976, we completed construction of a strain, designated χ1776, that seemed to possess a sufficient number of safety features to satisfy our goals and meet the standards of an EK2 host for cloning with nonconjugative plasmid vectors as set forth in the NIH Guidelines for Research Involving Recombinant DNA Molecules (16). We thus subjected this strain to an exhaustive series of tests to evaluate its safety and utility for recombinant DNA molecule experiments and to uncover new information that would hopefully permit future construction of a completely fail-safe *E. coli* strain. This chapter reviews what we have learned during our more than year-long education by *E. coli*.

RESULTS

Mutations that Increase the Usefulness of Strains for Recombinant DNA Molecule Research

As a standard necessary attribute, all strains being constructed must have *hsdR*[1] or *hsdS* alleles to abolish restriction thereby making it possible to introduce foreign DNA sequences into the strain. In studying transformability of strains, we found that transformation frequencies increased in strains with *endA* mutations (3- to 30-fold), *dapD*, and Δ[*bioH-asd*] mutations (approximately 3-fold), and with *galE* or Δ[*gal-uvrB*] mutations (4- to 10-fold). Although this information makes it possible to construct a strain that is considerably more transformable than wild-type K-12 strains, we have found that *rfb* mutations, which cause the deletion of carbohydrates from the lipopolysaccharide (LPS) core, reduce transformability 3- to 7-fold. This is unfortunate since these mutations have several desirable features in that they reduce recipient ability for many conjugative plasmid types and confer increased sensitivity to bile salts, detergents, and antibiotics. In this latter regard, strains with *dapD*, Δ[*bioH-asd*], and *rfb* mutations lyse easily, thus facilitating isolation of chimeric plasmid DNA.

[1] All gene symbol designations are those used and defined by Bachmann et al. (1).

Many of the strains being constructed possess mutations that lead to the continuous production of chromosome-deficient minicells during growth of the culture. Minicells result from abnormal cell divisions occurring near the polar ends of cells. Minicells are particularly useful in studies on the expression of DNA cloned on plasmid vectors, since minicells produced by plasmid-containing strains possess plasmid DNA which can be transcribed and translated (11,12,18). Consequently, such minicell-producing strains have already been useful in studying the expression of eukaryotic DNA sequences cloned on the pSC101 plasmid (5,15).

Mutations that Preclude Colonization of and Survival of Strains in the Intestinal Tract

Unpublished experiments done 10 years ago at Oak Ridge National Laboratory indicated that both *pur* and *thy* mutations caused *E. coli* strains to survive at reduced frequencies during passage through the intestinal tract of mice. We were also interested in both of these mutations since *thy* mutants should undergo thymineless death with concomitant degradation of DNA; *pur* mutations should also contribute to the further avirulence of *E. coli* K-12 since it is known that *pur* mutations in pathogenic microbial species cause avirulence because of interference with the ability of cells to grow intracellularly (3). In numerous tests in which approximately 10^{10} bacteria were suspended in milk and introduced into the esophagus of rats, we could obtain no evidence that *pur* mutations affected the intestinal titers or the duration of excretion of the *E. coli* K-12 strains before their elimination from the intestinal flora. We have thus discontinued including a purine requirement in our strains since it could impede the growth of strains in certain environments, thereby interfering with the phenotypic expression of other mutations that abolish synthesis of the rigid layer of the cell wall. Strains with mutations in the *thyA* gene, however, survive passage through the rat's intestinal tract at somewhat reduced frequencies. It is interesting to note that secondary mutations in the *deoC* gene but not the *deoB* gene usually cause a substantially lower survival and shorter duration of excretion of the *thyA* strain. Since *deoC* mutants lack deoxyriboaldolase, we can surmise that their lower survival rate in the rat intestine is owing to the presence of high concentrations of purine deoxyribonucleosides and/or thymidine, which are toxic to cells that possess such mutations (4). Further tests are underway to elucidate the basis for this behavior.

Mutations in the *rfbA* or *rfbB* cistrons confer increased sensitivity to a large number of antibiotics, drugs, ionic detergents, and, of greatest importance, bile salts. We have not yet tested the effect of an *rfb* mutation alone on colonization and survival of *E. coli* K-12 strains in the rat intestinal tract, but we know that this mutation, in conjunction with *thyA* and mutations that abolish synthesis of the rigid layer of the cell wall, has resulted in no recoverable survivors after feeding of rats. So far we have tested over 40 rats with an inoculation dose of approximately 10^{10} *E. coli* cells per rat. It is expected, although unconfirmed as yet, that other mutations conferring bile salts sensitivity such as *lpcA, lpcB,* and *rfa* would similarly lead to the inability of strains to survive in or colonize the intestinal tract of animals, provided that bile was produced and delivered to the duodenum.

Mutations that Preclude Biosynthesis of the Rigid Layer of the Cell Wall in Nonlaboratory Controlled Environments

Our initial goal was to construct strains that would require diaminopimelic acid (DAP), the precursor in a biosynthetic pathway unique to prokaryotic organisms leading to the synthesis of lysine. Since DAP is an unusual amino acid, we reasoned that it would not be prevalent in nature. Because DAP is an essential constituent of the murein layer of the *E. coli* cell wall, strains possessing defects in its synthesis would form spheroplasts in the absence of DAP and thereby lyse. Among numerous *dap* mutations tested, only the *dapD8* allele isolated and characterized by Bukhari and Taylor (2) was found to be genetically stable (reversion frequency of approximately 10^{-9}), and thus this allele was introduced into some of our strains for initial tests. We soon found, to our dismay, that strains possessing this allele did not always undergo DAP-less death and indeed were capable of growth, not only in liquid media but also when replica plated on solid media lacking DAP. The ability to survive in the absence of DAP was dependent on the presence of certain cations (Na^+, K^+, Mg^{2+}, and/or Ca^{2+}) and was also temperature dependent with survival being noted at 37°C or lower but not at 42°C. Upon microscopic observation, surviving cells growing in DAP-deficient liquid media appeared as spheroplast-like bodies surrounded by a capsular material and colonies forming on agar medium lacking DAP were very mucoid. These colonies were also composed of spheroplast-like bodies. It was thus evident that these *E. coli* strains, when placed under adverse circumstances, were able to synthesize colanic acid, a polysaccharide composed of galactose, glucose, glucuronic acid, and fucose, and whose synthesis is regulated by the *lon* gene (14). Consequently, it was necessary to introduce one of a variety of mutations such as *galE, galU, man,* or *non* that would abolish the ability of cells to make colanic acid and thus minimize, if not preclude, survival of spheroplast forms resulting from cessation of synthesis of the rigid cell wall layer.

During these studies we also investigated properties of other mutations that would confer a Dap^- phenotype, such as deletions of the gene for aspartic acid semialdehyde dehydrogenase. Ultimately, a strain with the *dapD8*, Δ[*bioH-asd*], and Δ[*gal-uvrB*] mutations was constructed and found to undergo DAP-less death under a variety of conditions. This strain was also shown to be unable to synthesize colanic acid. Such a constellation of mutations also promoted more rapid rates of death during passage through the intestinal tract of rats. The rates of DAP-less death, however, depend on a number of parameters: cells exhibiting faster rates of protein synthesis yield more rapid rates of death, and cells inoculated from log-phase cultures die more rapidly than cells inoculated from stationary-phase cultures. The cell density at the inception of DAP-less growth is also important in that cells starting at low density die more rapidly than cells inoculated at high density. Indeed, when one starts with cell densities of $>5 \times 10^7$ cells/ml, the culture never completely dies since survivors apparently can scavenge DAP released from lysed cells.

We have also observed that the presence of other microorganisms, whether introduced into the culture deliberately or accidently, accelerates the rate of DAP-less death. Since DAP-less death is dependent on concomitant protein synthesis and *dap* mutations confer an obligate requirement for lysine, and *asd* mutations confer an obligate requirement for both threonine and methionine in addition to DAP, DAP-less death can be achieved only when *dap asd* strains grow in the presence of threonine, methionine, and lysine. Because of this, we are currently exploring the behavior of mutations that abolish alanine racemase and D-alanyl-D-alanine ligase activities, which are necessary for the synthesis of D-alanyl-D-alanine, another unique component of the rigid layer of the *E. coli* cell wall. Since such mutants would be able to synthesize L-alanine and carry out protein synthesis in media containing a metabolizable energy source but devoid of amino acids, they should offer some safety advantages over use of strains with *dap* and/or *asd* mutations.

It is evident from our studies that strains defective in the synthesis of the rigid layer of the cell wall offer a significant safety advantage, provided that they contain additional mutations that abolish the synthesis of the extracellular capsule which balances the osmotic pressure differential between the medium and the inside of the bacterial cell. It appears, therefore, that this experimental approach to safer strain construction is not only highly efficacious for *E. coli* but conceivably valid for application to a wide range of microbial species for future development of safer host-vector systems.

Mutations that Facilitate Degradation of DNA in Nonlaboratory Controlled Environments

We expected that the presence of *thyA* mutations, which lead to thymineless death, would also lead to degradation of DNA because of the accumulation of unrepaired single-strand breaks (13). Although thymineless death does occur in thymine-deficient medium and this is accompanied by degradation of DNA, we have not observed a significant decrease in the molecular weight of single-stranded DNA as determined on alkaline sucrose gradients.

Since we have observed a loss of covalently closed circular pSC101 molecules in this strain during thymine starvation, it is possible that DNA degradation is closely associated with appearance of single-strand breaks such that we could not expect to detect the decrease in DNA molecular weights because of the formation of single-strand breaks, which have been estimated to occur at a rate of 0.98 per chromosome per minute (13). As an additional or alternative explanation, it should be noted that the strains in which DNA degradation has been investigated most extensively also possess mutations that abolish dark repair and restriction, either of which in the mutant or wild-type state might affect detection of single-strand breaks and/or subsequent DNA degradation during thymineless death. These possibilities are currently under study.

We have also investigated the behavior of strains with a *polA* (TS) allele such as *polA214* or *polA12* in combination with a *recA* allele such as *recA200* (TS) to see if these combinations of mutations in the presence of a *thyA* mutation would lead to an increased rate of DNA degradation in the presence or absence of thymidine. Although the presence of *polA* (TS) and *recA* (TS) in *thyA* strains causes an accelerated rate of DNA degradation during thymineless death at 42°C, the presence of these mutations has little noticeable effect at 37°C. The presence of a *polA* (TS) allele does, however, cause *thyA* strains to commence thymineless death immediately at 37°C with no lag in loss of colony-forming ability as is usually noted for *thyA* or *thyA recA* (TS) strains. We have not, however, observed that *polA* (TS) and/or *recA* (TS) alleles alter either the rates of death or the time for clearance of *thyA* strains from the rat intestinal tract. Based on these results, it is evident that if *polA* and *recA* mutations are to be introduced into future safer strains to accelerate the rates of DNA degradation at temperatures usually encountered by *E. coli,* it will be necessary to isolate additional temperature-sensitive alleles of the *polA* and/or *recA* genes. Since we already know what mutations eliminate survival of strains *in vivo,* it would seem judicious to isolate conditional alleles that confer cold rather than heat sensitivity. Indeed, the use of *polA* (CS) alleles should also abolish replication and transmission of ColE1 cloning vectors at temperatures below 30°C.

Mutations that Would Permit Replication of Cloning Vectors to Be Dependent on the Host

Since we have not undertaken the development of cloning vectors, no work at present has been attempted in this area other than to include a *supE* allele in all strains. The *supE* allele was maintained under the expectation that most safer cloning vectors would have some critical function that would be dependent on the presence of an amber suppressor.

Mutations that Preclude or Minimize Transmission of Recombinant DNA to Other Bacteria

Mutations that cause resistance to all of the familiar specialized and generalized transducing phages of *E. coli* are well known and indeed have been introduced into some of our strains. Nevertheless, it is highly likely that alterations in the cell surface which result in resistance to these known phages will endow the strain with sensitivity to other phages that lurk in the sewers and polluted rivers. It thus may be difficult, if not impossible, to ensure that a strain could not be infected by a potential transducing phage in nature that would act to transmit a cloned DNA fragment to some other robust organism. The probabilities for such occurrences, however, would be extremely small, especially if the disarmed host is unable to synthesize DNA because of a *thyA* mutation or is otherwise in a metabolically inactive state because of growth requirements or is in the process of dying. It should be noted in this regard that *E. coli* K-12 cells suspended in phosphate-buffered saline and then infected with T6 bacteriophage have been observed to yield a small burst of liberated phage several hours after infection, especially after prior starvation of the cells for 4 hr. Whether some of the transducing phages which are more host dependent than T6 can replicate or not is as yet unknown. This observation, coupled with our observation that *E. coli* can transfer conjugative plasmids at low frequency after long periods of starvation on nongrowth media at 37°C, underscores the necessity of ultimately introducing mutations that accelerate the rate of the death of *E. coli* under conditions of starvation when thymineless and DAP-less death do not occur.

To block the transmission of nonconjugative plasmid cloning vectors, we have been isolating and characterizing mutants with various cell surface defects that would contribute to a conjugation-deficient (Con^-) recipient phenotype. The rationale for this approach is based on the fact that a cell possessing a nonconjugative, plasmid vector–harboring, recombinant DNA would first have to acquire a conjugative plasmid in order to transmit that cloned DNA to some other microorganism. Although a considerable amount of work has been done on the isolation and characterization of Con^- recipients with regard to their ability to mate with Hfr and F′ donors (8,9,17, 19), there has been no previous work on the isolation of Con^- mutants defective in inheriting the 18 to 20 other conjugative plasmid types found in gram-negative enteric microorganisms. Because of the diversity of plasmid types and of mutational lesions that could potentially reduce the recipient ability of safer host strains, we have spent considerable time during the past year in isolating and characterizing Con^- mutants. For example, we have found that the presence of deletions of the *gal* operon reduces the recipient ability 1,000-fold or more for plasmids in the M, O, W, X, and 10 incompatibility groups with smaller yet significant reductions noted for plasmids in the C, I (only some of those tested), J, L, T, and 9 incompatibility groups. The addition of an *rfb* mutation to a strain with a Δ*gal* lesion either does not alter or further reduces recipient ability for the aforementioned plasmid types but more importantly reduces recipient ability approximately 1,000-fold for plasmids in the F and N incompatibility groups, which are quite prevalent in *E. coli* strains in nature, as are I-type plasmids (10). Thus, a strain such as χ1776 with both Δ*gal* and *rfb* mutations, when mated for 90 min under optimal laboratory conditions with donors possessing repressed conjugative plasmids, gives transconjugant frequencies of 10^{-4} to 10^{-5} for L, P, and some I group plasmids; 10^{-5} to 10^{-6} for J, 9, and other I group plasmids; and less than 10^{-7} for C, FII, H, M, N, O, W, X, and 10 group plasmids. Other mutants with different cell surface defects are currently being isolated and characterized with regard to their ability to either receive or donate the various types of conjugative plasmids.

Since little is known about plasmid transfer between gram-negative microorganisms in soil, water, and sewage, we investigated conjugational plasmid transfer at temperatures likely to occur in these environments. At 27°C we were unable, even when using wild-type Con^{+} recipient strains of *E. coli,* to detect plasmid transfer at frequencies in excess of 10^{-8} except for transfer of plasmids in the T and P incompatibility groups. Thus, if we assume that laboratory conditions reflect those found in nature, it would be unlikely that a significant amount of mobilization of nonconjugative plasmid cloning vectors would occur in natural environments other than in the intestinal tract because of temperatures unsuitable for the expression of the donor and/or recipient phenotypes necessary for conjugative plasmid transmission. A more complete treatment of the probabilities of transmission of recombinant DNA contained on nonconjugative plasmid cloning vectors has been presented and discussed elsewhere (6,7).

Mutations that Permit Monitoring of Strains

We have introduced mutations that confer high-level resistance to nalidixic acid as a means to recover strains during rat feeding experiments. This particular genetic marker is suitable for such experiments and routine monitoring since resistance to nalidixic acid is not common among naturally occurring gram-negative organisms and has not been observed to be plasmid mediated and since mutations to resistance occur at a very low spontaneous frequency. In addition, nalidixic acid has a high efficiency of killing such that it can be used to detect a single Nal^{r} isolate in the presence of 10^{8} to 10^{9} Nal^{s} cells. Many of our strains also possess mutations to resistance to cycloserine, which also has not been observed to be plasmid mediated. It is thus possible to use double selection for resistance to cycloserine and nalidixic acid to detect very low numbers of cells when testing strains in a variety of environments in and outside of the laboratory.

One of the chief concerns expressed by many who are apprehensive about the safety of recombinant DNA molecule research is the inadvertent contamination of a culture of a disarmed strain with cells of a robust transformable strain during transformation with recombinant DNA molecules. We have found that the addition of nalidixic acid and/or cycloserine to the culture during its growth before transformation and to the selective medium for plating transformants greatly minimizes, if not precludes, transformation of such contaminants. In this regard, the use of cycloserine is somewhat preferable to nalidixic acid since the latter is difficult to dispose of in a safe way, being stable to autoclaving, whereas the former has a reasonably short half-life at neutral or acidic pHs and is rapidly destroyed in nature.

Construction, Properties, and Testing of χ1776

χ1776 was constructed in 13 steps from χ1276 (12) and possesses 15 different mutational lesions. Table 1 lists most of the phenotypic properties

TABLE 1. *Phenotypic properties of χ1776*

Phenotype	Responsible mutation(s)[a]
Requires diaminopimelic acid	*dapD8* Δ29[*bioH-asd*]
Requires threonine	Δ29[*bioH-asd*]
Requires methionine	*metC65* Δ29[*bioH-asd*]
Requires biotin	Δ40[*gal-uvrB*] Δ29[*bioH-asd*]
Requires thymidine	*thyA57*
Cannot use galactose for growth	Δ40[*gal-uvrB*]
Cannot use maltose for growth	Δ29[*bioH-asd*]
Cannot use glycerol for growth	Δ29[*bioH-asd*]
Cannot synthesize colanic acid	Δ40[*gal-uvrB*]
Sensitive to UV and defective in dark repair	Δ40[*gal-uvrB*]
Cannot undergo photoreactivation	Δ40[*gal-uvrB*]
Sensitive to glycerol (aerobic)	Δ29[*bioH-asd*]
Sensitive to bile salts, ionic detergents, and antibiotics	*rfb-2*
Resistant to nalidixic acid	*nalA25*
Resistant to cycloserine	*cycA1 cycB2*
Resistant to chlorate (anaerobic)	Δ40[*gal-uvrB*]
Resistant to trimethoprim	*thyA57*
Resistant to T1, T5, and φ80	*tonA53*
Resistant to λ and 21	Δ29[*bioH-asd*]
Partially resistant to P1	*rfb-2*
Conjugation defective	Δ40[*gal-uvrB*] *rfb-2*
Produces minicells	*minA1 minB2*
Temperature sensitive at 42°C	One mutation that is partially responsible for phenotype is linked to *thyA*
Cannot restrict foreign DNA	*hsdR2*
Allows for vector to be host dependent	*supE42* (amber)

[a] See Bachmann et al. (1).

of this strain along with the responsible mutations for the designated phenotype. A more complete description of the properties associated with some of these mutations and data verifying its safety features will be presented elsewhere. In summary, χ1776 and its derivative χ1876, which possesses the pSC101 nonconjugative plasmid, (a) cannot survive passage through the intestinal tract of rats; (b) die, partially degrade their DNA, and lyse in growth media lacking DAP and thymidine; (c) cannot grow and die at variable rates after drying or when suspended in water or other nongrowth media; and (d) are unable to transmit genetic information contained on pSC101 to other bacteria under any of the above-described nonpermissive conditions at measurable frequencies. It should be pointed out that by factoring the various parameters necessary for a successful triparental mating to allow for receipt and then mobilization of pSC101 by a conjugative plasmid, we can predict that this series of events could occur in nature with a probability of approximately 10^{-22} per surviving bacterium (7).

DISCUSSION AND CONCLUDING REMARKS

Our future goals include the construction of χ1776 derivatives with additional mutations that would further reduce its recipient ability in matings with donors possessing various conjugative plasmids, increase its transformability, and increase its rate of demise in nonlaboratory controlled environments. At the same time, we are constructing other sublines of *E. coli* K-12 with defects in D-alanine metabolism as a means to block cell wall biosynthesis. The latter, together with the other mutations as described in this chapter, should confer the necessary attributes required for safer and more useful *E. coli* strains for recombinant DNA molecule research.

Since it is generally acknowledged that the cloning of DNA in bacteria, especially *E. coli,* is not without potential hazard, it is imperative that investigators attempting to clone DNA in these disarmed host-vector systems undertake the responsibility to verify the relevant phenotypic properties of this system immediately after isolating a clone containing recombinant DNA. With regard to χ1776 and its derivatives, the traits to be verified should include the requirements for diaminopimelic acid and thymidine, inability to utilize or ferment galactose or maltose, inability to produce colanic acid or other capsular material, and sensitivity to UV and bile salts. These tests can be easily done by either replica plating or streak testing on appropriate agar media. Clones that are not tested or show an alteration in even one phenotypic trait should be destroyed immediately. We also believe that those who clone DNA should devote some portion of their time to more extensive tests to determine the effects of cloned DNA on the survival and transmissibility characteristics of the host-vector system. We will be more than pleased to offer advice on the design of these tests and provide any strains that might be necessary to perform them.

Soon after the discovery of recombinant DNA molecule technology using *E. coli,* it became obvious that numerous potential biohazards were associated with this research. This realization was in large part based on the facts that *E. coli* is a normal intestinal inhabitant of humans and warm-blooded animals, is sometimes a severe opportunistic pathogen, and possesses the potential to exchange genetic information, especially plasmid DNA, with representative strains of over 30 bacterial genera. It is equally obvious that the development of biological containment systems and their usage in conjunction with physical containment facilities and procedures are, therefore, imperative for continued safe recombinant DNA molecule research. Although we tend to think primarily of hazard to the human species as a direct consequence of recombinant DNA molecule research, it must be remembered that other potential biohazards in recombinant DNA molecule research can exist that could cause perturbations of the ecosystem. Also, the extent of the possible damage to the biosphere and ultimately to humans when one organism acts on another to either displace it from or interfere with its normal functioning in its ecological niche is unpredictable but potentially catastrophic. These points should therefore serve to underscore the importance for developing, testing, and using disarmed microbial host-vector systems in all experiments in which the foreign cloned DNA might contribute to such potentially biohazardous conditions. This admonition is as applicable for use of microbial hosts other than *E. coli* for recombinant DNA molecule experiments as it is when using *E. coli.* Such disarmed microbial host-vector systems should minimize if not prevent the survival and transmission of cloned DNA if the chimeric host was inadvertently released from its rigorously controlled test tube habitat.

ACKNOWLEDGMENTS

Research was supported by grants from the National Science Foundation (GB-37546) and the National Institutes of Health (DE-02670, AI-11456, and 5 P02 CA 13148) and by an NIH Postdoctoral Fellowship (F32-AI-05222) to S.C.H., an NIH Postdoctoral Traineeship (T32-GM-7090) to J.E.C., and an NIH Predoctoral Traineeship (T32-GM-07164) to D.A.P.

REFERENCES

1. Bachmann, B. J., Low, K. B., and Taylor, A. L. (1976): Recalibrated linkage map of Escherichia coli K-12. *Bacteriol. Rev.,* 40:116–167.
2. Bukhari, A. T., and Taylor, A. L. (1971): Genetic analysis of diaminopimelic acid- and lysine-requiring mutants of Escherichia coli K-12. *J. Bacteriol.,* 105:844–854.
3. Burrows, T. W. (1955): The basis of virulence for mice of Pasteurella pestis. *Symp. Soc. Gen. Microbiol.,* 5:151–175.
4. Buxton, R. S. (1975): Genetic analysis of thymidine-resistant and low-thymine-requiring mutants of Escherichia coli K-12 induced by bacteriophage Mu-1. *J. Bacteriol.,* 121:475–490.
5. Chang, A. C. Y., Lansman, R. A., Clayton, D. B., and Cohen, S. N. (1975): Studies of

mouse mitochondrial DNA in Escherichia coli: structure and function of eucaryotic-procaryotic chimeric plasmids. *Cell,* 6:231–244.
6. Curtiss, R., III (1976): Genetic manipulation of microorganisms: potential benefits and biohazards. *Annu. Rev. Microbiol.,* 30:507–533.
7. Curtiss, R., III, Clark, J. E., Goldschmidt, R., Hsu, J. C., Hull, S. C., Inoue, M., Maturin, L. J., Moody, R., and Pereira, D. A. (1976): Biohazard assessment of recombinant DNA molecule research. *Proceedings of the Third International Symposium on Antibiotic Resistance* (*in press*).
8. Curtiss, R., III, Fenwick, R. G., Jr., Goldschmidt, R., and Falkinham, J. O. (1975): The mechanism of conjugation. In: *Transferable Drug Resistance Factor R,* edited by S. Mitsuhashi. University Park Press, Tokyo (*in press*).
9. Falkinham, J. O., III, and Curtiss, R., III (1976): Isolation and characterization of conjugation-deficient mutants of Escherichia coli K-12. *J. Bacteriol.,* 126:1194–1206.
10. Falkow, S. (1975): *Infectious Multiple Drug Resistance.* Pion Limited, London.
11. Frazer, A. C., and Curtiss, R., III (1973): Derepression of anthranilate synthase in purified minicells of Escherichia coli containing the Coltrp plasmid. *J. Bacteriol.,* 115:615–622.
12. Frazer, A. C., and Curtiss, R., III (1975): Production, properties and utility of bacterial minicells. *Curr. Top. Microbiol. Immunol.,* 69:1–84.
13. Freifelder, D. (1969): Single-strand breaks in bacterial DNA associated with thymine starvation. *J. Mol. Biol.,* 45:1–7.
14. Markovitz, A. (1976): Genetics and regulation of bacterial capsular polysaccharide biosynthesis and radiation sensitivity. In: *Surface Carbohydrates of Procaryotic Cells,* edited by W. Sutherland. Academic Press, New York.
15. Morrow, J. F., Cohen, S. N., Chang, A. C. Y., Boyer, H. W., Goodman, H. M., and Helling, R. B. (1974): Replication and transcription of eukaryotic DNA in Escherichia coli. *Proc. Natl. Acad. Sci. U.S.A.,* 71:1743–1747.
16. National Institutes of Health Guidelines for Research Involving Recombinant DNA Molecules (1976): U.S. Department of Health, Education and Welfare, Public Health Service, National Institutes of Health, Bethesda, Maryland.
17. Reiner, A. M. (1974): Escherichia coli females defective in conjugation and in absorption of a single-stranded deoxyribonucleic acid phage. *J. Bacteriol.,* 119:183–191.
18. Roozen, K. J., Fenwick, R. G., Jr., and Curtiss, R., III (1971): Synthesis of ribonucleic ιcid and protein in plasmid-containing minicells of Escherichia coli K-12. *J. Bacteriol.,* 107:21–33.
19. Skurray, R. A., Hancock, R. E. W., and Reeves, P. (1974): Con$^-$ mutants: class of mutants in Escherichia coli K-12 lacking a major cell wall protein and defective in conjugation and adsorption of bacteriophage. *J. Bacteriol.,* 119:726–735.

Recombinant Molecules: Impact on Science and Society, edited by R. F. Beers, Jr. and E. G. Bassett. Raven Press, New York © 1977.

7. Use of the T4 Ligase to Join Flush-Ended DNA Segments

*V. Sgaramella, H. Bursztyn-Pettegrew, and S. D. Ehrlich

**Laboratorio di Genetica Biochimica ed Evoluzionistica, C.N.R., 27100 Pavia, Italy; and Department of Genetics, Stanford University Medical School, Stanford, California 94305*

INTRODUCTION

The work published so far on the ability of T4 ligase to join "flush"-ended (fully base paired) DNA duplexes ["terminal" joining (13,15)] still leaves unanswered a variety of questions about the enzymatic and structural features of this reaction:

1. Is the enzyme involved in the terminal joining reaction the same as that involved in the nick-sealing or "cohesive" joining?
2. How do the intermediate stages of the two reactions compare?
3. Do other enzymes exhibit this property?
4. Is it a peculiarity of the ATP-dependent ligases?
5. What is its physiological role?

The relevance of these questions has been accentuated by the recent expansion of research in the *in vitro* recombination of specific DNA segments resulting from the action of type II restriction endonucleases (9). A sizeable fraction of these produces flush ends (see R. Roberts, *this volume*).

In this chapter we report experiments bearing on the first of the above questions and describe a new assay for the terminal ligation that is based on the intramolecular circularization of appropriately dilute solutions of flush-ended DNA duplexes. The circles formed in the presence of the T4 ligase have been chemically and biologically characterized as being covalently closed via phosphodiester bonds. The efficiency of the reaction can be expressed as the ratio of circular to linear molecules and can be easily determined by electron microscopic analysis (7).

Using this assay we have obtained preliminary evidence that the terminal and the cohesive reactions involve the same enzyme, and that it is possible to obtain a sufficiently pure ligase using a shortened version of the published procedure (16).

Some of the possible applications of this reaction to the splicing of DNA segments produced by different restriction endonucleases, or by altogether different kinds of cleavage, are examined. We propose a general procedure aimed at the splicing of long DNA segments by means of short, easily prepared molecular joints.

MATERIALS AND METHODS

DNA. Plasmid DNAs were prepared essentially according to published procedures (3). SPP1 phage DNA was prepared as described (11) or was a gift of J. Lu.

Enzymes. The T4 was purified according to the procedure of Weiss et al. (16) up to Fraction V. The *E. coli* ligase was a gift of Dr. I. R. Lehman. The *Bsu*R endonuclease was purified as described by Bron, Murray, and Trautner (1), with an additional DEAE-column to remove traces of exonuclease. *Eco*RI endonuclease was purified from Fraction I (18) by an ammonium sulfate precipitation followed by a DEAE-cellulose chromatography (Dr. T. Landers, *personal communication*). *Hin*dII was a gift of Dr. G. Bernardi, *Sma* was a gift of Drs. H. Kopecka and J. Lis, and *Hpa*I was a commercial enzyme.

Gel electrophoresis. Agarose gel electrophoresis was performed on vertical slabs using Tris-borate-EDTA buffer at pH 8.3 (10). SDS-polyacrylamide gel assays were run by Dr. D. Uyemura.

Electron microscopy. Aqueous spreading was performed according to Inman and Schnös (7).

Transfection assay. The recipient strain, W5449 (r^-m^+ recB recC F^- $\lambda^S thr^-$ leu^- arg^- his^- pro^- thi^- ara^- lac^- gal^- trp^- mtl^- xcl^- Str ^{R}tsx), was brought to competence essentially as described by Bursztyn et al. (2).

RESULTS AND DISCUSSION

Intramolecular Circularization of DNA Duplexes as an Assay for the Polynucleotide Ligase

Of the several restriction endonucleases known to generate flush-ended DNA segments, we have used *Hin*dII (6), *Hpa*I (M. Green and C. Mulder, *personal communication*), *Sma* (14) and *Bsu*R (1) with the DNA substrates listed in Table 1. The extent of circularization ranged from 5 to 75%. These differences can be probably ascribed to the varying levels of purity of the endonucleases used, as well as by the length of the resulting linear segments (see also "terminally" ligated DNA, *following*).

The greatest extent of circularization could be obtained by using *Bsu*-generated segments; EM counting of the circles formed allows quantifica-

TABLE 1. *Circularization of flush-ended DNA segments produced by restriction endonucleases*

Enzyme	DNA	Cleavage sites	Average segment length (kilobases)	Circularization (%)
*Hpa*I	pSC101	1	9	5
"	pMB9	1	6	5
Sma	pSC101	1	9	20
"	pMB9	1	6	20
*Hind*II	SPP1	~10	4	25
*Bsu*R[a]	pSC101	~10	~0.9	75
"	SPP1	~60	~0.8	75

[a] *Bsu*R cleavage was stopped before completion (see text).

tion of the joining reaction. However, it should be noted that this can lead to an underestimation of the enzymatic activity, since linear and circular molecules can still be the result of more than one joining event. The use of homogeneous samples of DNA molecules of optimal length precludes these limitations. Even under the described conditions, the circularization assay appeals to us because of its simplicity and rapidity.

Figure 1 shows the results of a reaction in which *Bsu*R-generated SPP1 DNA segments have been incubated at 20°C for 15 hr in the presence of T4 ligase and cofactors. Approximately 75% of the molecules can be scored as circles; this value does not exceed 2% if the ligase is omitted or if the *E. coli* enzyme is used. These circles appear to be covalently closed and, if exposed to denaturing conditions of temperature or high pH and brought back to a nondenaturing environment, they can still be visualized as relaxed circles. On the contrary, the linear molecules denature into single strands, the majority of which collapse into the so-called bushes (Fig. 2) or, because of the incomplete digestion with *Bsu*R endonuclease, renature into partially double-stranded molecules with single strands at one or both ends, identifiable as "flowers" (Fig. 3).

The time course of a joining reaction is shown in Fig. 4. Under the conditions used, a very rapid initial rate is followed by a slower, steady increase up to 75% circularization of SPP1 DNA segments that had been generated with *Bsu*R endonuclease.

Purification of the T4 Ligase "Terminal" and "Cohesive" Joining Activities

The T4 ligase can join the linear molecules produced by restriction endonucleases which introduce staggered cuts and thus originate "cohesive" ends. In the previous section we showed that it can join flush-ended segments. We decided to use the circularization of *Eco*RI- and *Bsu*R-generated DNA segments to monitor the purification of the "cohesive" and the

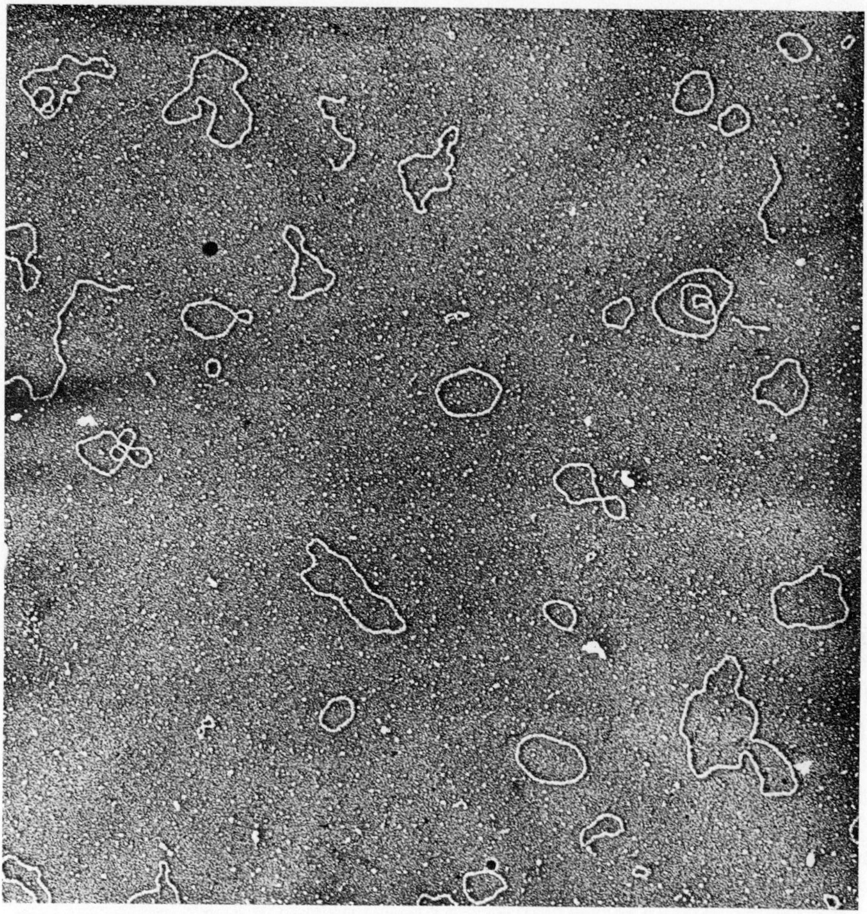

FIG. 1. Electron micrograph of flush-ended DNA segments produced by *Bsu*R endonuclease from SPP1 DNA and circularized in the presence of T4 ligase (see text).

"terminal" joining activities, respectively, of the T4 ligase. The purification procedure detailed by Weiss et al. (16) was followed until the second DEAE-column chromatography. The enzyme was assayed for ATP-^{32}PPi exchange, circularization of *Eco*RI-produced segments, and circularization of *Bsu*R-generated segments. Both substrates were obtained from SPP1 DNA, the former as limit digest and the latter after partial digestion, in order to keep the sizes of the segments within a comparable range. Figure 5 gives the elution profile of the enzyme on the second DEAE-column. Although ATP-PPi exchange activity is present in two large peaks, plus an intermediate minor one, only the first large peak contains joining activity. We have not investigated further the ATP-PPi exchange activities in the last two peaks.

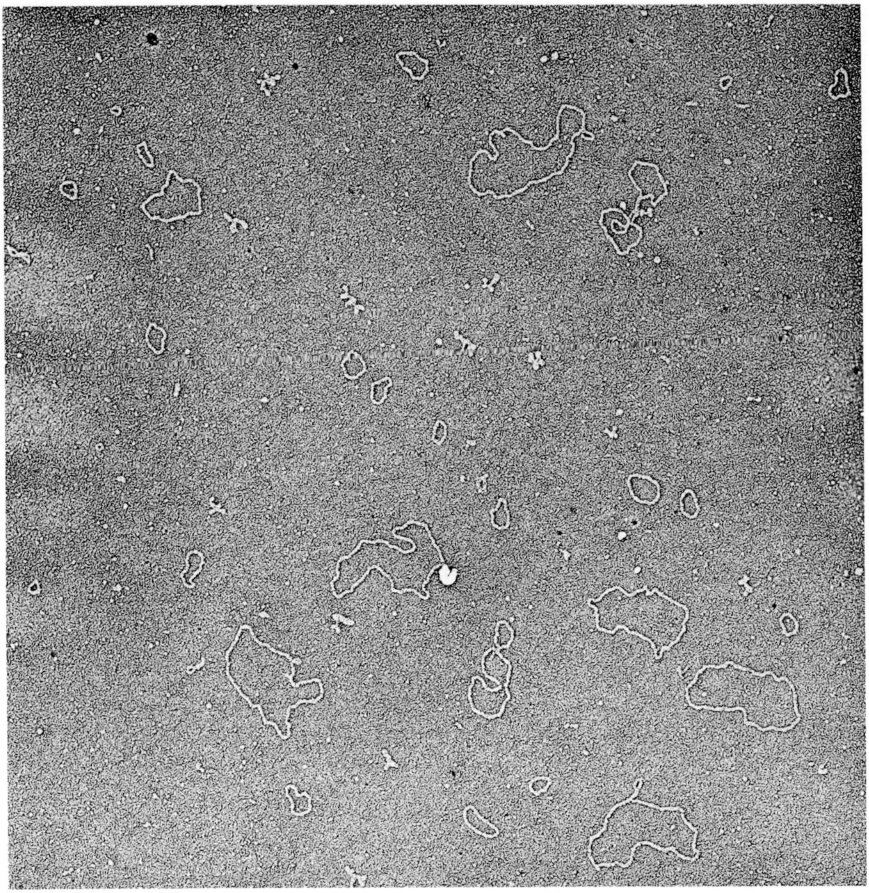

FIG. 2. Same as Fig. 1, after denaturation at pH 12.5 and renaturation.

The two joining activities appear to elute together, coinciding with the first ATP-PPi exchange peak; however, the "terminal" activity displays a broader peak than the "cohesive" one. The observed joining activity provides evidence that, at this point in the purification, there is little if any interference by contaminating phosphatases, endo-, and exonucleases. The level of endonuclease contamination was estimated by scoring the fraction of superhelical pSC101 DNA converted into relaxed circles using 10 times the amount of enzyme necessary for optimal joining; only traces of endonucleases are present (Fig. 5, *insert*) as shown by the finding that only approximately one nick per 9,000 base pairs of pSC101 was introduced during 3 hr incubation. An enzyme preparation was taken through the entire purification procedure described by Weiss et al. (16), shown to possess terminal joining activity, and then converted into the ^{32}P-AMP-intermediate

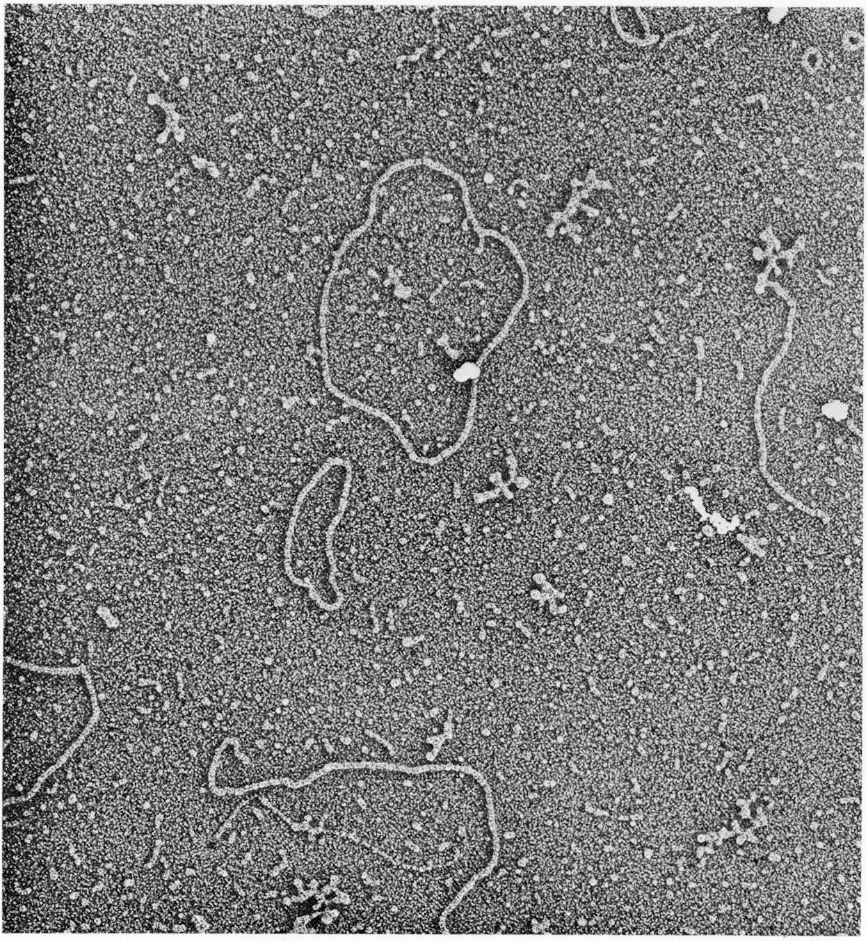

FIG. 3. Same as Fig. 2, with a detail of a circle and a partially double-stranded linear molecule.

(17). The activated and labeled enzyme was analyzed by SDS-polyacrylamide gel electrophoresis. The radioactivity associated with AMP traveled as a single peak, as shown in Fig. 6. All these results lend strong support to the interpretation that the terminal joining reaction is mediated by the same enzyme which catalyzes the cohesive reaction and that very likely a single polypeptide chain of a molecular weight approximating 52,000 is involved.

Biological Activity of "Terminally" Ligated DNA

Limit digestion with *Bsu*R endonuclease cuts pSC101 into probably more than 10 segments, of which two measuring 1×10^6 and 3×10^5 daltons

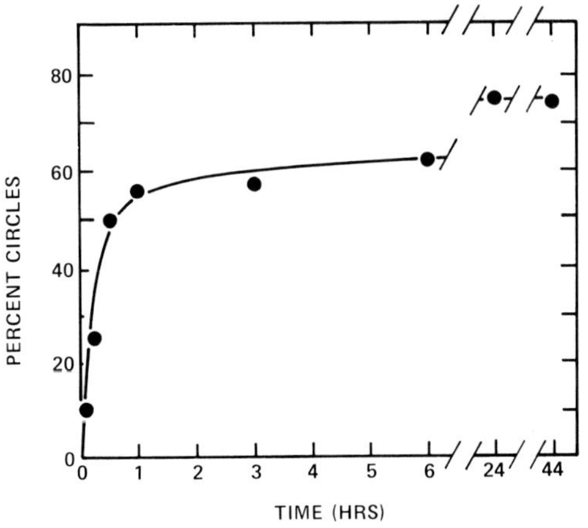

FIG. 4. Kinetics of the joining of *Bsu*R endonuclease-generated segments in the presence of T4 ligase. The digestion mixture contained: 10 mM Tris-Cl, pH 7.4; 10 mM $MgCl_2$; 150 mM NaCl; 60 μg/ml SPP1 DNA and 50 μl/ml *Bsu*R endonuclease. The ligation mixture contained: 50 mM Tris-Cl, pH 7.6; 5 mM $MgCl_2$; 1 mM 2-mercaptoethanol; 10 μM (nucleotides) tRNA; 60 μM ATP; 1 μg/ml segmented DNA and 10 units/ml T4 ligase. The ligation was run at 20°C and aliquots of 20 μl were taken at various times and heated at 65°C for 5 min before spreading.

are the longest; the remaining segments are too small for accurate measurement by agarose gel electrophoresis. The whole mixture can be ligated in the presence of the T4 ligase to produce circular structures of different sizes (see Fig. 1). We wanted to investigate whether this enzyme could allow the reconstitution of active plasmids carrying the tetracycline resistance determinant, and, given the shortness of the limit digestion products, we decided to use pSC101 segments resulting from incomplete cleavage by *Bsu*R endo. By varying the length of the digestion and electrophoresing samples taken at the various times, we observed a different spectrum of

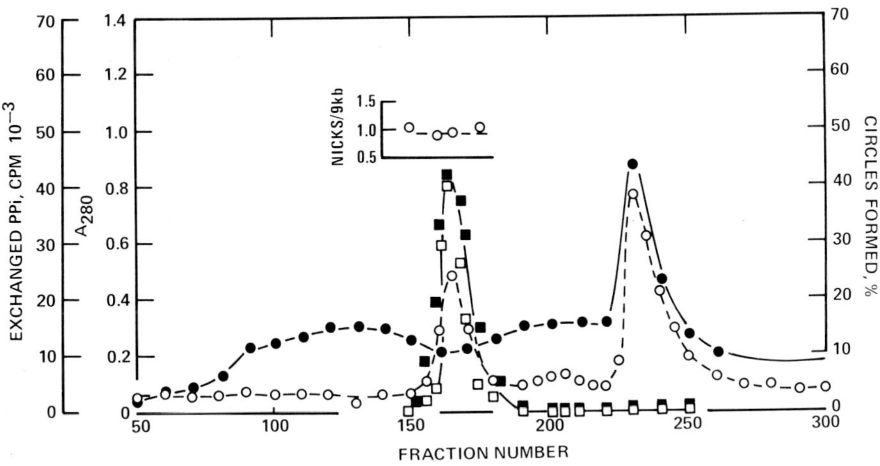

FIG. 5. DEAE-cellulose chromatography of the T4 ligase. For the details of purification, see Weiss et al. (16). ATP-^{32}PPi exchange (○); cohesive joining of *Eco*RI-SPP1 DNA (□); terminal joining of *Bsu*R-SPP1 DNA (■); A_{280} (●). The insert gives an estimate of the contaminating endonuclease (see text).

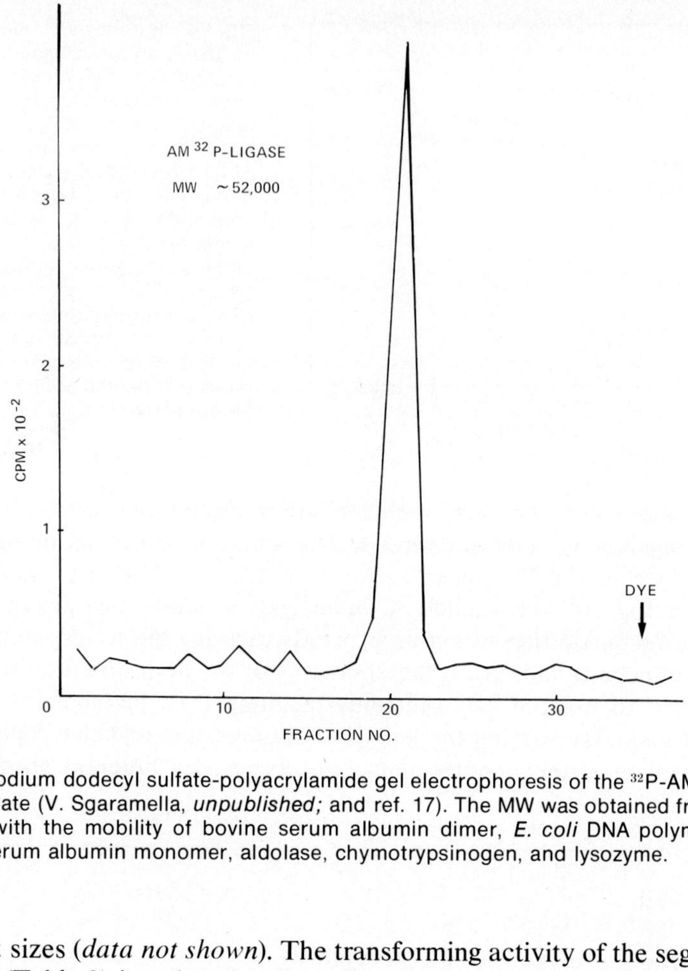

FIG. 6. Sodium dodecyl sulfate-polyacrylamide gel electrophoresis of the ^{32}P-AMP-ligase intermediate (V. Sgaramella, *unpublished;* and ref. 17). The MW was obtained from comparison with the mobility of bovine serum albumin dimer, *E. coli* DNA polymerase 1, bovine serum albumin monomer, aldolase, chymotrypsinogen, and lysozyme.

segment sizes (*data not shown*). The transforming activity of the segmented plasmid (Table 2) is reduced at the indicated times to approximately 5, 0.1, and 0.01% of that displayed by the uncut plasmid. Of the segments produced at the three stages of the digestion, those present after 60 min of incubation can be ligated by the T4 ligase in such a way as to recover a transforming activity about 15-fold higher than that of the unligated segments. Thus, it seems likely that 60 min digestion produced segments of length appropriate for the reconstitution of functional plasmids. Shorter incubation periods generated segments that were too long for optimal reassembly, and retained biological activity too high to allow the detection of an increase similar to that observed in the 60-min sample. Longer incubation, on the contrary, led to extensive cutting and thus to an essentially irreparable loss of activity.

An analysis of the size of both the starting segments and the resulting circles is under way, and we hope to obtain from it evidence supporting our interpretation of the data of Table 2. Moreover, we plan to observe the upper

TABLE 2. *Tc^r transformation with* Bsu*R endonuclease-treated pSC101 after T4 ligase circularization*[a]

Digestion time (min)	T4 ligase	% Circles	Tc^r colonies/μg/DNA
0	−	−	27,000
30	+	61	1,500
	−	2	1,400
60	+	57	400
	−	1	24
90	+	66	7.5
	−	2	2.5

[a] Ligation performed at 20°C for 1 hr, using a DNA level of 5.5 μg/ml solution. Transformation and selection for tetracycline resistance by method of Cohen et al. (4).

limit of the circles which can be formed through the terminal ligation. Also in progress is a study to determine the structure and the restriction enzymes pattern of the plasmids extracted from the tetracycline-resistant clones obtained with the ligated segments; the 10 analyzed so far have a size similar to pSC101 and retain one *Eco*RI site.

Joining of Different Ends

The recombination of DNA segments produced by restriction endonucleases entails using the same endonuclease to generate complementary ends. This requirement can be alleviated by the use of homopolymeric joints introduced with the terminal transferase (8), or by the use of flush ends, as described in this chapter. The scope of the terminal joining approach can be broadened by the conversion of single-stranded termini into fully base-paired ones with DNA polymerase (19) or reverse transcriptase (12). With regard to this possibility, it is useful to note that most of the ends produced by restriction endonucleases are either flush or longer at their 5′ terminus (R. Roberts, *this volume*), and thus repairable. An alternative solution to the problem of joining different ends is the use of small segments produced by the action of two different restriction endonucleases on the same DNA. A certain proportion of these segments contains ends of the two specificities, which could be inserted between the two termini of the DNAs to be spliced. The combination of any pair of endonucleases could produce all sorts of desirable joints, and their use might be competitive with that of chemically synthesized segments, as exemplified by the self-complementary octanucleotide containing the *Eco*RI target (5).

Figure 7 gives an example of the production of such linkers or joints carrying one flush end and the other of the *Eco*RI-cohesive type. They have been produced by incubating SPP1 DNA with *Eco*RI followed by *Bsu*R

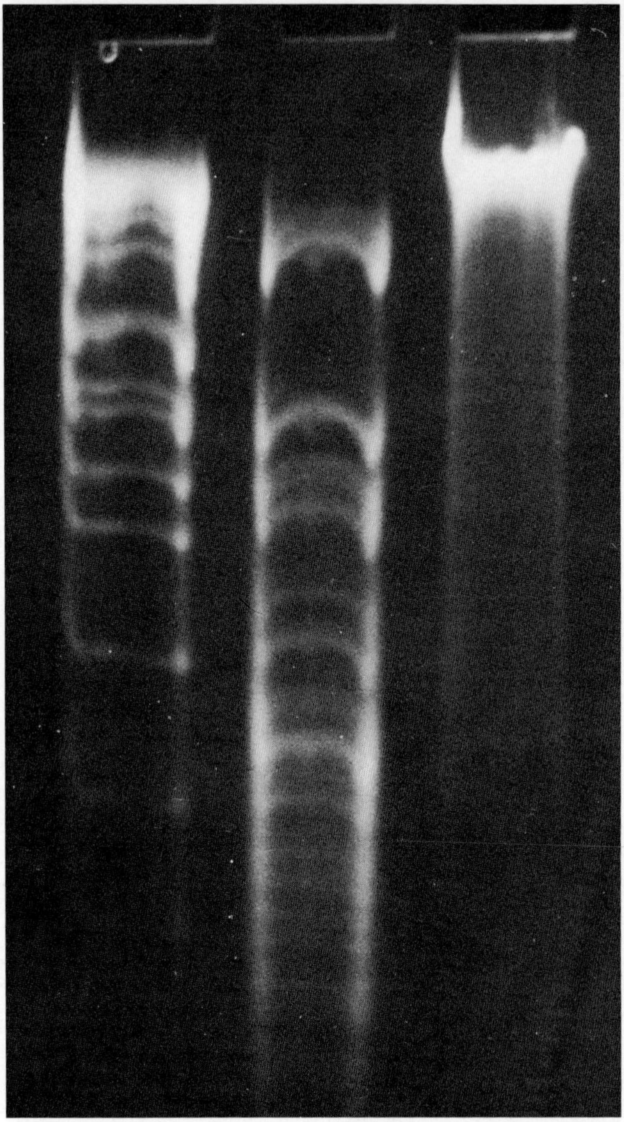

FIG. 7. Agarose gel electrophoresis of *Eco*RI-*Bsu*R joints (see text). A 120-μl mixture was prepared with: 6 μg of SPP1 DNA; Tris-Cl pH 7.4, 10 mM; NaCl 100 mM; $MgCl_2$, 5 mM; *Eco*RI, 8 μl. Incubation was at 37°C for 2 hr. To 90 μl of this mixture, 10 μl of a 10-fold concentrated cocktail was added, as required for the *Bsu*R endo digestion (see Fig. 4 legend for details), plus the enzyme. In parallel, intact SPP1 DNA was similarly digested with *Bsu*R endo. The incubation was at 37°C for 3 hr. Electrophoresis in 1.5% agarose gel was for 14 hr at 25 V. **Left:** *Eco*RI digest. **Center:** double digest. **Right:** *Bsu*R digest.

endonuclease digestion. Approximately 30% of the segments of the middle pattern are carrying different ends: appropriate selections could then be imposed for the enrichment of the most useful joints.

ACKNOWLEDGMENTS

It is a pleasure to acknowledge the stimulating participation of Dr. J. Lederberg. We want to thank Drs. K. Murray, S. Bron, G. Bernardi, L. Lis, and H. Kopecka for gifts of enzymes and suggestions on their purification and use. This work has been supported by grants from NIH to Dr. J. Lederberg and from the USA-Italian National Council of Research (C.N.R.) Cooperative Program.

REFERENCES

1. Bron, S., Murray, K., and Trautner, T. A. (1975): Purification and general properties of a restriction endonuclease from B. subtilis strain R. *Mol. Gen. Genet.,* 143:13–23.
2. Bursztyn, H., Sgaramella, V., Ciferri, O., and Lederberg, J. (1975): Transfectability of rough strains of S. typhimurium. *J. Bacteriol.,* 124:1630–1634.
3. Clewell, D. B., and Helsinki, D. R. (1969): Supercoiled circular DNA-protein complex in E. coli: purification and induced conversion to an open circular form. *Proc. Natl. Acad. Sci. U.S.A.,* 62:1159–1169.
4. Cohen, S. N., Chang, A. C. Y., Boxer, H. W., and Helling, R. B. (1973): Recircularization and autonomous replication of a sheared R-factor DNA segment in E. coli transformants. *Proc. Natl. Acad. Sci. U.S.A.,* 70:3240–3244.
5. Green, P. S., Poonian, M. S., Nussbaum, A. L., Tobias, L., Garfin, D. E., Boyer, H. W., and Goodman, H. M. (1975): Restriction and modification of a self-complementary octanucleotide containing the Eco RI substrate. *J. Mol. Biol.,* 99:237–261.
6. Gromkova, R., and Goodgal, S. H. (1972): Action of haemophilus endodeoxyribonuclease on biologically active DNA. *J. Bacteriol.,* 109:987–997.
7. Inman, R. B., and Schnös, M. (1970): Denaturation of thymine- and 5-bromouracil-containing lambda DNA in alkali. *J. Mol. Biol.,* 49:93–98.
8. Lobban, P. E., and Kaiser, A. D. (1971): Enzymatic end-to-end joining of DNA molecules. *J. Mol. Biol.,* 78:453–471.
9. Nathans, D., and Smith, H. O. (1975): Restriction endonucleases in the analysis and restructuring of DNA molecules. *Annu. Rev. Biochem.,* 44:273–293.
10. Peacock, A. C., and Dingman, C. W. (1967): Resolution of multiple RNA species by polyacrylamide gel electrophoresis. *Biochemistry,* 6:1818–1827.
11. Riva, S., Polsinelli, M., and Falaschi, A. (1969): A new bacteriophage of B. subtilis with infectious DNA having separable strands. *J. Mol. Biol.,* 35:347–361.
12. Rougeon, F., Kourilsky, P., and Mach, B. (1975): Insertion of rabbit beta-globin genes sequence into an E. coli plasmid. *Nucleic Acids Res.,* 2:2365–2378.
13. Sgaramella, V. (1972): Enzymatic oligomerization of bacteriophage P22 DNA and of linear Simian virus 40 DNA. *Proc. Natl. Acad. Sci. U.S.A.,* 69:3389–3393.
14. Sgaramella, V. (1976): Characterization of B. subtilis phages DNAs with restriction endonucleases. *Atti. A. G. I.,* 21:17–18.
15. Sgaramella, V., and Khorana, H. G. (1972): Enzymic joining of the chemically synthesized polydeoxynucleotides to form the DNA duplex representing nucleotide sequence 1 to 20. *J. Mol. Biol.,* 72:427–444.
16. Weiss, B., Jacquemin-Sablon, A., Live, R. T., Fareed, G. C., and Richardson, C. C. (1968): Further purification and properties of polynucleotide ligase from E. coli infected with bacteriophage T4. *J. Biol. Chem.,* 243:4543–4555.
17. Weiss, B., Thompson, A., and Richardson, C. C. (1968): Property of an enzyme-adenylate intermediate in the polynucleotide ligase reaction. *J. Biol. Chem.,* 243:4556–4563.

18. Wickner, W., Brutlag, D., Scheckman, R., and Kornberg, A. (1972): RNA synthesis initiates in vitro conversion of M13 DNA to its replicative form. *Proc. Natl. Acad. Sci. U.S.A.*, 69:965–969.
19. Wu, R., and Taylor, E. (1971): Complete nucleotide sequence of the cohesive ends of lambda bacteriophage DNA. *J. Mol. Biol.*, 57:491–511.

Recombinant Molecules: Impact on Science and Society, edited by R. F. Beers, Jr. and E. G. Bassett. Raven Press, New York © 1977.

8. Cloning of the Thymidylate Synthetase Gene of the Phage Phi-3-T

*S. D. Ehrlich, H. Bursztyn-Pettegrew, I. Stroynowski, and J. Lederberg

Department of Genetics, Stanford University Medical School, Stanford, California 94305

INTRODUCTION

Phi-3-T is a *Bacillus subtilis* temperate bacteriophage, isolated by Tucker (11) from the soil. Tucker has shown that infection by this phage causes, by lysogenic conversion of thy$^-$ *B. subtilis* clones, prototrophy. The phage carries genetic information specifying the thymidylate synthetase activity, the *thyP* gene.

Recently, we have reported purification of *B. subtilis* DNA segments achieved by *Eco*RI cleavage followed by gel electrophoresis (6). Since a still further purification can be achieved by cloning the segments (8), we have chosen to use that approach for the isolation of the *thyP* gene of the Phi-3-T. We describe here the successful cloning of that gene on the *Escherichia coli* plasmid pSC101 (3). The gene complements the thymine deficiency in *E. coli,* indicating its correct transcription and translation in this new host. The promoter of the gene is likely to be contained within the cloned segment. The hybrid plasmid transforms *B. subtilis* 100-fold less efficiently than the intact phage DNA. The excised segment, however, displays the same transforming activity (1,000 times less than the intact phage DNA) whether it comes from the phage or from the hybrid plasmid. *B. subtilis* clones transformed with the hybrid plasmid DNA do not contain detectable pSC101 sequences, but do show sequences homologous to part of the Phi-3-T genome.

METHODS AND RESULTS

Isolation of Tc^r thy$^+$ *E. coli* Transformants

The *Eco*RI cleavage pattern of the Phi-3-T DNA is shown in Fig. 1. In agreement with the data of Wilson et al. (12), 20 bands can be seen; four

* Present address: Institut de Biologie Moleculaire, Faculté de Science, Paris 75005, France.

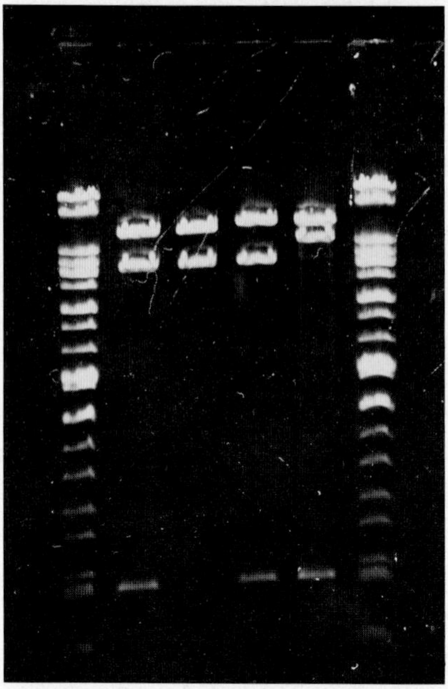

FIG. 1. Agarose gel electrophoresis of *Eco*RI-cleaved DNA. From *left to right:* Phi-3-T, pFT23, pFT24, pFT25, pFT33, and Phi-3-T. Five-hundred nanogram-cleaved DNA in 15 μl of digestion mixture containing 10% sucrose and 0.1% bromphenyl blue dye was loaded on a 4-mm thick 0.6% agarose slab gel. Electrophoresis was performed for approximately 8 hr at 2 V/cm in the TRIS-borate-EDTA buffer containing 500 ng/ml of ethidium bromide (6). Photograph was taken under the short wavelength ultraviolet light.

additional bands have been allowed to migrate out of the gel. A more detailed study has shown that approximately 30 segments can be resolved. Their sizes, determined using *Eco*RI-cleaved SPP1 DNA as a standard (4), are displayed in Table 1. The sum of their molecular weights approximates 7.2×10^7 daltons and is somewhat lower than the $82.6 \pm 5.3 \times 10^6$ daltons measured by electron microscopy that used pSC101 as a standard. The discrepancy probably results from one or more of the following factors: repetition within the phage genome (see below), incomplete resolution during gel electrophoresis, and the use of different molecular weight standards for the two measurements.

Kinetics of inactivation of *thyP* gene-transforming activity on cleavage with *Eco*RI enzyme is shown in Fig. 2. Nearly a 1,000-fold inactivation of fully cleaved DNA was observed. This finding prompted us to attempt the cloning of the *thyP* gene from an incompletely digested DNA sample, in which only 10% of the original transforming activity was lost. Such a sample was mixed with *Eco*RI-cleaved pSC101 DNA in a ratio 10 to 1, with a total DNA concentration of 11 μg/ml. Treatment with the T4 ligase resulted in appearance of approximately 40% circles in addition to a number of long, linear molecules. Ligated samples were used to transform thymine-requiring *E. coli* cells, selecting either for tetracycline resistance or thymine independence. The results of the transformation are displayed in Table 2.

Approximately 8% of the Tc^r clones contain sequences that are complementary to Phi-3-T cRNA. This value probably underestimates the pro-

TABLE 1. *Molecular weight of ϕ3T DNA segments after* EcoRI *cleavage*

Band	MW (10^{-6})
1	9.4
2	7.9
3	5.4
4	4.9
5a,b	4.5, 4.4
6	3.9
7a,b	3.2, 3.1
8	2.9
9	2.5
10a,b,c	2.07, 2.05, 2.05
11	1.96
12a,b	1.70, 1.70
13	1.45
14	1.25
15	1.05
16	.90
17	.70
18	.67
19	.65
20	.46
21	.41
22	.26
23	.24
24	.19
Total	71.86

portion of hybrid plasmids, since the very small inserts might have been undetected by the technique used (5). Two of the Tc^r clones hybridizing Phi-3-T cRNA also showed a thy^+ phenotype. Two more thy^+ clones have been obtained by selecting that phenotype directly; it is interesting to note that the frequency of occurrence of the directly selected thy^+ character is lower than when the Tc^r is selected first (Table 2).

Molecular Structure of Hybrid Plasmids

Plasmids from the clones hybridizing cRNA are denoted as "pFT," followed by the clone number. The *Eco*RI restriction pattern of the four DNAs extracted from the Tc^r thy^+ transformants is shown in Fig. 1. Molecular weights of *Eco*RI-released segments, together with those resulting from the action of *Bam* and *Sma* restriction enzymes, are shown in Table 3.

A paradox is immediately apparent: although the four plasmids all carry the thy^+ character, the only *Eco*RI segment that all four have in common is the pSC101 moiety of the hybrid (6.2×10^6 dalton segment, Table 3). The 4.5×10^6 dalton segment (A, Table 3) of the plasmids pFT23, 24, and 25, with a mobility identical to that of the segment 5a of the Phi-3-T, is missing

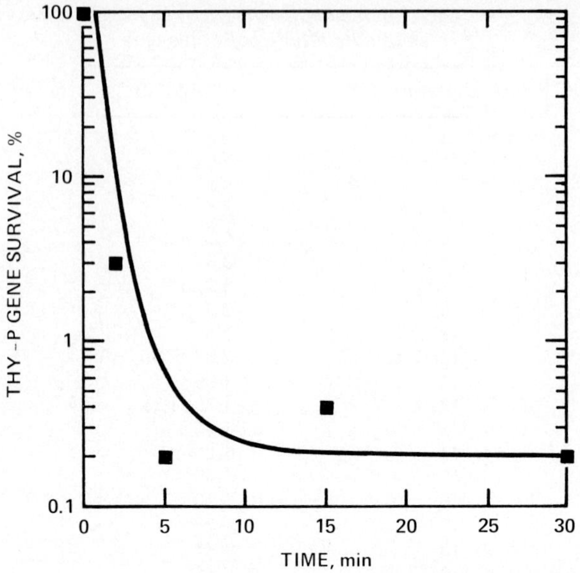

FIG. 2. Kinetics of inactivation of the *thyP* gene on cleavage of Phi-3-T DNA with *Eco*RI endonuclease. Enzymatic degradation at a DNA concentration of 100 μg/ml and transformation of *B. subtilis* strain SB591 (thy^-) were performed as described previously (6).

in the plasmid pFT33. The latter contains instead a 5.4×10^6 dalton segment A′, which differs from the segment A not only in size but also by embodying a *Bam* and *Sma* site. The two small segments (B and C, Table 3), which co-electrophorese with the segments 18 and 23 of the phage, are also not present in all of the plasmids. Which of the segments, then, carries the *thyP* gene?

An answer was obtained by electron microscope analysis of the heteroduplexes between different hybrid plasmid DNAs. The 5.4×10^6 dalton segment A′ of the pFT33 contains a 4.5×10^6 dalton stretch of DNA that is homologous to the total length of segment A of pFT23, 24, and 25. In

TABLE 2. *Isolation of* Thy^+ Tc^r E. coli *transformants*[a]

Selection	Tc^r	Thy^+	Thy^+Tc^r
Plated cells	10^7	10^8	10^8
Transformants	447	1	2
Hybridize φ3T cRNA	31	N.D.[b]	2
Tc^rThy^+ phenotype	2	0	2

[a] Recipient *E. coli* strain was W5443 (see legend of Fig. 3 for the genotype).
[b] N.D. = not determined.

TABLE 3. *Molecular weights of DNA segments from the hybrid plasmids*

Plasmid	*Eco*RI	*Bam*	*Sma*	Intact
pFT23	6.2	11.7	11.7	11.7
	A 4.5			
	B 0.67			
	C 0.24			
pFT24	6.2	10.8	10.8	10.8
	A 4.5			
pFT25	6.2	11.7	11.7	11.7
	A 4.5			
	B 0.67			
	C 0.24			
pFT33	6.2	8.3	8.0	12.6
	A′ 5.4	4.4	4.6	
	B 0.67			
	C 0.24			

addition, it contains a 9×10^5 dalton insert, which on denaturation and renaturation appears as a hairpin, with the stem approximately 100 base pairs long. Such structure can arise from the presence of palindromic sequences bordering the 9×10^5 dalton insert. We are presently investigating the origin of that insert. Thus, the sequences of segment A appear common to all of the hybrid plasmids, and presumably contain the *thyP* gene.

The thy$^+$ Character is Plasmid-borne

Three lines of evidence demonstrate that the thy$^+$ gene is harbored on a plasmid. First, DNA prepared by the standard, clear lysis procedure (2), containing less than 1% of linear molecules, can transform Tc^{s}thy$^-$ *E. coli* to Tc^r or Tc^{r}thy$^+$ phenotype in a manner indistinguishable from that displayed by pSC101 transforming to Tc^r character (Fig. 3). Approximately 1,000 transformants, selected either for Tc^r or thy$^+$, were tested by replica plating and found always to carry the unselected marker as well. The efficiency of transformation, calculated from the linear part of the curve (slope = 1) shown in Fig. 3, is displayed in Table 4.

Second, *E. coli* transformants of the Tc^{r}thy$^+$ phenotype can eliminate the plasmid using ethidium bromide treatment, followed by the ampicillin selection in the presence of tetracycline. All the Tc^s clones lost the thy$^+$ character simultaneously. Both markers could be reintroduced into the so-treated strain, at a frequency identical to that of the parent, by transformation with the hybrid plasmid DNA. Since the Tc^r character is known to be plasmid-borne, and our experiments have established linkage of the thy$^+$ marker

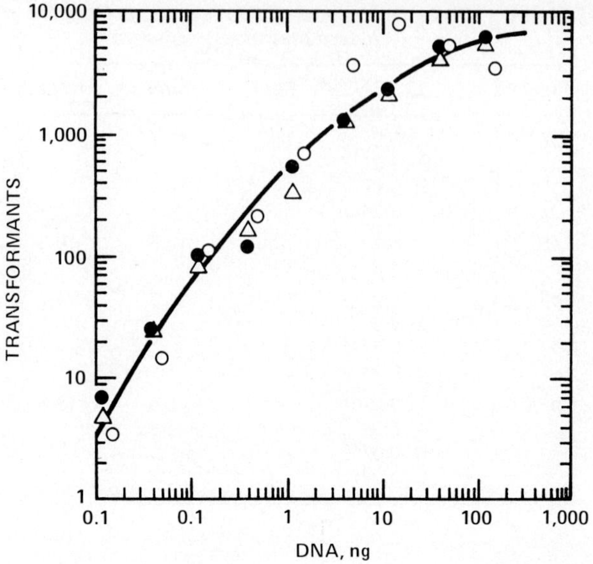

FIG. 3. Transformation of *E. coli* with pSC101 [Tc^r (○)] and pFT24 [Tc^r (△); Tc^{r}thy$^+$ (●)]DNA. *E. coli* strain W5443 (*thr-1 leu-6 thi-1 supE44 lacY1 r$^-$ m$^-$ thy$^-$ Str^r tonB tryp$^-$*) was rendered competent and transformed as described previously (7).

with the antibiotic resistance, then the thy$^+$ character must be carried by the plasmid.

Third, all the hybrid plasmid DNAs transform thymine-requiring *B. subtilis* strains to prototrophy. A representative experiment is shown in Fig. 4. Efficiencies of transformation, averaged from a number of experiments, are presented in Table 4. Intact Phi-3-T DNA was used as a control and displayed approximately a 100-fold higher efficiency than did pFT25 DNA. This low efficiency of plasmid DNA is not due to its supercoiled configura-

TABLE 4. *Transforming efficiencies of hybrid plasmids*

DNA	Cells	Efficiency (cge)[a]
pSC101	*E. coli* W5443	10^{-5}
pFTs		10^{-5}
ϕ3T	*B. subtilis* SB591	10^{-4}
ϕ3T/*Bam*		10^{-5}
ϕ3T/RI		10^{-7}
pFTs		10^{-6}
pFTs/*Bam*		10^{-6}
pFTs/RI		10^{-7}

[a] Colonies per genome equivalent.

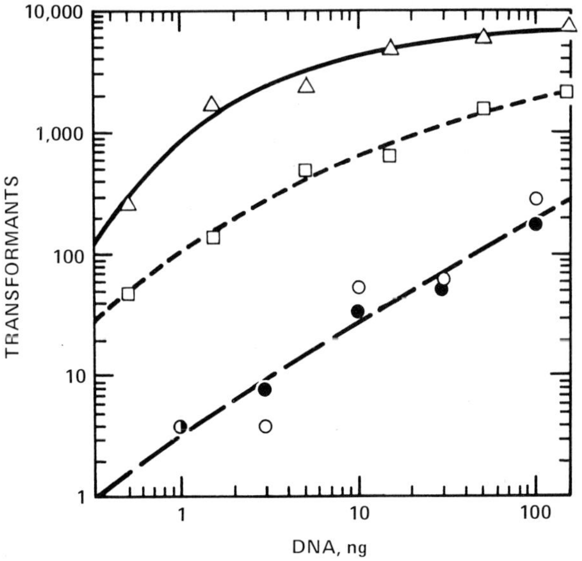

FIG. 4. Transformation of *B. subtilis* to thymine independence with Phi-3-T (△), pFT24 (○), *Bam*-cleaved Phi-3-T (□), and pFT24 (●) DNAs. The lines represent exponential functions, best fitting by the least squares test, to the experimental points.

tion: the cleavage with *Bam* endonuclease, which linearizes the molecule by introducing a cut about 2×10^5 daltons from the *Eco*RI site in the pSC101 part of the hybrid, does not change its biological activity. Similar treatment of the Phi-3-T DNA leads to a 10-fold decrease in transforming activity. Interestingly, the *Eco*RI treatment decreases biological efficiency of hybrid plasmids approximately 10-fold while the activity of the phage DNA is impaired 1,000-fold (Fig. 2). Therefore, the *Eco*RI segment embodying the *thyP* gene displays the same transforming efficiency regardless of the vector (Phi-3-T phage or the hybrid plasmid) by which it was carried or its previous host.

Hybrid Plasmids Contain Phi-3-T Sequences

Several lines of evidence demonstrate the presence of Phi-3-T sequences in the hybrid plasmids. As mentioned previously (Table 2), *E. coli* clones that harbor these plasmids hybridize RNA synthesized *in vitro* by the *E. coli* RNA polymerase on the Phi-3-T DNA template (Phi-3-T cRNA). In addition, all except one of the segments released by *Eco*RI cleavage from hybrid plasmids show electrophoretic mobility identical to some Phi-3-T segments (see above).

Further evidence for the presence of the *thyP* gene on the hybrid plasmids is that DNA of the latter can transform thy$^-$ *B. subtilis* strains to a state of thymine independence. These transformants acquire not only the thy$^+$ character, but also the ability to hybridize Phi-3-T cRNA.

Figure 5 shows that the colony hybridization technique (5) can be used with *B. subtilis* clones: SB168 (a standard *B. subtilis* strain), following lysis by Phi-3-T, hybridizes the Phi-3-T cRNA. Specificity of staining is demonstrated by the absence of hybridization with the pSC101 cRNA. Interestingly, SB168 (not carrying the prophage) also hybridizes the phage cRNA; the strain is not immune to the phage and does not release it on mitomycin induction (not shown). SB591, a thymine-requiring derivative of SB168, has lost the capacity to bind the Phi-3-T cRNA; the mutagenesis has apparently deleted sequences complementary to the phage. However, SB591 transformed with either the Phi-3-T DNA or the hybrid plasmid DNA does hybridize Phi-3-T cRNA. The low extent of hybridization displayed by clones of Phi-3-T DNA-transformed SB591 and their sensitivity to Phi-3-T phage indicate that only a limited amount of the phage genome (including the *thyP* gene) has been inserted in the chromosome. It is interesting to note that SB591 transformed to thymine prototrophy with DNA extracted from SB19 (a strain related to SB168) also hybridizes Phi-3-T cRNA.

Additional evidence for the presence of Phi-3-T sequences in the hybrid plasmids has been obtained using the hybridization technique developed by Southern (10). Phi-3-T cRNA binds to all *Eco*RI-cleaved segments of hybrid plasmids except the one corresponding to the pSC101 part of the molecules. Conversely, the cRNAs synthesized on each of the hybrid plasmid templates (including pFT33) hybridize to segment 5a of the phage. They also stain Phi-3-T segments 18 and 23, but not pFT24 cRNA; this is expected since pFT24 does not carry these latter segments (Table 3).

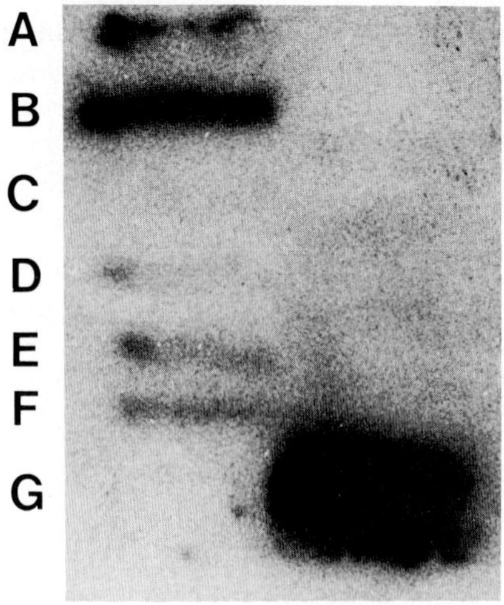

FIG. 5. Autoradiograph of *B. subtilis* colonies to which Phi-3-T (*left*) or pSC101 (*right*) cRNA was hybridized. An *E. coli* strain harboring pSC101 was used as control. Bacteria were streaked on nitrocellulose filter disks, incubated overnight, and prepared for hybridization according to the procedure of Grunstein and Hogness (5), as modified by J. Lis and L. Prestidge (*personal communication*). The filter was then cut into halves, one half was hybridized with Phi-3-T cRNA, the other with pSC101 cRNA. Hybridization and autoradiography were performed as described previously (5). **A:** SB168. **B:** SB168 (Phi-3-T). **C:** SB591*thy*$^-$. **D:** SB591 transformed with Phi-3-T DNA to *thy*$^+$. **E:** SB591 transformed with pFT23. **F:** SB591 transformed with pFT24 DNA. **G:** *E. coli* C600(pSC101).

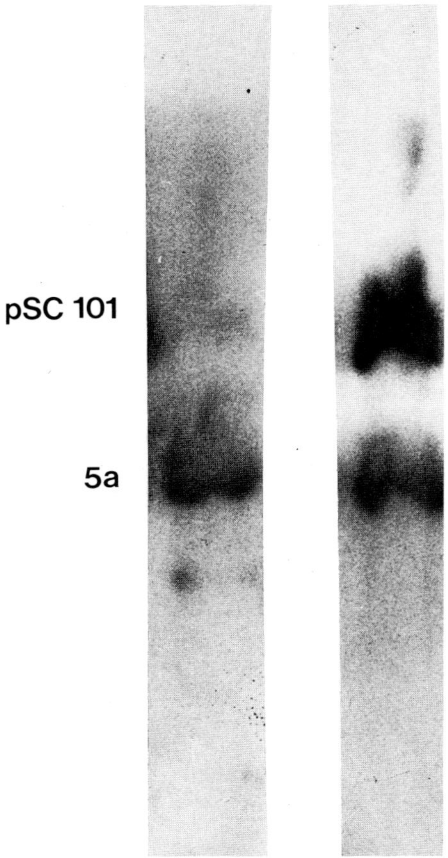

FIG. 6. Autoradiograph of *Eco*RI-cleaved pFT24 (*right*) and Phi-3-T (*left*) DNA, separated by gel electrophoresis and hybridized with the pFT24 cRNA. Electrophoretic separation was as described in the legend of Fig. 1, except that the segments smaller than 3.5×10^6 daltons have been allowed to migrate out of the gel. Hybridization and autoradiography were performed as described previously (10).

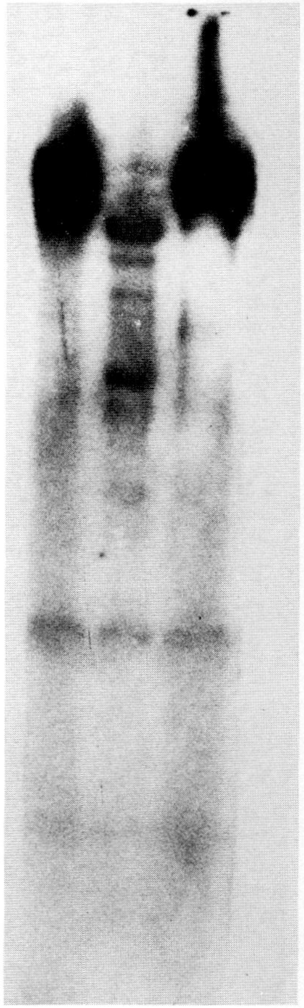

FIG. 7. Autoradiograph of *Eco*RI-cleaved pFT23 (*left*), Phi-3-T (*center*), and pFT23 plus carrier calf thymus (*right*) DNA, separated by gel electrophoresis and hybridized with pFT23 cRNA. Film was overexposed in order to show staining of shorter segments.

Hybridization of pFT24 cRNA to the phage and the plasmid DNA is displayed in Fig. 6.

A somewhat puzzling observation is that all pFT cRNAs hybridize to several additional bands in the Phi-3-T restriction pattern: 3, 6, 7, 10, 12, and 14 (Fig. 7 shows an example), and suggest to us the existence of several regions of internal homology within the phage genome. Incomplete degradation, another possible explanation, seems to be ruled out by the fact that a 10-fold increase in either the amount of *Eco*RI or the time of incubation did not change the described pattern. Similarly, the presence of multiple species in each of the plasmids is ruled out since the cleaved plasmids appear to hybridize both their own and Phi-3-T cRNA exclusively to segments detected by electrophoresis (Table 3), even when cleaved calf thymus DNA is added as a carrier prior to electrophoresis so that trace amounts of postulated contaminating segments are not lost (Fig. 7).

Still further evidence for the presence of Phi-3-T sequences in hybrid plasmids is obtained by electron microscope inspection (not shown) of the heteroduplex molecules between the phage and the plasmid DNAs. A double-stranded region without any detectable imperfection is formed between the pFT24 and the Phi-3-T DNAs: its length is 4.5×10^6 daltons—exactly that measured by gel electrophoresis (Table 3) for the insert.

An important conclusion can be drawn from these data. The inserts are homologous with the Phi-3-T DNA (except for pFT33) along their entire length. Therefore, the hybrid plasmids contain only the phage sequences added to the pSC101. Since these plasmids confer the thy$^+$ phenotype to *E. coli,* the gene that functions in the *E. coli* environment is the *thyP* gene of the Phi-3-T.

Promoter of the *thyP* Gene in Hybrid Plasmids

Is the *thyP* under control of a Phi-3-T promoter within the cloned segment in the hybrid plasmids, or does it depend on the read-through initiated on the promoter of the vector? The following evidence indicates that the former might be true:

1. The 4.5×10^6 dalton segment of the pFT23, which contains the *thyP* gene, is oriented in a direction opposite to that of the corresponding segment of pFT24, as revealed by the heteroduplex analysis (not shown). Nevertheless, both plasmids display the same thy$^+$ character.
2. The insert from the pFT24 was cloned in the ColE1-amp plasmid (9) and its biological activity was not impaired, regardless of its orientation.

DISCUSSION

*Eco*RI cleavage of the Phi-3-T DNA leads to a 1,000-fold decrease of the transforming activity of the *thyP* gene. Nevertheless, the cleavage does not occur within the gene, since segment 5a of the phage can transform both the

E. coli and the *B. subtilis* thymine-requiring strains to prototrophy; therefore, its genetic information must be intact. It seems more likely that the loss of biological activity is caused by the decrease in size, on cleavage, of the DNA carrying the *thyP* gene, a phenomenon we have observed for chromosomal markers of *B. subtilis* (6). This conclusion is supported by the fact that the cleavage of hybrid plasmids with the *Bam* restriction enzyme (which does not change the size of DNA) does not affect its transforming activity; on the other hand, treatment with the *Eco*RI enzyme (decreasing the size of DNA) also provokes a loss of biological activity.

In each of approximately 100 *B. subtilis* clones transformed to thymine independence by the hybrid plasmid DNA, Phi-3-T, but no pSC101, sequences were detected by hybridization. In addition, attempts to transform *B. subtilis* to Tc^r using the hybrid DNA, selecting directly for the resistance or by replica plating of thy^+ transformants, have consistently failed. It can therefore be concluded that there is excision of DNA sequences contained in the hybrid plasmids during transformation of *B. subtilis* cells. Experiments addressing that question in more detail are in progress.

We have demonstrated that a gene which functions in a Gram-positive organism (*B. subtilis*) can also function in a Gram-negative one (*E. coli*). Only one example of this nature has been previously reported: the ampicillin gene of the *Staphylococcus aureus* plasmid conferred resistance to *E. coli* when cloned in the pSC101 plasmid (1). The *thyP* gene of the phage Phi-3-T does not belong to the category of antibiotic resistance genes (known to be promiscuous), but is rather a gene involved in the biosynthetic processes of its host. Our finding has, therefore, a more general implication: it is quite likely that most of the genes remain functional on exchange between two prokaryotic species. This notion is supported by the fact that two other *B. subtilis* genes, *leu* and *ura,* have been demonstrated to be functional in *E. coli* (N. Y. Chi, *manuscript in preparation*). In addition, our evidence suggests that the *thyP* gene uses its own promoter in *E. coli.* This might make possible the experimental assessment of an analog of the universality of the genetic code: namely, the universality of genetic control.

ACKNOWLEDGMENTS

We thank J. Feitelson for help in preparing the manuscript and graphs via the SUMEX computer facility, P. Evans and L. Chaverri for technical help, Dr. D. Brutlag for a gift of RNA polymerase, J. Lis and H. Kopecka for a gift of *Sma* endonuclease, and D. Finnegan for the *E. coli* W5443 strain. This work was supported by NIH Grant 5 RO1 CA 16896–18.

REFERENCES

1. Chang, A. C. Y., and Cohen, S. N. (1974): Genome construction between bacterial species in vitro: replication and expression of Staphylococcus plasmid genes in Escherichia coli. *Proc. Natl. Acad. Sci., USA,* 71:1030–1034.

2. Clewell, D. B., and Helinski, D. R. (1969): Supercoiled circular DNA-protein complex in Escherichia coli: purification and induced conversion to an open circular DNA form. *Proc. Natl. Acad. Sci. USA,* 62:1159–1166.
3. Cohen, S. N., Chang, A. C. Y., Boyer, H. W., and Helling, R. B. (1973): Construction of biologically functional bacterial plasmids in vitro. *Proc. Natl. Acad. Sci. USA,* 70:3240–3244.
4. Ganesan, A. T., Anderson, J. J., Luh, J., and Effron, M. (1977): (*manuscript submitted*).
5. Grunstein, M., and Hogness, D. R. (1975): Colony hybridization: a method for the isolation of cloned DNAs that contain a single gene. *Proc. Natl. Acad. Sci. USA,* 72:3961–3965.
6. Harris-Warrick, R. M., Elkana, Y., Ehrlich, S. D., and Lederberg, J. (1975): Electrophoretic separation of Bacillus subtilis genes. *Proc. Natl. Acad. Sci. USA,* 72:2207–2211.
7. Mandel, M., and Higa, A. (1970): Calcium-dependent bacteriophage DNA infection. *J. Mol. Biol.,* 53:159–162.
8. Morrow, J. F., Cohen, S. N., Chang, A. C. Y., Boyer, H. W., Goodman, H. M., and Helling, R. B. (1974): Replication and transcription of eukaryotic DNA in Escherichia coli. *Proc. Natl. Acad. Sci. USA,* 71:1743–1747.
9. So, M., Gill, R., and Falkow, S. (1975): Generation of a ColE1-Ap^r cloning vehicle which allows detection of inserted DNA. *Mol. Gen. Genet.,* 142:239–249.
10. Southern, E. M. (1975): Detection of specific sequences among DNA fragments separated by gel electrophoresis. *J. Mol. Biol.,* 98:503–517.
11. Tucker, R. G. (1969): Acquisition of thymidylate synthetase activity by a thymine-requiring mutant of Bacillus subtilis following infection by temperate phage ϕ3. *J. Gen. Virology,* 4:489–504.
12. Wilson, G. A., Williams, M. T., Baney, H. W., and Young, F. E. (1974): Characterization of temperate bacteriophages of Bacillus subtilis by the restriction endonuclease EcoRI: evidence for three different temperate bacteriophages. *J. Virology,* 14:1013–1016.

Recombinant Molecules: Impact on Science and Society, edited by R. F. Beers, Jr. and E. G. Bassett. Raven Press, New York © 1977.

9. Discussion

Moderator: Frank E. Young

Dr. S. D. Ehrlich: Dr. Roberts, you referred to some difference that you could find adjoining the ends that were extending the nucleotides without sequence. It would seem that flush-ended molecules have a different efficiency of joining attributed mainly to the different state of purity of the enzymes used. For example, we can get 20% joining with *Sma*-cut DNA, 60% joining with *Bsu*-cut DNA, but only 5% with *Hpa*. How do you explain this situation?

Dr. R. J. Roberts: In the case we looked at we had in fact done terminal analysis on the DNA—which looked okay. We think in that particular case this was not the problem. This observation has been made in several other labs with different enzyme preparations. I think it is a real effect. However, we have not done a very detailed kinetic analysis of the reaction.

Dr. R. Sager: I would like to ask Dr. Curtiss to elaborate on some of the bouts that he has had with *E. coli.* In particular, I would ask him if he carefully refrained from stating in his chapter whether he feels he can or has outmaneuvered *E. coli* sufficiently so that there are strains he would consider safe for cloning mammalian DNAs.

Dr. R. Curtiss, III: I think the biggest problem that faced us was an unexpected ability of *E. coli,* under certain conditions, to survive in the absence of a cell wall. This necessitated introduction of additional mutations. We know how to do this now by a variety of means; thus, this poses no particular problem. The ability to "de-sexify" *E. coli* by making it either receive or donate genetic information through conjugation is going to be considerably more difficult in view of the large number of plasmids that promote conjugation and their different requirements for cell-cell interaction leading to DNA transfer. The strain that I described has two defects that make it conjugation-deficient. I did not mention that it is also resistant to most of the known transducing phages of *E. coli.* When the bacterial cell surface is altered, the real problem is the potential for endowing sensitivity for phages that may lurk in the sewers and about which we know nothing.

I can find no data in the literature that tell the quantitative titer of phages plated on a given strain per milliliter sewage. Everybody does enrichment. You know the next day but you do not know at the inception. We have not tested the sensitivity of this strain directly, say, to a sample of a phage that might be obtained from a sewage treatment plant. However, the frequency of transductional transmission, either by generalized or specialized phages, is 10^{-8} as required by the NIH definition of an EK2 host. As I understand, certain mammalian DNAs could be cloned in an EK2 host provided one had the suitable physical containment. Therefore, I think this strain meets the requirements; otherwise we would not have submitted it for consideration. We certainly can do better on the basis of what we have learned during the past 6 months and hope to have still better strains later this or early next year.

Dr. F. R. Blattner: Dr. Sgaramella, what do you make of the increased enzyme requirement for flush-end joining as opposed to ligation of an overlapping sequence? I gather from what you say that if the end concentrations are the same and you let the reaction go to infinity, you get a different percentage of joining. If this were simply a catalytic process, you would think that that would not be the case.

Dr. V. Sgaramella: There is a possible explanation which has to be looked into carefully and which somehow cancels the work I did with well-defined duplexes in Khorana's laboratory. We found that whenever a cohesive joining takes place close to the flush ends, the data are consistent. If we do cohesive joining four or five nucleotides from the flush end, we are able to detect piling up in the reaction of the intermediate AMP-5′ phosphoryl DNA. One possibility we looked into was whether the terminal ligation is such that large amounts of this intermediate are really accumulated, and whether this accumulation of the AMP-activated phosphorylated end is higher with flush ends than with cohesive ends. It would be nice to see if it is possible to detect this 5′-adenylate in the circles of DNA by phosphatase action, and whether the accumulation is dependent on the length of the termini.

This is a working hypothesis. I have no other explanation of why the reaction stops early. But we have found that structures have somehow frozen depending on subtle differences in the reaction conditions, such as pH, ionic strength, and temperature. The plateaus that one usually obtains even with well-characterized systems such as synthetic segments of genes really do not give rise to unequal final products. It might be an important line to explore further the possibility of frozen intermediates.

Dr. D. R. Helinski: I would like to ask Dr. Young a question with regard to data on the ϕ3T transformants. Although it is preliminary, has your lab made any attempts to isolate a plasmid from those ϕ3T transformants?

Dr. F. E. Young: Our work on the question is still preliminary. We did a few experiments that gave unclear results. On cesium chloride–ethidium bromide gradient centrifugation we obtain a peak that appears to be CCC DNA and has thy^{+} transforming activity. However, it contains chromosomal markers as well. Other preliminary data indicate that in some of the transformants *thy*P is integrated into the *B. subtilis* chromosome. Experiments are in progress to resolve this question. I should have pointed out in the chapter that the *B. subtilis* thy$^+$ transformants, unlike the *E. coli* thy$^+$ transformants, do not express Tet^r. Therefore, if *E. coli* DNA is taken up by *B. subtilis*, the Tet^r is not expressed.

Dr. J. Carbon: I would like to ask Dr. Curtiss about the relative transformation frequencies of χ1776 versus the parent and if he feels lower frequencies would be a detriment in working with χ1776.

Dr. R. Curtiss, III: Well, yes. Let me explain. Compared with the original parental strain, which is also a mini-cell producer, χ1776 in the right conditions transforms at approximately threefold higher frequency per microgram of DNA. It actually transforms better with pMB9 DNA than with pSC101. I should point out that during our endeavors we found that we obtained about a threefold increase in transformability because of the double diaminopimelic acid defect. We also showed that the introduction of the *gal*E mutation increases transformability three- to tenfold. We also know that endonuclease-I deficiency increases transformation three- to tenfold, although it is not in this strain. We actually have some strains in our lab that are increased 50-fold in transformation over the standard K-12 strains used by many workers.

The trouble with χ1776 is that it has a mutation that knocks out much of the lipopolysaccharide. This mutation has a multitude of phenotypes. First, it confers sensitivity to bile salts, which is important with respect to survival in the intestine. Second, it has increased sensitivity to detergents, which lessens its survival should it go down the drain. Third, it facilitates DNA isolation. Fourth, it causes the strain to be conjugation-deficient. Finally, it inhibits transformation by approximately tenfold over the strains found earlier in the family tree. Thus, the strain has a transformation frequency of about three times higher than the strain we started with. The problem is that one has to use new glassware. If old glassware that has any scratches or detergent residue is used, transformation is abolished. Thus, we have had to change the transformation procedure. We use 75 mM instead of 25 mM calcium chloride. After a 5-min shock at 42°C, we return the cells to 0°C in ice bucket for 20 min. We plate immediately on the medium containing the antibiotic. If the cells are allowed to grow in broth before plating, they tend to clump and lyse.

So these are a few changes over the standard procedure. With those techniques we achieve a frequency as good as we can obtain with other K-12 strains. I know many labs have not found that to be so, but they have yet to explore the effects of cell density and calcium concentration.

Dr. M. P. Kraus: This question is addressed to anyone who wants to answer it. A number of articles discussing the hazards of artificial maneuvering of recombinant DNA suggest that these maneuvers can be done only in the laboratory. Do you really believe that these various recombinant mechanisms are not also occurring in various ways in nature?

Dr. H. W. Boyer: I think they do.

Dr. R. Curtiss, III: I think that we get into a question of not whether or not they do, but probably of the frequency with which they do. Some can argue that 1 out of 10^9 bacteria in the intestinal tract is taken up by the mesenteric lymph node. I suppose on occasion the lysis of the bacteria being phagocytized or in lymph nodes gets taken up by eukaryotic cells. The fate of that DNA is not known.

Certainly DNA crosses eukaryotic-prokaryotic boundaries. The problem is that technology now allows one to do this with high frequency and fidelity and to select out a clone where one has the desired DNAs.

Dr. F. E. Young: I also think that recombination is occurring in nature. In the heterospecific transformation work, I have shown that it is possible to exchange DNA from closely related species. This had been done in *Pneumococcus* and *Streptococcus* and has been shown to occur *in vivo* in mice.

Dr. W. Szybalski: Dr. Curtiss, in considering possible hosts for phage vector, you did not stress the problem of sensitivity or resistance of your host to bacterial phages. As far as I know, it is resistant to lambda. Does it retain any receptor sites for any other phages, or because of its surface properties is it resistant to everything?

Dr. R. Curtiss, III: Yes, the strain is partially sensitive to 434. We have a derivative of 434 which now plates on the strain very well. It should be possible, using this modified 434 and recombining it with one of the safer lambda vectors now available, to make an appropriate lambda 434 hybrid which would have the 434 J-gene.

Dr. P. H. Kourilsky: Dr. Sgaramella, it seems to me that the circle formation is not a real assay for recombination because it is an *intra*molecular process. In what you describe, it seems to me, for instance, that fusion of the ends might be a limiting process in the reaction. I would like to know if there are any real data on the *inter*molecular process.

Dr. V. Sgaramella: Yes, I think if you raise the concentration of DNA you sometimes can observe the extent of this reaction by looking at the increase in length of segments you are using. In the extreme case, on sucrose gradients with the oligomerization of ^{32}P-DNA molecules, you could convert approximately 30 to 40% of the molecules into higher oligomers. So I am sure there is some variability in the *intra*molecular versus the *inter*molecular reaction, but I still think that it is a useful tool to monitor the joining between the two different ends. I am not sure that the ligase knows whether the two ends come from the same molecule or from two different molecules.

Mr. P. Youderian: I have two questions specifically for Dr. Boyer and Dr. Curtiss. I would appreciate comments from the rest of the group.

First, Dr. Curtiss presented some data in his discussion on χ1876 fed to rats in conjunction with tetracycline. The data show that the strain could survive in the presence of tetracycline, that is, the strain could survive under a specified set of conditions during passage through the gut, overcoming all the limitations that have been imposed on the strain.

Now, in conjunction with this, Dr. Curtiss raised the issue that people working with antibiotic resistance factors should not use antibiotics themselves. I was wondering, first, if Dr. Boyer could comment on what safety procedures are employed in his lab, which is concerned with the production of vectors involving multiple drug-resistance factors.

Second, a question to Dr. Curtiss: shouldn't his strain, χ1876, which under certain specified sets of conditions can survive unexpectedly, be tested in a wider variety of environmental conditions before it is certified as an EK2 host because the NIH Guidelines require 10^{-8} survival under any naturally occurring environmental conditions?

Dr. H. W. Boyer: We use the same physical containment that any other lab uses working with plasmids and drug-resistant markers, that is, P2 and occasionally P1.

Dr. R. Curtiss, III: I think it is a question I have to answer. Let me point out something. A level of 0.5 mg/ml tetracycline in the drinking water for the rat drinking 30 ml water a day represents an intake of 15 mg tetracycline. That huge amount would have adverse side effects. The point is that at this level of tetracycline, 1876 can not grow. It is resistant only up to 12.5 μg/ml. Obviously we are much higher.

When the metabolic activity of 1876—or any bacterium, for that matter—is completely abolished, you may not completely eliminate but certainly may reduce the probability that that cell can undergo conjugation, propagation of a transducing phage, or anything that would lead to the transmission of the microbe. These strains possess a number of mutations such that even if they do pass all the way through the intestinal tract in the feces and are distributed, they do not encounter thymidine and/or diaminopimelic acid. This is a nonpermissive condition under which the cells ultimately die. So I do not see this as a particular problem.

The other thing is that rats, when fed a microorganism, pass it in feces in 6 hr, whereas in humans the time span is closer to 24 hr. In addition, as I understand, rats have no gallbladder or bile. This strain is acutely sensitive to bile and, therefore, should have a lower survival rate when tested in humans or primates that have a gallbladder.

In meetings during the past 3 months more tests have been suggested. We have done them all, including determining the amount of nuclease in the intestines which have not previously been explored. I would be happy for anybody to do more tests if they wish. I think in our own lab group we are now endeavoring to make better strains rather than do more tests.

Dr. F. E. Young: Might I comment on that, Roy? One of the things I have seen through hospital surveillance of infections is that our environment contains large amounts of antibiotics. Thus, one of the gravest problems we face in hospitals is the overprescription of antibiotics. We have recently seen that *Haemophilus influenzae* has naturally incorporated an ampicillin-resistant gene on an endogenous plasmid. I believe, therefore, that the abuse of these chemotherapeutic agents must be curtailed and that the workers in recombinant DNA laboratories should not do experiments if they are on antibiotic therapy.

Dr. A. H. Deutch: One of our goals is to clone unique mammalian genes. Is it possible to have a test tube containing a pure gene, so that we could take a nanogram of such DNA through the restriction and ligation steps and use it in transformation?

Dr. F. E. Young: In the *Bacillus subtilis* system, transformants were obtained with 10 pg DNA per milliliter. The system still is far from developed in comparison with the sophistication of the *E. coli* system.

Dr. R. Curtiss, III: We ordinarily use 50 ng in transformation of χ1776. That gives us, say, 100 transformants. So if you went down to 5 ng you would have 10 transformants. I suppose with other strains one should be able to achieve that range eventually.

Dr. Deutch: So the enzymes could be active on a nanogram of DNA?

Dr. R. Curtiss, III: Yes.

Dr. F. E. Young: If I calculate it correctly, about 10 pg would be the limit of activity from the *B. subtilis* data that I presented.

Dr. M. H. Richmond: I would like to come back to Roy Curtiss' point on the resistance of the strain 1876 that was excreted in the presence of antibiotics. It seems to me that there must have been some multiplication of that strain. Otherwise, why is it excreted for so much longer? The antibiotic is not going to slow it down in passage through the gut.

Dr. R. Curtiss, III: I did not quite understand that. It is not multiplying.

Dr. M. H. Richmond: Why is it excreted for longer times?

Dr. R. Curtiss, III: It is cleared; we detected the strain 12 and 24 hr after feeding. The titers that we detected were about 10^4 to 10^2 in those two samples. We fed over 10^{10} microorganisms. To clear completely a strain that can not grow in the gut probably takes a couple of days just because some of the cells are lodged in the villi and things of this sort.

Dr. M. H. Richmond: That is fine; I want to check that. The other question was: did you check the organisms to see if they reverted?

Dr. R. Curtiss, III: Yes, we tested genotypic or at least the phenotypic properties of those microbes that passed through. They are identical to the original strain with respect to bile salt sensitivity, dap^-, gal^-, thy^-, and mol^-. We always test a reasonable sample.

Dr. P. H. Kourilsky: Dr. Curtiss, I think you wrote somewhere that you estimated the probability of dissemination of the strain of something such as 10^{-20}. What do you think is the probability that the work you did was with the wrong strain from the refrigerator? I am raising a more general question. Wouldn't it be better in general to emphasize the safety of the vector rather than the host? People usually prepare vectors in large amounts and can check the vector. I feel that the probability for somebody to pick up the wrong strain from the refrigerator and do an experiment with it is significantly high.

Dr. R. Curtiss, III: I suppose you are correct that somebody could pick up the wrong strain. That is a potential human error.

In the strains we have constructed, we have told people the strain is resistant to cycloserine. We suggest that one add cycloserine to the culture plating medium to preclude growth of any inadvertent microbe that gets in during the transformation as a contaminant, just as mammalian cell biologists use antibiotics to keep bacteria out of the mammalian cell cultures.

I should point out, though, that most of the contaminants that would be encountered in the lab would be *Pseudomonas* or *Staphylococcus,* which would probably not propagate *E. coli* cloning vectors, although that has not been looked at too thoroughly yet.

Dr. G. Wilcox: I would like to ask Dr. Roberts if one can draw any generalization about the storage and stability of the restriction endonucleases.

Dr. R. J. Roberts: I could give you a list. In general, they seem to be rather stable. Most of the enzymes we have worked with can be stored in 50% glycerol at −20°C for periods up to 6 months, 1 year, or 2 years without much loss in activity. This is true even though they have been only partially purified—through only two or three chromatographic steps. I think the probable reason for this is that the enzymes are fairly stable and do not inactivate very easily; and secondly, that the steps usually employed eliminate most of the proteases. I think the stability of the enzymes themselves is best demonstrated by the fact that digestions are linear for long periods of time, even after 24 hr.

Recombinant Molecules: Impact on Science and Society, edited by R. F. Beers, Jr. and E. G. Bassett. Raven Press, New York © 1977.

10. Introduction to Section B: Development of Plasmid Vectors

Stanley Falkow

Department of Microbiology and Immunology, School of Medicine, University of Washington, Seattle, Washington 98195

I confess that for roughly 17 years I thought that plasmids (or, as we used to call them, episomes) were doing pretty well as vectors and did not need much help in the way of development. In fact, my experience in the clinical laboratory has been that plasmids have been developing all too well!

The term plasmid means literally a thing in a cell's cytoplasm, although it is commonly defined as a nonessential, autonomously replicating extrachromosomal element of bacteria. Plasmids come in a wide variety of sizes and DNA compositions. It is probably fair to say that plasmids have been found in every bacterial species in which they have been sought. The presence of a plasmid often permits its bacterial host to better survive in specialized environments or to better compete with other microorganisms for a specific ecological niche. Since plasmids may be transmitted directly or indirectly among a variety of bacterial species, they have been often termed promiscuous.

But we have not compiled this volume to discuss plasmids and bacteriophages in general. Rather, we shall concentrate primarily on two plasmids, pSC101 or derivatives of the plasmid ColE1, as they relate to molecular cloning. In addition, we shall give at least equal time to the bacteriophage λ. (The distinction between a conjugative plasmid and a bacteriophage has always been a source of some puzzlement to me. I have always considered λ to be a plasmid that went wrong. Or perhaps it is the other way around.) In any event, we will consider the potential biohazards associated with splicing foreign DNA into pSC101, λ or other vectors.

I think it is fair to say that most scientists are unnerved by molecular cloning in that it represents yet another step in man's unnatural domination over his environment. I think it is also fair to assert that most of the contributors to this volume were and probably still are more than a bit uncomfortable about dealing objectively with the subject of biohazards. Indeed, those who attended the Asilomar Meeting on the Potential Hazards of Recombinant DNA Molecules may recall that on the first day there was a heated discussion of whether *E. coli* K-12 need be handled with aseptic technique. "You mean it isn't all right to pour *E. coli* K-12 down the drain?" By the time Asilomar was over, most people in attendance were convinced not only that they should not pour *E. coli* K-12 down the sink but that molecular cloning, even in *E. coli* K-12, posed some terribly complex biological as well as ethical questions. One of the many fascinating facets of the Asilomar meeting was that the standard practice of handling biohazards, sterilization, and physical containment, were considered not to be fully adequate for dealing with some of the potential hazards that were perceived for some molecular cloning experiments. So the call went forth to take *E. coli* K-12 and somehow "defuse the bug." The fascinating account of the attempts by Dr. Curtiss and his associates to meet this end was presented earlier in this volume. Plasmids and phage are molecular parasites but they can also "jump ship"—remember they are promiscuous, autonomously replicating elements. So it is not enough to defuse the bug, we must also defuse the plasmid and phage. The plan then is to develop a host strain/plasmid or host cell/phage combination that poses no threat to the investigator and would have an exceedingly low (theoretically) probability of escape and survival away from the confines of the laboratory.

I can assure you that the contributors to this section will not tell you about the perfectly "defused" plasmid or phage. They will attempt to provide you with an overview of molecular cloning with plasmids and phage as well as the first attempts at and tactics for developing defused plasmid and phage vectors. It is our hope that this information will give you some insight not only into the advantages, limitations and, indeed, hazards of these vectors of cloned DNA but also into the basic biology of these genetic vectors.

Recombinant Molecules: Impact on Science and Society, edited by R. F. Beers, Jr. and E. G. Bassett. Raven Press, New York © 1977.

11. DNA Cloning as a Tool for the Study of Plasmid Biology

Stanley N. Cohen, Felipe Cabello, Annie C. Y. Chang, and * Kenneth Timmis

Stanford University School of Medicine, Stanford, California 94305

INTRODUCTION

During the past several years, the development of DNA cloning procedures has made possible the introduction of prokaryotic and eukaryotic genes from diverse sources into *Escherichia coli* and other microorganisms. The earliest DNA cloning experiments (9) involved linkage of *Eco*RI restriction endonuclease-generated DNA fragments to a bacterial plasmid replicon, and subsequent introduction of the composite molecules into *E. coli* by transformation. Using this general method, investigators have propagated segments of bacterial plasmids (4,24,42), bacterial viruses (16), and eukaryotic genes from *Xenopus laevis* (3,8,27), *Drosophila melanogaster* (11,15,39), sea urchins (22), and mouse mitochondria (1,5) in bacteria using plasmid cloning vehicles such as pSC101 and ColE1. Plasmid vectors have also been used for the cloning of cDNA synthesized from partially purified mRNA of eukaryotes (30,34). Eukaryotic segments of chimeric plasmids are transcribed in the prokaryotic host (11,22,27), and polypeptide synthesis dependent on the eukaryotic DNA fragment has been observed (5).

The advances that led to genetic manipulation of microorganisms were made in several different laboratories in the late 1960's and early 1970's. There are four requirements: (a) a replicon (cloning vehicle or vector) able to propagate itself in the recipient organism, (b) a method of joining another DNA segment to the cloning vector, (c) a procedure for introduction of the composite molecule into a biologically functional recipient cell, and (d) a method of selection, from a large population of potentially recipient bacterial cells, of those microorganisms that have acquired the hybrid DNA species. Figure 1 illustrates use of these requirements in the initial cloning procedure.

---

* Present address: Max-Planck-Institut für Molekulare Genetik, 1 Berlin 33 (Dahlem), West Germany.

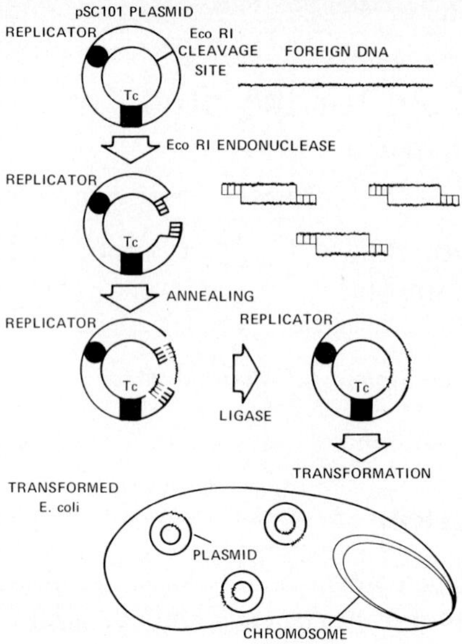

FIG. 1. DNA cloning procedure. The pSC101 plasmid, which carried replicative functions, a Tc-resistance gene, and a single-cleavage site for the *Eco*RI restriction endonuclease, is cleaved by the *Eco*RI enzyme. Cleavage of the plasmid, which occurs at a site that does not interfere with replicative functions or Tc-resistance, results in short, complementary, single-stranded ends. Similarly-cleaved foreign DNA is annealed with pSC101 plasmid DNA, and nicks are sealed by the enzyme DNA ligase. The resulting molecule is transformed into calcium chloride–treated *E. coli,* and selection is carried out for Tc-resistant cells. The pSC101 cloning vehicle and the reinserted foreign DNA fragment are propagated in the progeny of the original transformed cell.

During the past several years, it has become clear that a wide variety of experimental alternatives are available for the cloning of foreign DNA fragments in *Escherichia coli,* and a number of these are described in detail elsewhere in this volume. As shown in Table 1, various natural and constructed plasmid vectors and appropriate mutants of bacteriophage λ have been used; the essential requirement is that the cloning vector be a replicon able to propagate itself and an attached segment of foreign DNA in the recipient organism. Most conveniently, the construction of biologically functional, recombinant plasmids can be carried out using cohesive ends (26,35) generated by restriction endonucleases that cleave the plasmid at a single site (Table 2): some of the plasmids commonly used as vectors are cleaved once by each of several endonucleases, making these plasmids useful in the cloning of DNA fragments generated by different enzymes. Although λ mutants suitable for cloning of either *Hin*dIII or *Eco*RI-generated DNA fragments have been isolated (28,31,40), suitable λ mutants are not currently

TABLE 1. *Replicons capable of propagating themselves and attached genes*

A. Plasmid cloning vectors
pSC101
ColE1, pML21, ColE1-Ap, mini-ColE1
RSF1010
B. Hybrid plasmid replicons
pSC101-ColE1, RSF1010-ColE1, pSC101-RSF1010, pMB9, pBR313, and cloned replication region fragments of large plasmids (e.g., mini-R6-5, mini-F)
C. Bacteriophage vectors
λgt, λdv

TABLE 2. *Methods of joining DNA segments to plasmid replicons*

A. Cohesive termini generated by restriction endonucleases	
	e.g., *Eco*RI cleavage of plasmids pSC101, ColE1, or pMB9
B. Polynucleotide projections (e.g., dA-T)	
	Can be used at any restriction endonuclease cleavage site (blunt or cohesive end)
	Also used for sheared DNA or cDNA
C. Blunt-end ligation	
	Accomplished with T4 ligase
	Can be used to add endonuclease-generated adapter fragments to sheared DNA or cDNA

available for use with other enzymes. However, the use of "adapter fragments" as described below potentially circumvents this problem and allows considerable flexibility.

Biologically functional DNA molecules can also be constructed by addition of a series of identical deoxyribonucleotides (e.g., dA) to the ends of one DNA species and addition of complementary deoxynucleotides (e.g., dT) to the ends of a second species (6,43). The dA-T "terminal transferase" procedure (18,23) prevents the joining together of different fragments derived from the same DNA molecule, which may interfere with the joining of such fragments to the cloning vector (6). However, fragments of DNA joined by the dA-T method cannot be recovered conveniently from chimeric molecules, whereas inserted fragments are readily recovered when restriction endonuclease–generated cohesive ends are used for linkage (9). A method that involves the addition of dC-G termini to DNA molecules lacking cohesive ends, plus utilization of the short cohesive ends generated by the *Eco*RI restriction endonuclease under conditions of low salt potentially, enables the reconstruction of *Eco*RI-cleavable termini on cloned fragments of cDNA (34). Variations of this method seem likely to enable coupling of dA-T termini to cohesive ends generated by certain other restriction enzymes—with resulting reconstruction of endonuclease cleavage sites. However, it is important to point out that cohesive termini are not required for the joining of DNA fragments as reviewed elsewhere in this volume; blunt-ended DNA molecules can be joined using the T4 DNA ligase (36).

Introduction of chimeric plasmids into recipient cells is accomplished using a plasmid DNA transformation procedure (Table 3) (10); this procedure has enabled the cloning of individual plasmids, making it possible to study the progeny of single DNA molecules in ways previously feasible only with viral genomes by using the infective process. Through the transformation procedure, the progeny of single molecules of plasmid DNA can be amplified manyfold in bacterial cultures and can be propagated indefinitely. The procedure employed depends on the ability of divalent cations such as calcium chloride (25) and barium chloride (38) to alter membrane permeability of bacteria, thereby enabling them to take up plasmid DNA

TABLE 3. *Recombinant DNA procedures*

Introduction of recombinant DNA molecules into bacterial cells
A. Plasmid transformation
B. Phage transfection
Selection of microorganisms that acquire recombinant DNA molecules
A. Use of phenotypic property of vector and/or inserted DNA (e.g., antibiotic resistance, colicin resistance)
B. Insertional inactivation of gene on vector (e.g., colicin production, Tc resistance, Km resistance)
C. Insertional activation (e.g., phage λ, inserted fragment is required for plaque production)
D. Radioactive probe To identify cells carrying hybrid molecules by subculture cloning or *in situ* hybridization
E. Immunological probe

(10). Bacteriophage vectors are also taken up by appropriately treated cells, but in this case the cloning of inserted DNA fragments does not require survival of recipient cells – merely the ability of such cells to yield viable viral particles. Introduction of phage chimeras into bacterial cells potentially can be accomplished by using one of several transfection procedures employing calcium chloride treatment (25), spheroplast production (14), or a "helper phage" assay (21).

Antibiotic resistance genes carried by both the plasmid cloning vehicle and the inserted DNA fragments were initially used to select for cells that had acquired the chimeric molecules. Other genetic markers on cloning vectors such as colicin immunity or metabolic characteristics have since been used (Table 3). Inactivation of a cloning vehicle gene by insertion of a foreign DNA fragment at a restriction endonuclease cleavage site of the cloning vehicle has also been employed to select for chimeric plasmids (i.e., insertional inactivation) (41). A wide variety of other methods have been used to enable selection of plasmids carrying *specific* DNA fragments, such as subculture cloning procedures (22) and membrane hybridization methods (13,20); these methods use radioactively labeled probes to identify clones carrying the desired chimeric plasmid or phage.

USE OF DNA CLONING PROCEDURES IN THE STUDY OF PLASMID REPLICATION AND INCOMPATIBILITY

As noted above, plasmids have proved to be highly useful as gene carriers in DNA cloning experiments; conversely, DNA cloning has proved to be a useful technique for the study of plasmid biology. This has been especially true in studies of plasmid replication.

Although naturally occurring plasmids carry a wide variety of genes ranging from metabolic degradative pathways to antibiotic resistance traits (10), the essential feature common to all plasmids is a system that accomplishes autonomous replication of the plasmid as an extrachromosomal element. Thus, plasmids can be considered as replication systems that have acquired additional genes conducive to propagation of the plasmids. Although the molecular biology of small plasmids is readily amenable to study, the genetic complexity and structural fragility of large plasmid DNA molecules present potential problems in the study of DNA replication. In addition, large plasmids are known to carry more than one origin of replication (33), and replication may proceed differently at each origin; little has been known about the functional interaction of multiple replication systems that may coexist on a single DNA molecule.

DNA cloning procedures have enabled us to use two novel approaches to the study of plasmid replication: the first has been the construction of model replicons in which two physically and functionally distinct, well-characterized replication systems have been fused by joining them at restriction endonuclease cleavage sites. The second is the "paring-down" of complex replicons by specific selection of those DNA sequences that carry replication functions. Figure 2 diagrams both of these approaches.

The pSC134 composite plasmid was constructed by *in vitro* linkage of the colicin-producing plasmid ColE1 and the tetracycline-resistance plasmid pSC101 at their unique *Eco*RI restriction endonuclease restriction sites (41). Since the ColE1 plasmid cannot replicate in bacteria deficient in DNA polymerase I (12), and the pSC101 plasmid cannot replicate in bacteria that are not able to synthesize protein (41), the two components of the pSC134 have functionally distinguishable replicative properties. It was possible, therefore, to show that each component replicative system of the hybrid plasmid could accomplish replication of the entire molecule; however, studies of the origin of replication by the hybrid plasmid indicated that under normal conditions it uses only its ColE1 replication (41). When function of the ColE1 replication system is prevented (for example, in a host strain that synthesizes a temperature-sensitive DNA polymerase), the pSC101 replication origin and associated functions are used, and the copy number remains at the level characteristic of that pSC101. When the pSC134 composite plasmid uses exclusively the ColE1 replicative system, the plasmid is nevertheless incompatible with *both* of the parent replicons (2). Such studies support the view that a *trans*-dominant gene product is involved in the determination of plasmid incompatibility.

The replicon model of Jacob et al. (19) postulates that replication takes place at specific cell membrane attachment sites, and that incompatibility between plasmids is the result of competition between similar replicative systems for the same attachment site. This model would predict that each of the component replicative systems of pSC134 should be able to replicate some plasmid molecules, since both the ColE1 and pSC101 replication origins and functions have been shown to be available for propagation of

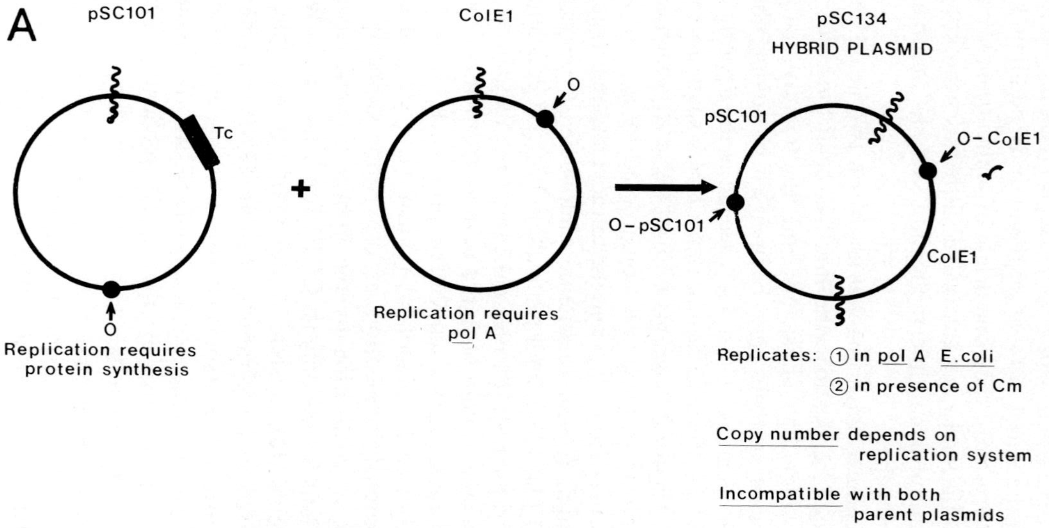

FIG. 2. Utilization of DNA cloning in studies of plasmid replication.

2A: Construction of hybrid plasmid replicons. The pSC101 and ColE1 plasmid are joined at the single *Eco* RI restriction endonuclease cleavage site of each plasmid to form the pSC134 hybrid replicon. Since the component replicative systems of pSC134 have distinguishing properties, it can be determined which origin (0) of replication and which set of replicative functions is being used by the hybrid plasmid.

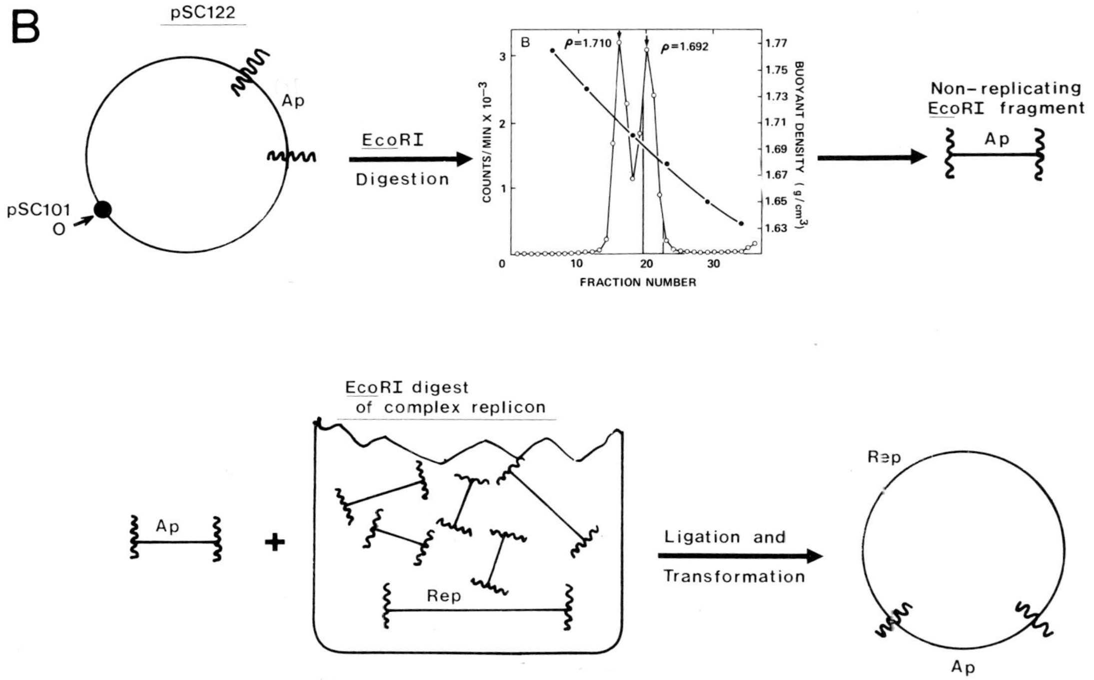

2B: Isolation of replicative regions of complex genomes. An Ap-resistant, *Eco*RI-generated fragment of the pSC122 plasmid, which bands at a different buoyant density ($\rho = 1.692$) than the pSC101 component ($\rho = 1.710$), can be separated from the pSC101 (replicative system) fragment by *Eco*RI digestion and centrifugation in cesium chloride gradients. The Ap-resistant gene segment is then used as a probe for picking out the *Eco*RI-generated fragment(s) carrying the replicative functions of a complex replicon.

the entire molecule. The model would also predict that pSC134 should have a copy number similar to the additive copy number of the independently replicating ColE1 and pSC101 plasmids, since each component replicative system should be capable of producing its normal complement of copies. As summarized above, studies of the pSC134 plasmid constructed by recombinant DNA techniques have yielded data inconsistent with these predictions and have provided evidence to support the view that a *trans*-dominant, plasmid-specified gene product is involved in plasmid incompatibility.

CLONING OF REPLICATION REGIONS OF COMPLEX GENOMES

The ability to select and clone restriction endonuclease–generated fragments of larger genomes has also made possible the isolation of specific segments of plasmids that carry replicative functions. The experimental plan for such studies is shown in Fig. 2B; the procedure involves the use of a selectable but nonreplicating endonuclease-generated DNA fragment as a probe to enable the identification and isolation of DNA segments capable of autonomous replication. In our studies (42), a fragment of staphylococcal plasmid DNA carrying a gene for ampicillin (Ap) and penicillin (Pc) resistance was used as a probe. Although this fragment was derived from another bacterial species, it codes for an antibiotic resistance trait indigenous to *E. coli,* and this trait is expressed in the *E. coli* host; the "probe" fragment had previously been cloned in *E. coli,* using the pSC101 plasmid as a cloning vehicle (4). Since the buoyant density of the staphylococcal DNA Ap-resistance fragment differs substantially from the buoyant density of the pSC101 vector, the nonreplicating staphylococcal DNA probe fragment can easily be separated from the pSC101 plasmid in cesium chloride gradients. A Km-resistance DNA fragment originally derived from the R6-5 plasmid (9) can also be used as a probe for the isolation of plasmid replication regions (24).

Using the procedure outlined in Fig. 2B, we were able to isolate *Eco*RI-generated DNA fragments carrying the replication region of the R6-5 and F*lac* plasmids. Although the replication regions of these plasmids contain only about 10% of the nucleotide sequences of the parent genomes, they were shown to specify both the incompatibility and copy number control properties of the parent plasmids. In addition, they express other replication-related plasmid properties, such as the ability to integratively suppress the *dna* A mutation (29) on the chromosome of bacteria grown under nonpermissive conditions. The characteristic acridine orange sensitivity of the F*lac* plasmid (17) also appears to be determined by the replication region segment of the plasmid.

The methods used for isolation and characterization of replication regions of the R6-5 and F*lac* plasmid genomes are potentially applicable for the isolation of DNA segments containing the replication origin and/or genes of any complex replicon capable of functioning in microorganisms, and they may be useful in the study of chromosomal replication. The sequestration of replicative functions of large genomes onto small plasmid DNA molecules should potentially facilitate *in vitro* and *in vivo* investigations of gene products involved in DNA replication. Moreover, the use of multiple restriction enzymes should enable fine-structure mapping of specific replicative functions within the isolated fragment; the cloning of a *Hin*dIII-generated replication region fragment of the R6-5 plasmid has already enabled identification of a region within the *Eco*RI-generated fragment that carries replicative functions. With appropriate modification, the principle of using a restriction endonuclease–generated "probe" fragment that lacks a particular genetic function to select for another fragment carrying the function may also permit identification and isolation of other phenotypically defined regions of complex genomes. Conversely, the phenotypic properties associated with specific segments of complex plasmid genomes can be studied by cloning the various restriction endonuclease–generated fragments of the genome (K. Timmis, F. Cabello, and S. N. Cohen, *in preparation*).

USE OF ADAPTER FRAGMENTS FOR DNA CLONING

The pSC101 plasmid contains a series of cleavage sites for different restriction endonucleases within close proximity to each other (15,32; also A. C. Y. Chang and S. N. Cohen, *unpublished observations*). Because the pSC101 plasmid contains single *Eco*RI, *Hin*dIII, *Bam*I, and *Sal*I cleavage sites within a region about 800 nucleotides in length, the plasmid serves as a source of "adapter fragments" which can be produced by treatment of pSC101 DNA with different combinations of these enzymes (Fig. 3A,B). Such adapter fragments are potentially useful for the conversion of a cleavage site generated by one of the restriction enzymes into a terminus that can be joined to a cohesive end produced by digestion of DNA with another enzyme. The utility of adapter fragments in DNA cloning experiments is enhanced substantially by the observation that blunt-ended DNA molecules can be joined together by the bacteriophage T4 ligase (36). The combined use of blunt-end ligation and adapter fragments from the pSC101 plasmid enables the cloning of blunt-ended DNA in cohesive-ended sites generated by the *Eco*RI, *Hin*dIII, *Bam*I, or *Sal*I restriction endonucleases. The procedures outlined in Fig. 3C are applicable for the cloning of cDNA segments or short polynucleotide fragments synthesized *de novo;* moreover, use of the adapter fragment technology readily enables excision of these inserted DNA pieces from chimeric molecules.

INSTABILITY OF CERTAIN CHIMERIC PLASMIDS

During studies of chimeric plasmids containing various kinds of eukaryotic or prokaryotic DNA, we observed that some chimeric plasmids were

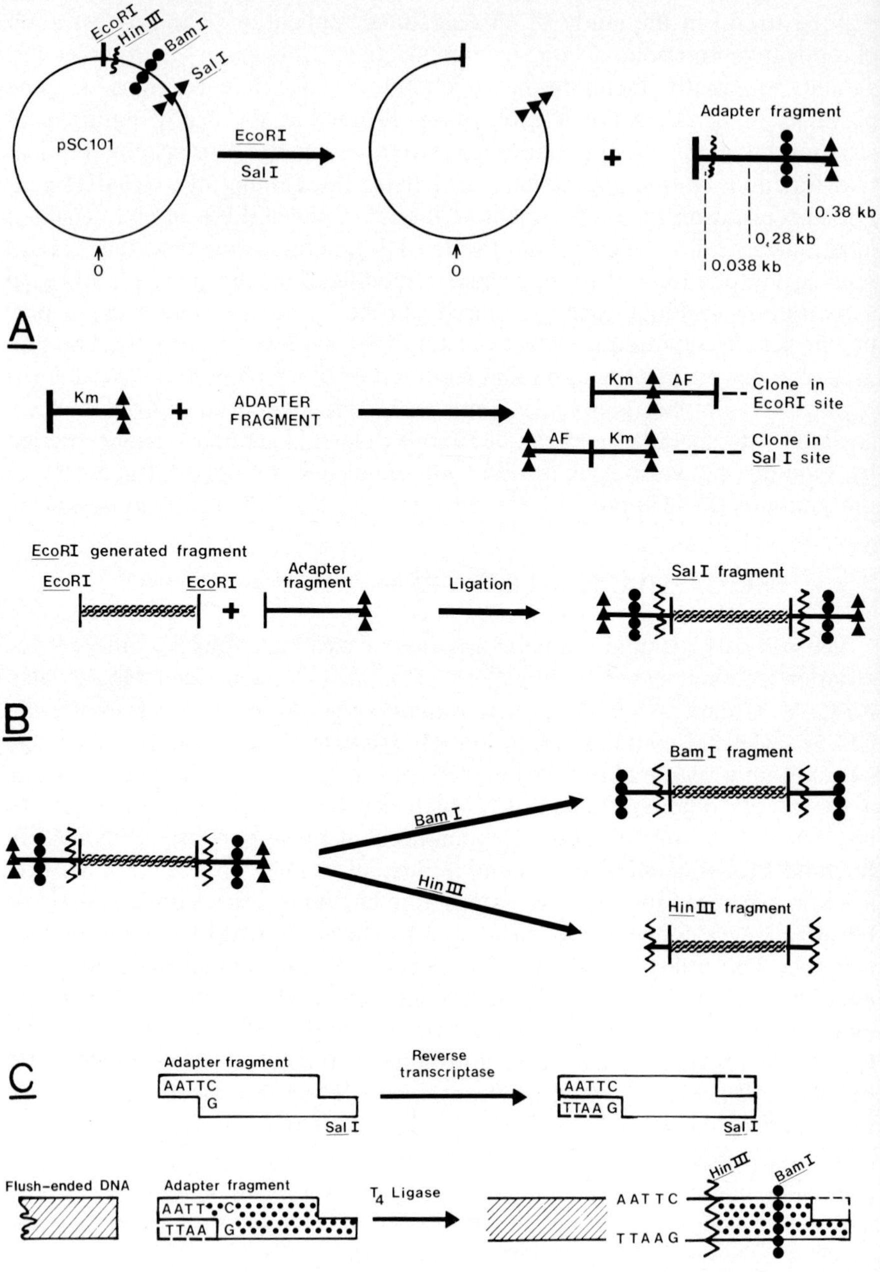

FIG. 3. A: Use of adapter fragments in DNA cloning. The pSC101 plasmid contains *Eco*RI, *Hin*dIII, *Bam*I, and *Sal*I restriction endonuclease cleavage sites in close proximity to each other. The location of these sites is shown in relation to the pSC101 origin of replication (0). Treatment of pSC101 plasmid DNA with both the *Eco*RI and *Sal*I enzymes leads to formation of the structures shown. The adapter fragments resulting from cleav-

lost from bacterial cells in the absence of continued selection for an antibiotic-resistance gene carried by the cloning vehicle component of the plasmid. Moreover, when continued selection for a cloning vehicle gene was carried out, the inserted foreign gene fragment appeared, in at least some cases, to have been lost.

Figure 4 shows the relative stability of three different composite plasmids constructed entirely of prokaryotic DNA segments; bacteria were grown in the absence of antibiotics for the number of generations shown and were then tested for expression of antibiotic-resistance determinants as indicated. The pSC134 plasmid, which is a hybrid DNA molecule constructed from ColE1 and pSC101 components as indicated above, was stable in the ab sence of selection—as was the pSC122 plasmid, which contains an ampicillin-penicillin resistance fragment of staphylococcal plasmid DNA inserted into pSC101. However, another *E. coli*–staphylococcal hybrid plasmid (pSC112), which contains another Ap-resistance gene derived from the pI258 plasmid (4), appeared to be *unstable;* growth of bacteria containing pSC112 in the absence of antibiotics led to a gradual loss of the tetracycline resistance specified by the pSC101 cloning vehicle, and a much more rapid loss of the Ap-resistance trait carried by the inserted DNA fragment. The observation that Tc-resistant, Ap-sensitive cells were produced in cultures of bacteria carrying a plasmid, which was shown by heteroduplex analysis and agarose gel electrophoresis to originally contain pSC101 and staphylococcal DNA fragments (4), suggested that preferential loss of the staphylococcal DNA Ap-resistance gene component of the chimeric plasmid is occurring in some cells. Isolation and characterization of DNA from Tc-resistant, Ap-sensitive cells (F. Cabello, K. Timmis, and S. N. Cohen, *in preparation*) has confirmed this hypothesis.

It has previously been reported (5) that mouse mitochondrial DNA-pSC101 hybrid plasmids appear to undergo a rearrangement of plasmid DNA under certain conditions. Recombination, DNA inversion, and transposition (translocation) of DNA segments also appear to occur within other chimeric plasmids, even in $recA^-$ hosts (A. C. Y. Chang and S. N. Cohen, *unpublished observations*). In more recent studies, growth of different chimeric plasmids containing various eukaryotic DNAs in the absence of

---

age can be used to convert a terminus produced by one restriction endonuclease to an end that can be cloned at a cleavage site generated by another enzyme.

B: Use of adapter fragments for substitution of restriction endonuclease sites. The illustration indicates that adapter fragments derived from the pSC101 plasmid can be used for the insertion of *Eco*RI-generated fragments of DNA into cleavage sites generated by the *Sal*I, *Bam*I, or *Hin*dIII restriction endonucleases.

C: The use of adapter fragments to add cohesive termini to blunt-ended DNA fragments. The cohesive ends of the adapter fragments are first filled in by the enzyme reverse transcriptase, and then are joined to the blunt-end termini of other DNA fragments by the bacteriophage T4 ligase. Subsequent introduction of cohesive termini within the adapter fragment is accomplished by treatment by an appropriate restriction endonuclease.

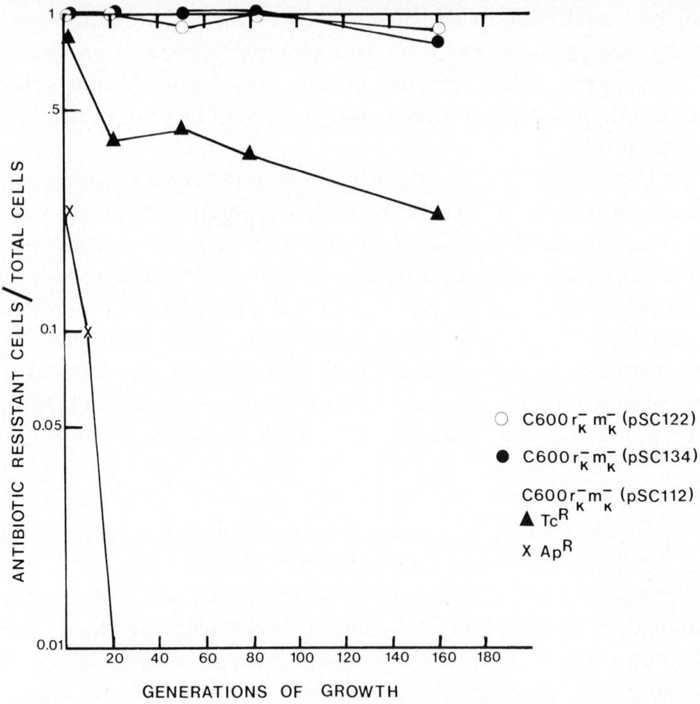

FIG. 4. Instability of certain chimeric plasmids in the absence of selective pressure. Culture of *E. coli* strain C600 rk⁻mk⁻ carrying the *E. coli* hybrid plasmid pSC134, or the *E. coli–S. aureus* hybrid plasmids pSC112 or pSC122 were grown in media containing antibiotics and were then diluted at time zero into culture media lacking drugs. Serial passage was carried out for 160 generations of growth in the absence of antibiotics. Samples were removed as indicated, and the fraction of cells expressing the Tc-resistance of the pSC101 component of each plasmid, and the Ap-resistance of the inserted staphylococcal DNA fragment, was determined.

antibiotics appeared to result in loss of the entire plasmid; however, in the presence of continued antibiotic selection, bacteria carrying certain of the prokaryotic-eukaryotic chimeras were able to compete effectively with bacteria carrying the pSC101 plasmid alone. These studies will be reported in detail elsewhere (F. Cabello et al., *in preparation*). Although the mechanism for reassortment and/or excision of DNA fragments in chimeric plasmids is not known, we have observed inverted repeat DNA sequences during electron microscopic heteroduplex examination of certain chimeric plasmids. Recent observations (7,37) indicating that inverted repeats or palindromes are implicated in the translocation of at least some discretely defined segments of plasmid DNA suggest the possibility of involvement of these structures in the "illegitimate" recombinational events described here.

SUMMARY

This brief review has outlined the various requirements for the cloning of foreign DNA in microorganisms and has considered the various alternatives currently available for each step in the procedure. The use of adapter fragments derived from the pSC101 plasmid by cleavage with various combinations of restriction endonucleases has been described, along with experiments that have employed the DNA cloning procedure for study of the molecular biology of plasmid DNA replication. Preliminary findings indicating the instability of certain chimeric plasmids have also been reported.

ACKNOWLEDGMENTS

The investigations from the laboratory of the authors have been supported by Grant AI 08619 from the National Institute of Allergy and Infectious Diseases, Grant PCM 75–14176 from the National Science Foundation, and by American Cancer Society Grant VC139A to Stanley N. Cohen. Kenneth Timmis was a Postdoctoral Fellow of the Helen Hay Whitney Foundation.

REFERENCES

1. Brown, W. M., Watson, R. M., Vinograd, J., Tart, K. M., Boyer, H. W., and Goodman, H. M. (1976): The structures and fidelity of replication of mouse mitochondrial DNA-pSC101 EcoRI recombinant plasmids grown in E. coli K-12. *Cell,* 7:517–530.
2. Cabello, F., Timmis, K., and Cohen, S. N. (1976): Replication control in a composite plasmid constructed by in vitro linkage of two distinct replicons. *Nature,* 259:285–290.
3. Carroll, D., and Brown, D. D. (1976): Adjacent repeating units of Xenopus laevis 5S DNA can be heterogeneous in length. *Cell,* 7:477–486.
4. Chang, A. C. Y., and Cohen, S. N. (1974): Genome construction between bacterial species in vitro: replication and expression of Staphylococcus plasmid genes in Escherichia coli. *Proc. Natl. Acad. Sci. U.S.A.,* 71:1030–1034.
5. Chang, A. C. Y., Lansman, R. A., Clayton, D. A., and Cohen, S. N. (1975): Studies of mouse mitochondrial DNA in Escherichia coli: structure and function of the eukaryotic-prokaryotic chimeric plasmids. *Cell,* 6:231–244.
6. Clarke, L., and Carbon, J. (1975): Biochemical construction and selection of hybrid plasmids containing specific segments of the Escherichia coli genome. *Proc. Natl. Acad. Sci. U.S.A.,* 72:4361–4365.
7. Cohen, S. N. (1976): Transposable genetic elements and plasmid evolution. *Nature,* 263: 731–738.
8. Cohen, S. N., and Chang, A. C. Y. (1974): A method for selective cloning of eukaryotic DNA fragments in Escherichia coli by repeated transformation. *Mol. Gen. Genet.,* 134: 133–141.
9. Cohen, S. N., Chang, A. C. Y., Boyer, H. W., and Helling, R. B. (1973): Construction of biologically functional bacterial plasmids in vitro. *Proc. Natl. Acad. Sci. U.S.A.,* 70:3240–3244.
10. Cohen, S. N., Chang, A. C. Y., and Hsu, L. (1972): Nonchromosomal antibiotic resistance in bacteria: genetic transformation of Escherichia coli by R-factor DNA. *Proc. Natl. Acad. Sci. U.S.A.,* 69:2110–2114.
11. Glover, D. M., White, R. L., Finnegan, D. V., and Hogness, D. S. (1975): Characterization

of six cloned DNAs from Drosophila melanogaster, including one that contains the genes for rRNA. *Cell,* 5:149–157.
12. Goebel, W. (1972): Replication of the DNA of the colicinogenic factor E1 (ColE1) at the restrictive temperature in a DNA replication mutant thermosensitive for DNA polymerase 3. *Nature* [*New Biol.*], 237:67–70.
13. Grunstein, M., and Hogness, D. S. (1975): Colony hybridization: a method for the isolation of cloned DNAs that contain a specific gene. *Proc. Natl. Acad. Sci. U.S.A.,* 72:3961–3965.
14. Guthrie, G. D., and Sinsheimer, R. L. (1960): Infection of protoplasts of Escherichia coli by subviral particles of bacteriophage $\phi\chi$174. *J. Mol. Biol.,* 2:297–305.
15. Hamer, D. H., and Thomas, C. A., Jr. (1976): Molecular cloning of DNA fragments produced by restriction endonucleases SalI and BamI. *Proc. Natl. Acad. Sci. U.S.A.,* 73:1537–1541.
16. Hershfield, B., Boyer, H. W., Yanofsky, C., Lovett, M. A., and Helinski, D. R. (1974): Plasmid Co1E1 as a molecular vehicle for cloning and amplification of DNA. *Proc. Natl. Acad. Sci. U.S.A.,* 71:3455–3459.
17. Hirota, Y. (1960): The effect of acridine dyes on mating type factors in Escherichia coli. *Proc. Natl. Acad. Sci. U.S.A.,* 46:57–60.
18. Jackson, D. A., Symons, R. H., and Berg, P. (1972): Biochemical method for inserting new genetic information into DNA of simian virus 40: circular SV40 DNA molecules containing lambda phage genes and the galactose operon of Escherichia coli. *Proc. Natl. Acad. Sci. U.S.A.,* 69:2904–2909.
19. Jacob, F., Brenner, S., and Cuzin, F. (1963): On the regulation of DNA replication in bacteria. *Cold Spring Harbor Symp. Quant. Biol.,* 28:329–343.
20. Jones, K. W., and Murray, K. (1975): A procedure for detection of heterologous DNA sequences in lambdoid phage by in situ hybridization. *J. Mol. Biol.,* 96:455–460.
21. Kaiser, A. D., and Hogness, D. S. (1960): The transformation of Escherichia coli with deoxyribonucleic acid isolated from bacteriophage λdg. *J. Mol. Biol.,* 2:392–415.
22. Kedes, L. H., Chang, A. C. Y., Houseman, D., and Cohen, S. N. (1975): Isolation of histone genes from unfractionated sea urchin DNA by subculture cloning in E. coli. *Nature,* 255:533–538.
23. Lobban, P. E., and Kaiser, A. D. (1973): Enzymatic end-to-end joining of DNA molecules. *J. Mol. Biol.,* 78:453–471.
24. Lovett, M. A., and Helinski, D. R. (1976): Method for the isolation of the replication region of a bacterial replicon: construction of a mini-F′km plasmid. *J. Bacteriol.,* 27:982–987.
25. Mandel, M., and Higa, A. (1970): Calcium-dependent bacteriophage DNA infection. *J. Mol. Biol.,* 53:159–162.
26. Mertz, J. E., and Davis, R. W. (1972): Cleavage of DNA by R_1 restriction endonuclease generates cohesive ends. *Proc. Natl. Acad. Sci. U.S.A.,* 69:3370–3374.
27. Morrow, J. F., Cohen, S. N., Chang, A. C. Y., Boyer, H. W., Goodman, H. M., and Helling, R. B. (1974): Replication and transcription of eukaryotic DNA in Escherichia coli. *Proc. Natl. Acad. Sci. U.S.A.,* 71:1743–1747.
28. Murray, N. E., and Murray, K. (1974): Manipulation of restriction targets in phage λ to form receptor chromosomes for DNA fragments. *Nature,* 251:476–481.
29. Nishimura, Y., Caro, L., Berg, C. M., and Hirota, Y. (1971): Chromosome replication in Escherichia coli. IV. Control of chromosome replication and cell division by an integrated episome. *J. Mol. Biol.,* 55:441–456.
30. Rabbits, T. H. (1976): Bacterial cloning of plasmids carrying copies of rabbit globin messenger RNA. *Nature,* 260:221–225.
31. Rambach, A., and Tiollais, P. (1974): Bacteriophage having EcoRI endonuclease sites only in the nonessential region of the genome. *Proc. Natl. Acad. Sci. U.S.A.,* 71:3927–3930.
32. Rodriguez, R. L., Bolivar, F., Goodman, H. M., Boyer, H. W., and Betlach, M. (1976): Construction and characterization of cloning vehicles. In: *Molecular Mechanisms in the Control of Gene Expression* (*ICN-UCLA Symposium on Genetic Regulatory Mechanisms*), *Vol. 5,* edited by D. P. Nierlich, W. J. Rutter, and C. F. Fox. Academic Press, New York.

33. Rownd, R. H., and Perlman, D. (1976): Two origins of replication in composite R plasmid DNA. *Nature,* 259:281–284.
34. Ruogeon, F., Kourilsky, P., and Mach, B. (1975): Insertion of a rabbit beta globin gene sequence into an Escherichia coli plasmid. *Nucleic Acids Res.,* 2:2365–2378.
35. Sgaramella, V. (1972): Enzymatic oligomerization of bacteriophage P22 DNA and of linear simian virus 40 DNA. *Proc. Natl. Acad. Sci. U.S.A.,* 69:3389–3393.
36. Sgaramella, V., and Khorana, H. G. (1972): Studies on polynucleotides. XII. Total synthesis of the structural gene for an alanine transfer RNA from yeast: enzymic joining of the chemically synthesized polydeoxynucleotides to form the DNA duplex representing nucleotide sequence 1 to 20. *J. Mol. Biol.,* 72:427–444.
37. Starlinger, P., and Saedler, H. (1976): *Curr. Top. Microbiol. Immunol.,* 75:111–152.
38. Taketo, A. (1975): Sensitivity of Escherichia coli to viral nucleic acid. X. Ba^{++}-induced competence for transfecting DNA. *Z. Naturforsch.,* 30C:520–522.
39. Tanaka, T., Weisblum, B., Schnös, M., and Inman, R. B. (1975): Construction and characterization of a chimeric plasmid composed of DNA from Escherichia coli and Drosophila melanogaster. *Biochemistry,* 14:2064–2070.
40. Thomas, M., Cameron, J. R., and Davis, R. W. (1974): Viable molecular hybrids of bacteriophage lambda and eukaryotic DNA. *Proc. Natl. Acad. Sci. U.S.A.,* 71:4579–4583.
41. Timmis, K., Cabello, F., and Cohen, S. N. (1974): Utilization of two distinct modes of replication by a hybrid plasmid constructed in vitro from separate replicons. *Proc. Natl. Acad. Sci. U.S.A.,* 71:4556–4560.
42. Timmis, K., Cabello, F., and Cohen, S. N. (1975): Cloning, isolation, and characterization of replication regions of complex plasmid genomes. *Proc. Natl. Acad. Sci. U.S.A.,* 72: 2242–2246.
43. Wensink, P. C., Finnegan, D. J., Donelson, J. E., and Hogness, D. S. (1974): A system for mapping DNA sequences in the chromosomes of Drosophila melanogaster. *Cell,* 3:315–325.

Recombinant Molecules: Impact on Science and Society, edited by R. F. Beers, Jr. and E. G. Bassett. Raven Press, New York © 1977.

12. Molecular Cloning as a Tool in the Study of Pathogenic *Escherichia coli*

Magdalene So and Stanley Falkow

Department of Microbiology and Immunology, School of Medicine, University of Washington, Seattle, Washington 98195

INTRODUCTION

The natural habitat of *Escherichia coli* is the alimentary tract of man and warm-blooded animals. In addition to their usual habitat, *E. coli* are also found widely distributed in nature in soil, insects, and vegetation, as well as in birds and fish which have been in close proximity with man or other animals (20). The genus *Escherichia coli* and the "species" *E. coli* were defined on biochemical and serological criteria (12), but, one may ask, in what sense do the populations of cells which we conveniently call *E. coli* diverge in their overall genetic organization? If one examines a reasonably wide variety of independent *E. coli* isolates, it is found (4,5) that strains may differ by as much as 25% in their nucleotide sequences, and the genome size varies from a low of about 2.3×10^9 daltons to a high of about 3.0×10^9 daltons. Although *E. coli* strains share a significant proportion (about 75%) of their polynucleotide sequences with *Shigella spp.,* no strains from any other genus of the Enterobacteriaceae show greater than 25% relatedness to *E. coli* (4,5). Not unexpectedly, specific strains (clones?) of *E. coli* have been associated with ecological preferences or other specializations, including the ability to cause disease. This latter capability has recently been of considerable concern to those contemplating work with DNA recombinant molecules; the role of *E. coli* as a pathogen has been a concern to those interested in infectious diseases for a much longer period of time. In the following sections we shall attempt to present an overview of the pathogenicity of *E. coli* as well as how molecular cloning may be employed as a valuable tool for understanding the epidemiology and pathogenesis of *E. coli* infection and for developing means of preventing infection by this organism.

E. COLI AS A PATHOGEN

Extraintestinal Infection

The role of *E. coli* as a pathogen has been well documented for many years. The most common *E. coli* infection of man is that of the urinary tract. Organisms generally enter the bladder by an ascending route; the source of infection is usually the patient's own fecal flora. For a time, it was felt that the *E. coli* strains causing urinary tract infection were simply opportunists which had more or less accidentally been introduced into the right place at the right time. It is becoming increasingly clear (7,21,32,33), however, that *E. coli* isolated from urinary tract infections often have a distinct phenotype which may include one or all of the following attributes: a specific K (capsular) antigen, a hemolysin, and the ability to produce colicin V. By the same token, among hospitalized patients, *E. coli* is not only an important cause of urinary tract infection but also a leading cause of postoperative wound infections and serious septicemia. In the latter two instances, infections generally arise in patients with already compromised health (19). But here again, *E. coli* isolated from wounds and the bloodstream often are hemolytic or produce colicin V, whereas the incidence of such strains in the normal fecal flora is generally quite low (7,32,33; Table 1). These data suggest, therefore, that colicin V and hemolysin-producing strains are associated with increased pathogenicity.

The role of colicin V is of particular interest. Smith and Huggins (33) have shown that invasive human, bovine, ovine, and avian *E. coli* strains often produce colicin V and that this property is almost certainly plasmid-mediated (ColV). If the ColV plasmid is eliminated from wild-type invasive strains by a "curing" agent, the strains become significantly less virulent. Smith (32) transferred the ColV plasmid to a variety of laboratory strains, including *E. coli* K-12, and found that these strains became more pathogenic for experimental animals. In contrast, ColE plasmids have no detectable effect on pathogenicity (33).

The role of colicin V in causing increased pathogenicity is still not

TABLE 1. *Hemolysin (Hly) and colicin V (ColV) in clinical isolates of* E. coli[a]

Source of *E. coli* strain	No. tested	Hemolytic (%)	Col^+ (%)	ColV (%)
Stool	33	6	30	0
Urinary tract infection	59	49	30	7
Blood	53	34	21	11
Enteropathogenic	19	0	22	0
Miscellaneous extraintestinal infections	43	50	32	14

[a] J. Jorgensen, B. Minshew, and S. Falkow (*unpublished observations*).

clear. It appears that colicin V confers on organisms an increased ability to survive in the alimentary tract as well as increased protection against the normal defense mechanisms of the body. The selective advantage enjoyed by $ColV^+$ organisms does not appear to be associated *per se* with the inhibitory effects of colicin V on colicin-sensitive bacteria (33). Moreover, although increased pathogenicity is associated with ColV plasmids, it is not certain whether activity of colicin V or a related substance is responsible for increased pathogenicity or if another plasmid gene is the culprit.

The recent discovery that the ColV plasmid carries a determinant which can contribute significantly to the pathogenicity of *E. coli* has several important features that are relevant to the question of recombinant DNA molecules and the potential biohazards associated with this research. *E. coli* K-12 $ColV^+$ strains have been used in the microbial genetics laboratory for many years without the knowledge that they were of measurably increased virulence as compared to *E. coli* K-12 F^- strains. However, pathogenicity is a relative term. *E. coli* K-12 by all criteria is not a natural pathogen. Consequently, if one infects *E. coli* K-12 with ColV, its increased virulence can be shown only by injecting large doses ($>10^9$ cells/ml), usually intravenously, into experimental animals (32). By contrast, naturally occurring $ColV^+$ invasive strains usually produce disease at lower doses by more natural routes of infection, including the oral route, and with lesions resembling those present in natural disease (33). This simply underscores the fact that, in general, the pathogenicity of *E. coli* (indeed, of most microorganisms) cannot be attributed to a single determinant, but more often than not is a reflection of a constellation of unlinked bacterial genes (both chromosomal and extrachromosomal) as well as a multitude of specific and relatively nonspecific animal host factors. Thus, certain *E. coli* strains may have the inherent potential for invasiveness or may acquire accessory genetic information (often plasmid-mediated) that is sufficient to tip the balance from a strain that is usually a commensal to one capable of initiating overt disease.

From the standpoint of recombinant DNA molecule research in which *E. coli* is employed, the documentation of the effects of plasmid-mediated determinants on pathogenicity must be viewed as one of the most cogent arguments for potential biohazard. One may take this a step further. Presumably purified ColV plasmid DNA, like F and R plasmids, could easily be cleaved by restriction endonucleases and cloned within *E. coli* K-12 with conventional plasmid vectors such as pSC101 or a ColE1 derivative. Indeed, in light of the genetic and molecular work that has already taken place with ColV (reviewed in ref. 16), this might be viewed as a highly desirable experiment. Prior to 1974, it might have been assumed that in cloning the structural gene for colicin V biosynthesis, only an antibiotic determinant was being isolated; now we would consider that (potentially) a virulence gene was being isolated. For illustrative purposes, therefore, it may be useful to see that genes that might logically be considered "harmless" could, in fact, have more significance than expected. But having discovered that ColV is a "virulence" plasmid, should one take the view that the cloning of the ColV gene(s) not be done because it is a potential biohazard? It seems to us that cloning would, in fact, be a most useful tack to take for those interested in determining the precise role of ColV plasmids in the pathogenesis of extraintestinal infection by *E. coli.* One could, for example, determine if the increased pathogenicity bestowed by a ColV plasmid was due to the structural gene for colicin biosynthesis or if some other gene was responsible. The argument that the molecular cloning of determinants of virulence would be useful does not rest on the idea that no biohazard is involved, since working with clinical isolates or laboratory derivatives of pathogens is in itself a biohazard. Rather, the guiding principle should be that genetic determinants which are native to the species in question, *E. coli,* were being studied within *E. coli,* and with *E. coli* plasmids.

Enteropathogenic *E. coli*

Enteric pathogens can be divided into three groups depending on their characteristic interaction with the intestinal mucosa. The first group of pathogens, exemplified by *Shigella* and certain *E. coli* strains, invade the intestinal mucosa and subsequently multiply, leading to extensive inflammation, necrosis, and ulceration. The organisms are localized almost exclusively in the colon where the intense mucosal inflammation and destruction lead to fever and bloody, mucoid stools (dysentery) which are hallmarks of the disease. The second group of pathogens is typified by virulent *Salmonella* strains and certain *E. coli.* These organisms also penetrate the eipthelial lining of the intestinal tract, yet there is little mucosal destruction. They are usually locally contained, leading to a polymorphonuclear response, clinical gastroenteritis, and, occasionally, bacteremia. The third group of pathogens, of which certain *E. coli* strains are the most common, do not invade the intestinal mucosa. Infection occurs in the small bowel and illness is due to enterotoxin release and subsequent secretion of fluid and electrolytes leading to watery diarrhea. The *E. coli* strains of this latter group, the enterotoxigenic *E. coli,* have received the most attention in recent years and we shall confine our remarks to them.

Enterotoxigenic *E. coli* have been implicated in diarrheal disease of man, pigs, cattle, and lambs (16,26,28,34,35). Generally, the young are affected more seriously than adults. The toxigenic *E. coli* isolated from pigs have been particularly well studied and serve as a useful general model (26,34). A large proportion of these porcine strains produce enterotoxin, possess a common surface antigen K88 *ab* or *ac* (K88), produce a hemolysin, and ferment raffinose. Subsequent genetic and molecular studies coupled with animal studies have shown that all of these properties are plasmid mediated (1,16,24,34,35,37). Plasmids (Ent) encode for the production of heat-labile (LT) and/or heat-stable (ST) toxin, whereas hemolysin is carried on a separate plasmid species, Hly. The ability to use raffinose (Raf) and the biosynthesis of K88 are generally linked and appear to be usually mediated by yet another independent plasmid species (P. L. Shipley, C. L. Gyles, and S. Falkow, *unpublished observation*). The Ent and K88 plasmids act in concert to play the most critical role in the pathogenesis of toxigenic diarrhea (Table 2). The toxins are effective only on small bowel epithelium, whereas the K88 antigen, which is expressed as proteinaceous pili on the bacterial surface, facilitates adherence of bacterial cells to the small bowel mucosa. Thus, these plasmids permit their host bacterial cells to circumvent one of the most important normal animal host defense mechanisms, intestinal mobility.

The mobility of the small bowel is quite different from that of the colon. The small bowel undergoes continual movement, and food as well as ingested bacteria are continually propelled from proximal to distal parts. Bacterial proliferation in the small bowel is relatively unusual, and the mobility pattern helps to explain this fact. Yet the Ent and K88 plasmids act together to permit the offending organism to adhere to the small bowel where it can proliferate and deliver enterotoxin to its specific target tissue. It is supposed that these plasmid-mediated properties more often aid colonization rather than cause disease. Significantly, strains which are Ent^+ $K88^-$ or Ent^- $K88^+$ generally do *not* cause diarrheal disease; only when they coexist together is overt disease commonly produced (34). Moreover, not all strains which possess Ent and K88 are pathogenic. For example, *E. coli* K-12 carrying Ent and K88 does not cause overt disease even if very large doses ($>10^{11}$ cells) are fed (34). However, *E. coli* K-12 strains which are Ent^+ and/or carry a plasmid-mediated K antigen do colonize animals significantly better than *E. coli* K-12 F^- cells (17).

Ent plasmids and plasmids which mediate intestinal colonization have now also been definitively demonstrated in calf and human enteropathogenic strains (15,16,35,37). Nevertheless, a number of puzzling features remain. For example, although K88 permits effective colonization within the intestines of piglets, it does not appear to act in other animal species. Simi-

TABLE 2. *Effects of* E. coli *containing K88 and Ent plasmids on piglets*[a]

Plasmid	$K88^-$	$K88^+$
Ent^-	No effect	Mild diarrhea (33%)
Ent^+	No effect	Severe diarrhea (89%)

[a] From ref. 34.

larly, other plasmid-mediated colonization factors appear to have narrow specificities. Are these colonization factors a single molecular theme or are they as distinctly different as sex pili, which they resemble superficially? Why are the Raf and K88 properties so often linked together on the same plasmid? Is Raf important for K88 expression? What is the basis of toxigenicity? These and other questions important to our understanding of the pathogenesis of the toxigenic diarrhea can probably be most directly attacked by using molecular cloning to isolate each of the specific determinants. In the remaining sections, we describe our use of molecular cloning to focus attention more precisely on one genetic determinant, namely, the sequences involved in the biosynthesis of the *E. coli* heat-stable enterotoxin, ST.

THE MOLECULAR CLONING OF A PLASMID DETERMINANT WHICH ENCODES FOR THE PRODUCTION OF THE *E. COLI* HEAT-STABLE ENTEROTOXIN

The *E. coli* enterotoxins have been divided into two general classes (14,16,23,28,37). The heat-labile class, called LT, is an immunogenic protein whose molecular weight has been variously estimated to be from 24,000 to 100,000 (10). The LT toxin shares partial antigenic identity with the well-characterized enterotoxin of *Vibrio cholerae* (18). Cholera toxin has been shown to exert its effect on small bowel epithelial cells via stimulation of the adenylate cyclase–cyclic AMP system, and it is now apparent that the effects of the *E. coli* LT toxin are also mediated through this same mechanism (9,13,22).

The other class of *E. coli* enterotoxin, ST, is a heat-stable substance which is nonantigenic and estimated to be of low molecular weight, about 1,000 daltons (25,28). The response of the small bowel to ST is characterized by an immediate accumulation of fluid in the bowel lumen, but the duration of its action, unlike that observed with LT, is short-lived and not mediated through the adenylate cyclase system (14). The *E. coli* ST enterotoxin remains a perplexing problem. The observed differences between ST and LT with respect to their immunogenicity, mode of action, and apparent differences in molecular size support the concept that these two enterotoxins are distinctly different. Yet, whereas some *E. coli* Ent plasmids encode for only ST or only LT, by far the most common Ent plasmids isolated from *E. coli* encode for both ST and LT (24,37). The LT toxins are a homogeneous class. However, one finds considerable differences in toxicity and animal specificity when dealing with ST toxin preparations (31). In large part, the enigma of the ST enterotoxin is compounded by the fact that ST may be reliably assayed only in a ligated intestinal loop or by intragastric injection of suckling mice (8). In contrast, the LT enterotoxin may be reliably assayed in several tissue culture systems (9,15,22). *E. coli* which produce only ST were known pathogens in piglets and calves, but their role in human illness had been uncertain until 1975, when *E. coli* producing only ST were shown to be associated with diarrhea in travelers to Mexico (28,29) and incriminated in a hospital outbreak of infantile diarrhea (27). In every case in which strains have been carefully examined, ST biosynthesis has been unequivocally shown to be associated with a single plasmid species. Here again, however, heterogeneity is the rule. Ent plasmids encoding for ST + LT are generally conjugative, are 6.0 to 9.5 × 10^7 daltons in mass, possess a mol fraction G + C of 0.5, and belong to the F incompatibility complex (37,38). In contrast, Ent plasmids which encode for ST, although conjugative, range in mass from 2.0 to 8.0 × 10^7 daltons, possess a mol fraction of G + C of 0.41 to 0.5, and, except for isolates from calves, do not belong to the F incompatibility complex (37,38). Table 3 summarizes some of the molecular properties and polynucleotide sequence relationships of representative Ent plasmids.

We had previously characterized a 6.5 × 10^7 dalton conjugative Ent ST plasmid from calves of the F incompatibility complex in some detail (17,37). This plasmid had been transmitted to *E. coli* K-12, which, in turn, now gave a strongly positive suckling mouse test. Retransfer of this plasmid, ESF0041, to various other nontoxigenic K-12 sublines was always associated with the acquisition of the ability to synthesize ST toxin. *E. coli* K-12 (ESF0041) strains were tested in calves at the large animal isolation facility at Fort Collins, Colorado. The strains were innocuous when fed orally in doses in excess of 10^{12} cells, although *E. coli* K-12 (ESF0041) colonized the calves to a significantly greater extent than did *E. coli* K-12 F^- strains (17). Because ESF0041 was the best characterized plasmid available to us which

TABLE 3. *Polynucleotide sequence relationships between Ent plasmid ESF0041 and other plasmids*

Plasmid	Phenotype and source	Molecular mass (× 10^6 daltons)[a]	Mol fraction, guanine + cytosine[a]	Relatedness[b] with ESF0041 (%)	Relatedness[b] with P307 (%)
ESF0041	ST[c] (bovine)	65	0.65	100	45
P95	ST (porcine)	20	0.41	3	7
P2176	ST (porcine)	80	?	?	—
P16	ST (porcine)	25	?	1	1
P307	ST + LT[c] (porcine)	60	0.50	36	100
ESF119	ST + LT (human)	60	?	36	45
F	ST + LT (human)	60	0.49	33	21
R144	Inc I Tc Km	63	0.50	3	5

[a] Molecular weight and mol fractions guanine + cytosine were determined as described earlier (37,38).

[b] The degree of relatedness was measured by DNA-DNA duplex formation using the S1 endonuclease method (38).

[c] ST, heat-stable enterotoxin; LT, heat-labile enterotoxin.

encoded for ST, it appeared to be the logical choice from which to clone the gene(s) encoding for the heat-stable enterotoxin.

Figure 1 shows the migration in 0.7% agarose of *Eco*RI-cleaved ESF0041 plasmid DNA relative to similarly cleaved pSC101, a 5.5×10^6 dalton R plasmid encoding for tetracycline resistance (Tc) and λ bacteriophage DNA. ESF0041 possessed 11 sites susceptible to *Eco*RI cleavage. *Eco*RI-cleaved ESF0041 and pSC101 DNA were mixed, ligated by *E. coli* polynucleotide ligase (11), and the products of ligation sedimented through a linear 5 to 20% neutral sucrose gradient. Fractions from the gradient were used to transform an *E. coli* K-12 F^- strain with selection for Tc^r. In the absence of a direct selection for ST, or alternatively, a means to distinguish clones harboring only pSC101 from those harboring the pSC101 that carried one or more ligated ESF0041 fragments, random Tc^r clones from each sucrose gradient fraction were tested for the production of ST using the suckling mouse assay. Only one of the 72 Tc^r clones tested was positive for ST.

Plasmid DNA was prepared from the Tc^r ST^+ clone examined in the electron microscope and, after *Eco*RI cleavage, was analyzed by agarose gel electrophoresis. The Tc^r ST^+ clone harbored a single plasmid species of 1.15×10^7 daltons, and agarose gel electrophoresis (Fig. 2, c–e) confirmed that this plasmid, hereafter called ESF3000, possessed two fragments of DNA, one migrating at the position of pSC101 and one at the position of a 5.7×10^6 dalton fragment of ESF0041. Heteroduplex molecules (Fig. 3)

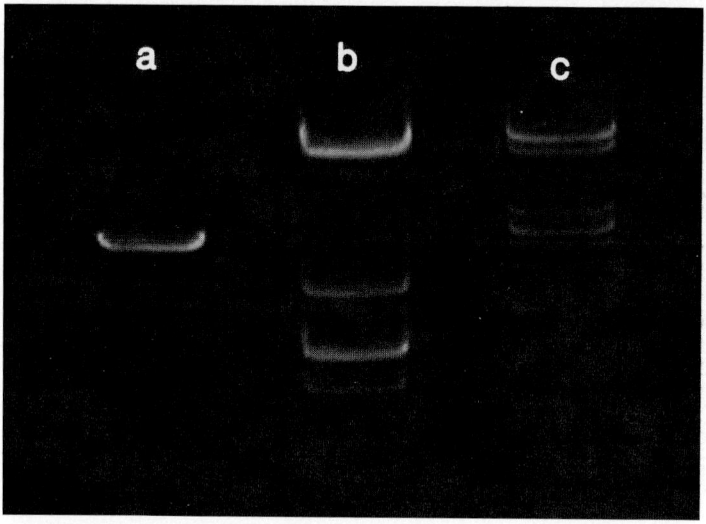

FIG. 1. Migration in 0.7% agarose gel of *Eco*RI cleaved **a:** pSC101 DNA (5.8×10^6 daltons); **b:** λ bacteriophage DNA (from top to bottom: 13.7, 4.71, 3.7, 3.5, 3.03, and 2.09×10^6 daltons); **c:** ESF0041 DNA (extrapolated molecular weights, from top to bottom: 15.5, 13.2, 7.3, 6.2, 5.6, 4.3, 3.4, 2.4, and 0.4×10^6 daltons. (See also Fig. 2.)

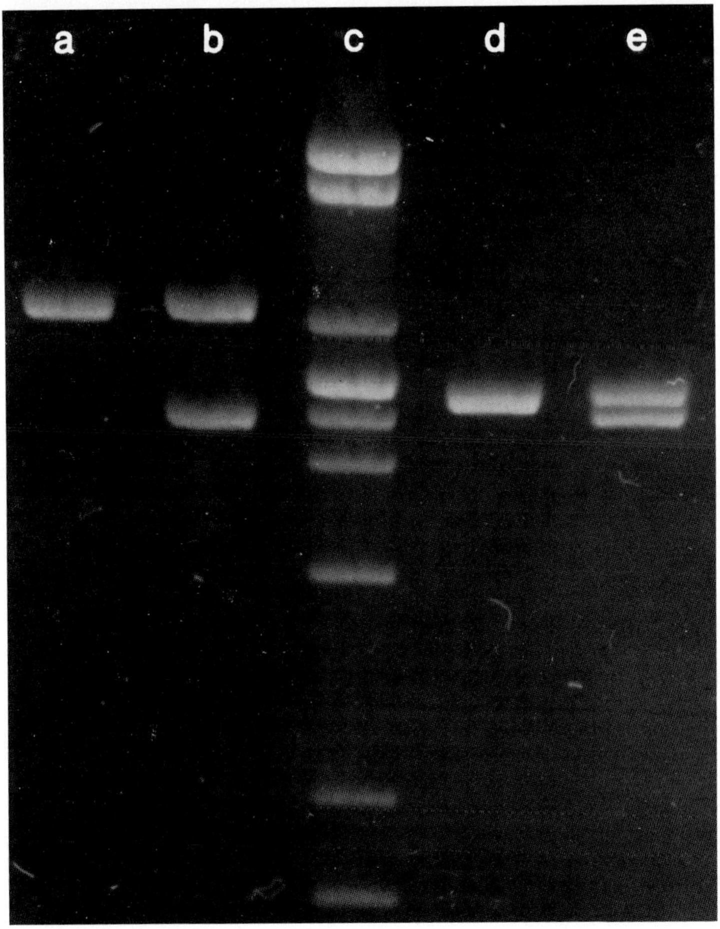

FIG. 2. Migration patterns in 0.7% agarose gel of *Eco*RI-cleaved DNA. **a:** RSF2124; **b:** ESF3001; **c:** ESF0041; **d:** pSC101; and **e:** ESF3000. The DNA fragments were electrophoresed in a horizontal gel apparatus at 60 V, 15 mA, for 20 hr.

show, as expected, that ESF3000 contained a complete pSC101 genome and the additional segment of DNA contributed by ESF0041.

The ESF3000 plasmid was transformed into a variety of *E. coli* K-12 sublines to confirm that a structural gene for ST biosynthesis resided on this plasmid. In every instance (Table 4), the acquisition of ESF3000 was associated with the biosynthesis of ST which was produced at a level comparable to that of the enterotoxigenic wild-type *E. coli* strain isolated from calves. We concluded from these data that the structural gene(s) associated with ST biosynthesis had been cloned as part of a 5.7×10^6 dalton fragment into pSC101. In order to simplify the analysis of the ST gene product(s) as well

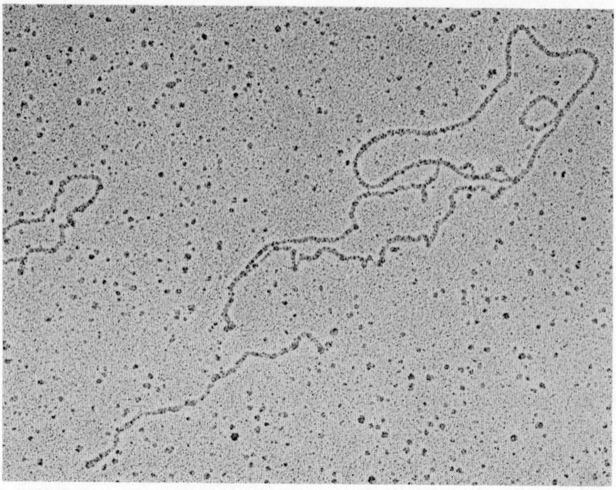

FIG. 3. Heteroduplex molecule of pSC101 and ESF3000 showing a complete duplexed pSC101 genome and the additional segment of single-stranded DNA cloned from ESF0041.

as provide ample DNA for *in vitro* studies, we subsequently cloned the 5.7×10^6 dalton fragment with a replicon which is normally maintained in cells as a multicopy pool. RSF2124 is a 7.3×10^6 dalton derivative of ColEl which has received the translocation sequence, TnA, carrying the structural genes for ampicillin resistance (39). The 5.7×10^6 dalton fragment was ligated to RSF2124 and the resulting 1.3×10^7 dalton plasmid, ESF3001 (Fig. 2, a–c), was found to be present in approximately 20 to 30 copies per cell. Consequently, we expected that the levels of ST toxin

TABLE 4. *Toxicity of* E. coli *derivatives for suckling mice*[a]

Strain	Derivation	Gut weight / Whole body weight
E. coli B41	Clinical isolate from calf	0.126
E. coli K-12	Plasmid⁻, Tox⁻	0.064
E. coli K-12 (pSC101)	By transformation with pSC101	0.065
E. coli K-12 (ESF0041)	By conjugation from *E. coli* B41	0.122
E. coli K-12 (ESF3000)	By transformation of ligated *Eco*R1 fragments of pSC101 and ESF0041	0.130
E. coli B (ESF3000)	By transformation	0.124

[a] From ref. 36.

produced by *E. coli* (ESF3001) cells might be significantly increased. It is unfortunate that the measurement of ST toxin is, at best, semiquantitative. Nevertheless, when comparable populations of *E. coli* (ESF0041), *E. coli* (ESF3000), and *E. coli* (ESF3001) were prepared, the supernatant fluid from *E. coli* (ESF3001) appeared to contain about three times the amount of toxin (Table 5).

As noted in Table 3, there was no direct relationship between ESF0041 and porcine plasmids encoding for ST. ESF0041 did share a significant proportion of its sequences with Ent ST + LT plasmids, although it seemed likely that this reflected shared F-like transfer genes rather than being concerned with other functions. Obviously, use of the large (6.5×10^7 dalton) plasmid as a molecular probe cannot be expected to provide the necessary information concerning the relationship between the actual structural gene(s) for ST, which comprise but a small proportion of the plasmid genome. However, the cloned ST fragment can be usefully employed in nucleic acid hybridization experiments to examine the origin of the ST gene(s) found in various toxigenic isolates. Preliminary experiments along this line suggest that a sequence of approximately 1 to 2×10^6 daltons is often (but not always) common to ST-encoding plasmids from porcine and human isolates. These data must be confirmed, but initially it appears that there may be a limited pool of ST enterotoxin genes which are available to a variety of heterogeneous plasmids. Such a finding is not unlike the recent observation that antibiotic resistance determinants are often transposed from replicon to replicon as a discrete DNA sequence which is bounded by an inverted-repeated sequence (6,30). In this context, examination of the cloned ST fragment in the electron microscope reveals that a prominent inverted-repeated sequence bounding a short sequence of DNA is present on the cloned ST fragment (Fig. 4). The transposability of the ST enterotoxin gene(s) is currently under investigation. Hopefully, further studies of this nature should materially aid our understanding of the epidemiology of the *E. coli* implicated in toxigenic diarrheal disease.

TABLE 5. *Amplification of ST biosynthesis in strains harboring ESF3001*[a]

Strain	Relative toxicity[b]
E. coli K-12	0
E. coli B41	20
E. coli K-12 (ESF0041)	20
E. coli K-12 (ESF3000)	30
E. coli K-12 (ESF3001)	70

[a] Derivations of the strains are given in Table 4.
[b] Given as the reciprocal of the dilution of culture supernatant giving a suckling mouse gut to whole body weight ratio ≤0.08 (average of 4 tests).

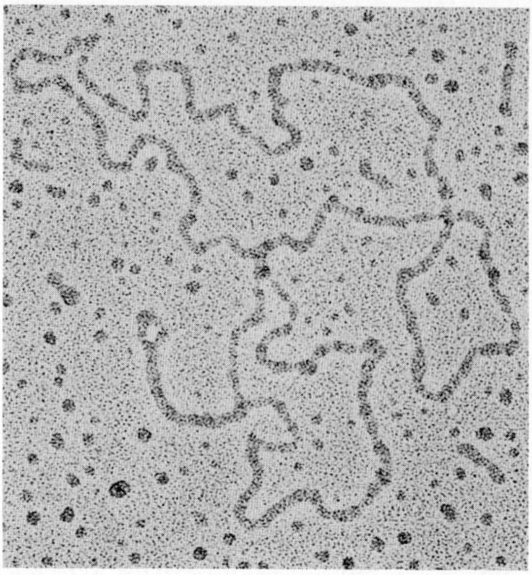

FIG. 4. Single-stranded molecule of ESF3000 showing presence of prominent inverted-repeated DNA sequence cloned on DNA fragment containing ST gene(s).

DISCUSSION

We have successfully isolated a 5.7×10^6 dalton segment of DNA containing a genetic determinant(s) essential for the biosynthesis of an *E. coli* heat-stable enterotoxin. The fragment was isolated from a large conjugative Ent plasmid of a bovine *E. coli* strain and the fragment joined to the nonconjugative plasmids pSC101 and RSF2124. The isolation of these hybrid plasmids containing ST gene(s) provides a useful tool with which to study the nature and possible origin of the *E. coli* ST toxin. These hybrid plasmids readily segregate into *E. coli* minicells (M. So and S. Falkow, *unpublished observations*), providing an excellent model system for examining the *in vitro* biosynthesis of the ST toxin as well as the requirements for its biosynthesis. It is still not certain that the plasmid directly encodes for the small (1,000 dalton) toxin moiety or if it encodes for an enzyme which converts some normal cell constituent to a toxic molecule. The amplification of the ST DNA fragment and the coincident increase in toxin levels should not only permit us to answer this question but should, in addition, simplify toxin isolation and at least indirectly help determine the mode of action of ST as well as the nature of the toxin receptor on the epithelial cells of the small bowel. As already noted, the cloned ST fragment can also be usefully employed in nucleic acid hybridization experiments to examine the origin(s) of the ST gene(s).

The research presented here is, to our knowledge, the first attempt to clone a genetic determinant concerned with bacterial pathogenicity. Given the concerns raised by Berg et al. (2) and later voiced at the Asilomar Conference on potential biohazards associated with the construction of DNA recombinant molecules (3), it is useful to examine the biohazards associated with our research. As noted earlier, working with bacterial pathogens is itself an acknowledged biohazard, so that the question is whether the molecular cloning procedure we employed presents an additional even greater hazard. Many of the genetic determinants of virulence of *E. coli* such as ST, LT, K88, Hly, ColV are normally carried by plasmids, most of which may be transmitted by conjugation (1,16,24,32–35,37). Indeed, H. Williams Smith and his associates (34,35) have shown that Ent plasmids may be transmitted *in vitro* to *Salmonella spp.* and *Shigella spp.* That ESF0041, as well as other Ent plasmids, and ColV are members of the F incompatibility complex (16,37,38) certainly indicates that the plasmids mediating these determinants of virulence are not unique plasmid species. In some cases, K88 and certain Ent plasmids may be found as multicopy pools within their naturally occurring bacterial hosts. Consequently, we feel that the *in vitro* linkage of an Ent ST fragment to the nonconjugative plasmids pSC101 and RSF2124 and their transfer to the relatively innocuous *E. coli* K-12 strain do not represent significantly increased hazard as compared to naturally occurring enterotoxigenic isolates. This is not to say that the *E. coli* K-12 (ESF3000) and *E. coli* K-12 (ESF3001) strains should be treated haphazardly; we have treated these strains with the same procedures we normally apply to known enteric pathogens.

One of the earliest fears voiced about DNA recombinant molecules concerned the dissemination of antibiotic resistance genes that might be part of the carrier replicon or the carried cloned fragment (2,3). In the present study, the carrier replicons employed for the ST fragment encoded for tetracycline resistance (pSC101) or ampicillin resistance (RSF2124). Although the toxigenic diarrheas are not generally treated with antibiotics, should one, nevertheless, be concerned about the added biohazard of the combination of antibiotic resistance and the ST enterotoxin gene on a single plasmid? ESF0041 was, in fact, isolated from a host cell which carried an independent R plasmid encoding for tetracycline and streptomycin resistance (17). More recently, an Ent ST plasmid in *E. coli* from a nursery epidemic was described (40), which coexisted with an R plasmid encoding for sulfonamide, streptomycin, tetracycline, ampicillin, chloramphenicol, and kanamycin resistance. Indeed, we find among Ent^+ bacteria of animal origin that the coincidence of R plasmids is >50%, whereas among human isolates the coincidence of R plasmids is about 25%. As noted earlier, antibiotic resistance determinants such as those encoding for tetracycline and ampicillin resistance are often found on transposable DNA sequences (6,30). Recently, we (F. L. Heffron, M. So, P. Bedinger, and S. Falkow, *manuscript submitted*) have shown that the Ent plasmids receive transpositions of antibiotic resistance genes at a relatively high frequency when they are co-resident in the same cell as R plasmids. These Ent^+-R^+ plasmids are thereafter transferred as a single unit. Given the enormous reservoir of antibiotic resistance genes in naturally occurring *E. coli* (16), it seems likely that the plasmids we have constructed in the laboratory have counterparts in nature.

In our judgement, the use of molecular cloning to isolate specific determinants of bacterial pathogenicity holds considerable promise for the better understanding of bacterial pathogenesis and may be extremely useful at the practical level for the construction of vaccine strains and/or the identification of specific therapeutic measures to treat disease. For example, we (M. So and S. Falkow, *unpublished observation*) have recently cloned the LT structural gene in much the same way as described here for ST. The structural genes for toxin biosynthesis appear to reside on the plasmid. Since LT is immunogenic, we are currently isolating deletions of the toxin gene to obtain a mutant protein which retains the ability to induce an immune antitoxin response, while being nontoxic to the animal cell. We feel this approach may be useful for the eventual use of the cloned fragment in a vaccine strain and may, in addition, be similarly applicable to other determinants of virulence such as the K colonization antigens. Of course, we do not advocate that the molecular cloning of microbial determinants of virulence be taken lightly. Such cloned determinants should not be introduced into bacterial species which do not normally carry them. However, so long as the cloned genetic determinants of virulence are preserved within their natural host species, molecular cloning would appear to provide an extremely valuable research tool and to pose no more risk to the investigator or the environment than naturally occurring pathogenic species studied in the laboratory.

ACKNOWLEDGMENTS

The work described in this report was supported by grant #AI10085–04 from the National Institutes of Allergy and Infectious Diseases, NIH, and Contract FDA 223–73–7210 from the Bureau of Veterinary Medicine, Food and Drug Administration. M. So was supported, in part, by a fellowship from Smith, Kline and French.

REFERENCES

1. Bak, A. L., Christiansen, G., Christiansen, C., Stenderup, A., Ørskov, I., and Ørskov, F. (1972): Circular DNA molecules controlling synthesis and transfer of the surface antigen (K88) in Escherichia coli. *J. Gen. Microbiol.,* 73:373–385.
2. Berg, P., Baltimore, D., Boyer, H. W., Cohen, S. N., Davis, R. W., Hogness, D. S., Nathans, D., Roblin, D. O., Watson, J. D., Weisman, S., and Zinder, N. D. (1974): Potential biohazards of recombinant DNA molecules. *Science,* 185:303.
3. Berg, P., Baltimore, D., Brenner, S., Roblin, R. O., and Singer, M. F. (1975): Summary statement of the Asilomar Conference on recombinant DNA molecules. *Proc. Natl. Acad. Sci. U.S.A.,* 72:1981–1984.
4. Brenner, D. J., and Falkow, S. (1971): Molecular relationships among members of the Enterobacteriaceae. *Adv. Genet.,* 16:81–118.
5. Brenner, D. J., Fanning, G. R., Skerman, F. J., and Falkow, S. (1972): Polynucleotide

sequence divergence among strains of Escherichia coli and closely related organisms. *J. Bacteriol.,* 109:953–965.
6. Cohen, S. N., and Kopecko, D. (1976): Structural evolution of bacterial plasmids: Role of translocating genetic elements and DNA sequence insertion. *Fed. Proc.,* 35:2031–2036.
7. Cooke, E. M., and Ewins, S. P. (1975): Properties of strains of Escherichia coli isolated from a variety of sources. *J. Med. Microbiol.,* 8:107–111.
8. Dean, A. G., Ching, Y., Williams, R. G., and Handen, L. B. (1972): Test for Escherichia coli enterotoxin using infant mice: application in a study of diarrhea in children in Honolulu. *J. Infect. Dis.,* 125:407–411.
9. Donta, S. T., King, M., and Sloper, K. (1973): Induction of steroidogenesis in tissue culture by cholera enterotoxin. *Nature* [*New Biol.*], 243:246–247.
10. Dorner, F. (1975): Escherichia coli enterotoxin: partial purification and characterization. In: *Microbiology—1975,* edited by D. Schlessinger, pp. 242–251. American Society for Microbiology, Washington, D.C.
11. Dugaiczyk, G., Hedgpath, J., Boyer, H. W., and Goodman, H (1974): Physical identity of the SV40 deoxyribonucleic acid sequence recognized by the EcoRI restriction endonuclease and modification methylase. *Biochemistry,* 13:503–512.
12. Edwards, P. R., and Ewing, W. H. (1972): *Identification of Enterobacteriaceae,* Ed. 3. Burgess Publishing Co. Minneapolis, Minnesota.
13. Evans, D. J., Chen, L. C., Curlin, G. T., and Evans, D. G. (1972): Stimulation of adenylcyclase by Escherichia coli enterotoxin. *Nature* [*New Biol.*], 236:137–138.
14. Evans, D. G., Evans, D. J., and Pierce, N. F. (1973): Differences in the response of rabbit small intestine to heat labile and heat stable enterotoxins of Escherichia coli. *Infect. Immun.,* 7:873–880.
15. Evans, D. G., Silver, R. P., Evans, D. J., Chase, D. G., and Gorbach, S. L. (1975): Plasmid-controlled colonization factor associated with virulence in Escherichia coli enterotoxigenic for humans. *Infect. Immun.,* 12:656–667.
16. Falkow, S. (1975): *Infectious Multiple Drug Resistance.* Pion Ltd., London.
17. Falkow, S., Williams, L. P., Seaman, S. L., and Rollins, L. D. (1976): Increased survival in calves of Escherichia coli K-12 carrying an Ent plasmid. *Infect. Immun.,* 13:1005–1007.
18. Finkelstein, R. A. (1975): Cholera enterotoxin. In: *Microbiology—1975,* edited by D. Schlessinger, pp. 236–241. American Society for Microbiology, Washington, D.C.
19. Finland, M. (1971): Changing prevalence of pathogenic bacteria in relation to time and the introduction and use of new microbial agents. In: *Bayer Symposium III: Bacterial Infections: Changes in Their Causative Agents, Trends and Possible Reasons,* edited by M. Finland, W. Marget, and K. Bartmann, pp. 4–23. Springer-Verlag, New York.
20. Geldreich, E. E. (1966): *Sanitary Significance of Fecal Coliforms in the Environment.* U.S. Department of the Interior, Washington, D.C.
21. Glynn, A. A., Burmfitt, W., and Howard, C. J. (1971): "K" antigens of Escherichia coli and renal involvement in urinary tract infections. *Lancet,* 1:514–516.
22. Guerrant, R. L., Brunton, L. L., Schnaitman, C., Rebhun, L. I., and Gilman, A. G. (1974): Cyclic adenosine monophosphate and alteration of Chinese hamster ovary cell morphology: a rapid, sensitive in vitro assay for the enterotoxins of Vibrio cholerae and Escherichia coli. *Infect. Immun.,* 10:320–327.
23. Gyles, C. L. (1971): Heat labile and heat stable forms of the enterotoxin from E. coli strains enteropathogenic for pigs. *Ann. N.Y. Acad. Sci.,* 176:314–322.
24. Gyles, C., So, M., and Falkow, S. (1974): The enterotoxin plasmids of Escherichia coli. *J. Infect. Dis.,* 130:40–49.
25. Jacks, T. M., and Wee, B. J. (1974): Biochemical properties of Escherichia coli low-molecular weight, heat stable enterotoxin. *Infect. Immun.,* 9:342–347.
26. Moon, H. W. (1974): Pathogenesis of enteric diseases caused by Escherichia coli. *Adv. Vet. Sci. Comp. Med.,* 18:179–211.
27. Ryder, R. W., Wachsmuth, I. K., Buxton, A., Evans, D. G., DuPont, H. L., Mason, E., and Barrett, F. F. (1976): Heat-stable enterotoxigenic Escherichia coli in a newborn nursery: relation to infantile diarrhea. *N. Engl. J. Med.* (*in press*).
28. Sack, R. B. (1975): Human diarrheal disease caused by enterotoxigenic *Eschericha coli. Annu. Rev. Microbiol.,* 29:333–353.
29. Sack, D. A., Merson, M. H., Wells, J. G., Sack, R. B., and Morris, G. K. (1975): Diarrhea

associated with heat-stable enterotoxin-producing strains of Escherichia coli. *Lancet,* 1:239–241.

30. Saunders, J. R. (1975): Transposable resistance genes. *Nature,* 258:3844–3845.
31. Smith, H. W. (1971): The bacteriology of the alimentary tract of domestic animals suffering from Escherichia coli infection. *Ann. N.Y. Acad. Sci.,* 176:110–125.
32. Smith, H. W. (1974): A search for transmissible pathogenic characters in invasive strains of Escherichia coli: the discovery of a plasmid-controlled toxin and a plasmid-controlled lethal character closely associated, or identical, with colicine V. *J. Gen. Microbiol.,* 83:95–111.
33. Smith, H. W., and Huggins, M. B. (1976): Further observations on the association of the colicine V plasmid of Escherichia coli with pathogenicity and with survival in the alimentary tract. *J. Gen. Microbiol.,* 92:335–350.
34. Smith, H. W., and Linggood, M. A. (1971): Observations on the pathogenic properties of the K88, Hly and Ent plasmids of Escherichia coli with particular reference to porcine diarrhea. *J. Med. Microbiol.,* 4:467–485.
35. Smith, H. W., and Linggood, M. A. (1972): Further observations on Escherichia coli enterotoxins with particular regard to those produced by atypical piglet strains and by calf and lamb strains: the transmissable nature of these enterotoxins and of a K antigen possessed by calf and lamb strains. *J. Med. Microbiol.,* 5:243–250.
36. So, M., Boyer, H. W., Betlach, M., and Falkow, S. (1976): The molecular cloning of an E. coli plasmid determinant which encodes for the producing of heat stable enterotoxin. *J. Bacteriol.* (*in press*).
37. So, M., Crandall, J. F., Crosa, J. H., and Falkow, S. (1975): Extrachromosomal determinants which contribute to bacterial pathogenicity. In: *Microbiology—1974,* edited by D. Schlessinger, pp. 16–26. American Society for Microbiology, Washington, D.C.
38. So, M., Crosa, J. H., and Falkow, S. (1975): Polynucleotide sequence relationships among Ent plasmids and the relationship between Ent and other plasmids. *J. Bacteriol.,* 121:234–238.
39. So, M., Gill, R., and Falkow, S. (1976): The generation of a Co1E1-Ap^r cloning vehicle which allows detection of inserted DNA. *Mol. Gen. Genet.,* 142:239–249.
40. Wachsmuth, I. K., Falkow, S., and Ryder, R. W. (1976): Plasmid-mediated properties of a heat-stable enterotoxin producing Escherichia coli associated with infantile diarrhea. *Infect. Immun.,* 14:403–407.

Recombinant Molecules: Impact on Science and Society, edited by R. F. Beers, Jr. and E. G. Bassett. Raven Press, New York © 1977.

13. Expression of Bacterial Genes in Phage Lambda Vectors

Noreen E. Murray

Department of Molecular Biology, University of Edinburgh, Edinburgh EH9 3JR, Scotland, U.K.

INTRODUCTION

Our current understanding of the molecular genetics of *E. coli* has gained much from the use of specialized transducing phages. These phages not only augment genetic analyses but provide a ready means for the quantitation of specific messenger RNAs (19), the purification of proteins that bind to specific regions of DNA (15), the study of transcription and its regulation *in vitro* (10,36), and the amplification of gene products via increased gene dosage (23).

Transducing derivatives of bacteriophage λ are now readily made *in vitro* by the insertion of fragments of donor DNA, produced by restriction endonucleases, into appropriate phage receptor chromosomes or vectors (4,5, 24,25,29,33). These transducing phages are not only plaque-forming but, for a given vector, are of a predictable structure since the fragment of heterologous DNA is inserted, in either of two orientations, at a predetermined site within the phage chromosome. The optimal use of transducing derivatives of λ, in those experiments in which the transcription of the acquired genes is required, depends on our understanding of the transcriptional control of the λ genome.

In this chapter data concerning the expression of heterologous genes within λ transducing phages are presented. These data are from model systems in which genes from the *trp* operon of *E. coli* are included in the phage genome. λ*trp* phages provide good systems because of the ease with which they can be analyzed genetically (13) and because the *trp* enzymes can readily be assayed (8).

OBSERVATIONS AND DISCUSSION

Control Circuits of Phage λ

The essential genes of phage λ are transcribed from three *major* promoters, P_L, P_R, and P′R (Figs. 1 and 2). Following infection of a sensitive

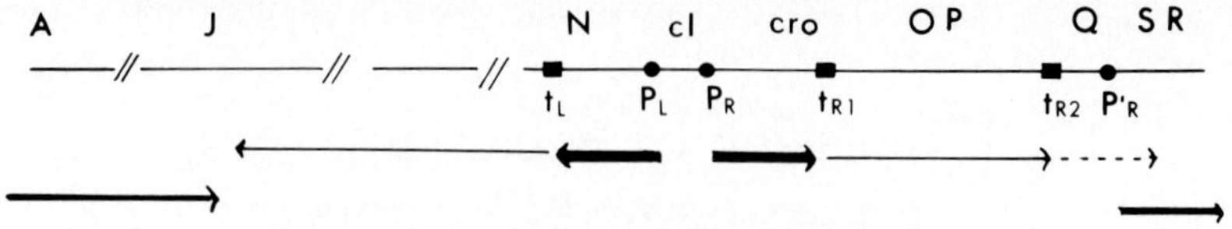

FIG. 1. Control circuits of phage λ. The heavy arrows from the promoters P_L and P_R to the termination sites t_L and t_{R1} indicate the early transcripts obtained in the absence of *N* protein; a transcript continuing beyond t_{R1} encounters a second terminator, t_{R2}. In the presence of *N* protein, transcription continues leftward through t_L, rightward through t_{R1}, t_{R2}, and gene *Q*. The product of gene *Q* is necessary for transcription through genes *S*, *R*, *A*, and *J*.

host, transcription, which is subject to repression by the product (repressor) of the *c*I gene, proceeds leftward from P_L through gene *N* and rightward from P_R through gene *cro*. In the absence of the product of gene *N*, most transcripts terminate just beyond these genes at the sites t_L and t_{R1}; a second termination site, t_{R2}, impedes those transcripts from P_R that escape the first termination signal, t_{R1}. Gene *N* protein exerts its positive regulatory role by interacting with RNA polymerase to permit transcription to ignore these "stop" signals (1,14,32). The *cro* gene product, when available in sufficient

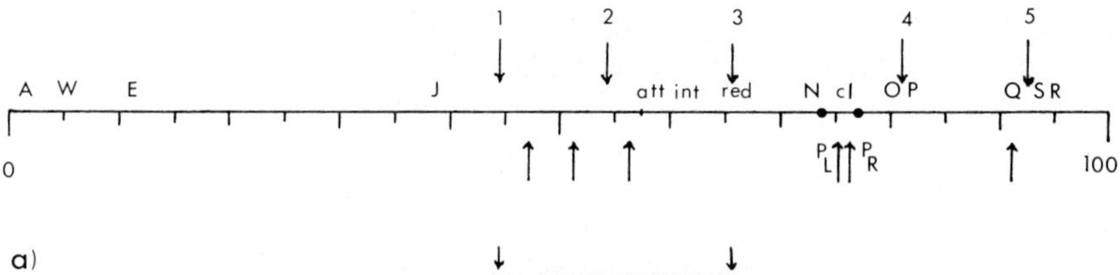

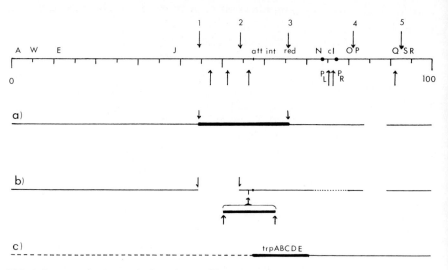

FIG. 2. Structures of transducing phages. The map at the top of the figure indicates some of the important genetic markers of phage λ; the scale 0 to 100 represents the length of λ^+DNA; arrows above the line indicate targets for endo R.*Eco*RI; arrows below the line those for endo R.*Hin*dIII. Below the map are the structures of three classes of transducing phages, wherein gaps represent deletions; . . . . . *imm*^{21}; – – – φ80 DNA; and heavy lines bacterial DNA. **a:** A transducing phage made by replacing the phage DNA between the *Eco*RI targets 1 and 3 with bacterial DNA. **b:** A transducing phage made by inserting a fragment of bacterial DNA at the *Hin*dIII target closest to the attachment region. **c:** A biologically made transducing phage, λ*trp*BG2 (9), derived from a φ80/λ hybrid phage.

concentrations, depresses transcription from P_L (13,27,28) and P_R (12). In the presence of *N* protein, transcription continues leftward beyond t_L into the *red* and *int* genes, and rightward through genes *O* and *P*, to permit efficient replication of phage DNA, and subsequently through t_{R2} and gene *Q*. The product of the latter gene is essential for the expression of the "late" genes of lambda. Since the linear chromosome of the phage circularizes on infection, *Q*-dependent transcription from P′R (located between genes *Q* and *S*) continues not only through genes *S* and *R* but into *A* and beyond gene *J*. Both genes *N* and *Q* play a positive regulatory role essential for plaque formation, but an N^- phage *can* produce a small plaque if the termination site t_{R2} is removed by an appropriate deletion, as in λN^-*nin* (7).

The λ genes *int* and *red* are not essential for plaque formation, but the product of the *int* gene is required for those special recombination events which integrate phage genomes into the bacterial chromosome. This integrative recombination is confined to a particular site, the attachment region (*att*), on both phage and bacterial chromosomes (16).

Expression of Bacterial Genes in Transducing Phages

Bacterial genes within phage λ can be expressed either from their own promoter or from a phage promoter (13,30,31). Whether the genesis of a *plaque-forming* transducing phage is biological (*in vivo*) or biochemical (*in vitro*), the incorporated genes are generally confined to the central region of the phage genome (see Fig. 2), where they replace some of the nonessential region of the genome between genes *J* and *N* (or *J* and P_R in those special cases where gene *N* may be deleted in the absence of t_{R2}). Transcription of the bacterial genes from a major λ promoter is possible either from P_L or from a rightward promoter if transcription through the late genes is not terminated prior to the heterologous DNA.

The most extensive information concerning the transcription of bacterial genes from the promoters of phage λ concerns the expression of the *trp* genes of *E. coli* from P_L (9,13). These data suggest that P_L and the *N* operon of phage λ offer especially useful features. Experiments with λ*trp* phages (14,32), as well as others (1), have shown that a transcript initiated at P_L in the presence of the *N* gene product will read through sequences that would otherwise impede transcription. This role of *N* protein, as an antiterminator, may enable the transcription of genes that are separated from the phage promoter by a sequence capable of deterring RNA polymerase. In addition to providing the peculiarly useful *N* protein, P_L itself promotes a very high rate of gene expression. The maximal rate of expression from P_L may be an order of magnitude higher than that from an efficient, derepressed bacterial promoter such as the *trp* promoter (9). The achievement of this expression does, however, depend on the loss of the negative control normally provided by the *cro* gene product (9,14).

Structure of Transducing Phages

The structures of transducing phages generated by *in vitro* recombination of DNA fragmented by endo R.*Eco*RI (Fig. 2a), or by endo R.*Hin*dIII (Fig. 2b), are compared with a biologically derived λ*trp* phage in which gene *N* and the termination site t_L are both retained (Fig. 2c). For a transducing phage generated by endo R.*Eco*RI, in which the bacterial genes are oriented so that they are transcribed from the same strand as the genes they have replaced, expression of the acquired genes will be *N*-dependent and subject to negative control by both λ repressor and *cro* gene product (13). Phages made via endo R.*Hin*dIII (Fig. 2b) have the bacterial DNA inserted to the left of the attachment region (4). These transducing phages remain integration proficient (4), and the site of insertion of the heterologous DNA differs from that of transducing phages derived *in vivo* (6). λ*trp* phages have been isolated in which fragments of the *trp* operon of *E. coli* have been incorporated into such a λ vector (Fig. 2b), and these phages have been used to study the expression of the *trp* genes from both P_L and the *trp* promoter (18).

λ*trp* Phages Made *in vitro*

The five genes of the *trp* operon of *E. coli* (Fig. 3) are contained in three fragments following digestion of the bacterial DNA with endo R.*Hin*dIII (18). These fragments were isolated as λ*trpA*, λ*trpC*, and λ*trpE* phages selected by their ability to complement Trp^- strains of bacteria. Complementation results in a "Trp^+ plaque" when, for example, a λ*trpE* phage is plated on $TrpE^-$ indicator bacteria in the absence of exogenous tryptophan (13). A λ*trpA* phage will *complement trpA* mutants of *E. coli* and will *recombine* with many *trpB* mutations, a *trpC* phage includes a complete *C* gene and parts of both the *B* and *D* genes, whereas the remainder of the operon is within the λ*trpE* phage.

The orientation of a fragment of the *trp* operon within a given λ*trp* phage was deduced from a cross to a well-characterized ϕ80 transducing phage containing the entire *trp* operon (11). The homology necessary to produce λ/ϕ80 recombinants from such a cross resides in the incorporated *E. coli* DNA. Crosses of this sort

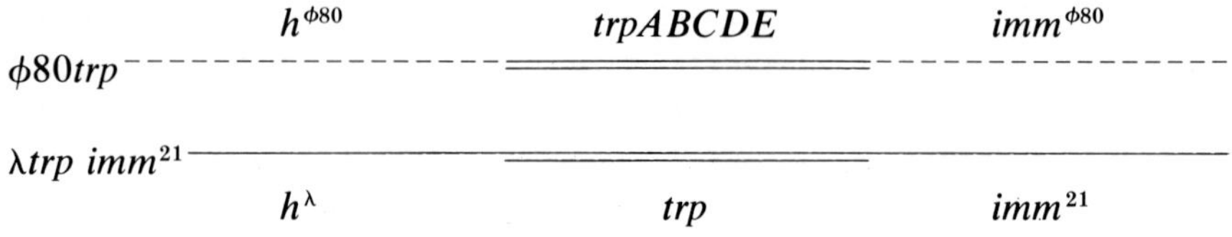

divided the λ*trp* phages into two classes. One, which permits the recovery of h^{λ} imm^{80} recombinants, contains its *trp* genes in the same orientation as

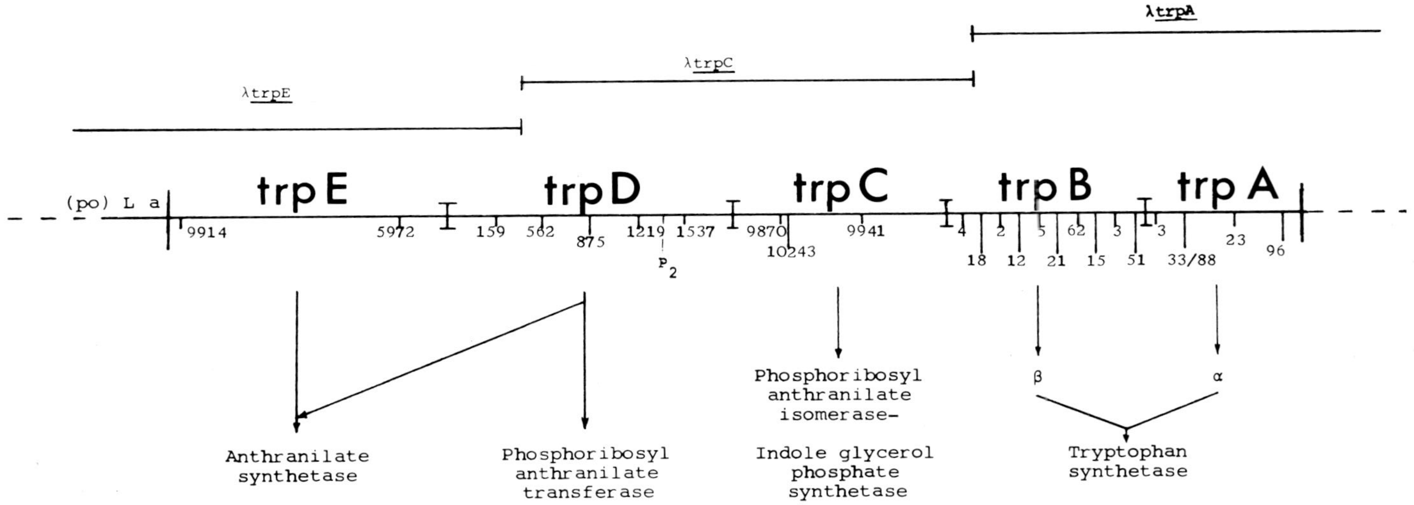

FIG. 3. The *trp* operon of *E. coli* and the content of λ*trp* phages. The extents of the fragments included in the λ*trp* phages are indicated above the genetic map of the *trp* operon of *E. coli* (35). (po), L, and a, left of the *trpE* gene, are the regulatory elements of the *trp* operon, the promoter (p), operator (o), leader region (L), and attenuator (a) (3); P2 within the *trpD* gene is an internal constitutive promoter (21). The targets for endo R.*Hin*dIII are between mutations *trpD*159 and D562 and between mutations *trpB*4 and B8 (18). The enzymes coded by the *trp* genes are given below the map.

those in the *ϕ80trp* phage. In this class of transducing phage, the *trp* genes are transcribed from P_L on the DNA strand defined as the *l*-strand, as are the *trp* genes in the *ϕ80trp* phage; in the alternative class, the *trp* genes must be transcribed in the opposing direction from the other strand, defined as the *r*-strand. We use the abbreviations $\lambda trp(A)^l$ and $\lambda trp(A)^r$ to distinguish these orientations. Both classes of λ*trpA*, λ*trpC* or λ*trpE* phages were isolated (18).

Recombination Between λ*trp* Phages

The *trpB* gene can be restituted by either *in vitro* or *in vivo* recombination between λ*trpA* and λ*trpC* phages; similarly, the *trpD* gene from λ*trpC* and λ*trpE* phages. Recombination *in vivo* requires the two phages to be of like orientation and is mediated via the homology of the host's *trp* operon (18).

Use of Recombinant Phages

The synthesis of $\lambda trp(ABC)^l$ and $\lambda trp(ABC)^r$, or $\lambda trp(CDE)^l$ and $\lambda trp(CDE)^r$, phages facilitates study of the expression of the *trp* genes since tryptophan synthetase is the product of the *trpA* and *B* genes and the assay for anthranilate synthetase is dependent on both the *trpE* and the *trpD* gene products (20). λ*trpCDE* phages are expected to contain the promoter and operator of the *trp* operon (Fig. 3). This was shown to be so since anthranilate synthetase production from λ*trpCDE* phages, in the absence of expression from a phage promoter, was repressible by high levels of tryptophan (18). λ*trpCDE* phages were used to study expression from the *trp* promoter. λ*trpABC* phages include only the weak constitutive promoter located in the *trpD* gene (21). This promoter is readily detected in complementation tests but is, in fact, sufficiently weak that tryptophan synthetase expression from λ*trpABC* phages can readily be used to monitor expression from the λ promoters.

Transcription from Phage Promoters

$\lambda trp(ABC)^l$ phages were used to quantitate the expression of tryptophan synthetase from transcripts initiated at P_L. A cro^- derivative of $\lambda trp(ABC)^l$ was used to derepress transcription from P_L, and expression of the *trp* genes was followed during infection of Trp^- bacteria in the presence of repressing concentrations of tryptophan. Tryptophan synthetase was detected in a λ-sensitive host but not in an immune host. The rate of expression of the λ*trpA* and *B* genes from $\lambda trp(ABC)^l$ (9) was similar to that from a classic λ*trp* phage, in which the *trp* genes are located close to gene *N* (Fig. 4). Our data suggest that transcription initiated at P_L proceeds unabated beyond *att* into the central region of the phage (18).

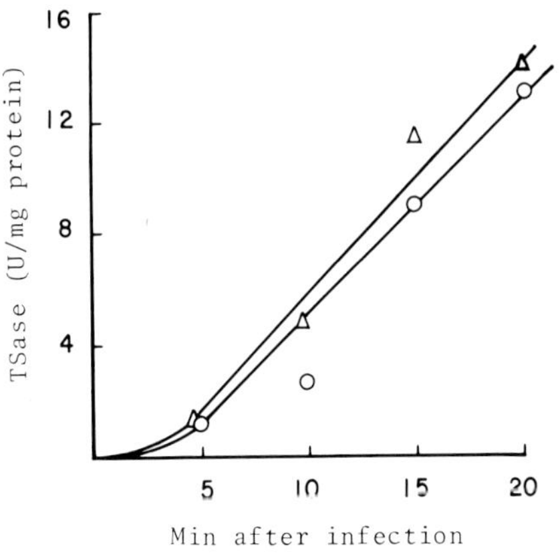

FIG. 4. *Trp* gene expression from P_L of λ*trp* (*ABC*)^l. λ*trp* (*ABC*)^l *cro*$^-$ and λ*trp*BG2 *cro*$^-$ (ref. 9; see also Fig. 2c) were used to infect a *trp*$^-$, *thy*$^-$ host at a multiplicity of 2 phages/cell. Replication of DNA was prevented by thymine starvation. Tryptophan synthetase (TSase) coded by the *trpA* and *B* genes of λ*trp*(*ABC*)^l *cro* (△) and λ*trp*BG2*cro* (○) was assayed (8). When the host strain is lysogenic for λ, no tryptophan synthetase is detected; neither phage includes the *trp* promoter.

λ*trp*(*A*)^r phages were isolated as "Trp$^+$ plaques" on TrpA$^-$ bacteria, and, since the DNA fragment containing *trpA* has no known promoter for this gene, *trpA* is probably expressed from a λ promoter. TrpA$^-$ bacteria lysogenic for λ*trp*(*A*)^r grow slowly in the absence of tryptophan but, in contrast to λ*trp*(*A*)^l lysogens, grow well when only one drop of broth is added to the top layer of a minimal plate. This very weak expression of the *trpA* gene may be from a constitutive promoter in the central region of the λ genome, and it may suffice to provide enough *trpA* product to detect complementation of a *TrpA*$^-$ host. Attempts to detect expression of *trpA* from a rightward promoter have been unsuccessful, but this does not rule out a low level of transcription from P_R or P'_R (18).

Convergent Transcription of the *trp* Genes

Since transcription from P_L can proceed beyond *att* into the *trp* genes, it follows that the *trp* genes in a λ*trp*(*E*)^r phage can potentially be transcribed in either direction. We have investigated this possibility by following the expression of the *trpD* and *E* genes, transcribed rightward from the *trp* promoter, in the face of leftward transcription from P_L.

When a λ-lysogenic host was infected with λ*trp*(*CDE*)^r *cro*$^-$, expression from the *trp* promoter proceeded at a high and constant rate (Fig. 5). In contrast, during infection of a λ-sensitive host, when transcription from P_L can proceed, *trp* gene expression started normally but was severely decreased after 5 min (18). Clearly, powerful leftward transcription from P_L interferes with productive expression of the *trp* genes in the opposing orientation. We can not be sure that the rightward transcription of the *trp* gene is prevented, since the effect could be due to an inhibition of translation of *trp* mRNA by the production of excess RNA of complementary sequence. We do not know whether the effect detected reflects the relative

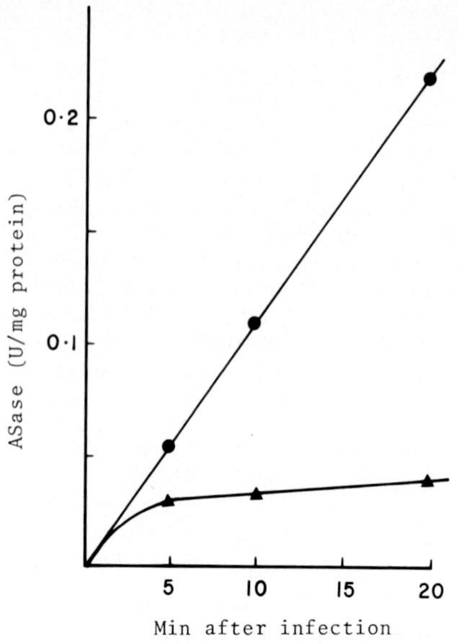

FIG. 5. Converging transcription of the *trp* genes of λ*trp*(*CDE*)[r]. λ*trp*(*CDE*)[r]*cro*$^-$ was used to infect the λ-sensitive, *trp*$^-$ host (▲) and λ-lysogenic, *trp*$^-$ host (●) at multiplicities of 2 phages/cell, followed by an assay for anthranilate synthetase (ASase) (8). In each host strain, expression of the *trp* genes of the phages was derepressed by the presence of a mutation in the *trp* repressor gene (*trpR*). Replication of the phages was prevented in the λ-sensitive host by thymine starvation and in the λ-lysogenic host by the presence of the lambda repressor.

efficiencies of the two opposing promoters or whether it is another feature of the action of lambda's *N* protein in overcoming barriers to transcription.

Convergent transcription has both academic and practical interest, for it illustrates the need to consider not only the positive effects of the promoters within the vector but also the potential inhibitory ones.

Amplification of Gene Expression from an Included Promoter

In order to maximize the yield of products from a bacterial promoter included in a transducing phage, it is necessary to delay cell lysis but permit DNA replication so that the number of gene copies is greatly enhanced. This end was originally achieved by the introduction of a mutation in lambda's gene *S* (23). A defect in gene *S* prevents cell lysis but permits DNA replication and protein synthesis to continue for some hours (2). Recently, Moir and Brammar (22) obtained significantly better amplification of gene products by using mutations in either gene *Q* (Fig. 6) or gene *N*. In the former case all late functions, including gene *S*, are blocked, and although lysis is prevented so too are other late functions including the consequent packaging of the replicated DNA. An N^- phage is, also, defective in all late functions, and although it replicates more slowly than N^+Q^- phage (26), the yields of enzyme achieved have been at least as high. In either case, using biologically derived *trp* phages as model systems, anthranilate synthetase, the product of the *trpE* and *D* genes, comprised more than 25% of the total soluble protein of infected cells, and an even higher proportion of

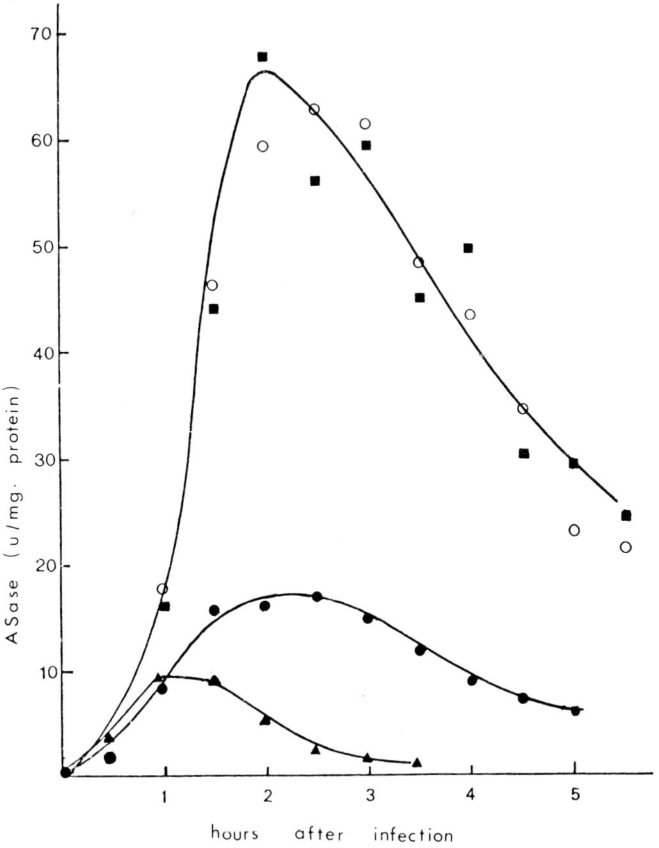

FIG. 6. The effect of *Q, S,* and *R* mutations on anthranilate synthetase (ASase) production from the *trp* promoter of λ*trp* phages (22). The trp^-, $trpR^-$ host strain was infected at a multiplicity of 2 phages/cell, and infected cells were incubated under conditions allowing normal replication of phage DNA. Values for anthranilate synthetase are λ$trpR^-$ (▲); λ$trpS^-$ (●); λ$trpQ^-$ (■); and λ$trpQ^-S^-$ (○).

the protein synthesized after infection. Amplification of the phage genomes was easily able to override tryptophan-mediated repression by titration of the *trp* repressor protein (22). Experiments with λ$trp(CDE)^{l}Q^-S^-$, λ$trp(CDE)^{l}N^-$, or λ$trp(CDE)^{r}N^-$ phages have been equally successful in achieving high levels of anthranilate synthetase (18). Amplification of the *trp* genes using the *Col*E1 plasmid (17) have produced similar yields of *trp* enzymes.

The potential of all these approaches is limited by the efficiency of the incorporated bacterial promoter and, in some cases, by other features of the normal control system. Transcription of heterologous genes from an efficient phage promoter, controlled by components of the phage regulatory system including lambda's *N* protein, offers a more generally useful approach.

Expression of Bacterial Genes from P_L of λ

USE OF *cro*⁻ MUTANTS

λ*trp* phages lacking the *trp* promoter have been used to investigate ways of optimizing gene expression initiated at the phage promoter, P_L (22). Derepression of transcription from P_L was obtained by the use of a mutation in gene *cro,* and cell lysis was delayed by blocks in genes *Q* and *S*. Cells infected with λ*trp* cro^- Q^- S^- phages in which the *trp* genes were expressed exclusively from P_L can contain about 10% of their soluble protein as anthranilate synthetase. Such levels of specific proteins provide an advantageous start in the purification of a protein, particularly one that is normally present in only small amounts. Encouragement for the general applicability of this approach derives from the special properties of lambda's *N* gene product in permitting RNA polymerase to read through normal barriers to transcription. Franklin (13) showed that transcription from P_L under the auspices of lambda's *N* protein is undeterred by repression of the *trp* operon.

The success in the use of cro^- derivatives of λ is mitigated by the finding that λ cro^- Q^- S^- phages are not easy to construct or propagate and the yields of enzyme achieved are very dependent on temperature.

USE OF "HYBRID IMMUNITY"

As an alternative to cro^- phages, in the absence of a mutation in P_L that would prevent the interaction of the negatively acting *cro* gene-product with this promoter but would, nevertheless, leave transcription unimpaired, we chose to investigate phages with "hybrid immunity" regions (N. E. Murray and W. J. Brammar, *unpublished observations*).

The *cro* gene is within the immunity region of λ; its product is immunity specific and fails to act on the promoter of the heteroimmune phage λimm^{434} (27). It follows from this specificity of interaction that a phage with a hybrid immunity region, having P_L from λ but P_R and *cro* from phage 434 (or vice versa) should be phenotypically cro^- for leftward transcription, while retaining normal moderation of rightward transcription from P_R. Such a phage has previously been isolated from an elegantly designed phage cross, but this phage included a defect in the leftward promoter to reduce transcription from P_L (34). Our objective was to maximize expression from P_L and we sought to control expression from this promoter by making a phage with a hybrid immunity region and a suppressible defect in gene *N*. An N^- phage having a hybrid immunity region, and a deletion (*nin*) removing t_{R2}, will propagate in a nonsuppressing host, but in an N^- suppressing host uncontrolled expression of *N*-dependent gene products is realized.

DNAs from phages λ and λimm^{434}, in which the only targets for endo R.*Hin*dIII were in the immunity regions, were used to make *in vitro* recombinants having hybrid immunities (Fig. 7). Derivatives of λ*trp* phages having hybrid immunity regions were made, and expression of the *trp* genes from both $P_L\lambda$ and P_L^{434} was quantitated following infection of a Trp^-, *N*-suppressing host. In each case, when cell lysis was prevented by a mutation in gene *S*, anthranilate synthetase (the products of the *trpD* and *E* genes) accounted for about 25% of the soluble protein of the cell. Hybrid immunity phages offer an efficient way of harnessing the powerful λ promoter. The dependence of transcription, but not replication, on suppression of gene *N* allows the propagation of the N^- *nin* phage in the absence of excessive expression from P_L. The ability to moderate expression during propagation is a necessary feature of a good vehicle for the amplification of gene products, since overproduction of a normally nontoxic product may be disadvantageous and lead to decreased stability of the clone. Helinski (*this volume*) has achieved this same moderation from lambda's promoter (P_L) cloned within a relaxed plasmid. Replication proceeds to produce about 20 copies of the plasmid in the absence of expression from P_L, but elevation of the temperature inactivates the heat-labile λ repressor leading to derepressed transcription in the absence of *cro* gene product.

Deletion derivatives of λ-transducing phages are readily isolated, and DNA between the bacterial genes and lambda's leftward promoter, including unwanted genes or restriction targets, can be removed. Bacterial genes fused to lambda's regulatory elements may be transposed *in vitro,* within a single fragment of DNA, to the Co1E1 vector. Such systems offer tremendous potential for the efficient production of bacterial enzymes by fermentation in the near future.

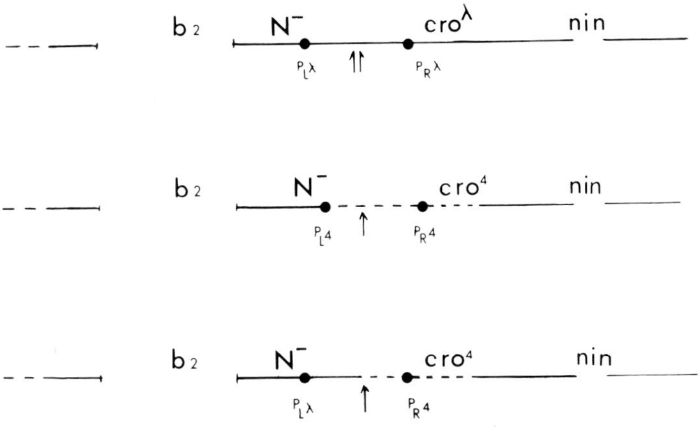

FIG. 7. Derivation of "hybrid immunity" phages by *in vitro* recombination. The DNAs of λ*b2N⁻imm*$^{\lambda}$*nin* and λ*b2N⁻imm*^{434}*nin* phages, each deprived of their targets for endo R.*Hin*dIII (other than those within the immunity region), were treated with endo R.*Hin*dIII and then T4 polynucleotide ligase. One of the two types of recombinants is shown in the diagram.

ACKNOWLEDGMENTS

The work described in this chapter has depended on the use of restriction enzymes, the tryptophan operon of *E. coli,* and continued interchange of ideas, strains, and enzymes within our laboratory. I wish to thank all my colleagues, particularly Kenneth Murray and Bill Brammar, for their contributions to this environment and the Medical Research Council for financial support.

REFERENCES

1. Adhya, S., Gottesman, M., and de Crombrugghe, B. (1974): Release of polarity in E. coli by gene N of phage λ: termination and antitermination of transcription. *Proc. Nat. Acad. Sci. U.S.A.,* 71:2534–2538.
2. Adhya, S., Sen. A., and Mitra, S. (1971): The role of gene S. In: *The Bacteriophage Lambda,* edited by A. D. Hershey, pp. 743–746. Cold Spring Harbor Laboratory, Cold Spring Harbor, New York.
3. Bertrand, K., Squires, C., and Yanofsky, C. (1976): Transcription termination in vivo in the leader region of the tryptophan operon of Escherichia coli. *J. Mol. Biol.,* 103:319–337.
4. Borck, K., Beggs, J. D., Brammar, W. J., Hopkins, A. S., and Murray, N. E. (1976): The construction in vitro of transducing derivatives of phage lambda. *Mol. Gen. Genet.,* 146:199–207.
5. Cameron, J. R., Panasenko, S. M., Lehman, I. R., and Davis, R. W. (1975): In vitro construction of bacteriophage λ carrying segments of the Escherichia coli chromosome: Selection of hybrids containing the gene for DNA ligase. *Proc. Nat. Acad. Sci. U.S.A.,* 72:3416–3420.
6. Campbell, A. (1962): Episomes. *Adv. Genet.,* 11:101–145.
7. Court, D., and Sato, K. (1969): Studies of novel transducing variants of lambda: dispensability of genes N and Q. *Virology,* 39:348–352.
8. Creighton, T. E., and Yanofsky, C. (1970): Chorismate to tryptophan (Escherichia coli)—Anthranilate synthetase, PR transferase, PRA isomerase, InGP synthetase, and tryptophan synthetase. *Methods Enzymol.,* 17A:365–380.
9. Davison, J. R., Brammar, W. J., and Brunel, F. (1974): Quantitative aspects of gene expression in a λ-trp fusion operon. *Mol. Gen. Genet.,* 130:9–20.
10. de Crombrugghe, B., Chen, B., Anderson, W., Nissley, P., Gottesman, M., Pastan, I., and Perlman, R. (1971): Lac DNA, RNA polymerase and cyclic AMP receptor protein; cyclic AMP, *Lac* repressor and inducer are the essential elements for controlled lac transcription. *Nature* [*New Biol.*], 231:139–142.
11. Deeb, S. S., Okamoto, K., and Hall, B. D. (1967): Isolation and characterisation of non-defective transducing element of bacteriophage ø80. *Virology,* 31:289–295.
12. Echols, H., Green, L., Oppenheim, A. B., Oppenheim, A., and Honigman, A. (1973): Role of the cro gene in bacteriophage λ development. *J. Mol. Biol.,* 80:203–216.
13. Franklin, N. C. (1971): The N operon of lambda: extent and regulation as observed in fusions to the tryptophan operon of Escherichia coli. In: *The Bacteriophage Lambda,* edited by A. D. Hershey, pp. 621–638. Cold Spring Harbor Laboratory, Cold Spring Harbor, New York.
14. Franklin, N. C. (1974): Altered reading of genetic signals fused to the N operon of bacteriophage λ: Genetic evidence for modification of polymerase by the protein product of the N gene. *J. Mol. Biol.,* 89:33–48.
15. Gilbert, W., and Müller-Hill, B. (1967): The lac operator is DNA. *Proc. Natl. Acad. Sci. U.S.A.,* 58:2415–2421.
16. Gottesman, M. E., and Weisberg, R. A. (1971): Prophage insertion and excision. In: *The Bacteriophage Lambda,* edited by A. D. Hershey, pp. 113–138. Cold Spring Harbor Laboratory, Cold Spring Harbor, New York.
17. Hershfield, V., Boyer, H. W., Yanofsky, C., Lovett, M. A., and Helinski, D. R. (1974):

Plasmid ColE1 as a molecular vehicle for cloning and amplification of DNA. *Proc. Natl. Acad. Sci. U.S.A.,* 71:3455–3459.
18. Hopkins, A. H., Murray, N. E., and Brammar, W. J. (1976): Characterization of λtrp-transducing phages made in vitro. *J. Mol. Biol.,* 107:549–569.
19. Imamoto, F., Morikawa, N., Sato, K., Mishima, S., Nishimura, T., and Matsushiro, A. (1965): On the transcription of the tryptophan operon in Escherichia coli. II. Production of the specific messenger RNA. *J. Mol. Biol.,* 13:157–168.
20. Ito, J., Cox, E. C., and Yanofsky, C. (1969): Anthranilate synthetase, an enzyme specified by the tryptophan operon of Escherichia coli: purification and characterization of component I. *J. Bacteriol.,* 97:725–733.
21. Jackson, E. N., and Yanofsky, C. (1972): Internal promoter of the tryptophan operon of E. coli is located in a structural gene. *J. Mol. Biol.,* 69:307–313.
22. Moir, A., and Brammar, W. J. (1976): The use of specialised transducing phages in the amplification of enzyme production. *Mol. Gen. Genet.,* 149:87–99.
23. Müller-Hill, B., Crapo, L., and Gilbert, W. (1968): Mutants that make more Lac repressor. *Proc. Natl. Acad. Sci. U.S.A.,* 59:1259–1264.
24. Murray, K., and Murray, N. E. (1975): Phage lambda receptor chromosomes for DNA fragments made with restriction endonuclease III of Haemophilus influenzae and restriction endonuclease I of Escherichia coli. *J. Mol. Biol.,* 98:551–564.
25. Murray, N. E., and Murray, K. (1974): Manipulation of restriction targets in phage λ to form receptor chromosomes for DNA fragments. *Nature,* 251:476–481.
26. Ogawa, T., and Tomizawa, J. (1968): Replication of bacteriophage DNA I. Replication of DNA of lambda phage defective in early functions. *J. Mol. Biol.,* 38:217–225.
27. Pero, J. (1971): Deletion mapping of the site of the tof gene product. In: *The Bacteriophage Lambda,* edited by A. D. Hershey, pp. 549–608. Cold Spring Harbor Laboratory, Cold Spring Harbor, New York.
28. Radding, C. M., and Shreffler, D. R. (1966): Regulation of λ exonuclease and a new λ antigen. *J. Mol. Biol.,* 18:251–261.
29. Rambach, A., and Tiollais, P. (1974): Bacteriophage λ having EcoRI endonuclease sites only in the non-essential region of the genome. *Proc. Natl. Acad. Sci. U.S.A.,* 71:3927–3930.
30. Sato, K., and Matsushiro, A. (1965): The tryptophan operon regulated by phage immunity. *J. Mol. Biol.,* 14:608–610.
31. Schleif, R., Greenblatt, J., and Davis, R. W. (1971): Dual control of arabinose genes on transducing phage λdara. *J. Mol. Biol.,* 59:127–150.
32. Segawa, T., and Imamoto, F. (1974): Diversity of regulation of genetic transcription. II. Specific relaxation of polarity in read-through transcription of the translocated trp operon in bacteriophage λtrp. *J. Mol. Biol.,* 87:741–754.
33. Thomas, M., Cameron, J. R., and Davis, R. W. (1974): Viable molecular hybrids of bacteriophage λ and eukaryotic DNA. *Proc. Natl. Acad. Sci. U.S.A.,* 71:4579–4583.
34. Wilgus, G. S., Mural, R. J., Friedman, D. I., Fiandt, M., and Szybalski, W. (1973): $\lambda imm^{\lambda 434}$: A phage with a hybrid immunity region. *Virology,* 56:46–53.
35. Yanofsky, C., Horn, V., Bonner, M., and Stasiowski, S. (1971): Polarity and enzyme functions in mutants of the first three genes of the tryptophan operon of Escherichia coli. *Genetics,* 69:409–433.
36. Zalkin, H., Yanofsky, C., and Squires, C. L. (1974): Regulated in vitro synthesis of Escherichia coli tryptophan operon messenger RNA and enzymes. *J. Biol. Chem.,* 249:465–475.

Recombinant Molecules: Impact on Science and Society, edited by R. F. Beers, Jr. and E. G. Bassett. Raven Press, New York © 1977.

14. Safety of Coliphage Lambda Vectors Carrying Foreign Genes

Waclaw Szybalski

McArdle Laboratory for Cancer Research, University of Wisconsin, Madison, Wisconsin 53706

INTRODUCTION

Since most major research breakthroughs have elicited both positive and negative responses from the scientific community and the general public, it is not surprising that the novel technologies of genetic engineering leading to the synthesis of new genomes have not only their enthusiastic proponents but also their vocal critics. The proponents stress the great potentials of this technology both in fundamental studies of gene function and in practical medical, agricultural, and industrial applications, whereas the critics stress the potential dangers of the cloned foreign DNA carried by the bacterial vectors. The early apprehensions about the dangers culminated in a letter published by Berg et al. (2) and the ensuing Asilomar Conference on Recombinant DNA Molecules (3) (see also publications cited in ref. 12).

One rational approach to this controversial area is to pose the following questions:

1. Are the dangers real, potential, or imaginary?
2. Do the potential benefits of the technique warrant exposure to the potential dangers? In short, is the benefit-to-risk ratio a favorable one?
3. Is it possible to decrease or eliminate the potential dangers without seriously limiting application of the technique?

In the absence of hard data, it is difficult to answer the first question: theoretically there is a potential danger, no matter how remote, whenever a new strain or new mutant is produced in the laboratory. However, from the practical point of view such danger is only an imaginary one. The best example from the past was the decision to test chemical mutagens that never preexisted in nature with the result that no pathogen has ever emerged among the myriads of laboratory-created *E. coli* or other mutants. On the other hand, the various beneficial effects of the research on mutagenicity are quite substantial.

The answer to the second question might be a subjective one. We simply do not know and cannot predict the unexpected deleterious or beneficial effects of some foreign DNA carried in association with the vector. Scenarios of a negative nature might include synthesis of a potent toxin, whereas a beneficial scenario might be represented by the adventitious immunization to a pathogenic agent by an antigen coded by a foreign DNA fragment. Due to these unknowns, the safest assumption is that the adventitious risks and benefits would cancel out, whereas the predictable scientific and practical benefits of the gene implantation technology amply warrant use of the technique. However, I sincerely hope that even the most beneficial foreign DNA construct, e.g., one that codes for a new "miracle hormone" but also shows pathogenic side effects, will not be regarded with the same leniency as another artificial construct, the automobile, which brings us many comforts at the price of about 50,000 American lives per year, not to mention environmental pollution. It was pointed out to me that automobiles do not self-replicate, but from the practical point of view their production can hardly be impeded for numerous economic, sociological, and political reasons.

The answer to the third question is yes, and in this presentation I plan to identify all the potential routes of escape for the λ phage vector and to specify means that will block these escape routes. However, before doing so, let me first compare the general safety features of phage vectors versus plasmid-host vectors, the two presently most commonly used prokaryotic vehicles for cloned foreign DNA.

COMPARISON OF SAFETY FEATURES OF PHAGE AND PLASMID VECTORS

The plasmid and its bacterial host constitute a *single*-component, self-replicating system, which after escape into the natural environment has the potential to propagate the inserted foreign DNA. The phage and its bacterial host, on the other hand, is a *two*-component system, in which each component alone could be considered as safe:

a. The host alone is safe, since it does not carry the cloned DNA.
b. The phage vector alone has no capacity to replicate after escaping into the natural environment, unless it first survives, finds a sensitive host, and infects it successfully (see Fig. 1).

This two-component feature of the phage-host system provides it with intrinsically much higher safety than that of most one-component systems, especially if all routes are blocked that would convert the phage-host vector into a single-component system, such as a lysogen, plasmid carrier, or some other appropriately balanced phage-host carrier system. Such undesirable conversion to a one-component system would decrease the safety of the phage vector to the level of the comparable plasmid vector, or in some cases even lower, since nondefective prophage and some prophage plasmids could mature into phage particles and thus become more infectious than conventional plasmids.

ESCAPE MODES OF PHAGE VECTORS

To evaluate the various safety features when designing a safer vector, it is helpful first to consider all the possible modes of vector escape from the laboratory where it is propagated in a specially designed host. There are two primary routes for the cloned foreign DNA carried by the phage to escape from the laboratory into the environment.

The first route involves the conversion of the two-component, phage-host combination into a stable one-component prophage or plasmid carrier system followed by escape of the laboratory host that now carries the cloned foreign genes.

The second route is escape of the phage particles that carry the cloned DNA, followed by their finding a suitable host in nature and a successful infection. These two routes of escape are depicted in Fig. 1. They will be

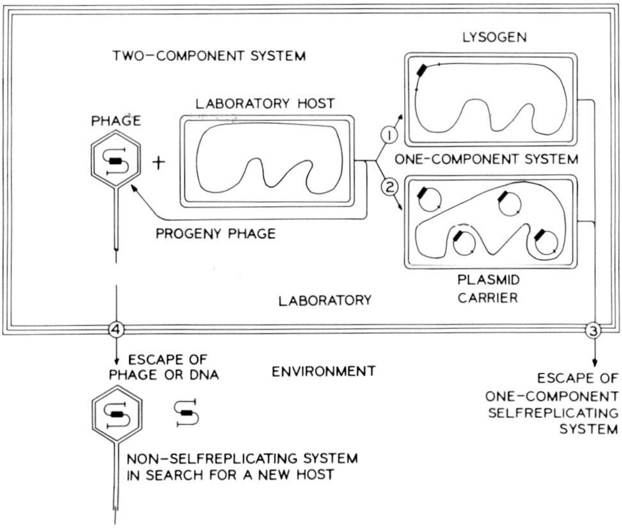

FIG. 1. Modes of possible escape routes for phage vectors. Within the laboratory the phage is replicating in the specially adapted bacterial host (two-component system) producing large yields of progeny phage. The phage and host genomes are indicated by thin lines, and the dark rectangle within the phage DNA represents the cloned foreign DNA fragment. There are two major escape routes. The phage vector could become either (1) prophage or (2) plasmid in the laboratory host, converting the two-component system to a less safe one-component, self-replicating system, which could (3) escape from the laboratory into the environment. Alternatively, (4) the phage could escape from the laboratory in the form of phage particles or their DNA, but, since they do not constitute a self-replicating system, their survival and further replication depend on finding a new suitable host in nature.

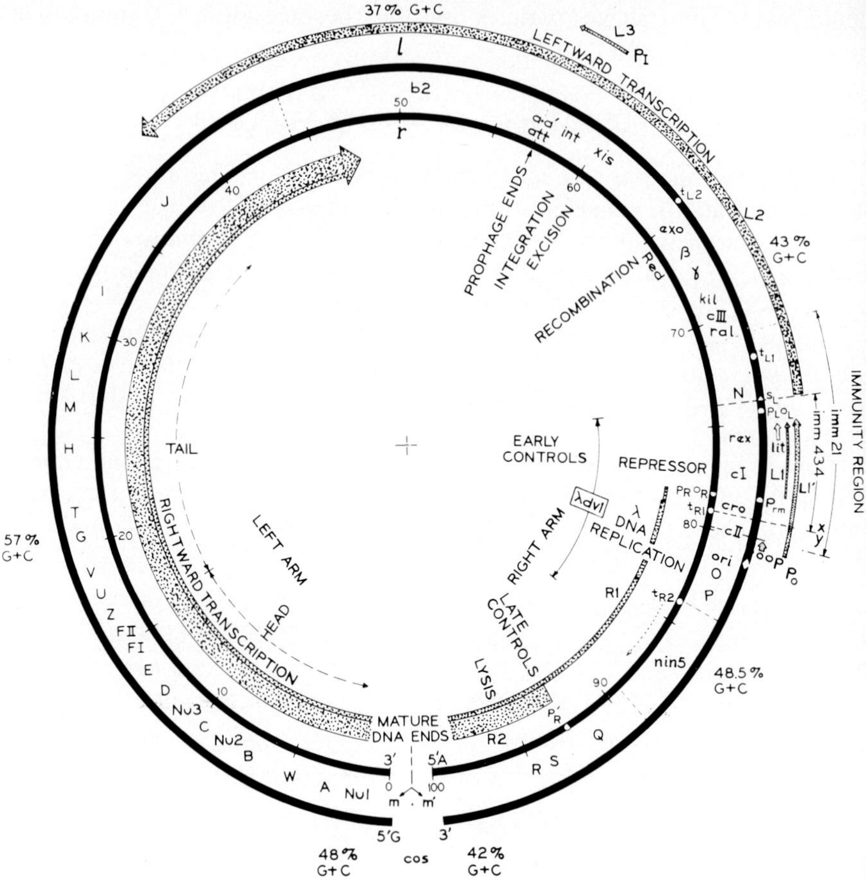

FIG. 2. A genetic and molecular map of bacteriophage lambda (λ) of *Escherichia coli.* The *vir* mutations are in the o_L and o_R operators. Specifically, mutations *v*1, *v*3, *vs,* and *virCc* are in the o_R operator region. Mutations *c*17 and *ri*^{c} are in the *y*-cII and *O* regions and create new promoters not regulated by the *cro* product. (Additional details in ref. 16.)

discussed in more detail in the following sections on biological containment. The λ phage genes, as related to this discussion, are shown in Fig. 2.

BIOLOGICAL CONTAINMENT OF PHAGE VECTORS

Preventing Formation of Prophage or Plasmids

During propagation of wild-type λ phage in the laboratory host, there is a high probability that a stable prophage state will be established, since λ is endowed with special mechanisms for that purpose. On the other hand, plasmid formation is a rare event, on the order of 10^{-6} to 10^{-8}. Both events could be blocked by: (a) genetic modification of the phage; (b) genetic modification of the host; (c) avoidance of any DNA sequence homologies between vector and host; (d) early harvesting of the lysate before the lysogens or plasmid-carrying bacteria have the chance to propagate and overgrow the lysate; and (e) selective inactivation of bacterial cells in the lysate.

PREVENTION OF LYSOGENY

For lysogeny to be established successfully, the repressor function is required along with a mechanism for stable integration of the phage genome into the host DNA. To abolish the repressor function, it is best to delete the repressor gene *c*I [e.g., one of the KH deletions (4)] and/or to inactivate the o_L and o_R operators, the sites of the repressor function, by the *vir* mutations. Moreover, deletion of the phage *int* gene together with the *att*P site on the phage and the *att*B site in the laboratory host will eliminate site-specific phage integration. Even if the phage should become integrated by some general or specific recombination event, the host bacterium will not survive the lethal effects of the phage, which cannot be repressed in the absence of the *c*I and operator functions. Stable integration of the phage and foreign cloned genes would thus be highly unlikely (probably less than 10^{-16}), requiring a very rare, illegitimate recombination event (estimated 10^{-8}) and deletion of two sets of lethal genes (estimated 10^{-8}).

PREVENTION OF PLASMID FORMATION

Formation of a plasmid from the λ phage requires establishment of a well-regulated balance between plasmid and host replication. Excessive replication of the plasmid will result in host death, whereas an insufficient level of plasmid replication will result in loss of the plasmid by dilution. Either the whole genome of phage λ, usually its N^- mutant (11), or only a part of it carrying the p_R-*cro-O-P* operon and denoted λdv (1,8), can establish a stable plasmid. In the case of the λN^- plasmid, the *nin*5 deletion (Fig. 2), which leads to *N*-independent expression of the late genes, prevents plasmid formation of the first kind. Furthermore, formation of this and of the λdv type of plasmid can be achieved through deregulation of the p_R-*cro-O-P* operon by mutations that affect the autocontrolling mechanism of the gene *cro* product (1,9,16). The so-called *vs* and especially *vir Cc* mutations in the o_R operator, which permit the phage to grow on λdv carrier cells and thus could be relatively easily selected (10), block establishment of λdv plasmids (K. Matsubara, *personal communication*). The same is true for the *c*17 and some ri^c type of mutants (K. Matsubara, *personal communication*), but unfortunately the vectors carrying a *c*17 mutation seem to produce low phage yields (F. R. Blattner, *personal communication*). The latter appears to be true also for the cro^- mutants, but it should be possible to use *cro*ts mutants

[K. Matsubara cited in (5)] and to propagate such vectors at permissive temperatures followed by heating at a nonpermissive temperature to inactivate the λdv*cro*ts carrier cells. Since the frequency of plasmid formation is inherently quite low (10^{-6} to 10^{-8}), the introduction of the *virCc* and other mutations cited above should reduce the overall frequency of stable plasmid-carrier cells to less than 10^{-14}. The observation of G. Kellenberger-Gujer and D. E. Berg (*personal communication*) that some *dnaA* ts mutations in the *E. coli* host block λdv formation even at permissive temperatures should also be explored when designing a proper laboratory host.

PREVENTION OF OVERGROWTH OF THE LYSATE BY RARE LYSOGENS AND PLASMID CARRIERS

Although formation of lysogens and plasmid carrier cells could be rendered quite improbable, as outlined above, one such cell could perhaps be formed when large volumes of lysate are prepared. Such cells could replicate in the lysate and produce up to 10^9 progeny per milliliter, especially if such a lysate would be left for a prolonged time under conditions permissive for growth. How could such overgrowth be prevented? One way is to harvest the culture as soon as lysis has occurred, a procedure which, however, is subject to human error. This error could be minimized by installing an automatic device to inject chloroform into the growth medium as soon as massive lysis is sensed by a special, e.g., photoelectric, sensor. Another way is to use S^- phage mutants, which prevent lysis of the host; in such cultures, overgrowth by lysogens and plasmid carriers cannot easily occur. The artificial lysis of S^- mutants by chloroform would result in death of any lysogens and plasmid carriers, as discussed below. However, use of λS^- mutants permits only one cycle of λ growth, and thus the method of phage propagation must be properly adjusted.

SELECTIVE INACTIVATION OF SURVIVING CELLS IN THE LYSATE

E. coli cells are extremely sensitive to shaking with chloroform (less than 10^{-8} survival), whereas λ phage vectors survive this treatment (with the exception of some λh mutants). Thus, prompt treatment of the lysate with chloroform will efficiently kill any of the surviving host cells, including already very rare lysogens or plasmid carriers. Thus, the procedure for propagation of phage vectors that carry cloned DNA should include addition of chloroform at the time of cell lysis and before any handling of the lysing culture which could lead to escape of lysogens or plasmid-carrying cells. A simple and practically foolproof procedure should be relatively easy to design.

One might question whether an essentially chemical mode of containment, such as chloroform treatment, could be considered a part of the biological containment. My feeling is that this step does constitute a part of the biological containment measure, analogous to the use of any chemical ingredient in the media and the proper microbiological techniques. Chemical containment is quite different from physical containment procedures depending on costly hardware, mainly cabinets and specially designed laboratory facilities.

PREVENTION OF SURVIVAL OF THE VECTOR-CARRYING LABORATORY HOST IN NATURE

In the very unlikely case that the lysogen or plasmid carrier should become established in the laboratory and then escape into the natural environment, the use of laboratory host mutants that cannot survive in nature could provide an additional level of containment. For instance, one could use as hosts mutants that require the presence of streptomycin or neamine (15) in the laboratory media. Several other types of mutations that would disarm the host have been discussed by Curtiss et al. [(5) and *this volume*].

The combination of approaches outlined above should reduce to undetectable levels the probability of stable association of the laboratory host with the vector carrying the foreign DNA and their subsequent escape.

Containment of Infective Phage Particles

As mentioned before, the phage particle by itself is not a self-replicating unit, and thus, its survival in nature would depend on finding a sensitive host and establishment of a stable association that would permit the replication and expression of the foreign DNA. I shall first discuss the natural safety features of phage λ and then outline measures that would further decrease the probability of escape of phage vectors.

NATURAL CONTAINMENT OF PHAGE λ

The phage could escape the laboratory environment by various routes, most important of which would be (a) ingestion, (b) the airborne route or transport on dry surfaces, and (c) the sewer system.

(a) Because of the sensitivity of λ phages to the low pH of the stomach, no surviving λ phages were detected in human feces after experimental ingestion of 10^8 (N. Murray, *unpublished observations*) or even 10^{11} (W. Szybalski, *unpublished observations*) λ phage particles.

(b) The high sensitivity of phage λ to desiccation gives it a low chance of escape by transport on dry surfaces or through the airborne route.

(c) The low survival of λ in the sewer, probably due to the presence of detergents and bile salts, further contributes to the natural containment of the phage.

All these features and some of those discussed below are documented by Blattner et al., *this volume*. The encounter with and infection of a new host cell in nature are hampered by two more factors. In a survey of over 2,000 *E. coli* strains isolated from natural sources, investigators found that not one was able to propagate the present "wild-type" version of phage λ, which is adapted to grow in laboratory strains of *E. coli* K-12 [R. W. Davis, cited in (5) and *personal communication*]. Moreover, when phage particles are released into the natural environment the dilution factor is enormous, which further decreases the chance for a successful encounter.

Even if infection should occur in nature, a stable association between the new host and the phage is required to successfully propagate and express the foreign genes. This, however, was rendered practically impossible by the features preventing lysogenization and plasmid formation, as discussed in previous sections. On the other hand, one might imagine that a new host in nature carries a lambdoid prophage that could recombine with the infecting λ vector and, as a result, a new phage would be produced which has lost all the safety features but still carries the foreign DNA. The presence of such a helper lambdoid phage could also mediate the stable association of the foreign DNA with the new host found in nature. This unlikely scenario points out the necessity of adding a new safety feature to the phage vector, namely the rapid loss of infectivity after it escapes from the laboratory environment.

PREVENTION OF PHAGE INFECTIVITY OUTSIDE THE LABORATORY ENVIRONMENT

There are several ways to abolish the infectivity of a phage, either by genetic or other approaches.

Laboratory-Dependent Infectivity

One of the simplest ways to construct such a conditionally stable phage would be to introduce either missense or suppressed nonsense mutations into the λ genes coding for the capsid components, so as to make the phage extremely unstable and/or noninfectious under conditions encountered in the natural environment and a standard laboratory media, but quite stable and able to propagate well under specially designed laboratory conditions (12). Vance Makin in my laboratory was attempting to isolate this kind of λ mutant, which, e.g., could grow well only in the presence of 10 mM putrescine (or other supplements including high concentrations of special salts), but in the absence of which would rapidly disintegrate and lose its infectivity. Such vectors would not be able to transfer their genome into any other host present in nature, since they would perish long before finding a suitable host.

A variation on this strategy is construction of phage mutants that are extremely sensitive to some agents frequently present in nature, e.g., DNAses or detergents, whereas these agents would be absent from laboratory media.

Tailless Mutants (Three-Component System)

Another way to eliminate the infectivity of phage particles is to delete or mutate the tail-coding genes and to propagate only the phage heads packed with the recombinant DNA. Only under controlled laboratory conditions could such heads be made infectious by adding separately prepared tails (66), and they then could be effectively propagated through a single infectious cycle, at quite high yields if they carry an S^- mutation. An additional safety factor of such head preparations is their extreme instability unless stored in 10 mM putrescine (66), a condition readily provided in the laboratory but rare in the natural environment. The disadvantages of this approach are the cumbersome procedure, the necessity to use only a single infectious cycle, and the "unsafe" phase when the tails are combined with the head preparation. It should be possible, however, to greatly alleviate or eliminate all these disadvantages, and thus to develop a safe three-component method (heads, tails, and host), which would be an improvement over the two-component method (phage, host) since heads, the only component that carries the cloned foreign DNA, are noninfectious (V. C. Bode, *personal communication*).

Phage Inactivation

Another way to abolish phage infectivity would be to treat the lysate, before further processing, with some inactivating agent, e.g., diethyl pyrocarbonate, which would not interfere with further phage processing or DNA isolation. This approach is being studied in our laboratory. Obviously, phages used for further propagation could not be treated in such a manner unless a method were devised to reactivate the inactivated phage particles.

Infectivity of Recombinant DNA

In the three procedures immediately above, the phage DNA either remains within the noninfectious phage particles or may be released into the natural environment. The probability of finding and infecting a susceptible host with such a DNA is nil, since (a) successful transfection requires special treatment of the host cell and even then its rate is very low, and because (b) free DNA would be rapidly degraded by omnipresent DNAses. Moreover, in the normal course of events we do ingest large quantities of foreign DNA in our daily diet, without any obvious ill effects. The same pertains to our normal environment.

PREVENTION OF PHAGE PROPAGATION OUTSIDE THE LABORATORY ENVIRONMENT

Regimens outlined in the previous section were designed to abolish the infectivity of the phage particles. As a further precaution, however, one might also introduce modifications that would prevent survival and replication of the cloned DNA should it be introduced into a sensitive cell encountered in nature.

Restriction

If the phage were grown in the r^-m^- (restriction and modification deficient) laboratory host, there is a high probability that the phage DNA would be restricted after entering the sensitive cell since most of the cells in nature carry some r^+m^+ system.

Conditional Mutants

If the phage carries mutations which permit its propagation only in the su^+ and gro^- type of hosts, there would be a very low chance that a sensitive cell encountered in nature would be of such complex genotype that would allow phage replication. A further improvement would be for the phage vector to carry several nonsense mutations requiring the presence of more than one kind of suppressor in the host cell. Also, thermosensitive (*ts*) mutations in the phage would be useful to prevent phage growth at elevated temperature, e.g., in the intestinal tract.

Host Lethality

It would be helpful to maximize the early lethal effect of the phage on its host so as to prevent phage propagation. The laboratory host would obviously have to be protected from this premature destruction.

Recombination Functions

There are several ways to minimize possible recombination events between the genomes of the phage vector and the host, which might lead to transfer of the foreign DNA into the host genome or loss of the safety features of the phage: (a) If the phage remains gam^+, this function usually depresses the recombination systems of the host; (b) the phage should be deficient in its Red recombination functions; (c) one should avoid introducing into the phage any DNA homologous to *E. coli* DNA, including the IS insertion sequences.

High Yields

High yields of phage permit using small volumes of culture media, which inherently increases the safety factor. Therefore, when introducing the above mutations into phage vectors and the laboratory host, one should always check the phage yield, which should approach or exceed 10^{11} particles/ml of raw lysate (see Blattner et al., *this volume*). It has been our experience that the phage yield could very frequently be improved by repeated cycles of phage growth and isolation of larger plaques plated under fully permissive conditions. For example, many of the λ*bio*(N^-)*nin*5 phages (which in earlier studies produced less than 10^9 particles/ml) are now routinely yielding over 10^{11} particles/ml without affecting the deletion or substitution, probably by accumulation of some *chi*-like mutations (7).

CONVENIENCE FEATURES FOR INCORPORATION OF FOREIGN DNA

These aspects are discussed by K. Murray and by F. R. Blattner et al. (*this volume*). Earlier references are cited in (12). The work in this area has been predominately concentrated on (a) deletion of nonessential DNA so as to increase the capacity of the vector, (b) elimination of the undesirable restriction endonuclease cuts within the vital parts of the genome, and (c) easy detection of the desired recombinant phages. Much of this early work will have to be reevaluated in view of the availability of many new types of restriction nucleases, some of which produce only one or two cuts in nonessential regions of λ, the observation of Sgaramella (*this volume*) that cohesive ends are not necessarily required for the fusion of fragments by ligase, and various new strategies discussed in this volume for attaching adapter cohesive ends and proper prokaryotic promoter and protein-initiation signals to the fragments of foreign DNA. The number and variety of new ingenious approaches is growing at a rapid pace. However, one should bear in mind that the expression of cloned genes could sometimes be considered as an additional hazard, especially if the product is known to be toxic.

TESTING OF THE SAFETY FEATURES

Simple and rapid tests should be designed to verify the presence of various genetic modifications which confer the safety features and to monitor the survival of the phage vector and retention of the cloned DNA under conditions representing the natural environment. For example, to test the first route of escape (Fig. 1), one should quantitatively assess the effect of various mutations and lysate treatment on the frequency of lysogen or plasmid formation in the laboratory host. The second route of escape could be tested by measuring the effect of various mutations on the kinetics of survival of phage infectivity, phage replication, integration, and recombination in various hosts or lysogens representing the natural environment. When designing tests one should also explore the "worst case," but only if it could represent, with reasonable probability, the natural conditions.

CONCLUDING REMARKS

When designing, testing, and using phage vectors, one should also consider the following aspects, some of which have been previously discussed (12,13).

1. The benefit-to-risk ratios for some recombinant molecules might be very high; for example, by cloning specific human genes one should be able to employ them for synthesizing in large quantities any human protein or peptide, possibly of great medical importance. Furthermore, by directed mutagenesis of the cloned DNA, investigators could easily produce many new variants of human proteins and test them for the desired pharmacological activities. This is an area of great opportunity.

2. One should also keep in mind that any organism, including man, throughout the eons has been exposed to all kinds of foreign DNAs, especially including those ingested in food, and that formation of recombinant DNA molecules might at all times be occurring in nature. There is no point in increasing the safety of the vector to exceed that of the natural spontaneous background.

3. Other host-phage systems should perhaps be explored as possible vectors, especially if they represent some rare and exotic ecological niche. However, developing such a new system might prove much more difficult than radical genetic modifications of the λ-*E. coli* system, which would make its survival fully dependent on unusual laboratory conditions while retaining the great advantages of the genetically and biochemically highly developed and well-understood system. One should never be unwisely tempted to replace well-known and easily minimized risks of a highly explored system with the uncertainties of some new undeveloped system.

3. Carefully designed λ lysogens and plasmids should also be considered as rather safe vectors; for example, λdv plasmids, which carry *cro*ts mutations and are grown in properly contained hosts, could be both practical and relatively safe since they can produce over 1,000 copies of cloned DNA per cell while killing their carrier cells at the human body temperature [K. Matsubara, cited in (5,12) and *personal communication*].

5. One should realize that the technology of recombinant DNA molecules lends itself to easy mechanical containment using a simple, inex-

pensive sealed glove box. This is possible since all the operations, including the creation, testing, and destruction of the vector which carries the cloned DNA, could be performed in such a completely closed system (12).

These are only the general outlines of the possible factors involved in the design of safer vectors, but further improvements and projections are governed only by the imagination and ingenuity of the designer.

ACKNOWLEDGMENTS

Studies leading to this chapter were supported by the Program-Project Grant CA-07175 of the National Cancer Institute. I am greatly indebted to Dr. F. R. Blattner for many stimulating discussions.

REFERENCES

1. Berg, D. (1974): Genes of phage λ essential for λdv plasmids. *Virology,* 62:224–233.
2. Berg, P., Baltimore, D., Boyer, H. W., Cohen, S. N., Davis, R. W., Hogness, D. S., Nathans, D., Roblin, R. O., Watson, J. D., Weissman, S., and Zinder, N. (1974): Potential biohazards of recombinant DNA molecules. *Science,* 185:303.
3. Berg, P., Baltimore, D., Brenner, S., Roblin, R. O., and Singer, M. F. (1975): Summary statement of the Asilomar Conference on recombinant DNA molecules. *Science,* 188:991–994. (Also, *Nature,* 255:442–444 and *Proc. Natl. Acad. Sci. U.S.A.,* 72:1981–1984.)
4. Blattner, F. R., Fiandt, M., Hass, K. K., Twose, P. A., and Szybalski, W. (1974): Deletions and insertions in the immunity region of coliphage lambda: revised measurements of the promoter-startpoint distance. *Virology,* 62:458–471.
5. Curtiss, R., III, Szybalski, W., Helinski, D. R., and Falkow, S. (1976): Workshop on design and testing of safer prokaryotic vehicles and bacterial hosts for research on recombinant DNA molecules. *Am. Soc. Microbiol. News,* 42:134–138.
6. Harrison, D. P., Brown, D. T., and Bode, V. C. (1973): The lambda head-tail joining reaction: purification, properties and structure of biologically active heads and tails. *J. Mol. Biol.,* 79:437–449.
7. Henderson, D., and Weil, H. (1975): Recombination-deficient deletions in bacteriophage λ and their interaction with chi mutations. *Genetics,* 79:143–174.
8. Matsubara, K., and Kaiser, A. D. (1968): λdv: An autonomously replicating DNA fragment. *Cold Spring Harbor Symp. Quant. Biol.,* 33:769–775.
9. Matsubara, K., and Takeda, Y. (1975): Role of the tof gene in the production and perpetuation of the λdv plasmid. *Mol. Gen. Genet.,* 142:255–230.
10. Ordal, G. W. (1973): Mutations in the right operator of bacteriophage lambda: physiological effects. *J. Mol. Biol.,* 79:723–729.
11. Signer, E. R. (1974): Plasmid formation: a new mode of lysogeny by phage λ. *Nature,* 223:158–160.
12. Szybalski, W. (1976): Safety of bacteriophage vectors for cloned foreign DNA. *Nucleic Acids Recomb. Scient. Mem.,* NAR-18:27–31.
13. Szybalski, W. (1977): Initiation and regulation of transcription and DNA replication in coliphage lamda. In: *Regulatory Biology,* edited by J. C. Copeland and G. A. Marzluff. Ohio State University Press, Columbus.
14. Szybalski, W. (1976): Genetic and molecular map of *Escherichia coli* bacteriophage lambda (λ). In: *Handbook of Biochemistry and Molecular Biology, Nucleic Acids, Vol. II,* edited by G. D. Fasman, pp. 677–685. CRC Press, Cleveland.
15. Szybalski, W., and Cocito-Vandermeulen, J. (1958): Neamine and streptomycin dependence in *Escherichia coli. Bacteriol. Proc.,* p. 37.
16. Szybalski, W., and Szybalski, E. H. (1974): Visualization of the evolution of viral genomes. In: *Viruses, Evolution and Cancer,* edited by E. Kurstak and K. Maramorosch, pp. 563–582. Academic Press, New York.

ADDENDUM

Several safer phage vector–host systems were recently developed (1–3). Of these λ phages, λgt*WES*·λB (3) and Charon 3A, 4A, and 16A (1) together with the DP50*supF*-enfeebled host were certified by the Director of the National Institutes of Health as EK2 systems.

1. Blattner, F. R., Williams, B. G., Blechl, A. E., Denniston-Thompson, K., Farber, H. E., Furlong, L.-A., Grunwald, D. J., Kiefer, D. O., Moore, D. D., Schumm, J. W., Sheldon, E. L., and Smithies, O. (1977): Charon phages: safer derivatives of bacteriophage lambda for DNA cloning. *Science,* 196:161–169.
2. Donoghue, D. J., and Sharp, P. A. (1977): An improved bacteriophage lambda vector: construction of model recombinants coding for kanamycin resistance. *Gene,* 1:209–227.
3. Leder, P., Tiemeier, D., and Enquist, L. (1977): EK2 derivatives of bacteriophage lambda useful in the cloning of DNA from higher organisms: the λgt*WES* system. *Science,* 196:175–177.

Recombinant Molecules: Impact on Science and Society, edited by R. F. Beers, Jr. and E. G. Bassett. Raven Press, New York © 1977.

15. Construction and Properties of Plasmid Cloning Vehicles

Donald R. Helinski, *Vickers Hershfield, David Figurski, and Richard J. Meyer

Department of Biology, University of California, San Diego, La Jolla, California 92093

INTRODUCTION

Since the initial demonstration of the utility of a plasmid element for the cloning of genes in *Escherichia coli* (14), a variety of plasmid elements have been developed as cloning vehicles in both *E. coli* and other bacteria. These newer cloning vehicles possess the following plasmid properties that are advantageous for the cloning of DNA.

1. Stable maintenance in the host bacterial cell;
2. Non-self-transmissibility;
3. Ease of genetic manipulation;
4. Ease of isolation;
5. The capacity of joining with and replicating foreign DNA of a broad size range;
6. Ease of introduction of the *in vitro* generated hybrid plasmid into a bacterial cell.

In addition to these basic properties, the more recently developed plasmid cloning vehicles have one or more of the following features:

1. Multiple copy number;
2. Amplification of the DNA upon incubation of the cells in the presence of chloramphenicol;
3. Reduced size;
4. Presence of additional antibiotic resistance genes or strong selective markers other than antibiotic resistance genes;
5. A broad host range within the Gram-negative bacteria;

---

* Present address: Department of Microbiology, Duke University Medical Center, Durham, North Carolina 27710.

6. Presence of prokaryotic regulatory genes that promote high-level expression of introduced foreign DNA;
7. Presence of mutations that inhibit the replication of the plasmid element at mammalian body temperature.

These various modifications of plasmid elements are directed toward increasing their efficacy and safety as cloning vehicles. This chapter is concerned with the cloning vehicle properties of ColE1 and its derivatives constructed *in vitro* and the development of the broad host range plasmid RK2 as a cloning vehicle.

OBSERVATIONS

Properties of ColE1 as a Cloning Vehicle

ColE1 is a non-self-transmissible plasmid of a molecular weight of 4.2×10^6 (2). The majority, if not all, of the mature molecules of this plasmid exist in the bacterial cell in the covalently closed, circular DNA form to the extent of 12 to 15 copies per copy of bacterial chromosome (13). A variable portion (20 to 90%, depending on growth conditions) of the covalently closed, circular DNA molecules of ColE1 are associated with protein when the cells are growing under logarithmic growth conditions (13). The DNA of this DNA-protein complex, designated relaxation complex, is converted to the open circular form in the presence of ionic detergents as sodium dodecyl sulfate (SDS) or the intercalating agent ethidium bromide (11,26). The properties of this reaction indicate that one of the proteins in the complex catalyzes a cleavage at a unique position—specifically in one of the two DNA strands—and covalently links to the 5′ terminus of the nicked strand (7,12,19,28). The position of this nick is at or very near the site of the origin of replication of the plasmid (30). Since ethidium bromide induces the relaxation event (26), the use of dye–cesium chloride equilibrium centrifugation for the isolation of plasmid DNA results in the conversion of the plasmid DNA to the open circular form from the form of relaxation complex. Despite the loss of the covalently closed, circular DNA that is in the form of relaxation complex, covalently closed circles of ColE1 DNA can be obtained in relatively pure form to the extent of 1 to 2% of the total cellular DNA by a single dye–cesium chloride equilibrium centrifugation of a cleared lysate prepared by lysis of the cells with a nonionic detergent (12,24) or by the SDS-salt method (18).

ColE1, unlike most plasmid elements, continues to replicate for up to 12 hr in the presence of chloramphenicol (3,10,13). Under these conditions 1,000 to 3,000 copies per cell of the covalently closed, circular DNA form of this plasmid can be obtained, equivalent in most cases to 40% of the total cellular DNA (10,22). A variable amount (10 to 90%) of the plasmid DNA generated under these conditions contains an RNA segment in one of the two DNA strands (8). Under usual growth conditions, the proportion of RNA-containing molecules produced in the presence of chloramphenicol is low. Cells incubated in the presence of chloramphenicol have proven to be an excellent source of Co1E1 DNA for use as a cloning vehicle in transformation experiments (22).

Plasmid Co1E1 possesses a single *Eco*RI site located 18% from the origin of replication (28). Insertion of DNA at this site does not affect the replication properties of the plasmid (22). One of the major reasons for the development of Co1E1 as a cloning vehicle is its relaxed mode of replication. This offered the possibility that inserted foreign DNA would be maintained as many copies per cell and the multicopies of the foreign DNA would result in greatly amplified levels of gene products specified by this DNA. In addition, the ability of Co1E1 to continue to replicate in the presence of chloramphenicol offered the possibility of generating large amounts of the cloned foreign DNA in the *E. coli* cell. Both of these expectations have been realized (22). As seen in Table 1, the insertion of the entire tryptophan operon in the plasmid Co1E1 and the introduction of this hybrid plasmid (Co1E1-trp) into *E. coli* resulted in a 20-fold increase in the tryptophan synthetase α level and a 50-fold increase in the anthranilate synthetase level over the normal induced cellular levels (results with W3110 trpR$^-$) upon induction of expression of the operon by the addition of 3-indolylacrylic acid (22). The presence of the entire tryptophan operon, consisting of the five structural genes A, B, C, D, and E and a regulatory operator-promoter region, on the multicopied plasmid Co1E1 has been estimated to result in the production of tryptophan enzyme levels equivalent to 20 to 40% of the total cellular

TABLE 1. *Tryptophan enzyme levels in plasmid-containing strains*[a]

Strain	Medium additions	Specific activity: ASase	Specific activity: TSase α
W3110 trpR$^+$	Trp	0.02	0.08
W3110 trpR$^-$	Trp	1.4	15.0
C600 trpR$^+\Delta$trpE5	Trp	0.77	14.7
recA/pVH5	IA	70	308

[a] Cultures of the various strains were grown with the indicated supplements. trpR$^+$ and trpR$^-$ indicate the presence and absence, respectively, of a functional repressor of the tryptophan operon. Repression can be relieved also by the addition of the tryptophan analogue, 3-indolylacrylic acid (IA). pVH5 is the designation for a Co1E1-trp hybrid plasmid containing the entire tryptophan operon of *E. coli*. The plasmid also contains segments of the transducing phage ϕ80-trp that was used as the source of the tryptophan operon. On cleavage with *Eco*RI, three DNA fragments are obtained with molecular weights of 4.2×10^6 (Co1E1 DNA), 8.5×10^6 (ϕ80-trp DNA fragment), and 1.6×10^6 (ϕ80 DNA fragment). (From ref. 22.)

protein (22). A similar amplification of the level of mRNA corresponding to the structural genes of the tryptophan operon has been found for the Co1E1-trp–containing strains (22).

As shown in Fig. 1, the incubation in the presence of chloramphenicol of cells carrying Co1E1, the hybrid plasmid Co1E1-trp, or Co1E1-kan (Co1E1 carrying a fragment of DNA determining resistance to the antibiotic kanamycin) resulted in the accumulation, in each case, of approximately 40% of the total cellular DNA as covalently closed, circular plasmid DNA. This is equivalent to 1,000 to 2,000 copies of Co1E1, 300 copies of Co1E1-trp, and 560 copies of Co1E1-kan per cell (22). Although the insertion of DNA at the single *Eco*RI site on the Co1E1 molecule does not alter the replication properties of the plasmid, it does result in a loss of the plasmid's ability to produce the antibiotically active protein colicin E1 (22). This effect of the insertional event at the *Eco*RI site facilitates the use of Co1E1 as a molecular vehicle in that a distinction can be readily made between cells that have been transformed with normal Co1E1 or Co1E1 hybrid DNA molecules.

Co1E1 Derivatives as Cloning Vehicles

Even though Co1E1 is a relatively low molecular weight plasmid, it has been found that this element can be reduced in size by at least one-half without loss of its self-replicating properties. For example, a mini-Co1E1 plasmid, designated pVH51, with a molecular weight of 2.1×10^6 arose

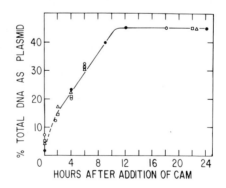

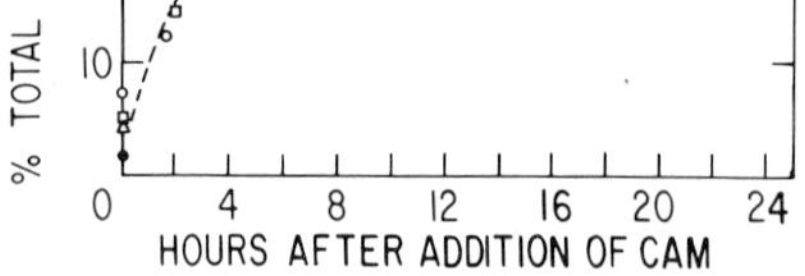

FIG. 1. Accumulation of Co1E1 and Co1E1 hybrid plasmids in cells incubated in the presence of chloramphenicol. *E. coli* cells carrying plasmids Co1E1 (●——●), pVH5 (□——□ and ○——○), and pML2 (△——△) were grown and samples removed at 0, 2, 4, 6, and 22 hr after the addition of chloramphenicol (CAM) (250 μg/ml). The amount of plasmid DNA was determined by ethidium bromide-CsCl equilibrium centrifugation of sarkosyl lysates of the cells. The Co1E1-trp plasmid, pVH5, is described in Table 1. pML2 is the designation for the Co1E1-kan plasmid. On *Eco*RI cleavage, pML2 yields Co1E1 DNA and a 4.5×10^6 dalton fragment that carries the kanamycin resistance gene. (From ref. 22.)

spontaneously from cells carrying the Co1E1-trp hybrid plasmid pVH5 (21). The mini-Co1E1 plasmid possesses a single *Eco*RI restriction site. Heteroduplex analysis showed that about 90% of the mini-Co1E1 plasmid hybridizes to approximately 50% of the parent Co1E1 plasmid (21). The remaining 10% of mini-Co1E1 presumably is a segment of the ϕ80-trp DNA that was the source of the tryptophan operon for the construction of plasmid pVH5. Mini-Co1E1 does not specify the production of colicin E1 but does confer immunity to this colicin (21).

The loss of a substantial segment of normal Co1E1 DNA apparently has resulted in the maintenance of the plasmid at a much higher level than normal Co1E1 in that 114 copies of mini-Co1E1 per cell are found (21). This loss of DNA, however, does not alter the ability of the plasmid to continue to replicate in the presence of chloramphenicol (21).

The mini-Co1E1 plasmid is an example of a Co1E1 derivative generated by a spontaneous deletion event in the host cell. Other derivatives of Co1E1 that have been developed for use as a cloning vehicle have been obtained by insertion *in vitro* of DNA fragments carrying genes that can serve as selective markers for transformation of the plasmid. For example, as shown in Table 2, a fragment (MW 4.5 × 10^6) of plasmid pSC105 that carries a kanamycin resistance gene was obtained by cleavage with the *Eco*RI restriction enzyme and inserted into the *Eco*RI-treated Co1E1 plasmid (22). One of the two *Eco*RI sites in the Co1E1-kan hybrid has been deleted yielding a plasmid, designated pCR1 (15), that possesses the desirable replication

TABLE 2. *Properties of Co1E1 derivatives*[a]

Plasmid	No. of copies per cell	Amplifiable with CAM addition	Molecular weight	Selection of transformants
Co1E1	25–30	+	4.2 × 10^6	Colicin E1 immunity
Co1E1-*kan* (pML2)	25–30	+	8.7 × 10^6	Kanamycin resistance
Co1E1-*trp* (pVH151)	25–30	+	8.9 × 10^6	Tryptophan independence
mini-Co1E1 (pVH51)	125	+	2.1 × 10^6	Colicin E1 immunity
mini-Co1E1-*kan* (pML21)	25–30	+	6.6 × 10^6	Kanamycin resistance
mini-Co1E1-*trp* (pVH153)	25–30	+	6.8 × 10^6	Tryptophan independence

[a] All of the hybrid plasmids were constructed by fusion of *Eco*RI-generated fragments. The kanamycin resistance fragment (MW 4.5 × 10^6) was derived from plasmid pSC105. The tryptophan fragment (MW 4.7 × 10^6) was obtained by *Eco*RI digestion of λ-trp bacteriophage λED10f (bot) and contains the poED region of the tryptophan operon. The host strain for the plasmids is *E. coli* C600 or a tryptophan-requiring mutant of this strain.

properties of Co1E1, the kanamycin resistance gene as a powerful selective marker and a single *Eco*RI site for the insertion of *Eco*RI-derived fragments of foreign DNA. The addition of the *Eco*RI-derived, kanamycin resistance fragment of pSC105 also provides single *Hin*dIII and *Sal*I sites for insertion of *Hin*dIII- and *Sal*I-derived fragments of DNA, respectively (15,20). A Co1E1-trp plasmid similarly has been constructed by the insertion of a fragment of DNA (MW 4.7×10^6) containing the poED region of the tryptophan operon (V. Hershfield, C. Yanofsky, and D. Helinski, *unpublished observations*). This trp fragment was produced by *Eco*RI cleavage of the DNA of the λ-trp bacteriophage λED10f(bot). The trp fragment provides a powerful selective marker in transformations using Co1E1-trp as the cloning vehicle and an *E. coli* recipient possessing a deletion within the *trp*E gene. The utilization of an auxotrophic marker as tryptophan for selection obviates the use of antibiotic resistance for this purpose. This may be desirable under circumstances where the employment of a cloning vehicle carrying an antibiotic resistance gene introduces into a bacterial species for the first time resistance to a theoretically useful antibiotic or increases the probability of survival of a cloned DNA segment that potentially is hazardous.

The *Eco*RI-derived kan and trp fragments also were inserted into *Eco*RI-cleaved mini-Co1E1 (Table 2). The resulting mini-Co1E1-kan and mini-Co1E1-trp plasmids, surprisingly, are maintained at a much lower copy number than the parental mini-Co1E1 (21). The transmissibility of the mini-Co1E1-kan and mini-Co1E1-trp plasmids, as promoted by the conjugative plasmids R100drd or R64drd11, is 10^4 to 10^5 times lower than that of Co1E1 (21). It has also been observed that the insertion of DNA into Co1E1 at the *Eco*RI site (in at least certain instances) reduces substantially the transmissibility of the plasmid. The decreased conjugal transfer of these plasmids when promoted by a conjugative plasmid is, of course, an important safety consideration with respect to the containment of these cloning vehicles in a disabled host strain.

In addition to the spontaneous generation of lower molecular weight derivatives of the Co1E1 plasmid, functional Co1E1 plasmids, reduced in size, have been obtained by cleavage of Co1E1 or its derivative with different restriction enzymes and subsequent transformation with the fragments (6).

Finally, a Co1E1 derivative possessing a specific transposable segment of DNA that contains an ampicillin resistance gene (designated TnA) has been isolated from cells harboring both the Co1E1 plasmid and the antibiotic resistance plasmid R1 (35). These Ap^r derivatives of Co1E1 most likely arose by the spontaneous translocation *in vivo* of the TnA segment from the R1 plasmid to Co1E1. The Co1E1-Ap^r hybrid plasmid, designated RSF2124, is particularly useful as a cloning vehicle in that transformants with this plasmid can be selected with ampicillin and those transformants possessing an insert of foreign DNA at the *Eco*RI site of Co1E1 are readily distinguished by their loss of ability to produce colicin E1.

Construction of a Co1E1-λ Hybrid Plasmid Cloning Vehicle

The insertion of a particular structural gene into a plasmid element does not assure expression of this gene in the *E. coli* cell in the form of a complete mRNA molecule or the corresponding intact protein. Expression of the inserted gene obviously depends on the presence of necessary regulatory elements and, in the case of an inserted eukaryotic gene, proper recognition of the DNA and the corresponding mRNA by the transcription and translation processes, respectively, of *E. coli*. To facilitate the production of mRNA from inserted prokaryotic or eukaryotic DNA in a Co1E1 hybrid plasmid, a Co1E1 derivative was constructed that carried the highly efficient promoters of the bacteriophage λ (V. Hershfield, C. Yanofsky, N. Franklin, and D. Helinski, *unpublished observations*). This cloning vehicle, generated *in vitro* by the fusion of *Eco*RI-cleaved Co1E1 and an *Eco*RI-generated fragment (MW 5.8 × 10^6) of the DNA of the transducing λ-trp phage pt190, carries as the essential λ elements the genes N, C_I, and cro and the P_L and P_R promoters (Fig. 2). (For a review of the regulatory elements of phage λ, see [23]). In addition, the structural genes B, C, and D of the tryptophan operon are present at the left end of the λ fragment. Since the principal operator-promoter region of the tryptophan operon is absent, expression of the D gene is under the control of the λ promotor P_L. Low-level expression of the tryptophan genes B and C, however, occurs due to the presence of the low-efficiency internal promoter p2 located before the C gene (for a recent review of the tryptophan operon, see [5]). As shown in Fig. 2, the C_1 gene is present as the temperature-sensitive mutant C_{IATS}. At 40°C, the thermosensitive repressor of the C gene is inactivated and transcription is initiated off of one strand from the P_L promoter and off of the opposite strand from the P_R promoter. The P_L promoter has been estimated to be 10-fold more efficient than the average *E. coli* promoter and in fact has been shown to promote 11-fold more transcriptions than the trp promoter under fully constitutive conditions (16). The cro gene is also in a mutant form to relieve the repressive effect of the cro gene product on transcription from the P_L and P_R promoters (see [23]). The initiation of transcription from the P_L promoter results in the high-level formation of the protein product of the N gene. The formation of the N protein is important to the ultimate effectiveness of the Co1E1-λ cloning vehicle in that it is expected that this protein will interfere with the functioning of transcription termination signals present on the plasmid and, thus, facilitate high-level production of fused mRNA units initiated from the P_L and P_R promoters (17). This may be particularly essential to the formation of a complete mRNA corresponding to inserted eukaryotic genes that most likely lacks a

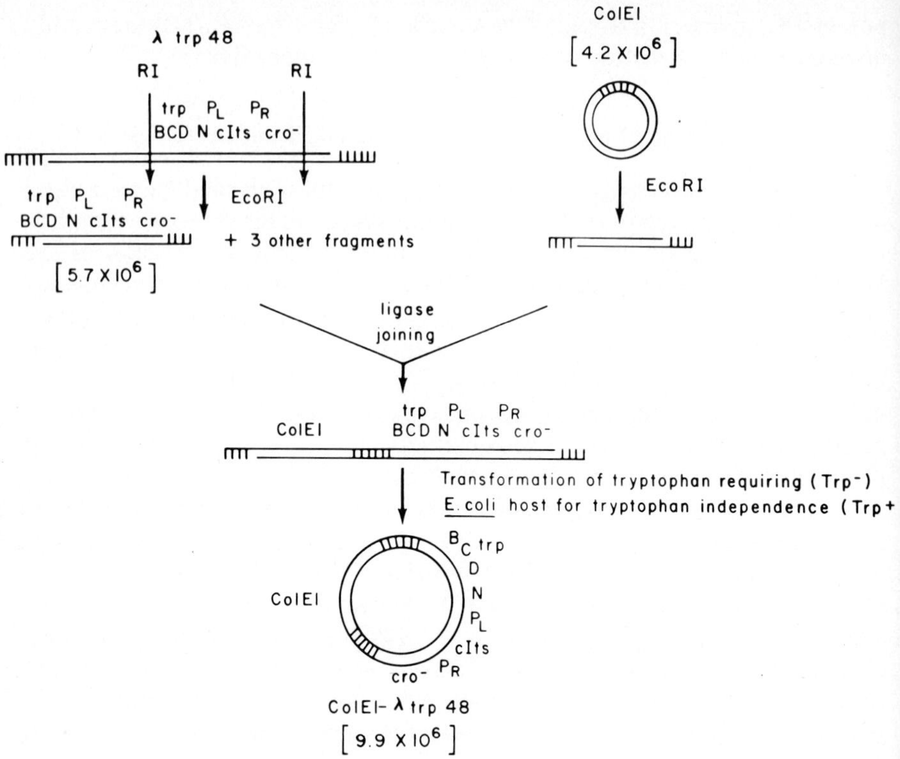

FIG. 2. Construction of the Co1E1-λ hybrid plasmid cloning vehicle. The 5.7×10^6 dalton fragment of DNA carrying the tryptophan genes B, C, and D and the promoter regions of λ was derived by *Eco*RI cleavage of the DNA of phage λtrp48. The various designations on the λ fragment appear in the text. Transformants carrying the Co1E1-λtrp48 plasmid will grow in the absence of tryptophan when the medium is supplemented with indole. The numbers in brackets refer to the molecular weight of the DNA.

promoter region recognized by the *E. coli* transcription system. As shown in Fig. 3, raising the temperature of the culture of cells carrying the Co1E1-λ cloning vehicle with a foreign DNA insert theoretically should result in the high-level expression of this foreign DNA in the form of mRNA. In the case of a prokaryotic foreign DNA insert, this also should result in the formation of high levels of protein product specified by this DNA. For eukaryotic foreign DNA, the greatly enhanced production of complete mRNA may compensate to some extent for the possible poor efficiency of translation of eukaryotic mRNA in *E. coli* (25), or at the least the Co1E1-λ hybrid plasmid can serve as a starting point for further biochemical or genetic modifications to obviate the complications of translation of the eukaryotic mRNA in the bacterial host. The multicopy nature of the Co1E1-λ plasmid is an additional important factor in the amplified produc-

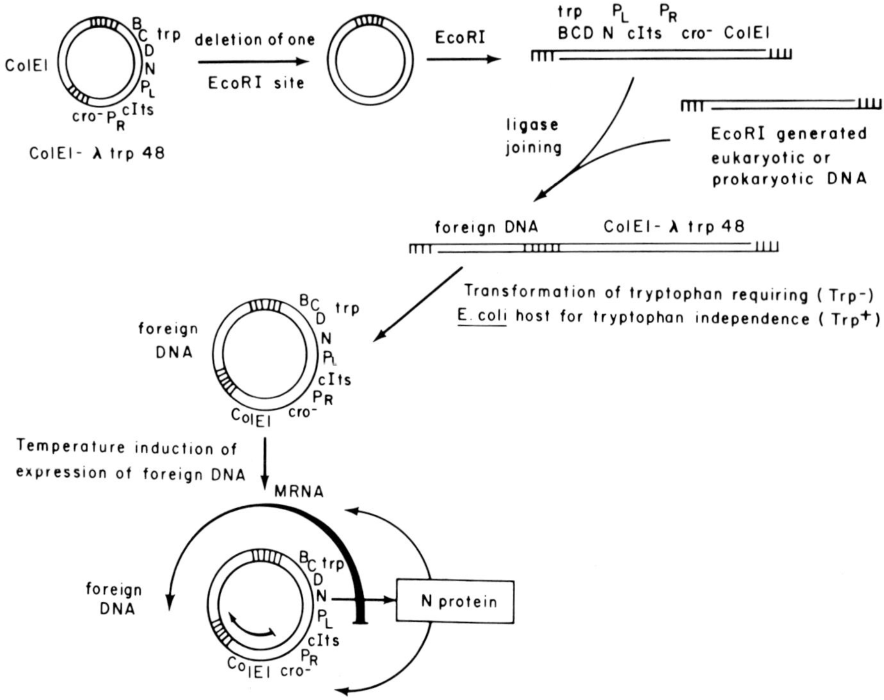

FIG. 3. Proposed use of the Co1E1-λ hybrid for the cloning of foreign DNA. The designations of the various genes on the hybrid plasmid appear in the text. Transformants for the plasmid will grow in the absence of tryptophan when the medium is supplemented with indole.

tion of products of foreign DNA present in the Co1E1-λ hybrids. To date, the constructed Co1E1-λ cloning vehicle exhibits expected properties with regard to temperature inducibility of the tryptophan operon structural genes B, C, and D (V. Hershfield, C. Yanofsky, N. Franklin, and D. Helinski, *unpublished observations*). Experiments presently are in progress to test the Co1E1-λ vehicle for its effectiveness in promoting with fidelity enhanced levels of expression of inserted foreign DNA. A similar approach to promote the expression of inserted DNA also is being carried out with phage λ cloning vehicle (33).

Construction of a Broad Host Range Cloning Vehicle

Co1E1, although effective as a cloning vehicle in *E. coli,* is not maintained stably in Gram-negative bacteria distally related to *E. coli* and, therefore, is of limited use for the cloning of genes in many bacterial species other than *E. coli.* The potential importance of augmenting or improving desirable characteristics of a variety of other Gram-negative bacteria, as, for example, nitrogen fixation by *Rhizobium* or hydrocarbon degradation by *Pseudomonas,* calls for the development of cloning vehicles with a broad host range. Plasmid RK2, a member of the P incompatibility group, has several properties that make it an attractive plasmid for developing as a broad host range cloning vehicle. It is a 3.76×10^7 MW plasmid (31) that is capable of replicating in a wide variety of Gram-negative bacteria (4,27); it will promote its own transfer across diverse Gram-negative genera (4,27); and it is cleaved once by each of the restriction enzymes *Eco*RI, *Hin*dIII, and *Bam*HI (31,32). Presently, efforts are being carried out to bring about several modifications of RK2 to improve its utility and increase its safety as a cloning vehicle. These modifications include: (a) reduction in size of the plasmid; (b) removal of antibiotic resistance genes (RK2 determines resistance to several antibiotics) and insertion of tryptophan operon genes for use as selective genes in transformations or conjugal transfer of the plasmid; and (c) removal or displacement of transfer genes to prevent the promiscuous transfer of the plasmid in the natural environment with retention of the ability to transfer the plasmid from *E. coli* to diverse species under controlled laboratory conditions. The following is a summary of progress toward these goals in our laboratory.

Initial efforts to reduce the size of RK2 were directed toward isolating a segment of the plasmid that contained all of the genes necessary for its replication in different host strains. Genes required for autonomous replication of the plasmids R6-5 and F*lac* have been isolated on a single, small *Eco*RI fragment derived from each of these two plasmids (29,37). A similar approach for RK2 was not possible using *Eco*RI, *Hin*dIII, or *Bam*HI since the plasmid possesses only one site sensitive to each of these restriction endonucleases. To facilitate attempts to construct a self-replicating, low molecular weight derivative of RK2, we obtained restriction fragments containing the poED region of the tryptophan operon of *E. coli* from Co1E1-trp hybrid plasmids and inserted them in the single *Eco*RI, *Hin*dIII, and *Bam*HI sites of RK2 (32). The trpED fragment chosen for insertion not only provides a selective marker other than antibiotic resistance but also provides the plasmid with additional *Bam*HI and *Hin*dIII cleavage sites (32). These restriction sites do not lie within the E gene of the tryptophan operon that is used for selection. The resulting RK2-trp plasmids replicate stably in *E. coli* and retain all of the characteristics of the parental RK2 plasmid with the exception of loss of kanamycin resistance in the case of insertion of the trp genes into the *Hin*dIII site (32). In addition to the sensitivity to the three restriction enzymes (*Eco*RI, *Hin*dIII, and *Bam*HI) known to generate cohesive termini, RK2 possesses two sites sensitive to the restriction enzyme *Sal*I, also shown to yield conhesive termini on cleavage of the DNA. Transformation with *Bam*HI, *Hin*dIII, or *Sal*I digests of the RK2-trp plasmid did not yield a plasmid of reduced size using the trp genes or antibiotic resistance for selection (32). Similarly, digestion of the

RK2 plasmid with an enzymatic activity designated *Eco*RI* (found associated with purified preparations of *Eco*RI [33]) and ligation of the fragments with a purified *Eco*RI fragment of DNA carrying a kanamycin resistance gene did not yield a low molecular weight derivative of RK2 upon transformation of *E. coli* with the mixture (32).

These results suggested the need for a different approach toward the goal of reducing the size of RK2 and deleting one or more of the antibiotic resistance genes and all or a part of the transfer function. It was decided to insert the bacteriophage Mu randomly along the RK2 molecule to provide additional restriction sites at a variety of locations on the RK2 molecule (32; D. Figurski, R. Meyer, and D. R. Helinski, *manuscript in preparation*). This phage lysogenizes *E. coli* DNA randomly (9,36), and Mu DNA has been mapped for several restriction enzymes including *Hin*dIII (1). Various RK2-Mu hybrids were obtained (D. Figurski, R. Meyer, and D. R. Helinski, *manuscript in preparation*) by the procedure of Razzaki and Bukhari (34). The purified covalently closed DNA of these hybrids was restricted with *Hin*dIII and the resulting digests used to transform *E. coli* cells for tetracycline or ampicillin resistance. A lower molecular weight derivative of RK2 was obtained by this procedure from each of these RK2-Mu hybrids (D. Figurski, R. Meyer, and D. R. Helinski, *manuscript in preparation*). Each of these RK2-Mu derivatives possessed a segment of RK2, a portion of Mu DNA, and a single *Hin*dIII site. In addition, all of the reduced plasmids lost kanamycin resistance and in several cases the plasmid was no longer self-transmissible. The region of RK2 successfully deleted by this approach is shown in Fig. 4. One of the RK2 derivatives obtained by this method that is being subjected to additional manipulation to further reduce the size of the DNA is the plasmid designated pRK214.1 with a MW of 28×10^6 and has lost the kanamycin resistance gene and self-transmissibility. On the basis of the results of these various manipulations of the RK2 genome to date, we have determined the locations of various restriction enzyme sites and the regions specifying kanamycin resistance and conjugal transfer (Fig. 4). The results of attempts to delete segments of RK2 also suggest that the genes essential for replication of this plasmid lie within the restriction fragments 2, 4, and 5 (Fig. 4). These various observations indicate that the genes essential for the maintenance of RK2 are not as tightly clustered as that found for the plasmids F*lac* and R6-5 (29,37). It remains to be determined whether or not the presence of a larger number and broader distribution of genes responsible for replication will be characteristic of all broad host range plasmids.

An attractive property of the RK2 plasmid is its ability to both enter and replicate in a wide variety of Gram-negative bacteria. This property, however, also confers the potential hazard of possible promiscuous spread of the genetic element containing cloned foreign DNA. One possible approach that may significantly reduce this potential problem is the construction of a

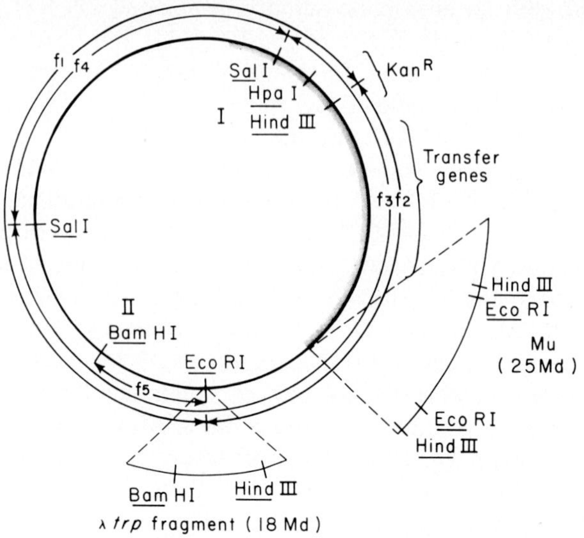

FIG. 4. The location of the cleavage sites on RK2 for various restriction enzymes. The sites for *Pst*I, *Bgl*II, and *Sma*I, which have not been precisely mapped, are located in regions indicated by I and II on the figure. Additional cleavage sites were introduced into RK2 by *in vitro* ligation of a λ *trp* fragment into the *Eco*RI site of RK2, and by lysogeny of RK2 with the phage Mu. The λ *trp* fragment shown here, containing the entire tryptophan operon, has been described previously (31). Since Mu lysogenizes randomly, different points of insertion were observed, and only one is shown. The cleavage sites shown on the λ *trp* piece and on Mu (1) were used in attempts to reduce the size of RK2. The region of RK2 successfully deleted by digestion of the various RK2-Mu hybrids with *Hin*dIII is indicated by the shaded area. The sizes (Md) of the RK2 restriction fragments are: f1: 24.5; f2: 13.2; f3: 25.2; f4: 12.8; f5: 5.1. (From ref. 32.)

binary vehicle system (32). One of the plasmid members (e.g., $Rep_{RK2}Tra^-$; RK2 replication functions but transfer is defective) would be capable of carrying foreign DNA and replicating in many different genera of bacteria but would be incapable of self-transmissibility. The other plasmid vehicle would have a more restricted host range (e.g., $Rep_{E1}Tra_{RK2}$; replication functions of ColE1 and transfer functions of RK2) but would be capable of mediating the transfer of the cloning vehicle $Rep_{RK2}Tra^-$. Thus, recombinant plasmids could be constructed in the well-characterized *E. coli* strain and then transferred to the target organisms by conjugal mating with the $Rep_{E1}Tra_{RK2}$ plasmid promoting the transfer. Since the $Rep_{E1}Tra_{RK2}$ member of this binary vehicle system is incapable of replicating in the target organism, the probability of subsequent transfer of the $Rep_{RK2}Tra^-$ cloning vehicle is greatly reduced. Some progress has been made toward the development of this binary vehicle system (R. Meyer, D. Figurski, and D. Helinski, *unpublished observations*). A region of RK2 (fragment 2 plus a short segment of DNA extending from *Hin*dIII to the neighboring *Sal*I site; see Fig. 4) known to contain genes responsible for transfer has been inserted into a ColE1-kan derivative. The resulting $Rep_{E1}Tra_{RK2}$ hybrid plasmid did not show self-transfer but mediated the transfer of two Rep_{RK2}-Tra^- plasmids generated by the reduction of the RK2-Mu hybrids with restriction enzymes. Further modifications are being carried out on this promising binary vehicle system.

CONCLUDING REMARKS

The multicopy ColE1 plasmid cloning system provides a rich source of a plasmid cloning vehicle and hybrid plasmid DNA and the means for the amplified production of the products of cloned foreign DNA. ColE1 and mini-ColE1 plasmids carrying a variety of auxotrophic and antibiotic resistance genes for selection in transformation have been derived both by spontaneous genetic events and by biochemical manipulations and offer advantages with regard to efficacy and potential safety as cloning vehicles. The development of plasmid cloning vehicles is clearly only in its early stage. Further genetic and biochemical manipulations will lead to even more effective and potentially safer vehicles for gene cloning in *E. coli.* In addition, the development of broad host range plasmid systems in the Gram-negative bacteria and plasmid cloning systems in a variety of species of Gram-positive bacteria offers the exciting prospect of applying the powerful technology of gene cloning to a variety of organisms of importance to human and veterinary medicine, agriculture, and the fermentation industry. To exploit fully these new developments for the benefit of society, however, similar advances must be made in our understanding of the judicious application of this rapidly expanding technology. Unfortunately, advances in this area may prove to be much more difficult to achieve than the technological improvements in gene cloning.

ACKNOWLEDGMENTS

We are indebted to Dr. Charles Yanofsky for his many contributions to the construction of the hybrid plasmids containing genes of the tryptophan operon. We are grateful to Dr. Naomi Franklin for her advice and assistance in the construction of the ColE1-λ hybrids. This research was supported by Public Health Service Research Grant AI-07194 from the National Institute of Allergy and Infectious Diseases and National Science Foundation Research Grant GB-29492. V. Hershfield and D. Figurski were supported by Public Health Service Postdoctoral Fellowships GB54149 and A101412, respectively. R. Meyer was supported by a Research Corporation Fellowship.

REFERENCES

1. Allet, B., and Bukhari, A. I. (1975): Analysis of bacteriophage Mu and λ-Mu hybrid DNAs by specific endonucleases. *J. Mol. Biol.,* 92:529–540.

2. Bazaral, M., and Helinski, D. R. (1968): Circular DNA forms of colicinogenic factors E1, E2 and E3 from E. coli. *J. Mol. Biol.,* 36:185–194.
3. Bazaral, M., and Helinski, D. R. (1970): Replication of a bacterial plasmid and an episome in Escherichia coli. *Biochemistry,* 9:399–406.
4. Beringer, J. E. (1974): R factor transfer in Rhizobium leguminosarum. *J. Gen. Microbiol.,* 84:188–198.
5. Bertrand, K., Korn, L., Lee, F., Platt, T., Squires, C. L., Squires, C., and Yanofsky, C. (1975): New features of the regulation of the tryptophan operon. *Science,* 189:22–26.
6. Betlach, M., Hershfield, V., Chow, L., Brown, W., Goodman, H. M., and Boyer, H. W. (1976): A restriction endonuclease analysis of the bacterial plasmid controlling the EcoRI restriction and modification of DNA. *Fed. Proc.,* 35:2037–2043.
7. Blair, D. G., Clewell, D. B., Sherratt, D. J., and Helinski, D. R. (1971): Strand-specific supercoiled DNA-protein relaxation complexes: comparison of the complexes of bacterial plasmids ColE_1 and ColE_2. *Proc. Natl. Acad. Sci. U.S.A.,* 68:210–214.
8. Blair, D. G., Sherratt, D. J., Clewell, D. B., and Helinski, D. R. (1972): Isolation of supercoiled colicinogenic factor E_1 DNA sensitive to ribonuclease and alkali. *Proc. Natl. Acad. Sci. U.S.A.,* 69:2518–2522.
9. Bukhari, A. I., and Zipser, O. (1973): Random insertion of Mu-1 DNA within a single gene. *Nature,* 236:240–243.
10. Clewell, D. B. (1972): Nature of ColE1 plasmid replication in Escherichia coli in the presence of chloramphenicol. *J. Bacteriol.,* 110:667–676.
11. Clewell, D. B., and Helinski, D. R. (1969): Supercoiled DNA protein complex in Escherichia coli: purification and induced conversion to an open circular DNA form. *Proc. Natl. Acad. Sci. U.S.A.,* 62:1159–1166.
12. Clewell, D. B., and Helinski, D. R. (1970): Properties of a supercoiled deoxyribonucleic acid-protein relaxation complex and strand specificity of the relaxation event. *Biochemistry,* 9:4428–4440.
13. Clewell, D. B., and Helinski, D. R. (1972): The effect of growth conditions on the formation of the relaxation complex of supercoiled ColE1 deoxyribonucleic acid and protein in Escherichia coli. *J. Bacteriol.,* 110:1135–1146.
14. Cohen, S. N., Chang, A. C. Y., Boyer, H. W., and Helling R. (1973): Construction of biologically functional bacterial plasmids in vitro. *Proc. Natl. Acad. Sci. U.S.A.,* 70:3240–3244.
15. Covey, C., Richardson, D., and Carbon, J. (1976): A method for the deletion of restriction sites in bacterial plasmid deoxyribonucleic acid. *Mol. Gen. Genet.,* 145:155–158.
16. Davidson, J., Brammar, W. J., and Brunel, F. (1974): Quantitative aspects of gene expression in a λ-trp fusion operon. *Mol. Gen. Genet.,* 130:9–20.
17. Franklin, N. C. (1974): Altered reading of genetic signals fused to the N operon of bacteriophage λ: genetic evidence for modification of polymerase by the protein product of the N gene. *J. Mol. Biol.,* 89:33–48.
18. Guerry, P., Le Blanc, D. J., and Falkow, S. (1973): General method for the isolation of plasmid deoxyribonucleic acid. *J. Bacteriol.,* 116:1064–1066.
19. Guiney, D. G., and Helinski, D. R. (1975): Association of protein with the 5′ terminus of the broken DNA strand in the relaxed complex of plasmid. *J. Biol. Chem.,* 250:8796–8803.
20. Hamer, D. H., and Thomas, C. A., Jr. (1976): Molecular cloning of DNA fragments produced by restriction endonucleases SalI and BamI. *Proc. Natl. Acad. Sci. U.S.A.,* 73:1537–1541.
21. Hershfield, V., Boyer, H. W., Chow, L., and Helinski, D. R. (1976): Characterization of a mini-ColE1 plasmid. *J. Bacteriol.,* 126:447–453.
22. Hershfield, V., Boyer, H. W., Yanofsky, C., Lovett, M. A., and Helinski, D. R. (1974): Plasmid ColE1 as a molecular vehicle for cloning and amplification of DNA. *Proc. Natl. Acad. Sci. U.S.A.,* 71:3455–3459.
23. Herskowitz, I. (1973): Control of gene expression in bacteriophage lambda. *Annu. Rev. Genet.,* 7:289–324.
24. Katz, L., and Helinski, D. R. (1974): Effect of inhibitors of ribonucleic acid and protein synthesis on the cyclic adenosine monophosphate stimulation of plasmid ColE1 replication. *J. Bacteriol.,* 119:450–460.
25. Kedes, L. H., Chang, A. C. Y., Houseman, D., and Cohen, S. N. (1975): Isolation of

histone genes from unfractionated sea urchin DNA by subculture cloning in E. coli. *Nature,* 255:533–538.
26. Kupersztoch-Portnoy, Y. M., Miklos, G. L. G., and Helinski, D. R. (1974): Properties of the relaxation complexes of supercoiled deoxyribonucleic acid and protein of the R plasmids R64, R28K, and R6K. *J. Bacteriol.,* 120:545–548.
27. Lai, M., Shaffer, S., and Panopoulos, N. J. (1976): Transfer of R plasmids among Xanthomonas spp. and other plant pathogenic bacteria. *J. Bacteriol.* (*in press*).
28. Lovett, M. A., Guiney, D. G., and Helinski, D. R. (1974): Relaxation complexes of plasmids Co1E1 and Co1E2: unique site of the nick in the open circular DNA of the relaxed complexes. *Proc. Natl. Acad. Sci. U.S.A.,* 71:3854–3857.
29. Lovett, M. A., and Helinski, D. R. (1976): A method for the isolation of the replication region of a bacterial replicon: construction of a mini-F′km plasmid. *J. Bacteriol.,* 127: 982–987.
30. Lovett, M. A., Katz, L., and Helinski, D. R. (1974): Unidirectional replication of plasmid Co1E1 DNA and relation of the origin-terminus to the nicked site in the Co1E1 relaxation complex. *Nature,* 251:337–340.
31. Meyer, R., Figurski, D., and Helinski, D. R. (1975): Molecular vehicle properties of the broad host range plasmid RK2. *Science,* 190:1226–1228.
32. Meyer, R., Figurski, D., and Helinski, D. R. (1976): Properties of the plasmid RK2 as a cloning vehicle. In: *DNA Insertion Elements, Plasmids and Episomes,* edited by A. I. Bukhari, J. Shapiro, and S. Adhya, Cold Spring Harbor Laboratory, Cold Spring Harbor, N.Y. (*in press*).
33. Murray, N. (1977): Expression of bacterial genes in phage lambda vectors (*this volume*).
34. Razzaki, T., and Bukhari, A. I. (1975): Events following prophage Mu induction. *J. Bacteriol.,* 122:437–442.
35. So, M., Gill, R., and Falkow, S. (1976): The generation of a Co1E1-Ap^r cloning vehicle which allows detection of inserted DNA. *Mol. Gen. Genet.,* 142:239–249.
36. Taylor, A. L. (1963): Bacteriophage-induced mutation in Escherichia coli. *Proc. Natl. Acad. Sci. U.S.A.,* 50:1043–1051.
37. Timmis, K., Cabello, F., and Cohen, S. N. (1975): Cloning, isolation, and characterization of replication regions of complex plasmid genomes. *Proc. Natl. Acad. Sci. U.S.A.,* 72: 2242–2246.

Recombinant Molecules: Impact on Science and Society, edited by R. F. Beers, Jr. and E. G. Bassett. Raven Press, New York © 1977.

16. Discussion

Moderator: Stanley Falkow

Dr. S. Falkow: We have heard that promising host-vector systems have been genetically constructed which exhibit a (theoretically) high level of biological containment. Certainly, one derivative of *E. coli* K-12 as well as several λ derivatives can meet the current NIH guideline criteria for an EK2 host-vector system.

But we do not appear to be so close in the development of an EK2 plasmid vector despite the efforts by Dr. Helinski's group. Dr. Helinski, would you care to comment on how close you think you are getting to something that would qualify as an EK2 plasmid?

Dr. D. R. Helinski: I guess our approach so far is to rely heavily on the Curtiss strain χ1776. We have isolated temperature-sensitive replication mutants of ColE1, which I think certainly helps but does not solve the whole problem.

Presently our approach is to build the suppressible mutations to better contain the plasmid within the disabled host strain that Curtiss has described, and other strains I am sure he is going to develop.

Dr. D. A. Gelfand: Two questions on the mini-ColE1 insertions. If you inserted a piece of DNA into the mini-ColE1 plasmid, did you observe increased transmissibility?

Dr. D. R. Helinski: We did not. The low level of transmissibility is retained even upon insertion of DNA.

Dr. D. A. Gelfand: Second question: On the ColE1-*trp* plasmid, is attenuator site activity preserved or is it titrated out?

Dr. D. R. Helinski: This is not entirely clear. The best guess is that attenuator site activity is there.

Mr. P. Youderian: I would like to address two comments to Dr. Szybalski. First, some facts argue against the safety of *any* lambda vector. The work of Ira Herschkovitz and colleagues has shown that many lambda genes are already present in *E. coli* and that lambda can recombine with those genes using the *rec*A-mediated host recombination systems. I remember a quote of Dr. Herschkovitz's to the effect that every gene is "somewhere" in lambda.

Second, could you address the question of the possible unintentional misuse of lambda vectors. For example, Dr. Leder's lambda derivative, which has been certified EK2, can be forced into lysogeny. Even lambda carrying a virulent mutation can be maintained in a strain carrying a λdv plasmid for a short period of time. One could "shotgun" DNAs from a higher organism and look for phenotypic properties under the impression that a safe lambda vector was in use. One would really be employing a lambda vector as a lysogen. In short, one could carry on this kind of "safe research" with "safe" vectors where no safety considerations were, in fact, being undertaken.

Dr. W. Szybalski: The first comment, I believe, you are addressing to the recombination event between the laboratory *E. coli* host and the lambda grown on it. This is a problem of probability. How high is the probability that some new genes could be acquired from the host to replace the defective genes in lambda? For the *Q* genes the probability was determined to be something on the order of 10^{-10}, which is well below the 10^{-8} level set for EK2. Obviously, the laboratory strain that is used and will carry lambda should be mutated, deleted, and selected so it would not substitute for any genes that were deleted for the purpose of safety. This is easy to assess in the laboratory and certainly should be done when changing from one host to another. I agree with you that this should be done.

In the second question you ask whether the certified strains of Leder are safe. This again is a question of probability. Remember that 10^{-8} was the level proposed by the NIH Advisory Committee, and the safety of this strain at 10^{-8} must be examined. Moreover, the committee thought that there was a safety factor at less than 10^{-8}.

Now, if you want to have something absolutely safe, it must be $10^{-\infty}$. I think such a thing does not exist in nature and the problem remains: where is the cutoff line for safety? That is something which, I guess, insurance companies can best answer. Is it safe if the dangerous event happens in a laboratory only once per hundred million years? It would be reasonably safe, safer than collision of the earth with some other flying object.

Dr. S. Lederberg: The question just asked is really impossible to answer by cloning even if you try very hard. The reason it is impossible is as follows. We know of the unlimited number of sites and of enzymes for these sites. Presumably what lambda has, at least, is a large number of recognition sites, which *E. coli* also has. So with respect to tetranucleotide homologies, pentanucleotide homologies, or hexanucleotide homologies, base pairs, etc., surely it is an impossible task to find a vector not matched by some types of sites and some type of host cell.

Dr. W. Szybalski: We asked the question: what happens if a hundred or a thousand of some host-vector combinations escape? If because of a spill a hundred or a thousand or even a million such combinations escape and the probability is 10^{-8}, it is still a low probability. The probability is what is important in this case to define the safety.

Dr. P. H. Kourilsky: I am afraid the people who heard this session have been left with the notion that it is easy to make *E. coli* a small factory specializing in the production of one protein.

I would like to argue against this attitude by saying that in another model system—that using the *lac* operon—it is clear that there are limiting factors in the production of beta-galactosidase. Even though you can pile up a thousand copies of the gene in a cell, very little enzyme is synthesized. You may try to put the enzyme under the control of an efficient PL promoter of lambda, but you will not obtain efficient enzyme synthesis because of the 5′ and of the messenger, as has also been shown by Beckwith's people. The 5′ and the messenger are quite important in the translation. Even though you transcribe the gene efficiently, you do not make active protein.

My question: do you people have information on the possible limiting factors that apparently do not exist in the case of tryptophan operon but might well exist in many other systems?

Dr. N. Murray: How many examples do you know of in which it has been difficult to obtain amplification from a phage promoter?

Dr. P. H. Kourilsky: It is difficult with the *lac* operon. It is apparently not that obvious with the first enzyme of the *gal* operon. My question is: is the tryptophan operon by any chance a particular case? It is quite possible.

Dr. T. Hirschfeld: The *trp* genes represented a test Co1E1 cloning system, and in some cases they have turned out to be the best system for maximal expression of protein.

In most other cases, at least those that have been brought to my attention, it seems there may be not as much as a 15-fold increase over the normal level, normal level being defined as one copy of the gene per chromosome versus 15 copies of the gene per chromosome when it is part of a Co1E1 plasmid. Whereas one may not find a 15-fold increase as was the case of *trp* products, probably one does obtain a several-fold increase with other cloned genes. What are the controls and limitations? Probably there are as many answers to these questions as there are number of genes that one can clone with Co1E1.

In other situations Co1E1 is not a particularly good plasmid cloning vehicle to use. When one inserts a gene onto the Co1E1 plasmid and enough product trickles through which is particularly bad for the cell, it is a lethal situation and one probably is not going to be able to use Co1E1.

Now, in the long run I think using promoter regions, as Dr. Murray described for her lambda vector and as I described for Co1E1, might overcome some of these problems. If one starts with the gene, perhaps somewhat defective in its transcription, under control of a temperature-sensitive lambda, then one can grow populations of cells up to a particular concentration. Increase the temperature and there is a burst of products. You do not have to worry about the viability problems with respect to the expressed product.

Dr. D. A. Gelfand: I would like to comment briefly on Dr. Kourilsky's remarks. When fully induced, part of a Co1E1 derivative plasmid, the beta-galactosidase gene, synthesizes up to 10 times the normal quantity of enzyme. When the plasmid has an additional mutation that confers overproduction, up to 60% of the total cellular protein can be beta-galactosidase. In that respect, it is analogous to Dr. Helinski's *trp* plasmid.

Dr. H. V. Aposhian: Fail-safe recognition, sooner or later, means one has to see any recombinant DNA attempting to depart. Has there been any effort to specify in recombinant DNA a function that could lead to easy recognition, for example, a fluorescent transfer RNA? One could then easily test anything going in and out for the presence of the culprit.

Dr. S. Falkow: Does anyone want to discuss that? The idea is to put a marker on the phage or the plasmid that will be so readily identifiable that it will be clear-cut for monitoring. Does that paraphrase it correctly?

Dr. H. V. Aposhian: Yes.

Dr. S. Falkow: Roy Curtiss has done it for *E. coli* in the sense that one can use nalidixic acid resistance or an arabinose/mannitol-negative phenotype. This makes the *E. coli* unlike any other coliform that you would find under natural circumstances, but what about "marking" the phage or the plasmid?

Dr. W. Szybalski: As far as the phage is concerned, I think Professor Blattner will discuss that in his chapter.

Dr. D. R. Helinski: Marking the phage is a difficult problem. The most suitable identifying genes to put on are antibiotic resistance genes, but, as repeatedly mentioned in this volume, the great frequency at which we find antibiotic-resistant intestinal organisms makes it difficult to develop a particular combination of resistance genes that would say, yes, indeed, that cell carries the plasmid that we are interested in. It is not impossible but much more difficult to do that.

Perhaps Dr. Falkow has something more to add to that.

Dr. S. Falkow: I would like to make one short comment. I think Dr. Richmond will discuss it more precisely. The reason that antibiotic resistance genes are so common in enteric bacteria is obvious: antibiotics are so commonly used. Sometimes they are used unwisely.

It is interesting to see the response of scientists to the idea of using antibiotic resistance genes on cloning vectors. Many are concerned about the inherent biohazards of using antibiotic resistance genes on cloning vehicles. Although I think this concern is proper, I have been disappointed that the same scientists have not before talked about the misuse of antibiotics in agriculture and in medicine. I hope that this forum will serve to identify hazards that we think are important not only in the kind of research we are doing but elsewhere as well.

Dr. H. V. Aposhian: Is there any information on the half-life or what happens to a naked plasmid, for example, *E. coli,* once it is lysed in an animal—not in an isolated mammalian cell but in the whole animal?

Dr. S. Falkow: I think Dr. Curtiss has the answer, or at least he may be one of the few, probably the only person, who did the experiment.

Dr. R. Curtiss, III: We have not checked the raw conditions. As I said previously (*this volume*), in the absence of any physical data about enzymes such as nucleases in the gut and based on a question I was asked at the meeting in February, we took the upper and lower intestinal contents and added radioactive DNA, including plasmid DNA, to see just how rapidly DNA is degraded. It is amazing, it just goes whiff, like that! In 5 min most of the DNA is solubilized. So the intestinal tract—at least that of rats—contains a large quantity of nucleases capable of degrading DNA from lysed bacterial cells.

Dr. H. V. Aposhian: But that is in the GI tract. The gut is loaded with bacteria, obviously. What would be the half-life in the body if the DNA were put in the tail of a rat?

The question is: what happens to the plasmid once it enters the cells of a whole animal, not just the intestinal tract? How long will naked plasmid DNA last? What happens to it once it enters the systemic circulation? Is there any information along those lines?

Dr. Curtiss: I do not think anybody has ever injected plasmid DNA intravenously or subcutaneously to see what its fate is. Some people have taken eukaryotic cells in culture and obtained uptake of, say, specialized transducing phage DNA or plasmid DNA. The results, in terms of the expression of that DNA, have not generally been successful.

Dr. S. Falkow: In vivo transformation has been described.

Dr. H. V. Aposhian: So in a sense you are saying the information is not available. People are not working with plasmid DNA once it enters the human or animal body outside the gut. I think that as molecular biologists, which I think we all are, we are not looking at it as a potentially toxic agent in the manner a toxicologist would. We just do not have that information available. It is fine to know what happens to cloned DNA of plasmid in the microbe; it is fine to set up a lambda system that may be safe. But the question still arises: what is the toxicology in the whole animal? I am surprised that this information does not seem to be available.

Dr. S. Falkow: Do you want to speak relative to that?

Dr. M. P. Gordon: I think some studies were done with polyoma and SV40 DNA showing they are infectious in whole animals. And I do not know if any studies on the half-life upon injecting the material were done, but you can certainly obtain tumors with free DNA. So I think that is worth considering here.

Dr. F. E. Young: I can speak to that point. That study was done by Aaron Bendich in the late 1950s or early 1960s and published, I believe, in *Science.* He studied the survival of a variety of DNAs, one of which, as you pointed out, was polyoma. He also examined pneumococcal transforming DNA in a variety of hosts. The original experiments of Griffith in 1928, published in the *Journal of Hygiene,* showed that it is dependent on the species being injected with the DNA. Fortunately, Griffith chose the mouse. Avery, in his hallmark paper of 1933, demonstrated that the rabbit, unlike the mouse, had larger amounts of circulating nuclease. If Griffith had selected the rabbit instead of the mouse, he might not have made that observation which he faithfully recorded but never returned to; and he was killed, as you know, when a V-2 bomb hit the laboratory in London.

Dr. J. King: I was quite struck, as I am sure many people were, by Dr. Falkow's presentation of the evidence that *E. coli* certainly is a pathogen in certain forms and that its pathogenicity, its ability to colonize, and its host range depend critically on the nature of the plasmids it contains. I do not remember hearing any evidence countering those observations.

On the other hand, we have had descriptions of experiments designed to determine in a specific case what the survival of this bacterial strain, *E. coli* K-12, is. It seems to me that what is needed is a monitoring of their working environment—we need to know if these cells are getting out and not rely on calculations from Vanderbilt as to what their escape probability is.

It seems to me that requires two things. One, it requires continuous monitoring of the laboratory and the work place; two, it requires medical surveillance of the personnel directly involved in the laboratory operations.

I wonder if any of the people who have spoken so far have such monitoring set-ups. If they do, would they describe to us how they monitor cell escape and how they monitor their personnel to make sure that they are not picking up laboratory-acquired infections?

Dr. S. Falkow: I think Dr. Richmond will probably discuss some of this; but we have for years, as have many people, used a nalidixic acid–resistant (nal^+) *E. coli.* It is relatively easy to plate undiluted feces on agar containing nalidixic acid and knock off the normal flora to see if nal^+ cells are being excreted. Over the years we have experienced no case in which any of the laboratory workers were harboring either a nalidixic acid–resistant *E. coli* K-12 or one carrying an R-plasmid that they were working with. However, we do not monitor daily but from time to time.

Some of the people we monitored certainly had R-factor–carrying *E. coli* in their guts, as I am sure most of you do, but none were the R-factors we were working with. In only one instance, which was myself, a blood sample was taken to look for antitoxic activity which might indicate that I had been exposed to the toxigenic *E. coli* we work with. It will be more significant when I bleed myself in future times to see if toxic neutralizing antibody levels increase. I certainly agree that monitoring must be part and parcel of recombinant DNA work. We have to build a monitoring system into this so it is as unequivocal as possible.

Dr. D. Hamer: First, to comment in response to Dr. King's question. We have also been cloning various DNA fragments with a number of different plasmid vehicles in our lab and have been routinely screening at least some of the people in the lab for the host strains of these plasmids. Although we have been monitoring only occasionally, usually once a month, we have found no examples yet either. Obviously, it is only a limited population.

The second comment I would like to make is that we have constructed a recombinant between Co1E1 and the suppressor tRNA gene SU-3; we obtained a deleted variant of this plasmid that has just one R-1 site. This can be used also as a cloning vehicle in the same sort of way that the Co1E1-*trp* plasmids which Dr. Helinski mentioned can be used.

Finally, cells harboring that particular plasmid produce the same amount of suppressor tRNA as strains with the suppressor mutation on the chromosome. So that is another example of having a gene present in approximately 20 copies per cell but having only the normal amount of product made.

Dr. A. B. Haberman: Dr. Szybalski, many of the genes in the safer lambda stages are amber genes. What is known about the distribution of amber suppressors in nature?

Dr. W. Szybalski: There are published studies on the presence of amber suppressors in nature, but I do not remember the exact figures at the present moment—the frequency was extremely low. Does anybody remember the figures?

Dr. R. Curtiss, III: I think Sidney Brenner, some years ago, screened *E. coli* strains for sensitivity to T4. Of many strains, he may have found 20 or so that were sensitive to T4. None of these was able to plate T4 amber or T4 ochre mutants. We really have no good data on it; they need to be obtained.

Dr. B. Lewin: Since it is important to know the details of vehicles we are using, is there any additional information on the origin of pSC101 than there was when it was isolated? The original report is not very detailed. What is known about the origin of pSC101 and what is its relation to other plasmids?

Dr. S. N. Cohen: The plasmid was isolated in experiments following shearing of the R6–5 plasmid. When it was isolated, we believed that it had formed by recircularization of a piece of that plasmid. Subsequent evidence obtained in my laboratory has led us to question our original view that formation of pSC101 simply involved recircularization of a sheared fragment of R6–5. There are a number of possibilities, some of which involve repeats of the insertion sequence element near the tetracycline gene of R6–5. It has been shown that the inverted repeats of this element and the tetracycline gene between them are subject to translocation and recombination with other DNA segments.

There are a number of small plasmids that have been derived from larger plasmids under various circumstances. In some instances, such "mini" plasmids appear to be formed by excision of certain plasmid DNA segments from the big plasmid.

pSC101 shows some homology with R6–5 by membrane hybridization. My best estimate of what is going on is that some of the sequences of pSC101 are similar to sequences in R6–5, but the entire plasmid does not appear to be homologous with R6–5. We are presently looking at this question in considerable detail, and we will have to wait and see what the data indicate.

Dr. S. Falkow: I will add a brief comment that plasmids which are not identical, but certainly very similar, to pSC101 are quite commonly found in nature. There are many small plasmids mediating tetracycline resistance.

Dr. B. Lewin: The point is that a number of these plasmids seem to be formed by popping out from larger plasmids. The question is whether the initial plasmid that was isolated following shearing of the R6–5 plasmid represented not a recircularization of the sheared molecule but perhaps a popping out of some transposable recombinational event.

Dr. S. D. Ehrlich: Dr. Cohen, you have presented interesting data on loss of plasmids from cells. What really struck me is the experiment in which you demonstrated that you started with one plasmid containing two *Eco*RI sites. From other later generations, you had almost entirely plasmid with only one *Eco*RI site.

If I understood correctly, segment loss was identical to the *Eco*RI segment that was inserted in the first place. I wonder if you have any evidence that you are not dealing with situations of mixed plasmids, one being lost faster than the other.

Dr. S. N. Cohen: That is, of course, the question that one must worry about. When one transforms initially in order to introduce plasmid into the cell, it is done under conditions for looking at these kinds of experiments with incorporation of small amounts of DNA in order to ensure that there is one plasmid per cell. Another point relevant to this is that the initial culture is cloned many times. If there were, in fact, two plasmids which had identical replication regions in the same cell or in the same clone, one would have to deal with questions of incompatibility and also differences in increased prominence of the particular band that would be represented by the component that is mixed.

Dr. S. N. Cohen: In some instances, the piece of DNA lost was larger than the fragment that had been inserted and an *Eco*RI cleavage site was removed during the deletion. However, the point you raise is a relevant consideration in some experiments in which it looks like the fragment deleted was precisely the *Eco*RI fragment that had been inserted in the first place. Prior to the growth of such cultures, the bacterial colonies were cloned in the presence of tetracycline. Any mixed plasmid situation would have to involve both plasmids being present in a single tetracycline-resistant clone. Since the pSC101 plasmid replication functions would be present on both molecules, I would expect incompatibility to prevent continuing coexistence of a mixed plasmid state. Whereas two plasmids might well have been taken up by the same cell initially at the time of transformation, they would be expected to quickly segregate. When cultures are started from a single clone it seems reasonable to expect that any heterogeneity observed at the end of the experiment occurred during the growing of the culture.

I do not know if that really answers your question. As you know, we and others have been observing this apparent recombination among cloned fragments for some time. Several years ago, Annie Chang found that some of the *Xenopus laevis*–pSC101 chimeras appeared to show intramolecular recombination. In the paper that we published with Bob Lansman and Dave Clayton on the mouse mitochondrial DNA–*E. coli* chimeras, similar observations are mentioned. Among the histone fragments that we have cloned, the pSP-2 and pSP-17 appear to be stable; however, there seems to be some instability in some other histone fragments.

What is going on here with respect to the molecular instability of certain chimeric plasmids I really cannot answer at this point. However, in some instances loss of the inserted fragment appears to allow the bacterial cell to grow better.

Dr. S. D. Ehrlich: If I may just continue my comment, it seems I observed essentially the same phenomenon you have described. Maybe one is dealing with basically two phenomena. One is rearrangement, which might arise as a product of an illegitimate recombination. In the other phenomenon you lose a specific segment, which really seems bizarre. I wonder if the data you have could be consistent with the hypothesis of having a mixture of plasmids, regardless of the fact that you take all the precautions to introduce just one.

Dr. S. N. Cohen: I really do not know how to answer that, Dr. Ehrlich. What we have at the end is a mixture of plasmids that appear to have been derived from a single clone. This is consistent with the view that some kind of recombination and/or deletion is occurring, but at this point, we do not know the precise nature of the process.

Dr. L. Villa-Komaroff: For the cloning experiments that many of us plan to do it would be useful to know if the pathogenic bit of information put into a new environment can in fact cause some kind of disease in an animal. I believe that Dr. Falkow addressed himself to that question, although I am not sure I heard all of the information.

My question is: does the chimera made with pSC101, the LT, or the ST cause any kind of disease in any animal that it might be put into?

Dr. S. Falkow: Well, in the first place I would not take one of the cloned DNAs and do the experiments. With the wild-type Ent plasmids which we have worked with in the contagious disease lab at Colorado State, we have never been able to duplicate disease in animals using *E. coli* K-12 cells. Using a plasmid-less typical *E. coli* strain from a calf, you can certainly put the Ent and K plasmids into that strain and duplicate the disease.

On the other hand, H. Williams Smith has shown that if the Ent plasmid is put into *Salmonella,* it has absolutely no effect on the pathogenesis of infection. The same is true with regard to *Shigella.* We have stayed strictly with *E. coli* for our cloning experiments, but it is certainly possible to transfer the wild-type Ent and K plasmids all around the enteric group.

I am struck, however, by the fact that one does not often find other microbial species producing enterotoxins typified by those of *E. coli.* I do not know what that means. It is an epidemiological observation, not necessarily a scientific one. Epidemiology is usually retrospective.

Recombinant Molecules: Impact on Science and Society, edited by R. F. Beers, Jr. and E. G. Bassett. Raven Press, New York © 1977.

17. Introduction to Section C: Practical and Potential Developments in Plant Genetics

E. W. Nester

This section entitled "Practical and Potential Developments in Plant Genetics" is concerned with the potential of genetics in improving the quality and quantity of food plants. It is also concerned with a system which naturally introduces and stably maintains bacterial genes in plants. In some respects, at the molecular level plants are much more difficult to work with than animal systems; however, they do have the important advantage that, in many cases, it is possible to regenerate whole plants starting from single cells. Techniques are now available for isolating and propagating haploid plant cells. In addition, techniques are now available for removing the tough outer wall from a cell to form a protoplast. These protoplasts, devoid of the normal cell wall, can be treated experimentally much as cultured mammalian cells. Several of our contributors have pioneered in the development and use of these technologies.

There is great anticipation that the use of modern techniques and thinking, in some part derived from microbial genetics, will lead to plants with increased food value. Such approaches have taken a number of forms. One investigator, J. Widholm, has isolated carrot and tobacco strains which are resistant to analogues of tryptophan and phenylalanine, as well as to those of methionine and lysine. Such mutants contain up to 20 times as much of the appropriate amino acid as does the wild-type plant. Since several of these amino acids are required by man and tend to be deficient in diets, the development of such plants conceivably could be a useful way of generating more nutritious plants. I am sure you will hear of other approaches from contributors to this section.

The isolation of mutants represents the first step in developing the system of genetic recombination. Considerable effort in a number of laboratories has gone into developing techniques for isolating desired mutants. In 1970 Peter Carlson reported the isolation of several leaky auxotrophic mutants by treating single haploid cells with ethyl methane sulfonate. A number of laboratories have reported the transformation of plants with prokaryotic DNA. A review of these data suggests that prokaryotic DNA can be taken up and expressed in plant cells, but that the effects are short-lived.

A serious question I believe exists of the validity of reports that bacterial

DNA can be integrated into the plant genome. However, there now may be an exception to this latter statement. Crown gall disease is caused by a bacterium that can transfer specific information to the plant—information which can be expressed for several decades at least. If the material transferred is DNA, this DNA apparently can function inside the plant cell for extended periods of time. This has ramifications not only for the mechanism by which a bacterium can alter the normal metabolic processes of eukaryotic cells, but it has potential application also for the introduction of other genes by using the elegant techniques that have already been discussed in this volume.

Therefore, I would like to begin this section by considering the crown gall system in some detail and presenting data which strongly suggest that bacterial DNA is transferred and stably maintained in plant tumors.

Recombinant Molecules: Impact on Science and Society, edited by R. F. Beers, Jr. and E. G. Bassett. Raven Press, New York © 1977.

18. Search for Bacterial DNA in Crown Gall Tumors

E. W. Nester, M.-D. Chilton, M. Drummond, D. Merlo, A. Montoya, D. Sciaky, and *M. P. Gordon

*Departments of Microbiology and *Biochemistry, University of Washington, Seattle, Washington 98195*

INTRODUCTION

Crown gall is a non-self-limiting neoplastic disease of dicotyledonous plants caused by *Agrobacterium tumefaciens* (26). When viable, actively metabolizing bacteria are inoculated into a fresh wound site of the plant, rapid growth ensues and the tissue continues to increase in size in an uncontrolled fashion. The tumor can be excised from the plant and grown on a solid sucrose-salts medium as a callus culture. Tumor cells differ from normal cells in a number of characteristics. Whereas normal cells require the plant hormones auxin and cytokinin for growth in culture, tumor cells require neither. Further, transformed cells undergo a profound alteration in their nitrogen metabolism. One obvious manifestation is that most tumor cells synthesize high levels of octopine or nopaline, two derivatives of arginine which do not appear in detectable levels in normal tissues. In addition, the tumor callus can be grafted onto a healthy plant and the tissue will develop into a tumor phenotypically identical to the original. All of these characteristics are stably inherited.

Ever since Smith and Townsend proved in 1907 that a bacterium is the causative agent of crown gall tumors (26), the most intriguing question has been: how does the bacterium induce these modifications? The elegant experiments of Braun and his collaborators (2) were the first to convincingly demonstrate that the continued presence of viable bacteria is not required for tumor formation. If the bacteria are heat killed about 2 days after being applied to a wounded plant, the tumor develops normally. Thus attention has focused on the identification of a putative "tumor-inducing principle" elaborated by the bacteria and transferred to the plant.

DETECTION OF PLASMID GENES IN TUMOR DNA

Because the unique characteristics of the transformed cell are stably inherited, numerous investigators have attempted to identify bacterial or viral nucleic acid as the tumor-inducing principle. Although many reports appear purporting to show the presence of both bacterial and viral genes in DNA isolated from the tumors (22,24,25,27), subsequent studies have indicated that all of these data were interpreted incorrectly (3,4,7–9,16,19,23). However, the possibility that a small fragment of bacterial DNA might be present in tumors was revived when Schell and his collaborators (31) reported that all virulent, but no avirulent, strains of *Agrobacterium* they examined contained large plasmids. Subsequent studies in several laboratories have convincingly demonstrated that a large plasmid is indeed essential for virulence (11,20,28–30). Therefore, for the past few years, our attention has focused on trying to determine whether plasmid genes are transferred to the plant and maintained in sterile tumor tissue.

Our approach has been to test for the presence of foreign DNA in tumor cells by measuring the kinetics of reassociation of small amounts of labeled plasmid DNA (probe) in the presence and absence of large amounts of tumor DNA (driver). If plasmid sequences are present in the tumor DNA, then the concentration of probe DNA is raised and it will reassociate more rapidly in the presence of tumor DNA. The kinetics of reassociation depend both on the number of copies of the plasmid sequences and on the fraction of the plasmid which is present in the tumor DNA.

Since the strain ($B_6$806) which induced the tumor under investigation contained two plasmids, only one of which is associated with virulence (11), it was necessary to transfer only the virulence-associated plasmid into a plasmidless strain by conjugation *in planta*. Pure virulence plasmid was isolated from the exconjugant bacterium and labeled with ^{32}P by the process of nick translation to a specific activity of approximately 10×10^6 cpm/μg DNA (13,17). DNA from cloned tobacco tumor line E9 was isolated free of significant amounts of polysaccharide and other interfering contaminants by pronase digestion, gel filtration on Sepharose 4B (12), RNase treatment and batch adsorption, and elution from hydroxylapatite (19).

The kinetics of reassociation of the plasmid in the presence and absence of tumor DNA are presented in Fig. 1 in the form of a Pot plot (3). There appears to be a perceptible increase in the rate of reassociation of the plasmid DNA in the presence of tumor DNA which is not observed when the probe DNA is incubated with either salmon DNA or DNA from normal tobacco callus. Significantly, this rate increase is abolished if the tumor DNA is incubated with DNase prior to incubation with the probe. Further, a heterologous DNA probe (tritiated DNA of *Serratia marcescens*) does not reassociate more rapidly in the presence of the tumor DNA (data not shown). These data were reproducible with four independently isolated preparations of tumor DNA. We conclude that tumor DNA contains plasmid sequences.

The data in Fig. 1 also include the reassociation of plasmid DNA at a concentration which corresponds to 0.8 copy of the plasmid per diploid

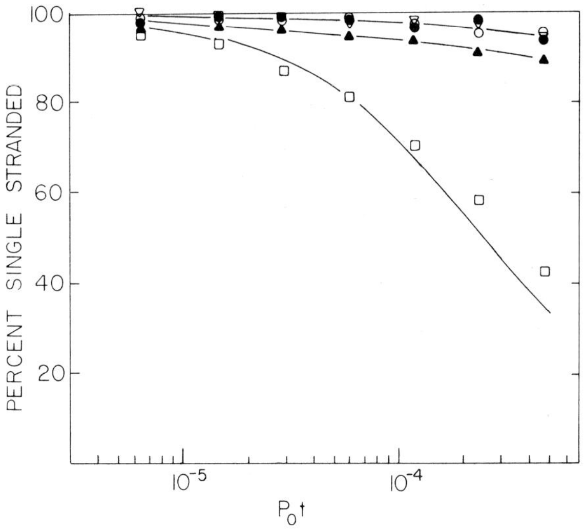

FIG. 1. Renaturation kinetics of *A. tumefaciens* A277 plasmid DNA in the presence of tumor and control DNAs. ^{32}P-labeled plasmid DNA (33×10^6 cpm/μg, 9×10^{-4} μg/ml) was allowed to reassociate in the presence of the following sheared DNA samples: 2.2 mg/ml salmon (○); 1.9 mg/ml normal tobacco callus (●); 2.3 mg/ml crown gall tumor E9 (△); DNase-treated crown gall tumor E9 DNA (originally 2.3 mg/ml) (▼); reconstruction: 1.9 mg/ml salmon DNA and 0.038 μg/ml A277 plasmid DNA, corresponding to 0.8 copy of plasmid per diploid tumor cell (□). Ideal second-order reaction curves are drawn for the reconstruction reaction only. All DNA concentrations were determined as described previously (4).

tumor cell. A comparison of the reassociation kinetics of this reconstruction mixture with the mixture of probe plus tumor DNA indicates that there cannot be as much as one copy of the plasmid in a diploid tumor cell. One reasonable explanation which could account for the slight shift in the kinetics of reassociation is that only a small region of the plasmid is represented in the tumor. Only this small fraction of the probe would renature more rapidly, and the rest of the probe (which is not represented) would renature at the same rate as the probe in the absence of the tumor DNA. To explore this possibility, investigators cleaved the labeled plasmid with the restriction enzyme *Sma*I (Lyle Brown, *personal communication*). The fragments were separated into 19 bands by agarose gel electrophoresis. Each radioactive band was then eluted from the gel, denatured, and its kinetics of reassociation followed in the presence and absence of tumor DNA. Results of such an experiment using the three bands of highest molecular weight derived from the cleavage are shown in Figs. 2, 3, and 4. It is clear that the rate of reassociation of DNA bands 1 and 2 is not accelerated in the presence of tumor DNA (Figs. 2 and 3). However, the DNA of band 3 does reassociate significantly more rapidly in the presence of tumor DNA (Fig. 4). This increased rate of reassociation is eliminated if the tumor

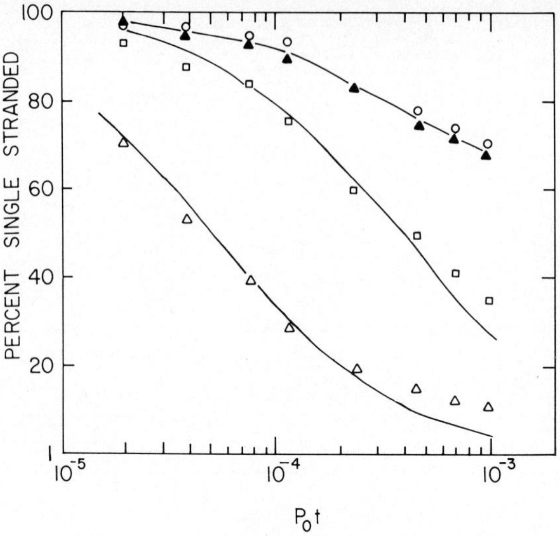

FIG. 2. Renaturation kinetics of *A. tumefaciens* A277 plasmid *Sma*I digest band 1 DNA in the presence of tumor DNA and control DNAs. ^{32}P-labeled DNA (22.8×10^6 cpm/μg, 1.3×10^{-3} μg/ml) was allowed to reassociate in the presence of the following sheared DNA samples: 2.2 mg/ml salmon (○); 1.45 mg/ml crown gall tumor E9 (▲); 1.74 mg/ml salmon DNA + 0.35 μg/ml A277 plasmid DNA (1.2 copy model) (□); 2.2 mg/ml salmon DNA + 0.44 μg/ml A277 plasmid DNA (15 copy model) (△).

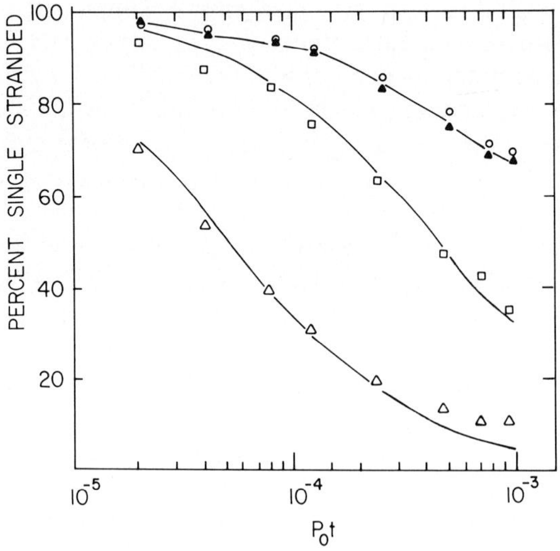

FIG. 3. Renaturation kinetics of *A. tumefaciens* A277 plasmid *Sma*I digest band 2 DNA in the presence of tumor and control DNAs. See Fig. 2 legend for details.

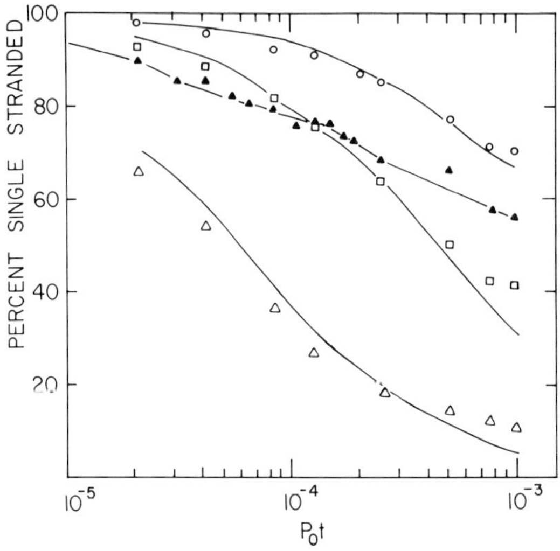

FIG. 4. Renaturation kinetics of *A. tumefaciens* A277 plasmid *Sma*I digest band 3 DNA in the presence of tumor and control DNAs. See Fig. 2 legend for details.

DNA is treated with DNase. When DNA eluted from each of the remaining bands was incubated with tumor DNA, ambiguous results were obtained. Several bands of DNA gave no evidence of being present in the tumor, whereas others appeared to be present but at a much lower concentration than the DNA of band 3. These data cannot be explained unequivocally at this time. One reasonable possibility is that the DNA of band 3 undergoes degradation as a result of radiation damage and perhaps nuclease treatment in the course of the nick translation reaction. These smaller fragments contaminate the other bands following electrophoresis and cause small artifactual changes in the renaturation kinetics.

The kinetics of reassociation of band 3 DNA in the presence of tumor DNA are clearly complex and deviate from what would be expected if all of the DNA of band 3 were present. Less than 18% of the labeled probe DNA is accelerated by tumor DNA. It is possible to estimate the number of copies of the plasmid fragment that are present in each tumor cell. One merely compares the $Pot^{1}/_{2}$ of the fast renaturing component with the $Pot^{1}/_{2}$ of the reconstruction mixture, which represents 15 copies of plasmid per tumor cell. The rate is between 1.4 and 2 times faster in the experiment than in the reconstruction. Therefore, we estimate that approximately 20 to 30 copies of the fragment are present per diploid tumor cell.

POSSIBLE EXPRESSION OF PLASMID SEQUENCES IN TUMOR

Since these DNA reassociation studies provide strong evidence for the presence of a fragment of the plasmid in tumor DNA, we attempted to determine whether some of these plasmid genes might be expressed in the tumor. The gene products we studied are involved in the synthesis of the unusual guanido amino acids, octopine and nopaline. These markers were chosen because tantalizing biochemical data had already accumulated which suggested that these markers were transferred from the bacterium to the plant. Most virulent strains of *Agrobacterium* degrade either octopine or nopaline (15), and sometimes both. Most tumors synthesize either octopine or nopaline (never both) (1,10,11). Morel and co-workers (21) were the first to draw attention to the correlation that octopine-degrading strains of *Agrobacterium* induce tumors that synthesize octopine and strains that degrade nopaline induce tumors that synthesize nopaline. They postulated that the genes coding for the enzymes mediating these reactions are transferred to the plant, and that the enzymes are reversible, functioning in the direction of degradation in the bacterium (oxidase activity) and in the biosynthetic direction (dehydrogenase activity) in the tumor. The data in Table 1 together with data from other laboratories (10,18,21) extend the number of bacterial strains and tumor lines examined. The correlation between the utilization and production traits is apparent. Several strains which did not degrade either octopine or nopaline induced tumors which did not synthesize either amino acid. However, there are several notable exceptions. Several bacterial strains utilize nopaline but the tumors they induce synthesize neither nopaline nor octopine.

TABLE 1. Agrobacterium *strains and their utilization patterns of octopine and nopaline*

Strain	Utilization by bacteria	Production by tumor	Utilization by progeny[a]	Production by tumor, induced by progeny
A6NC	Octopine	Octopine	Octopine (1,2)	Octopine
15955	Octopine	Octopine	Octopine (1,2)	Octopine
CGIC	Octopine	Octopine	Not tested	Not tested
B_6V87	Octopine	Octopine	Octopine (3)	Octopine
$B_6$806	Octopine	Octopine	Octopine (1,3)	Octopine
B_2A	Octopine	Octopine	Not tested	Not tested
140	Octopine	Octopine	Octopine (3)	Octopine
C58	Nopaline	Nopaline	Nopaline (3)	Nopaline
T37	Nopaline	Nopaline	Nopaline (3)	Nopaline
542	None	None	None (3)	None
AT1	None	None	Not tested	Not tested
AT4	None	None	Not tested	Not tested
EU6	Nopaline	None	Nopaline (1,3)	None
AT181	Nopaline	None	Nopaline (1)	None

[a] Progeny strains contained a single plasmid which was transferred from each of the strains listed in the first column to a plasmidless, avirulent strain. The numbers in parentheses refer to the mechanism of transfer.

1. DNA-mediated transformation (J. Schell, *personal communication*).
2. RP4-promoted conjugation (5).
3. *In planta* conjugation (14).

The data in Table 1 also verify that the ability to degrade octopine and nopaline is coded by plasmid-borne genes in those strains which can utilize only one or the other amino acid. Thus, these traits can be readily transferred to an avirulent plasmidless strain which cannot degrade either octopine or nopaline. This transfer can be mediated by mobilization by the promiscuous plasmid RP4 (5), by DNA-mediated transformation employing the plasmid DNA as donor (J. Schell, *personal communication*), or by conjugation *in planta* (14).

Another interesting conclusion that derives from these studies concerns strains that utilize both octopine and nopaline (Table 2). When plasmid isolated from these strains was used to transform the plasmidless strain, only the ability to degrade nopaline was transferred. Further, when the transfer of both markers was attempted by RP4 mobilization, again only the ability to degrade nopaline was transferred. Interestingly, the tumors induced by these strains synthesized only nopaline. One reasonable explanation for all of these observations is that the ability to degrade octopine and nopaline is generally a plasmid-borne trait, but in those strains that utilize both amino acids, the ability to utilize octopine is coded by a chromosomal gene. Further, it follows that only the information coded on the plasmid with regard to octopine/nopaline utilization is transferred to the plant.

From all of these data, it is tempting to speculate that the bacterial gene for octopine or nopaline utilization is transferred to the plant where it is expressed for octopine or nopaline synthesis. If this hypothesis is correct, then a mutation in the bacterial gene concerned with utilization should be manifested in the amount of the corresponding guanido amino acid synthesized in the tumor induced by the mutant. Several different arginine-requiring strains which can utilize octopine or nopaline as a source of arginine were treated with nitrosoguanidine (6), and mutants were selected

TABLE 2. Agrobacterium *strains which utilize both octopine and nopaline and the utilization patterns of progeny from these strains*

Strain	Utilization by bacteria	Production by tumor	Utilization by progeny[a]	Production by tumor, induced by progeny
27	Nopaline and octopine	Nopaline	Nopaline (1,3)	Nopaline
223	Nopaline and octopine	Nopaline	Nopaline (1,3)	Nopaline
2A	Nopaline and octopine	Nopaline	Nopaline (1)	Nopaline

[a] Progeny strains were derived as indicated in Table 1.
1. DNA-mediated transformation (J. Schell, *personal communication*).
2. RP4-promoted conjugation (5).
3. *In planta* conjugation (14).

which could not utilize these two compounds as a source of arginine. Each mutant was checked to verify that growing cells could still take up radioactive octopine or nopaline but could not degrade it, as measured by the inability of the breakdown product (arginine) to be incorporated into TCA-insoluble material. Enzymatic analysis of the membrane fractions of each of these mutants indicated that they possessed less than 1% of the wild-type activity of octopine or nopaline oxidase. All of these independently isolated mutants retained their virulence after the arginine locus was reverted to prototrophy. Tumors induced on *Kalanchoe* seedlings with each of these mutants as well as the wild-type parental strains were excised and analyzed for their content of octopine and nopaline (20). No significant differences in the level of octopine and nopaline synthesized by tumors induced by the wild-type and by the six mutants that we have examined could be detected. We cannot rule out the possibility that these mutants suffer a lesion in a regulatory gene which governs octopine or nopaline oxidase activity, but this seems a rather remote possibility for all of these mutants.

A more reasonable explanation is that the bacterial gene that specifies the bacterial oxidase activity is not the same gene specifying the analogous dehydrogenase activity in the tumor. This interpretation is consistent with the meager biochemical evidence which also suggests that the bacterial degradative enzyme and the tumor synthetic enzyme differ markedly. It is clear from the data of Table 1 that genetic information specifying which guanido amino acid will be synthesized by the tumor is encoded in the plasmid. However, if synthesis of these guanido amino acids in the tumor is due to the expression of transferred plasmid genes, these genes differ from those that code for the octopine/nopaline degradation pathways expressed in the causative organisms.

SUMMARY

Evidence is presented that a fragment of the virulence plasmid of the inciting bacterium is present in crown gall tumor cells. Attempts to gain evidence that this DNA is expressed by the synthesis of a plasmid-coded enzyme were unsuccessful. These studies represent a unique system in which prokaryotic genes are stably maintained in a eukaryotic cell.

Note added in proof: Matthysee and Stump (1976: *J. Gen. Microbiol.*, 95:9–16) have also reported that plasmid sequences are present in a *Vinca rosea* crown gall tumor. Their studies suggest that over half of the total plasmid sequences are present, considerably more than we found in our tobacco tumor studies.

ACKNOWLEDGMENTS

This work was supported by Public Health Service Grant CA-13015 from the National Cancer Institute and NP 194 from the American Cancer Society. D. Merlo was supported by Research Fellowship 1F32-CA05222 from the National Cancer Institute. We thank Prof. G. Melchers and Dr. Sacristan for providing the cloned line of tobacco tumor used in these experiments as well as numerous stimulating discussions.

REFERENCES

1. Bomhoff, G. H. (1974): Studies on crown gall—a plant tumor. Investigations on protein composition and on the use of guanidine compounds as a marker for transformed cells. Dissertation, University of Leiden, Leiden, the Netherlands.
2. Braun, A. C., and Stonier, T. (1958): Morphology and physiology of plant tumors. *Protoplasmatologia,* 10 (50):1–93.
3. Chilton, M.-D., Currier, T. C., Farrand, S. K., Bendich, A. J., Gordon, M. P., and Nester, E. W. (1974): Agrobacterium tumefaciens DNA and PS8 bacteriophage DNA not detected in crown gall tumors. *Proc. Natl. Acad. Sci. U.S.A.,* 71:3672–3676.
4. Chilton, M.-D., Farrand, S. K., Eden, F. C., Currier, T. C., Bendich, A. J., Gordon, M. P., and Nester, E. W. (1975): Is there foreign DNA in crown gall tumor DNA? In: *Second Annual John Innes Symposium,* edited by R. Markham, D. R. Davies, D. A. Hopwood, and R. W. Horne. North Holland Publishing Co., pp. 297–311. Amsterdam.
5. Chilton, M.-D., Farrand, S. K., Levin, R., and Nester, E. W. (1976): RP4 promotion of transfer of a large Agrobacterium plasmid which confers virulence. *Genetics,* 83:609.
6. Davis, C. H., and Rothman, R. H. (1973): Induction of crown gall by nitrosoguanidine-treated Agrobacterium tumefaciens. *Mutation Res.,* 20:283–285.
7. Drlica, K. A., and Kado, C. I. (1974): Quantitative estimation of Agrobacterium tumefaciens DNA in crown gall tumor cells. *Proc. Natl. Acad. Sci. U.S.A.,* 71:3677–3681.
8. Eden, F. C., Farrand, S. K., Powell, J. S., Bendich, A. J., Chilton, M.-D., Nester, E. W., and Gordon, M. P. (1974): Attempts to detect deoxyribonucleic acid from Agrobacterium tumefaciens and bacteriophage PS8 in crown gall tumors by complementary ribonucleic acid/deoxyribonucleic acid–filter hybridization. *J. Bacteriol.,* 119:547–553.
9. Farrand, S. K., Eden, F. C., and Chilton, M.-D. (1975): Attempts to detect Agrobacterium tumefaciens and bacteriophage PS8 DNA in crown gall tumors by DNA·DNA-filter hybridization. *Biochim. Biophys. Acta,* 390:264–275.
10. Goldman, A., Tempe, J., and Morel, G. (1968): Quelques particularites de diverse souches d'Agrobacterium tumefaciens. *C. R. Soc. Biol. (Paris),* 162:630–631.
11. Gordon, M. P., Farrand, S. K., Sciaky, D., Montoya, A. L., Chilton, M.-D., Merlo, D. J., and Nester, E. W. (1976): The crown gall problem. In: *Proceedings of a Symposium on the Molecular Biology of Plants,* edited by I. Rubenstein. University of Minnesota Press, Minneapolis (*in press*).
12. Heyn, R. F., Hermans, A. K., and Schilperoort, R. A. (1973): Rapid and efficient isolation of highly polymerized plant DNA. *Plant Sci. Lett.,* 2:73–78.
13. Kelly, R. B., Cozzarelli, N. R., Deutscher, M. R., Lehman, J. R., and Kornberg, A. (1972): Enzymatic synthesis of deoxyribonucleic acid XXXII. Replication of duplex deoxyribonucleic acid by polymerase at a single strand break. *J. Biol. Chem.,* 245:39–45.
14. Kerr, A. (1971): Acquisition of virulence by non-pathogenic isolates of Agrobacterium radiobacter. *Physiol. Plant Pathol.,* 1:241–246.
15. Lippincott, J. A., Biderbeck, R., and Lippincott, B. B. (1973): Utilization of octopine and nopaline by Agrobacterium. *J. Bacteriol.,* 116:378–383.
16. Loening, U., Butcher, D. N., Schuch, W., and Satirana, M. L. (1975): Experiments on the homology between DNA from Agrobacterium tumefaciens and from crown gall tumor cells. In: *Second Annual John Innes Symposium,* edited by R. Markham, D. R. Davies, D. A. Hopwood, and R. W. Horne, pp. 269–280. North Holland Publishing Co., Amsterdam.
17. Maniatis, T., Jeffery, A., and Kleid, D. G. (1975): Nucleotide sequences of the rightward operator of phage λ. *Proc. Natl. Acad. Sci. U.S.A.,* 72:1184–1188.
18. Menage, A., and Morel, G. (1964): Sur la presence d'octopine dans les tissus de crown-gall. *C. R. Acad. Sci. [D] (Paris),* 259:4795–5796.
19. Merlo, D. J., and Kemp, J. D. (1976): Attempts to detect Agrobacterium tumefaciens DNA in crown gall tumor tissue. *Plant Physiol.,* 58:100–106.

20. Montoya, A. L., Chilton, M.-D., Gordon, M. P., Sciaky, D., and Nester, E. W. (1977): Octopine and nopaline metabolism in Agrobacterium tumefaciens and crown gall tumor cells: role of plasmid genes. *J. Bacteriol.,* 129:101–107.
21. Petit, A., Delhaye, S., Tempe, J., and Morel, G. (1970): Recherches sur les guanidines des tissues de crown gall. Mise en evidence d'une relation biochimique specifique entre les sources d'Agrobacterium tumefaciens et les tumeurs qu'elles induisent. *Physiol. Veg.,* 8:205–213.
22. Quetier, F., Huguet, T., and Guille, E. (1969): Induction of crown gall: Partial homology between tumor-cell, bacterial DNA and the G + C-rich DNA of stressed normal cells. *Biochem. Biophys. Res. Commun.,* 32:128–133.
23. Schilperoort, R. A., Dons, J. J. M., and Ras, H. (1975): Characterization of the complex formed between PS8 cRNA and DNA isolated from A6-induced sterile crown gall tissue. In: *Second Annual John Innes Symposium,* edited by R. Markham, D. R. Davies, D. A. Hopwood, and R. W. Horne, pp. 253–286. North Holland Publishing Co., Amsterdam.
24. Schilperoort, R. A., Van Sittert, N. J., and Schell, J. (1973): The presence of both phage PS8 and Agrobacterium tumefaciens A6 DNA base sequences in A6-induced sterile crown gall tissue cultured in vitro. *Eur. J. Biochem.,* 33:1–7.
25. Schilperoort, R. A., Veldstra, H., Warnaar, S. O., Mulder, G., and Cohen, J. A. (1967): Formation of complexes between DNA isolated from tobacco crown gall tumors and RNA complementary to Agrobacterium tumefaciens DNA. *Biochim. Biophys. Acta,* 145:523–525.
26. Smith, E. F., and Townsend, C. O. (1907): A plant tumor of bacterial origin. *Science,* 25:671–673.
27. Srivastava, B. I. S. (1970): DNA:DNA hybridization studies between bacterial DNA, crown gall tumor cell DNA and normal cell DNA. *Life Sci. Part 2,* 9:889–892.
28. Van Larabeke, N., Engler, G., Holsters, M., Van den Elsacker, S., Zaenen, I., Schilperoort, R. A., and Schell, J. (1974): Large plasmid in Agrobacterium tumefaciens essential for crown gall inducing ability. *Nature* [*New Biol.*], 252:169–170.
29. Van Larabeke, N., Gentello, Ch., Schell, J., Schilperoort, R. A., Hermans, A. K., Hernalsteens, J. P., and Van Montagu, M. (1975): Acquisition of tumor-inducing ability by non-oncogenic agrobacteria as a result of plasmid transfer. *Nature* [*New Biol.*], 255:742–743.
30. Watson, B., Currier, T. C., Gordon, M. P., Chilton, M.-D., and Nester, E. W. (1975): Plasmid required for virulence of Agrobacterium tumefaciens. *J. Bacteriol.,* 123:255–264.
31. Zaenen, I., Van Larabeke, N., Teuchy, H., Van Montagu, M., and Schell, J. (1974): Supercoiled circular DNA in crown-gall inducing Agrobacterium strains. *J. Mol. Biol.,* 86:109–127.

Recombinant Molecules: Impact on Science and Society, edited by R. F. Beers, Jr. and E. G. Bassett. Raven Press, New York © 1977.

19. Genetics of Nitrogen Fixation: Some Possible Applications

Winston J. Brill

Department of Bacteriology and Center for Studies of Nitrogen Fixation, University of Wisconsin, Madison, Wisconsin 53706

INTRODUCTION

Nitrogen fixation is a key reaction in the N cycle on this planet. Of the N_2 fixed, most is fixed biologically; however, increasing requirements for protein are escalating the demand for industrially fixed N_2. Bulk fertilizer N, as ammonia, urea, or nitrate, is becoming quite expensive because of the tremendous energy demands of the Haber-Bosch process and transportation required to get the fertilizer to the farmer. A disadvantage with the use of bulk fertilizer N is that it is applied in large quantities and thus is readily brought into bodies of water by runoff after a rain. Biological N_2 fixation, on the other hand, occurs gradually and normally stops when sufficient fixed N is available. We hope that studies of biological N_2 fixation will yield information that can be applied to decrease our dependence on the Haber-Bosch process and take greater advantage of the natural biological process.

Only certain species of bacteria fix N_2. Examples of free-living N_2-fixing bacteria include *Azotobacter vinelandii,* which fixes N_2 aerobically, and *Klebsiella pneumoniae,* an organism closely related to *Escherichia coli,* which fixes N_2 only under anaerobic conditions. Some bacteria fix N_2 only when they are in a symbiotic association with a eukaryote. A very important bacterium of this class is *Rhizobium,* which normally only fixes N_2 within root nodules of specific legumes such as soybean, alfalfa, clover, and bean. This symbiosis is the reason why such plants require no fertilizer N.

Nitrogen fixation is catalyzed by the enzyme nitrogenase, which requires ATP and electrons. The enzyme is composed of two proteins, component I, containing Fe and Mo, and component II, containing Fe [reviewed in (6)]. Both components are extremely O_2 labile and thus need to be protected from air. *Azotobacter* (19) and *Rhizobium* (2) have evolved specialized systems to prevent O_2 from reaching nitrogenase. One of the most important discoveries was that nitrogenase, besides reducing N_2 to NH_4^+, also is capable of reducing acetylene to ethylene (8). Ethylene is a gas that can easily be measured by gas chromatography. The acetylene reduction assay allows the investigator to quantitate N_2 fixation readily. Another property common to all N_2-fixing systems is that nitrogenase synthesis is completely repressed in the presence of excess NH_4^+ (4). In fact, no nodules are seen in legumes that are fertilized with sufficient N.

RESULTS AND DISCUSSION

Studies with *Klebsiella pneumoniae* and *Azotobacter vinelandii*

The approach we have used to study N_2 fixation has been to obtain mutant strains (Nif$^-$) that grow well on fixed N but do not grow at all on N_2. Results obtained with such strains have given us insight into the regulation of nitrogenase synthesis (4,13), the order of *nif* genes on the chromosome (3,17), the active site of nitrogenase (18), and the existence of factors, other than nitrogenase components, that are specifically required if an organism is to fix N_2 (17).

Mutant strains that we obtained were first analyzed to determine which nitrogenase component is active. Inactive components can be detected serologically (18) or by specific staining techniques on polyacrylamide gels (5). A molybdenum cofactor that is a part of active component I also can be assayed with the use of mutant strains that lack this cofactor (17). We wanted to determine which mutations are in genes involved with the regulation of nitrogenase synthesis. Mutant strains lacking both components were reverted and then checked for their ability to fix N_2 in the presence of NH_4^+. The first such derepressed strain obtained was in *A. vinelandii* (13). Derepressed strains have potential agronomic importance because they continue to fix N_2 even when fertilizer N is present. We are attempting to determine conditions and construct strains that will optimize NH_4^+ excretion. Possibly, these derepressed strains will use cheap carbon compounds to produce NH_4^+ for crop growth.

Most of the genetic work on N_2 fixation has been performed with *K. pneumoniae*. Phage P1 that is normally used for transduction in *E. coli* can be used for genetic analysis in *K. pneumoniae* (20). The *nif*$^-$ mutations are closely linked to the genes specifying the enzymes of histidine biosynthesis (10,20). It has also been possible to obtain plasmids that carry the *nif* region (9,11). One of the most exciting experiments involved transferring of *nif*-containing plasmids, carrying drug-resistance transfer factors, to *E. coli* (11) and other bacteria that normally do not fix N_2 (7,9). It seems that these *nif* genes are expressed only when the cytoplasm is protected from O_2. Is it then possible to transfer *nif* genes from bacteria to corn or wheat to obtain plants that have no requirement for fertilizer N? Even if the genes are introduced into such plants, a mechanism for excluding O_2 must be developed. *Rhizobium* seems to take care of this problem by inducing the legume to produce specialized structures—root nodules that are filled with an O_2-scavenging protein, leghemoglobin (2).

The lesions in several Nif^- mutant strains were mapped by reciprocal transduction crosses in *K. pneumoniae* to yield an order of markers with respect to the *his* operon. A genetic map based on these crosses is seen in Fig. 1. Further, more detailed mapping requires that deletions that end within the *nif* region be found. The phage Mu is useful for obtaining deletions in *E. coli,* but wild-type *K. pneumoniae* is not lysed by Mu. We were able to isolate Mu-sensitive mutant strains of *K. pneumoniae,* however, by obtaining strains resistant to a phage that lyses *K. pneumoniae* (1). His^- mutant strains were obtained after Mu infection, and when Mu-sensitive isolates were isolated, some of them were Nif^-, suggesting that both Mu and a part or all of the *nif* genes had been deleted. Many deletions have been obtained and these should be useful for detailed mapping of the several hundred Nif^- point mutants that we have phenotypically analyzed.

It has been possible to get Mu integrated into a *nif*-containing plasmid (1). Again, some Mu-sensitive strains obtained from a Mu lysogen are Nif^-. The collection of *nif* deletions obtained in this manner are being used in conjugation experiments for mapping. However, the main use of these *nif* mutations in the plasmid will be for complementation analyses of the different genes. We hope that this work will result in an understanding of the order of genes, number of genes, and function of genes in this complex system.

We had earlier characterized many mutant strains of *A. vinelandii,* but a genetic system was not available. Recently, a transformation system was described for this organism (16), and we therefore began to map some of the nif^- mutations with respect to each other (3). The structural genes for components I and II are closely linked as they are in *K. pneumoniae,* but the gene coding for the molybdenum cofactor is not close to these structural genes. Also, regulatory genes seem to be distant from the structural genes. *Azotobacter* is an organism that has been grown in large scale for agricultural use in the Soviet Union and in India. Several commercial inocula of

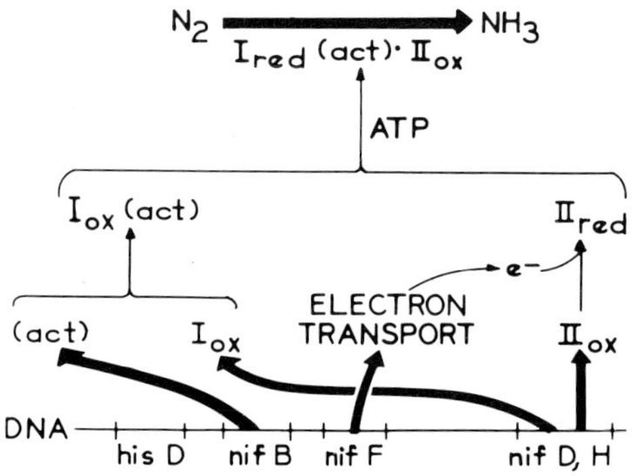

FIG. 1. Order of *nif* genes in *Klebsiella pneumoniae.* (act) represents the molybdenum cofactor. (Adapted from ref. 17.)

Azotobacter are currently being marketed in the United States. We feel that genetic manipulation of this organism might very well create strains that will further benefit agriculture.

Studies with *Rhizobium*

The *Rhizobium*-legume symbiosis is much more difficult to analyze genetically. Neither the bacterium nor plant can grow by itself on N_2. We wanted to isolate mutant strains that are arrested at various stages of the nodule-forming process as well as strains that are missing one of the nitrogenase components. No positive selection technique is available for obtaining such strains; therefore, we had to devise a rapid method for screening survivors of a mutagenesis for their ability to fix N_2 (21). An effectiveness assay for N_2 fixation that we used involved surface-sterilizing soybean seeds and placing one seed to a vial containing agar or vermiculite with an N-free salts medium. Each seed was inoculated with a colony of *Rhizobium,* a sterile plastic bag was placed over the vial, and the vials were incubated in a plant-growth chamber for 2 weeks. At the end of the incubation, the plants were cut at the base of the stem and N_2 fixation was measured by quantitating the amount of acetylene reduced per plant. Several thousand colonies could be screened in 1 month by this assay. We stocked mutant strains that were unable to reduce acetylene but would grow as well as the wild-type on a variety of rich and minimal media (15). Several mutant strains that have been characterized are described in Table 1. Strain SM1 is unable to nodulate soybean, whereas strain SM4 produces small, green nodules. Strain SM5 forms nodules that appear normal; however, when the *Rhizobium* in these nodules were examined for nitrogenase activities, it was found that active component I was present but component II activity absent. Serological analysis showed that component II protein was synthesized. Therefore, the mutation in strain SM5 is the first example of a structural mutation in a *Rhizobium nif* gene.

There are some indications that the *nif* genes are located on a plasmid in *Rhizobium*. For instance, prolonged subculture of *Rhizobium* on fixed N

TABLE 1. *Mutant strains of* Rhizobium japonicum *unable to form an N_2-fixing symbiosis with soybean*

		Description of nodules	
Strain	Size	Leghemoglobin	Acetylene-reducing activity
Wild-type	Large	Present	Present
SM1	No nodules	—	—
SM4	Small	Absent	Absent
SM5	Large	Present	Absent

TABLE 2. *Acetylene-reducing activity of mutant strains of the* Rhizobium japonicum–*soybean symbiosis*

Strain	nmoles C_2H_2 reduced per plant per hour	No. of nodules per plant
Wild-type	149	17
SM31	324	25
SM35	199	43

sometimes yields ineffective strains (12). Also, acridine orange has been shown to induce strains unable to form N_2-fixing nodules (14). We plan to isolate plasmids that are found in the wild-type and use them to attempt transformation experiments with the mutant strains. This type of experiment should identify which genes, if any, are plasmid-borne.

One surprising phenotype that we found during the screening experiments was the ability of two mutant strains to have activities greater than the wild-type, as tested by the effectiveness assay. Table 2 presents results with such strains. The two seem to be different because SM31 is twice as active as the wild-type but forms the same number of nodules per plant, whereas strain SM35 produces more nodules per plant. Will strains such as these be important in agriculture? It is possible that the strains will not successfully compete in the soil because the wild-type is a standard laboratory strain that has been maintained on slants for many years. However, these results have given us incentive to mutate strains that are successful in the field and to screen for superior mutants. It may also be possible soon to transfer "super" genes from one strain to another. At present, strains SM31 and SM35 are being tested in field experiments to determine if benefits to soybean do accrue from use of these strains.

We wondered whether the effectiveness assay (21) could be used to study plant genes that are important for N_2 fixation. Many cultivars of soybean have been tested with this assay and a great variation has been seen among them (21). The acetylene reduction values obtained with the assay do not bear a direct relationship to yield data of the cultivars. It is possible that the effectiveness assay demonstrates N_2-fixing potential—a potential that may not be realized in the field. Plant breeding experiments might be able to demonstrate plant genetic loci responsible for high N_2-fixing potential. These genes might then be bred into suitable plants to create possible new useful soybean cultivars.

ACKNOWLEDGMENTS

This work was supported by the College of Agricultural and Life Sciences, University of Wisconsin, Madison, by the National Science Foundation, and by the Cooperative State Research Service.

REFERENCES

1. Bachhuber, M., Malavich, T., and Howe, M. (1976): Use of bacteriophage Mu to generate deletions in Klebsiella pneumoniae. *Bacteriol. Proc.*, p. 108.
2. Bergersen, F. J., and Turner, G. L. (1975): Leghaemoglobin and the supply of O_2 to nitrogen-fixing root nodule bacteroids: Studies of an experimental system with no gas phase. *J. Gen. Microbiol.*, 89:31–47.
3. Bishop, P. E., and Brill, W. J. (1976): Genetic analysis of Azotobacter vinelandii mutants unable to fix nitrogen. *Bacteriol. Proc.*, p. 163.
4. Brill, W. J. (1975): Regulation and genetics of bacterial nitrogen fixation. *Annu. Rev. Microbiol.*, 29:109–129.
5. Brill, W. J., Westphal, J., Stieghorst, M., Davis, L. C., and Shah, V. K. (1974): Detection of nitrogenase components and other nonheme iron proteins in polyacrylamide gels. *Anal. Biochem.*, 60:237–241.
6. Burris, R. H., and Orme-Johnson, W. H. (1974): Survey of nitrogenase and its EPR properties. In: *Microbial Iron Metabolism*, edited by J. B. Neilands, pp. 187–209. Academic Press, New York.
7. Cannon, F. C., and Postgate, J. R. (1976): Expression of Klebsiella nitrogen fixation genes (nif) in Azotobacter. *Nature*, 260:271–272.
8. Dilworth, M. J. (1966): Acetylene reduction by nitrogen-fixing preparations from Clostridium pasteurianum. *Biochim. Biophys. Acta*, 127:285–294.
9. Dixon, R., Cannon, F., and Kondorosi, A. (1976): Construction of a P plasmid carrying nitrogen fixation genes from Klebsiella pneumoniae. *Nature*, 260:268–271.
10. Dixon, R. A., and Postgate, J. R. (1971): Transfer of nitrogen-fixation genes by conjugation in Klebsiella pneumoniae. *Nature*, 234:47–48.
11. Dixon, R. A., and Postgate, J. R. (1972): Genetic transfer of nitrogen fixation from Klebsiella pneumoniae to Escherichia coli. *Nature*, 237:102–103.
12. Fred, E. B., Baldwin, I. L., and McCoy, E. (1932): *Root Nodule Bacteria and Leguminous Plants*. University of Wisconsin Press, Madison, Wisconsin.
13. Gordon, J. K., and Brill, W. J. (1972): Mutants that produce nitrogenase in the presence of ammonia. *Proc. Natl. Acad. Sci. U.S.A.*, 69:3501–3503.
14. Higashi, S. (1967): Transfer of clover infectivity of Rhizobium trifolii to Rhizobium phaseoli as mediated by an episomic factor. *J. Gen. Appl. Microbiol.*, 13:391–398.
15. Maier, R. J., and Brill, W. J. (1976): Ineffective and non-nodulating mutant strains of Rhizobium japonicum. *J. Bacteriol.* (*in press*).
16. Page, W. J., and Sadoff, H. L. (1976): Physiological factors affecting transformation of Azotobacter vinelandii. *J. Bacteriol.*, 125:1080–1087.
17. St. John, R. T., Johnston, H. M., Seidman, C., Garfinkel, D., Gordon, J. K., Shah, V. K., and Brill, W. J. (1975): Biochemistry and genetics of Klebsiella pneumoniae mutant strains unable to fix N_2. *J. Bacteriol.*, 121:759–765.
18. Shah, V. K., Davis, L. C., Gordon, J. K., Orme-Johnson, W. H., and Brill, W. J. (1973): Nitrogenaseless mutants of Azotobacter vinelandii. Activities, cross-reactions and EPR spectra. *Biochim. Biophys. Acta*, 292:246–255.
19. Shah, V. K., Pate, J. L., and Brill, W. J. (1973): Protection of nitrogenase in Azotobacter vinelandii. *J. Bacteriol.*, 115:15–17.
20. Streicher, S., Gurney, E., and Valentine, R. C. (1971): Transduction of the nitrogen-fixation genes in Klebsiella pneumoniae. *Proc. Natl. Acad. Sci. U.S.A.*, 68:1174–1177.
21. Wacek, T. J., and Brill, W. J. (1976): A simple, rapid assay for screening nitrogen-fixing ability in soybean. *Crop Sci.* (*in press*).

Recombinant Molecules: Impact on Science and Society, edited by R. F. Beers, Jr. and E. G. Bassett. Raven Press, New York © 1977.

20. Plant Protoplast Fusion: Progress and Prospects for Agriculture

Edward C. Cocking

Department of Botany, University of Nottingham, Nottingham, NG7 2RD England

INTRODUCTION

In this volume on recombinant molecules, it is highly appropriate, and particularly relevant, for us to assess the extent to which the consequences of plant protoplast fusion are enabling the generation of new plant strains or new plant species. When considering bacteria, we are, in this volume, largely concerned with the mechanism of recombination of genomes at the molecular level. We are, however, primarily concerned, when considering bacteria, fungi, plants, and animal species, with the broader aspect of the ability of the genetic apparatus of these living cells to incorporate and substitute foreign genetic material between cells of the different species groups.

Plant protoplasts are plant cells from which the cell wall has been removed by suitable enzymatic degradation, leaving the homeostatic unit of the plant cell intact. Because of the absence of the cell wall, uptake of foreign genetic material is greatly facilitated (10). Fusion is also readily possible between isolated protoplasts, thus enabling the ready mixing together of the genomes of different species. Not only is interplant species fusion possible, but also interfungal (25), plant-fungal (12), and plant-animal (2). A significant feature of the development of isolated plant protoplasts is that when suitable cultural conditions have been established, they regenerate a new cell wall and divide to produce small cell aggregates. Such cell aggregates are sometimes capable of developing into an organized callus mass, producing roots and shoots and subsequently whole flowering plants. Progress in this field of investigation is therefore dependent on the extent to which the genetic consequences of protoplast fusion are stabilized and expressed in regenerated plants. The relevance of this progress to agriculture will necessarily depend on the extent to which it is possible to apply any generally applicable, somatic hybridization strategy to crop plants. The Agricultural Research Council in the United Kingdom is particularly interested in this potential, and my colleagues Drs. Power, Evans, and Bhojwani and I recently presented the general content of this article to that group.

ISOLATION, CULTURE, AND FUSION OF PLANT PROTOPLASTS

Protoplasts can be isolated from leaf mesophyll tissue, epidermal tissues, petals, roots and root nodules, germinating pollen grains, tetrads, fruits, and *in vitro* cultured plant tissues. Leaf material is the preferred source of protoplasts, since after removal of the lower epidermis by peeling or an enzyme treatment, large numbers of cells are accessible for conversion into protoplasts. This consideration also applies to cultured plant tissues, particularly cell suspensions, but protoplast release is often critically dependent on the stage in the growth cycle of such tissues. Yields are variable, but from leaf tissue an average of 5×10^5 protoplasts per gram of fresh weight can be expected.

Although the isolation of protoplasts presents no major problems, successful culture is still restricted to a relatively few species, and even to certain varieties of a given species.

Protoplasts of most species regenerate a new cell wall after 2 to 3 days, and, following dedifferentiation, those of a few species enter division. Division, if maintained, results in the formation of cell colonies after 3 weeks. At this point in the regeneration process, the concentration of the plasmolyticum is progressively reduced so that after 8 to 12 weeks callus can be transferred to media lacking added plasmolytica. The growth of protoplast-derived callus and its handling for plant regeneration may parallel that already established for callus obtained directly from the plant. The time course of events is dependent largely on the plant species in question, but flowering plants can be produced from individual protoplasts after 4 to 6 months. Organogenesis of callus produced from protoplasts has been described for an ever-increasing number of species and includes several *Petunia* species (37), *Nicotiana tabacum* (33), *Asparagus officinalis* (6), *Ranunculus sceleratus* (14), *Atropa belladonna* (17), *Brassica napus* (22), *Pharbitis nil* (31), *Antirrhinum majus* (35), *Datura innoxia* (42), and *Daucus carota* (18).

To this list can be added approximately 25 species whose isolated protoplasts, mainly of leaf origin, exhibit division with callus production and limited organogenesis. These species include important crop plants such as potato and rice. The vast majority of plants produced from protoplasts are normal and fertile, and it is becoming apparent that protoplast culture will play an important role in the future in the cloning of desired plant types. Certain features of protoplast growth and subsequent plant regeneration potential can be explained in simple genetic terms. For example, in a wide range of culture media, leaf protoplasts of *Petunia parodii* will stop dividing at the 50-cell colony stage, yet those of *P. hybrida* continue growth to form callus. Callus of *P. parodii* readily undergoes organogenesis, yet that of *P. hybrida* does not. Leaf protoplasts of the F_1 (*P. hybrida* × *P. parodii*) and the reciprocal cross not only produce callus, like *P. hybrida*, but this callus also undergoes organogenesis, like *P. parodii.* Protoplast division capability and plant regeneration are therefore heritable characteristics in this system.

The somatic hybridization of plant species can be divided into four main stages:

1. Isolation of protoplasts.
2. Fusion of protoplasts.
3. Selective culture of fusion products.
4. Their regeneration into plants.

As previously implied, isolation of protoplasts presents no major problems, likewise with their fusion. Three methods of fusion are now established and their respective merits can be summarized as follows: (a) $NaNO_3$ (36): low percentage fusion with high viability; (b) high pH and Ca^{2+} ions (24): good fusion (1 to 10%) with low viability; and (c) polyethylene glycol (PEG) (23): good fusion (2 to 20%) with moderate viability. It could be argued that, in the final analysis, the three methods would show a comparable production level of viable heterokaryons and ultimately somatic hybrids. Protoplast fusion involves the establishment of very close contact (adhesion) between the plasmalemmas of adjacent protoplasts. Simultaneously, the nature of the membrane is altered in such a way so as to initiate limited membrane fusion. The mechanism of membrane fusion in plant cells is not clear, but all three of the above fusion-inducing agents alter the charge on the membrane surface thus facilitating close contact. An expansion of localized cytoplasmic bridges (similar to plasmodesmata) results in the coalescence of the respective cytoplasms to form a cytoplasmic hybrid cell (cybrid). However, several factors will combine to prevent the production of hybrid cells proper. These must include:

1. Instability of one or both components following fusion treatment.
2. Inherently low plating efficiency of one or both parental protoplast systems.
3. Asynchronous nuclear division.
4. Incomplete mixing of the cytoplasms because of the presence of vacuoles.
5. Chromosome loss during successive mitoses.
6. The inability of hybrid cells to survive at low density.

Very little can be done to alter points 1 through 5, but the design and operation of selection methods can be such as to minimize the problem of low density survival. It seems that the minimal plating density for leaf protoplasts of *Petunia* and *Nicotiana,* in the absence of nurse tissue, is 6×10^3/ml. At this density the plating efficiency is less than 0.5%. It seems, therefore, that the recovery of somatic hybrids is going to be a low-frequency event, and that the extent of protoplast fusion achieved using any of the three methods may bear no relationship to the number of somatic hybrids recovered. If the selection is rigorous, then the extent of fusion may be of secondary importance, when considering the many factors operating at the cultural level, which will reduce the numbers of surviving hybrid cells.

DEVELOPMENT OF SELECTION PROCEDURES FOR SOMATIC HYBRIDS

In order to recover these somatic hybrids it is necessary to select a few colonies from among many thousands or even millions derived from the parental protoplasts. For this reason a great deal of effort is now being directed toward the development of stringent selection systems for the recovery of somatic hybrids.

Experience with the fusion of cultured animal cells has clearly shown it is possible to devise powerful selection procedures that will recover rare hybrid cells by using differing physiological and biochemical capabilities of these cells. The classic example of this approach is the selection technique developed by Littlefield in 1964 (27) for the isolation of mammalian somatic hybrid cells.

Another approach used for the selection of mammalian hybrid cells is the exploitation of dominant, rather than recessive, drug-resistant mutants. Actinomycin D resistance in hamster cells has been shown to be dominant or co-dominant as has the resistance to ouabain and vinblastine (13). The fusion of two cell lines, each containing a different dominant drug-resistant mutation, followed by culture with the addition of the appropriate drugs, would allow the recovery of hybrid cells.

Unfortunately, once again progress in the isolation of dominant, drug-resistant, mutant higher plant cell lines is, at present, not as advanced as in mammalian cell culture. However, there appears to be no reason why similar dominant, drug-resistant mutants should not be isolated in higher plants, and recently Marton and Maliga (28) reported the recovery of a semi-dominant BUdR-resistant tobacco mutant.

These selection systems, which have been briefly outlined, are what may be termed generally applicable in that they can be used to select out hybrids which cannot be achieved by other means (i.e., mouse × man) and in which the selection is not based in any way, other than on general genetic principles, on prior knowledge of the behavior of the hybrid cells in culture. Naturally, these generally applicable methods are also needed for plant cells. It is significant that the recovery of the first somatic higher plant hybrids (7) did in fact depend on a knowledge of the behavior in culture of the sexual counterpart.

The quest for more generally applicable selection methods has prompted Melchers and Labib (30) to advocate the exploitation of mutations that affect the color of leaves. Such chlorophyll and carotenoid mutations are not uncommon and can be quickly recognized, and, what is more, many are already well documented. Some of these mutations prove to be light-sensitive.

It is also likely that many of these recessive mutations will complement to normal green in the hybrid. Moreover, under the appropriate conditions of light, mineral salts, and growth regulators, callus cultures will turn green and it may be possible, therefore, to identify such chlorophyll-deficient mutations at the level of callus tissue. Therefore, Melchers and Labib (30) isolated mesophyll protoplasts from two haploid chlorophyll-deficient, light-sensitive varieties of *N. tabacum,* exposed the mixed population to high pH in the presence of calcium ions to induce fusion, and then plated the protoplasts in nutrient medium and subsequently incubated the cultures under high light conditions. After 2 months a green colony developed and plants regenerated from this colony were normal green. Analysis of the F_2 revealed segregation for mutant types, clearly establishing that this plant resulted from a somatic cell fusion. In subsequent experiments under altered conditions, more somatic hybrids were recovered. A difficulty which was encountered centered on the expression of the mutant phenotype at the callus level; under some conditions the light-sensitive mutant lines became green even under high light intensities. Reduction of the concentration of the organic components of the medium seemed to solve the problem.

A somewhat different approach but one which still uses chlorophyll-deficient mutations has been developed by Gleba, Butenko, and Sytnik (16). In their experiments they used a variegated variety of *N. tabacum,* which had a plastom mutation, and a *N. tabacum* variety, which was yellow-green, because it was homozygous with respect to a semidominant nuclear mutation. The interesting feature of this system is that if complementation takes place, it should be possible to identify not only the products resulting from nuclear fusion, but also hybrids that have arisen from heterokaryons in which nuclear fusion has not occurred, and where one of the nuclei has subsequently been eliminated and is, therefore, a cytoplasmic hybrid. After the induction of fusion in a mixed population of protoplasts by PEG, the protoplasts were cultured to form colonies and plants regenerated from them. Screening of these regenerated plants revealed the presence of yellow-green variegated plants and green variegated plants, both of which would be expected as a result of protoplast fusion, but the latter would be expected only from the development of a cytoplasmic and not a nuclear hybrid.

In this laboratory a different approach to the selection problem has been followed (11). Since mutant cell lines are at present difficult to obtain and maintain, and, as there are considerable advantages in isolating protoplasts directly from the diploid plant, it was decided that, instead of attempting to isolate mutants, we would exploit the natural variation in metabolic capabilities exhibited by preparations of protoplasts isolated from different species. As an example, mesophyll protoplasts isolated from *Petunia hybrida* cv comanche and *N. tabacum* cv xanthi nc were compared in their sensitivities to various compounds such as amino acid analogues, growth substances, and other drugs included in the nutrient medium. Some of the compounds showed no differential effect on the growth of the protoplasts. Several of the drugs, however, showed a pronounced differential effect, preventing the growth of tobacco protoplasts while at the same concentration allowing the growth of petunia protoplasts, and with certain other drugs the converse situation prevailed. Differences in sensitivity of 10-fold and sometimes 100-fold have been obtained, comparable to differences expected from mutant cell lines.

Similar differences were also observed in the ability of protoplasts isolated from different species to grow under different environmental conditions. Mesophyll protoplasts of *N. otophora,* for instance, show a marked sensitivity to light during the first few days of culture. Differences in the ability of the protoplasts to grow in various nutrient media have also been detected. *Petunia parodii* protoplasts were characterized by a marked inability to sustain cell division beyond the 50-cell stage in most media. This behavior is apparently related to levels of growth regulators in the medium and, as such, can be regarded as another expression of differential drug sensitivity. Clearly, then, careful study of the growth potential of the various protoplast systems under different environmental, nutritional, and chemical conditions reveals marked differences in their capabilities.

This naturally leads one to speculate as to whether such differences might form the basis for the complementation selection of somatic hybrids. The development of a successful selection system using these naturally occurring differences would hinge on the fact that the ability to grow under a particular cultural regime must be expressed in the hybrid and therefore be dominant or semidominant. Doubtless some of the differences we have observed will be recessive or may be unsuitable for other reasons such as cross-feeding. (A similar difficulty is likely to be encountered with mutant cell lines.) Furthermore, for success we must choose two properties both of which exhibit dominance.

There is, therefore, an element of chance as to whether we make the correct choice. It is possible, however, to narrow the odds considerably. For instance, it is known that certain drug resistances, for example, resistance to actinomycin D, are dominant in animal cells. It is also possible that this resistance is also dominant among plant cells. We are also in a position to check whether a property is dominant or recessive within a particular group by examining the behavior of protoplasts isolated from the F_1 hybrid between two species, one of which exhibits sensitivity and the other resistance. In this way we can catalog the various characteristics of a protoplast system as to whether they are recessive, dominant, or semidominant.

Recent results obtained in this laboratory (39) have justified the logic of this approach. Mesophyll protoplasts were isolated from *P. hybrida* and *P. parodii,* mixed together and treated with PEG to induce fusion, and then cultured on a nutrient medium in which *P. parodii* protoplasts fail to grow beyond the small colony stage. Actinomycin D, to which *P. hybrida* protoplasts are more sensitive than those of *P. parodii,* was included in the medium at a concentration sufficient to stop the growth of *P. hybrida* protoplasts. After some weeks of culture ten colonies were recovered. Some of these colonies have now produced plants and these plants are characteristic of the "doubled up" F_1 hybrid between *P. parodii* and *P. hybrida.* It is significant that this experiment was initiated before the sexual hybrid was produced in the greenhouse and before the behavior of its protoplasts in culture was determined.

We are confident, therefore, that in using natural differential sensitivities we have a generally applicable selection system.

CONSEQUENCES OF PROTOPLAST FUSION—THE RELEVANCE OF PROTOPLAST FUSION TO AGRICULTURE

The aim of this chapter is not to assess comprehensively the overall potential of genetic manipulation of plant cell cultures for plant breeding. This has already been done adequately recently by Bottino (5) and by Holl (20). Rather, the aim is to assess the relevance for agriculture of one of the major potentials of the manipulation and culture of plant protoplasts, namely, the induced fusion of protoplasts from differing genetic backgrounds and the regeneration of somatic hybrid plants. Although there is no problem isolating large numbers of healthy protoplasts in crop plants, their response in culture has been the barrier in subjecting them to somatic hybridization or improving their types through introduction of desirable genetic materials.

Leguminous protoplasts have shown some encouraging response in culture. They often divide and sustain divisions yielding a callus tissue. However, the callus fails to regenerate whole plants. The response of cereal protoplasts in culture is more discouraging. The mesophyll protoplasts of cereals may regenerate a wall and remain viable for weeks but they never divide regularly. Only sporadic divisions have been reported in *Triticum, Secale* (15), and *Avena* (P. K. Evans, *personal communication*). The protoplasts from endosperm callus of corn showed only nuclear divisions, but there was no indication of cytokinesis (32). Recently Chinese workers (1) have reported sustained divisions in protoplasts isolated from haploid rice callus on medium fortified with (2,4-dichlorophenoxy)acetic acid (2,4-D) and coconut milk. In these cultures 0.1% of the protoplasts divided once after 2 weeks and only some of these underwent repeated divisions forming multicellular colonies of up to 30 cells. The only other graminaceous crop plant which has exhibited sustained divisions in protoplast cultures is *Saccharum officinarum.* These protoplasts were also derived from cultured cells.

Surveying the literature on cell and tissue cultures of the cereals, one is increasingly convinced that a logical approach to overcoming the present cultural difficulties of protoplasts is to lay a sound foundation of cell and callus culture.

It would be naive to suggest, even with the somatic hybridization approach and methodology developed at Nottingham, that the consequence of fusion between protoplasts of different species would necessarily be the production of fertile, somatic amphidiploid hybrid plants. Many factors could operate to preclude this possibility. Throughout their developmental history plants have evolved a variety of protective mechanisms—morphological, physiological, and biochemical—to maintain the integrity of species. The plant breeder may circumvent some of these difficulties by embryo culture techniques. It has also been suggested that sexual incompatibility mechanisms in plants may be analogous to immunochemical systems in animals, and the use of immunosuppressant drugs has been recommended. Whatever the actual limitations of somatic hybridization eventually prove to be, it is clear that prezygotic and endosperm-embryo relationships will not be among the incompatibility factors operating in this new method of hybridization. Two considerations will be uppermost in the mind of the plant breeder: first, the extent to which any selected somatic hybrid plants will be sterile, and second, the extent to which directional chromosome elimination will take place. Following somatic hybridization between cells of different mammalian species, there is often an undirectional loss of chromosomes, probably as a result of genome incompatibility; and advantage has been taken of this feature to assign mammalian genes to syntenic groups. To what extent genome incompatibility will be a problem in plant cell somatic hybridization is not yet clear. Some recent work on barley and wheat sexual crosses has indicated that directional chromosome elimination in interspecific hybrids in angiosperms may be more frequent than has been generally realized (4). It may be significant that examination of the consequence of fusion of petunia and boston ivy protoplasts has suggested that such complete elimination of chromosomes from one of the parents may occur in somatic hybrid cells on prolonged culture. Numerous examples of interspecific hybrids in plant evolution suggest that genome incompatibility is not always a serious problem, but, of course, present investigations are restricted to sexually derived hybrids. Experimentally, the extent of chromosome elimination or repression of gene function in interspecific somatic plant hybrids can be determined only by the use of genetic markers, or specific chromosome identification. Although the outcome of nuclear gene interactions in interspecific somatic hybrids is largely unknown, it is even less clear what the extent of cytoplasmic interaction might be following fusion (40). Cytoplasmically inherited traits are of widespread use in plant breeding, particularly as a source of male sterility; and it is likely that somatic hybridization may be used to transfer cytoplasmic features (such as male sterility) from one species to another.

Normally the plant breeder produces amphidiploids (plants having a complete diploid chromosome set from each parent strain) by doubling the chromosome number of F_1 interspecific hybrids; and when the two genomes are highly divergent, such amphidiploids are often fertile and reasonably stable both cytologically and genetically. There may be considerable advantage to be gained by using haploids in plant cell genetics and for the production of homozygous lines. It could also be argued that the fusion of haploids would be analogous to sexual fertilization. Nevertheless, there appear to be no real advantages in the use of haploids in somatic hybridization, particularly since to achieve fertility, following somatic hybridization between haploids of diverse species, the chromosome complement of the fusion product would have to be doubled.

Our somatic hybridization procedure therefore advocates the use of plant material with naturally occurring ploidy levels. The disposition of plant cultures toward ploidy changes emphasizes the need for selection and regeneration of hybrids at an early stage. Our somatic hybridization procedure also emphasizes the use of protoplasts isolated either directly from the plant (including albino plants), or from cells cultured *in vitro* for minimal lengths of time. The extent to which the plant breeder will use this somatic hybridization technique will depend on a number of factors. One of the main considerations will be the extent to which it is not possible for the breeder to achieve hybridization sexually (including the use of embryo culture methods) when the breeder must produce a hybrid between two species. It would therefore be unrealistic to consider somatic hybridization as a panacea for plant breeding; rather, it should be viewed as an additional hybridization method to be attempted when the sexual method is impossible or undesirable. Some crops such as banana, potato, and sugar cane are, of course, propagated vegetatively, and a sexual cycle is used merely to increase genetic diversity in a program of crop improvement. Here somatic hybridization may be particularly useful, thereby advantageously bypassing the need for such a sexual cycle.

The relevance of protoplast fusion, culture, and the selection and regeneration of somatic hybrids to agriculture is therefore best assessed by examining in detail those situations in which hybrids are desired but cannot be obtained by sexual procedures. The relevance to agriculture of the somatic hybridization procedures developed at Nottingham will be briefly critically examined, with particular reference to desired interspecific crosses in cereals, legumes, brassicas, and other crop plants.

Desired Interspecific Crosses in Cereals, Legumes, Brassicas, and Other Crop Plants

Generally speaking, production of interspecific crosses in crop plants is desirable for two major reasons: the production of new amphidiploids, and the use of interspecific hybridization for gene transfer, for example, resistance to plant diseases or environmental stress. The breeding of *Triticale*, which is an amphidiploid between wheat and rye, resulting from the sexual crossing of these two species, illustrates the already established need of the plant breeder to combine the characteristics of these two important cereals. The classic amphidiploid *Raphanobrassica* arose by spontaneous chromosome doubling between *Raphanus sativus* and *Brassica oleracea,* and recently extensive breeding work has indicated that a range of morphologically distinct *Raphanobrassica* forms may be created which may have agronomic potential, and in particular resistance to infection by *Plasmodiophora* (the causal agent of club root disease) (29). Many crosses within the Brassicas are desired for agronomic reasons. Some of these crosses (such as the production of *B. napus*) are possible only when embryo culture is used. A general feature of the production of interspecific crosses is that the usual sterile hybrids become reasonably fertile when their chromosome number is doubled, usually achieved, where it is possible, as a result of colchicine treatment. Such amphidiploids may also become reasonably stable for the hybrid condition, and produce progeny which are more or less uniformly like the original hybrid. Consequently, the use of diploid somatic cells for fusion which subsequently yield amphidiploids directly without, of course, the reduction division of the usual sexual reproduction (necessitating either spontaneous doubling, or the subsequent artificial doubling, of the chromosome complement by colchicine treatment) may be a distinct advantage. Crosses desired by the plant breeder include those, such as maize × sorghum and barley × rye, in this instance, enabling the disease resistance and winter hardiness of rye to be transferred to barley.

Although rarely of direct value, amphidiploids may be used to introduce into cereals, such as wheat, particular features of related species, including the insertion of small segments of alien chromosomes into wheat chromosomes by induced translocation (41). This new method of somatic hybridization is particularly relevant within this context. Let me be more specific. In introgressive hybridization (another form of interspecific hybridization), hybridization is followed by recrossing with the parental species in such a way that certain features of one species become transferred to the other species without impairment of taxonomic integrity. Since the development of crop varieties resistant to diseases and insects is one of the major objectives of plant improvement, it would seem likely that somatic cell hybridization will come to play an ever-increasing role in this respect. Sexual hybridization is currently being exploited to the limit for introgressive hybridization in the cereals and other crop plants. A typical example is the transfer of mildew resistance from the wild oat *Avena barbata* into the cultivated oat (Table 1). When sexual hybridization is impossible, the ability to transfer the necessary resistance genes might be made possible by somatic hybridization (Fig. 1). Basically the same considerations apply to the question of the possibility of transferring nodulation ability from legumes to nonleguminous crop plants (9). Indeed, recent work on the induction of nitrogenase in *Rhizobium* (26,34) has served to emphasize the genetic control by the host plant of the initiation and general establishment of nitrogen fixation in the

TABLE 1. *Transfer of mildew resistance from the wild oat* (Avena barbata) *into the cultivated oat*[a]

1. Amphidiploid, *A. barbata* (2n = 28) × *A. sativa* (2n = 42) backcrossed to *A. sativa.*
2. From selfed progeny of second backcross, mildew resistant plants isolated (2n = 44), disomic addition line. *A. sativa* chromosomes + pair *A. barbata* chromosomes.
3. Seeds of this disomic addition line irradiated ^{60}Co (15,000 rad).
4. Seedlings produced by crossing irradiated resistant lines with *A. sativa.*
5. Some seedlings resistant: euploid (2n = 44), the *A. barbata* resistance gene had been incorporated into the *A. sativa* genome.

[a] From ref. 3.

symbiotic association. Holl (21) has recently shown in the genetic analysis of a mutant line of *Pisum,* resistant to nodulation, that at least two genes are involved. One affects nodulation and the other influences fixation. The two genes segregated independently as dominant mendelian characters, and effective symbiosis requires the presence of at least one dominant gene at each locus. If there were a better knowledge of the plant genes controlling nodulation and the establishment of nitrogen fixation in legumes, somatic hybridization could play a key role in the transfer of this ability of nitrogen-fixing symbiotic association from legumes to nonleguminous crop plants by

A. Sexual hybridization

Vicia faba (Broad bean) x Vicia narbonensis (French Bean)

Breeders have long been interested in producing hybrids between these two species to transfer the resistance to "chocolate spot" disease from French bean to Broad bean.

B. Somatic hybridization

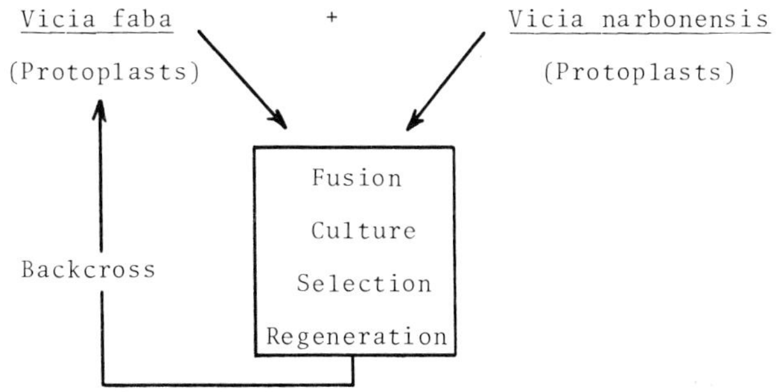

A similar transfer of resistance genes from pigeon pea to cowpea is also required.

FIG. 1. Somatic hybridization requirements in the Leguminosae.

enabling the production of legume/nonlegume amphidiploid somatic hybrids. Suitable selections could be made during the backcross to the nonlegume.

Somatic hybridization may also be particularly useful in introgressive hybridization. Sometimes the sexual hybrid is not fertile in the backcross to the desired parental species, and somatic hybridization could be usefully employed when this is the case. Somatic hybridization could also be useful for the crossing of vegetatively propagated plants. The magnitude of importance of extrachromosomal inheritance in crop plants is steadily increasing (19). Cytoplasmic markers, including the use of Fraction 1 (ribulose diphosphate carboxylase) analysis, are increasingly being used to decipher these effects (8). Protoplast fusion produces cytoplasmic hybrids [for a detailed discussion see (38)], and somatic hybridization could greatly facilitate the rapid production of cytoplasmic male sterile lines.

ACKNOWLEDGMENT

Original work described herein was supported by a grant from the Agricultural Research Council.

REFERENCES

1. Anonymous (Acad. Sin. Peking Inst. Botany) (1975): Isolation and culture of rice protoplasts. *Scientia Sinica,* 18(6):779–784.
2. Ahkong, Q. F., Howell, J. L., Lucy, J. A., Safwat, F., Davey, M. R., and Cocking, E. C. (1975): Fusion of hen erythrocytes with yeast protoplasts induced by polyethylene glycol. *Nature,* 255:66–67.
3. Aung, T., and Smith, H. (1976): Transfer of mildew resistance from the wild oat Avena barbata into the cultivated oat. *Nature,* 270:603–604.
4. Bennett, M. D., Finch, R. A., and Barclay, I. R. (1976): The time rate and mechanism of chromosome elimination in Hordeum hybrids. *Chromosoma,* 54:175–200.
5. Bottino, P. J. (1975): Potential of genetic manipulation in plant cell cultures for plant breeding. *Radiat. Bot.,* 15:1–16.
6. Bui-Dang-Hui, D., and Mackenzie, I. A. (1973): Division of protoplasts from Asparagus officinalis L. and their growth and differentiation. *Protoplasma,* 78:215–221.
7. Carlson, P. S., Smith, H. H., and Dearing, R. D. (1972): Parasexual interspecific plant hybridization. *Proc. Natl. Acad. Sci. U.S.A.,* 69:2292–2294.
8. Chen, K., Gray, J. C., and Wildman, S. G. (1975): Fraction 1 protein and the origin of polyploid wheats. *Science,* 190:1304–1305.
9. Cocking, E. C. (1974): Plant cell somatic hybridisation—its role and prospects in research and in practice. In: *Enciclopedia Della Scienza e Della Tecnica, Vol. 74,* edited by E. Macorini, pp. 199–208. Mondadori.
10. Cocking, E. C. (1977): Uptake of foreign genetic material by plant protoplasts. *Int. Rev. Cytol.,* 48:323–343.
11. Cocking, E. C., Power, J. B., Evans, P. K., Safwat, F., Frearson, E. M., Hayward, C., Berry, S. F., and George, D. (1974): Naturally occurring differential drug sensitivities of cultured plant protoplasts. *Plant Sci. Lett.,* 3:341–350.
12. Davey, M. R., and Power, J. B. (1975): Polyethylene glycol-induced uptake of microorganisms into higher plant protoplasts: Ultrastructural study. *Plant Sci. Lett.,* 5:269–274.
13. Davidson, R. L. (1974): Gene expression in somatic cell hybrids. *Annu. Rev. Genet.,* 8:195–218.
14. Dorion, N., Chupeau, Y., and Bourgin, J. P. (1975): Isolation culture and regeneration into plants of Ranunculus sceleratus L. leaf protoplasts. *Plant Sci. Lett.,* 5:325–331.

15. Evans, P. K., Keates, A. G., and Cocking, E. C. (1972): Isolation of protoplasts from cereal leaves. *Planta,* 104:178–181.
16. Gleba, Y. Y., Butenko, R. G., and Sytnik, K. M. (1975): Fusion of protoplasts and parasexual hybridization in Nicotiana tabacum L. *Dokl. Akad. Nauk S.S.S.R.,* 221:1196–1198.
17. Gosch, G., Bajaj, Y. P. S., and Reinert, J. (1975): Isolation, culture, and fusion studies on protoplasts from different species. *Protoplasma,* 86:405–410.
18. Grambow, H. J., Kao, K. N., Miller, R. A., and Gamborg, O. L. (1972): Cell division and plant development from protoplasts of carrot cell suspension cultures. *Planta,* 103:348–355.
19. Harvey, P. H., Levings, P. S., and Wernsman, E. A. (1972): The role of extra chromosome inheritance in plant breeding. *Adv. Agronomy,* 23:1–27.
20. Holl, F. B. (1975*a*): Innovative approaches to genetics in agriculture. *Can. J. Genet. Cytol.,* 17:517–524.
21. Holl, F. B. (1975*b*): Host plant control of inheritance of dinitrogen fixation in Pisum-Rhizobium symbiosis. *Euphytica,* 24:767–770.
22. Kartha, K. K., Michayluk, M. R., Kao, K. N., Gamborg, O. L., and Constabel, F. (1974): Callus formation and plant regeneration from mesophyll protoplasts of rape plants (Brassica napus l cv zephyr). *Plant Sci. Lett.,* 3:265–271.
23. Kao, K. N., and Michayluk, M. R. (1974): Method for high frequency intergeneric fusion of plant protoplasts. *Planta,* 115:355–367.
24. Keller, W. A., and Melchers, G. (1973): Effect of high pH and calcium on tobacco leaf protoplast fusion. *Z. Naturforsch.* [*C*], 28:737–741.
25. Kevei, F., and Peberdy, J. F.: *J. Gen. Microbiol.* (*in press*).
26. Kurz, W. G. W., and LaRue, T. A. (1975): Nitrogenase activity in rhizobia in absence of plant host. *Nature,* 256:407–409.
27. Littlefield, J. (1964): Selection of hybrids from matings of fibroblasts in vitro and their presumed recombinants. *Science,* 145:709–710.
28. Marton, L., and Maliga, P. (1975): Control of resistance in tobacco cells to 5-bromodeoxy uridine by a simple Mendelian factor. *Plant Sci. Lett.,* 5:77–81.
29. McNaughton, I. H. (1973): Resistance of Raphanobrassica to clubroot disease. *Nature,* 243:547–548.
30. Melchers, G., and Labib, G. (1974): Somatic hybridization of plants by fusion of protoplasts. I. Selection of eight resistant hybrids of "haploid" light sensitive varieties of tobacco. *Mol. Gen. Genet.,* 135:277–294.
31. Messerschmidt, N. (1974): Callus formation and differentiation of protoplasts isolated from cotyledons of Pharbitis nil. *Z. Pflanzenphysiol.,* 74:175–178.
32. Motoyoshi, F. (1972): Protoplasts isolated from callus cells of maize endosperm: Formation of multinucleate protoplasts and nuclear division. *Exp. Cell Res.,* 68:452–456.
33. Nagata, T., and Takeba, I. (1971): Plating of isolated tobacco mesophyll protoplasts on agar medium. *Planta,* 99:12–20.
34. Pagan, J. D., Child, J. J., Scowcraft, W. R., and Gibson, A. H. (1975): Nitrogen fixation by Rhizobium cultured on a defined medium. *Nature,* 256:406–407.
35. Poirier-Hamon, S., Rao, P. S., and Harada, H. (1973): Culture of mesophyll protoplast and stem segments of Antirrhinum majus (snapdragon): Growth and organization of embryoids. *J. Exp. Bot.,* 25:752–760.
36. Power, J. B., Cummins, S. E., and Cocking, E. C. (1970): Fusion of isolated plant protoplasts. *Nature,* 225:1016–1018.
37. Power, J. B., Frearson, E. M., George, D., Evans, P. K., Berry, S. F., Howard, C., and Cocking, E. C. (1976): The isolation, culture and regeneration of leaf protoplasts in the genus Petunia. *Plant Sci. Lett.,* 7:51–55.
38. Power, J. B., Frearson, E. M., Hayward, C., and Cocking, E. C. (1975): Some consequences of fusion and selective culture of Petunia and Parthenocissus protoplasts. *Plant Sci. Lett.,* 5:197–207.
39. Power, J. B., Frearson, E. M., Hayward, C., George, D., Evans, P. K., Berry, S. F., and Cocking, E. C. (1976): Somatic hybridisation of Petunia hybrida and P. parodii. *Nature,* 263:500–502.
40. Power, J. B., and Scowcroft, W. R. (1975): Somatic cell genetics and plant hybridisation.

In: *Form, Structure and Function in Plants,* edited by H. Y. Mohan-Ram, J. J. Shah, and C. K. Shah, pp. 88–93. Sarita Prakashan, India.

41. Riley, R., and Bell, G. D. H. (1959): The evolution of synthetic species. In: *Proceedings of the First International Wheat Genetics Symposium,* pp. 161–179. Winnipeg Public Press, Winnipeg.
42. Schieder, O. (1975): Regeneration of haploid and diploid Datura-innoxia Mill mesophyll protoplasts to plants. *Z. Pflanzenphysiol.,* 76:462–466.

Recombinant Molecules: Impact on Science and Society, edited by R. F. Beers, Jr. and E. G. Bassett. Raven Press, New York © 1977.

21. Plant Hybrids by Fusion of Protoplasts

Georg Melchers

Max-Planck-Institut für Biologie, D-74 Tübingen, West Germany

INTRODUCTION

This chapter reports the results of prior publications and some unpublished studies done by my group and by guests in my laboratory. I will discuss the implications of these findings, and their relationships to material already published and to the findings of other investigators. No attempt will be made to do complete justice to scientific developments less than a few years old because, it seems to me, it is still too early for such an attempt. Unfortunately, the reviews presently available [Cocking (14) and Chaleff and Carlson (9)] do not fill this need in a completely satisfactory manner. This is even more unfortunate as a thorough reading of the original literature is becoming more and more uncommon.

This section deals, in the main, with applications of plant breeding. Two new and unconventional methods of plant breeding that have shown practicality are the following:

1. *Using the techniques of Guha and Maheswari (23), it has been possible in some plants to shorten considerably the time for the breeding of a new cultivar by the recombination of traits from present varieties.* Starting with F_1 plants, one does not need to obtain an F_2 generation by self-fertilization or by sibling crosses; one can generate plants directly from the products of meiosis. These contain the recombinations either as haploids or as homozygotes, obtained either spontaneously or by treatment with colchicine. In Japan (48) and China (17) new cultivars of tobacco have been bred in this fashion; thus, there is no rationale for continuing to use the conventional procedures in tobacco breeding. In the case of the major crop plants such as rice, maize, wheat, barley, rye, potatoes, and rape seed, these new methods of breeding will become applicable only after considerable experimental effort. Only time will tell how successful and how important these unconventional breeding techniques will be for these plants. So far as I am aware, the greatest efforts in this area are taking place in China (17,34, 35,53,76–78), where hundreds of scientific and technical workers are engaged in such programs. At the Max-Planck Institute for Plant Genetics,

Drs. Hoffmann, Thomas, and Wenzel (72–74,79,80) direct three projects that attempt to apply this technique and that cited below to rye and rape seed.

2. *Employing the techniques of microbial genetics in the selection of mutants during the cell culture of haploid plants is certainly more effective than the prior methods of selection and propagation of mutants.* I am not aware of any new variety of crop plant that has been bred by the latter technique, although demonstrative examples have been described (5,6,36, 38,41).

Later we will hear about some more (and some less) successful and reproducible experiments leading to the formation of somatic hybrids. Up to now, only hybrids that can be sexually generated have been obtained by fusion. At the present time it is not possible to appraise the importance somatic hybridization will have for plant breeding, as compared to the methods discussed under *1.* and *2.* above. Possibly, the incompatibility between two types of plants may occur only at sexual fertilization, and if this obstacle is avoided by fusion of somatic cells, no other incompatibilities will occur during the division of cells, morphogenesis, or the next sexual phase. However, there are no guarantees: a number of cases are known in which two types of plants can be crossed, but the hybrid embryos or the plants do not develop normally (24,25,36,67). My question for colleagues who believe that incompatibility of genomes is limited mainly or exclusively to fertilization is, What basis can you cite for this optimism?

I think that it is quite dangerous in the case of plant breeders, and even more dangerous in the case of politicians, to hold forth hope for early spectacular results from somatic hybridization in order to help feed the hungry world. We should remember that a sexual hybrid between wheat and rye has existed for decades in many laboratories. It has been worked on by commercial plant breeders at great expense, and, nevertheless, today we have no important advances for the nourishment of mankind. One can say only that this may be accomplished within a perceptible time.

Colleagues who want to introduce cell organelles or pure DNA into protoplasts or into cells and whole plants, and thereby wish to "manipulate" the genotype of the plants, might ask themselves critically how high, with good conscience, they should raise the expectations of plant breeders.

As stated earlier, two unconventional methods are available for use in plant breeding. These methods would already have given more practical results if the work in our institutes were more practically (and less academically) oriented and if the work had been more strictly organized and performed with more technical help. Above all, in applied research, money is wasted if too little is invested for definite, clearly defined projects and if too many ill-defined investigations are poorly supported. The procedures of applying microbiological methods to higher plants mentioned in *2.* should be handled as has been done with fungal genetics. The Proceedings of the Guelph Symposium (1974) provides a review of the recent situation (31).

THE PROTOPLASTS

I will not review what has been reported or reviewed in the last few years about the preparation of naked plant cells, the "protoplasts." References are found in the chapter of Cocking (*this volume*). Many different aspects of protoplasts are of interest. The criteria by which the quality of protoplasts are judged vary widely. They include the appearance in the light microscope of the so-called indications of their viability, the ultramicroscopic structure, the occurrence of metabolic reactions, and so forth. If the protoplasts can resynthesize a cell wall, this indicates that they possess functions which are also characteristic of intact cells in tissues. Mature plant cells that have reached their final stage of differentiation normally do not divide further. If these cells then form protoplasts that are able to divide, this is a sign of their special vitality, but it is not self-evident and this is certainly not currently possible for all plants and all tissues. In the case of rice, for example, protoplasts from the mesophyll of mature green leaves have not yet been made to divide, whereas protoplasts from white leaf sheath or from callus cultures of this plant show cell division and regeneration of roots (18).

There are numerous media for the culture of protoplasts. It has rarely been shown that only one specific medium is suitable for the culture of a definite type of protoplast. The condition of the plant materials from which the protoplasts are derived is very important. The plants should be healthy and sufficiently moist; overexposure to the sun is to be avoided during their culture. Callus cultures grown under completely controlled conditions, or better, submerged cell cultures, are valuable [for example, the soybean culture of Gamborg and co-workers (30)]. By complete control of the cultural conditions, one can obtain material that is phenotypically homogenous. However, a disadvantage of such cultures is that they may no longer have the chromosomal composition of the original plant material. In other cases, the cells of such cultures are not genetically homogenous, principally because of varying numbers of chromosomes (37,38,55–62). One should use his judgment and choose his starting material according to the task at hand. One has the impression that in many instances the experimental materials have been chosen not on the basis of their suitability but because of the background and experience of the investigators designing the experiment.

For some purposes, such as studies of viral replication, protoplast cultures need live only for a few days; it is not necessary that cell division occur (71). Although the problem of virus infection and the multiplication of viruses in protoplasts lie somewhat outside of my topic, I would like to mention them because the technical side of this matter is within the main theme of this volume. It is reasonable to use protoplasts as receptors for the introduction of foreign genetic material into a plant cell. The simplest case, which occurs in nature, is infection by a virus. Before investigators were able to infect protoplasts with virus or introduce pure viral RNA, it was virtually impossible to study the kinetics of replication of plant viruses. Studies have shown that only about 10^3 of the 10^8 cells of a leaf could be synchronously infected by rubbing a viral solution on the surface of the leaf. The infection gradually spread hemispherically throughout the leaf from foci of infected cells. Although viral replication in the earliest infected cells was not yet complete, infection of neighboring cells was observed. Shortly thereafter, asynchronous replication occurred at indeterminable proportions. The situation is quite different in viral infection of a protoplast suspension. Nearly all the cells are simultaneously infected, and upon reaching a population of 10^6 to 10^7 virus particles per cell an almost synchronous replication of the virus takes place. Under the most favorable conditions, about 10^5 RNA molecules per protoplast are available (63,65). In transfection experiments with phage RNA, a similar ratio is observed. Because of the possibility of infecting protoplasts with viruses, the plant virologist can now conduct experiments that bacterial virologists have been able to perform for decades. It is astounding how little use is made of the new methods.

The pioneering work on viruses and protoplasts by Cocking and coworkers (11–13,15,16,55,56) goes back to the 1960s, and the precise, quantitative investigations of Takebe et al. (1,49,68,70,71) date back 6 to 8 years ago. This approach has yet to be used widely in the study of the localization of viral multiplication in or on various cell organelles. Otsuki and coworkers (50–52) applied these methods to the problem of how different viruses, but not variants of the same virus, exert interference.

In the presence of Ca^{2+}, tobacco protoplasts are able to withstand a pH of 11 for 30 min at 37°C (32). This implies that one can adjust the external medium of protoplasts to such unusual conditions that, for example, RNases are not active. This stability makes it possible to induce infection using tobacco mosaic virus RNA (very sensitive to RNase) at a concentration only $^1/_{1,000}$ of that used in more acidic solutions, and to achieve approximately 100% infection of the protoplasts (65). Little use has been made of this technique in attempts to introduce other nucleic acids into protoplasts. The introduction of auxin autotrophy by DNA of *Agrobacterium tumefaciens* by this procedure has not been successful. Sarkar's review (64) is an excellent compilation of studies on protoplasts and viruses.

FUSION

Fusion of cells is the prerequisite for fertilization and occurs in all sexually propagated organisms. In somatic tissues of higher plants, the cells frequently communicate with one another by means of plasmodesmata; so to speak, they are "fused." During the dissolution of the cell walls with enzymes (pectinase, which dissolves the middle lamellae and thereby liberates the cells; and cellulase, which removes the cell walls), fusion of neighboring cells sometimes takes place. During these processes, the plasmodesmata are not torn off but become enlarged (58). This "spontaneous fusion" is of no help if the goal is fusion of cells of different tissues, or, indeed, of different plants.

Before practical fusion media were known, several attempts were made to carry out somatic hybridization. We (41) mixed pectinase-treated cells from two different plant varieties in the hope that fusion could be achieved from protoplasts *in statu nascendi.* After cellulase treatment, two of the five samples demonstrated coagulation of the protoplasts. One of the two coagulated samples did not give any living material; however, the other one gave many thousands of dividing protoplasts. Unfortunately, it could not be determined whether fusion had occurred in the material because the cell clumps were not transparent.

In this experiment we used protoplasts from the chlorophyll-defective tobacco varieties "yellow green" and "yellow crittenden." These varieties complement each other, resulting in a completely normal colored F_1 hybrid; 3,625 plants were regenerated and grown.

In addition to the above studies, progeny were prepared from several plants that contained 96 chromosomes and were, therefore, suspected of possibly being fusion hybrids. We did not find, however, a single fusion hybrid (43). We now know that spontaneous fusion of cells which are not immediately adjacent is very unlikely, even in the case of *statu nascendi.* Thus, spontaneous fusion of protoplasts appears to be possible only for cells that are neighbors. During the treatment with pectinase and cellulase, the cells are in a plasmolyzed condition; that is, the plasmodesmata of most of the cells are torn off. The surfaces of the plasmolyzed nascent protoplasts are probably similar to the properties of the free protoplasts. Protoplasts have a negative surface charge of around 30 mVζ potential, as determined electrophoretically by T. Nagata during his recent stay in our laboratory. Thus, on the basis of electrostatic considerations, a fusion of such protoplasts is impossible. The negative charge decreases in solutions containing Ca^{2+} ions. This finding now provides the explanation for the successful fusion experiment of Keller and Melchers (32). [See also Devor et al. (19) who, in their model for the lipid membrane, postulate that the presence of polyvalent cations is necessary for fusion to occur.]

It was found by Grout et al. (22) that monovalent cations can also reduce the surface charge. We, however, have never observed fusions of mesophyll protoplasts. The same phenomenon was observed in solutions of $Mg(NO_3)_2$ (32). The coagulation observed in two of the five samples in the above-mentioned experiment may have resulted from a residue of a surface-active cleaning agent in the culture vessels. If we had obtained fusion hybrids from this experiment, the results could hardly have been reproducible because a reliable fusion medium was not known at the time.

Another attempt at inducing somatic hybridization was that undertaken by Power et al. (56,57). In these studies, they observed fusion in a $NaNO_3$ solution of embryonic, weakly vacuolated cells, but none could be observed with terminally differentiated, strongly vacuolated cells. Carlson et al. (8) claimed to have observed "25% fusion events" in a $NaNO_3$ solution containing protoplasts. The latter were derived from mesophyll; that is, from differentiated, vacuolated cells. The question is, What is one fusion event and how does one calculate the percentage? A further question: Was fusion between mesophyll protoplasts of *Nicotiana glauca* and *N. langsdorffii* ever observed, except in the one published experiment using $NaNO_3$? To obtain approximate quantitative results, Keller and Melchers (32) used random samples and estimated the number of protoplasts undergoing fusion as compared to the total number of the protoplasts present in a given field of view; successful fusion events included the fusion of many protoplasts that never divided. Unfortunately, Carlson et al. (8) provided only a brief description of their experimental methods and quantitation of results.

If one wants to obtain callus and possibly plants out of protoplast fusion, then one should obtain as many one-to-one fusions of A + B as possible, A and B being the two types of plants of which somatic hybridizations are wanted. One must strive to make the ratio of A + B fusions as large as possible compared to the nonfused protoplasts A and B and the A + A and B + B fusions. This is especially true if a good selective system is not available for selecting the A + B combination over the undesired combinations. We are attempting to do this in preliminary investigations by changing the surface charge of protoplasts from −30 mVζ potential to a positive value. This procedure could thus create a type of "artificial sexuality." Nagata has succeeded in obtaining a charge of about +5 mVζ potential using a phospholipid synthesized by Hj. Eibl of the Max-Planck Institute of Biophysical Chemistry, Göttingen. This surface charge is so weak that the protoplasts still coagulate; however, no effect on their viability is seen. The repulsive forces are nearly compensated for by the van der Waals forces. The weakly positively charged protoplasts stick only very loosely to negatively charged glass or plastic surfaces. These investigations on surface charge alterations will be continued as new synthetic products are made available by Dr. Eibl.

Polyethylene glycol (PEG) is a useful agent for carrying out the aggregation of protoplasts (29,75). True fusion appears to occur chiefly during removal of the PEG. Especially good results are obtained by using an alkaline solution of high calcium content to remove the PEG (27). At the moment, habit rather than rational understanding seems to speak for the use of the Ca^{2+}-pH method or its combination with PEG to effect protoplast aggregation.

CULTURE OF FUSION PRODUCTS

If one wants to obtain hybrid plants from hybridized cells produced by fusion of protoplasts, one must carefully cultivate the fused protoplasts so that they will divide and form callus. Then one has to regenerate plants from such a callus, a technique described by Takebe et al. (69). After the fusion process, the protoplasts can be first cultured in liquid medium in small Petri dishes (or in droplets) or plated directly into liquid agar (2–4). The agar method is recommended if one wants to follow individual cells or their fusion products and avoid the complications of secondary coagulations. Many protoplasts divide much more easily in liquid droplets, and, therefore, one must sacrifice the advantages of agar culture. Unfortunately, the number of plants able to go through the entire cycle (plants → protoplasts → callus → plants) is still very small. We "plant people" cannot console ourselves with the fact that in animal and human cells somatic genetics cannot be performed without direct involvement of sexual genetics. Fortunately, the situation is beginning to change (26,45) and conventional sexual genetics in humans is so handicapped that human geneticists must be happy to have the newly developed somatic cell genetic techniques. Nevertheless, this is not the situation in conventional plant genetics. The application of somatic genetics to animal cell studies was successfully achieved years ago. As has been done in the case of fungi, one should attempt in studies of higher plants to combine somatic cell genetics with conventional sexual genetics.

As soon as is possible, one should distinguish and separate the fusion hybrids from the cells of the parents. Schieder (66) has used an especially elegant technique involving fused protoplasts from the liverwort *Sphaerocarpos donnellii*. One mutant required nicotinic acid and the other mutant required carbohydrate since it was defective in the formation of chlorophyll. Only the hybrid is able to grow autotrophically on a medium lacking nicotinic acid. Furthermore, the hybrids can be identified by karyotype in this case.

In our recent work, we used the tobacco varieties "sublethal" (*ss*) and "virescent" (*vv*), which are strongly light sensitive; this sensitivity was shown to be caused by recessive, nonallelic genes (43,44). The seedlings of these plants grow slowly in normal light (Fig. 1). The sexual F_1 hybrids develop normal leaves and exhibit good growth. The $v \times s$ and $s \times v$ hybrids show no difference; hence, the extranuclear genetic information for visible markers is not different in the two varieties. When the seedlings are given large amounts of a sugar-containing nutrient solution, they become green and lose their sensitivity to light. Reduction of the oxygen pressure of the atmosphere also reduces the light sensitivity. Especially for the "virescent" strain, it appears that under normal conditions the chlorophyll is destroyed by photo-oxidation.

If one wants to achieve good selection for hybrid calluses after a fusion experiment, one should illuminate the material with 10,000 lux intensity of fluorescent light (Osram L 65 W/32 Warmwhite de luxe fluorescent lamps), reduce the organic components of the medium to $\frac{1}{5}$ or $\frac{1}{10}$ of the normal content, and reduce the benzyladenine concentration by 50%. We start using "haploid" plants of *v* and *s* (actually these plants are dihaploids with 24 chromosomes). The hybrid callus tissues containing 48 chromo-

FIG. 1. Seedlings 1.5 months after germination in a greenhouse during winter. **1:** "Sublethal" = *ss;* **2:** $(s \times v)F_1$; **3:** $(v \times s)F_1$; **4:** "virescent." These specimens demonstrate the light sensitivity of *ss* and *vv*, the complementation in the hybrids and the identity of $(s \times v)$ and $(v \times s)$. Note that there are no differences in the plastom (the total genetic information of the plastids).

somes are in no way inferior in growth to *v* and *s* calluses. The hybrid calluses are not only distinguished by their dark green color (Fig. 2), but they also grow faster than the yellowish or occasionally light green calluses generated by *v* and *s*. However, confusion is possible with *v* or with *vv*. In every case, the regenerated plants in the greenhouse are easily distinguished (Fig. 3). In four independent experiments, 20 hybrid plants developed if only the numbers 1, 2, 3, etc. were counted. However, if 1a, 2a, 3a, etc. were counted, the population of hybrid plants was increased to 43 (see Fig. 4). It is improbable but not impossible that the calluses, being small and compact, are separated mechanically by the first diluting step (44).

Gleba et al. (21) used the white part of an albino variegated tobacco variety which was, in nuclear genes, normal and thus had the genotype *su su*. Its heterozygotic state (*Su su*) is a gold-yellow plant. The two types of protoplasts were mixed and allowed to fuse. Plants were obtained which could not have been derived from simple mixing of the protoplasts and must have arisen as a result of fusion, since variegated plants with yellow portions were obtained. This system is not particularly elegant, because it is difficult to foresee the karyological behavior of the primary products (having 96 chromosomes) in the culture medium.

We have determined the karyotype of all our hybrids in not just one specimen from each fusion, but often in plants from several calluses which arose from the same fusion (Fig. 5). In these investigations, it was shown that during culture some calluses acquired irregular chromosome numbers. Some fluctuations from the normal chromosomal number of 48 can readily be seen very early; so early, in fact, that we suspect, particularly in the

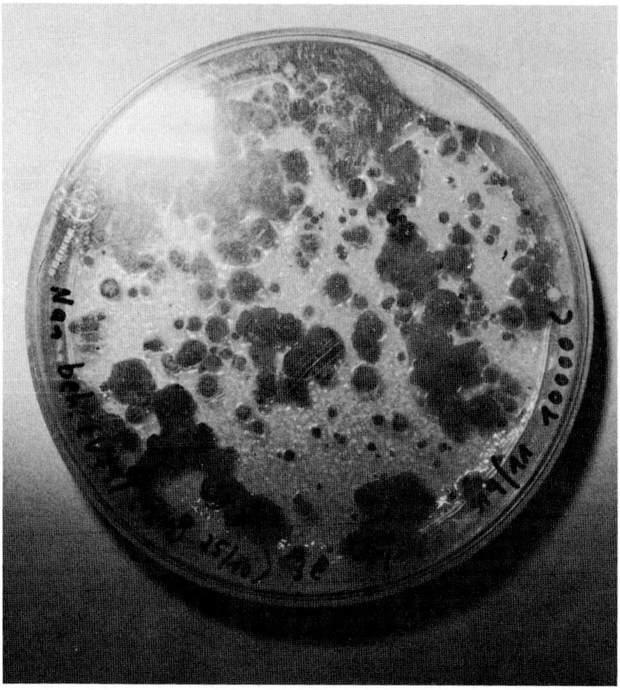

FIG. 2. Culture of calluses resulting from the first successful fusion trial. Cultivated in Nagata-Takebe medium without auxin and under high light intensity. Photograph taken approximately 2 months after fusion occurred. One callus developed a dark green regenerate that had 48 chromosomes, normal hybrid characteristics, and offspring with *s*- and *v*-like attributes.

case of triploids, that these have arisen as a result of the fusion of three protoplasts (44).

It has been known for a long time that hybrids between *Nicotiana glauca* and *N. langsdorffii* will form tumors either spontaneously or, more frequently, after wounding (81). These tumors are auxin autotrophic, as are the materials obtained either by habituation or after infection with *Agrobacterium tumefaciens* (20). These characteristics are exhibited not only by the sterile F_1 hybrids but also by the fertile amphidiploid hybrids that have arisen as a result of chromosome doubling. It should be easy to separate somatic hybrids with these characteristics from their parents by use of an auxin-free medium. In experiments of Carlson et al. (8), a plating efficiency of 0.01% was obtained using protoplasts prepared from amphidiploid hybrids. The medium used was that of Nagata and Takebe (47) containing auxin and kinetin. However, after the treatment of about 2×10^7 protoplasts in a 1:1 mixture, only 33 calluses were obtained. The results of Chupeau et al. (10), who found a 5% plating efficiency for *N. glauca* and

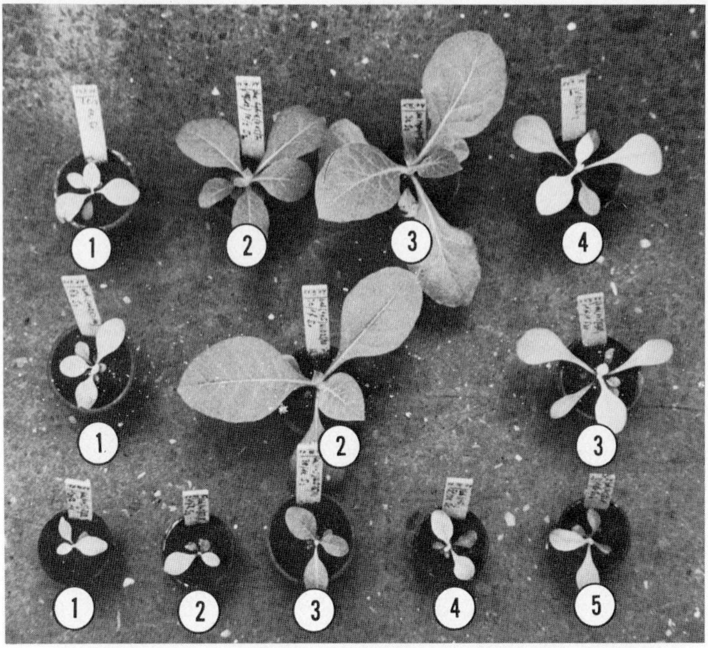

FIG. 3. Plants regenerated from callus cultures following protoplast fusion; grown in greenhouse under weak shade. **Upper, 1:** *s* (from control); **2:** ($v \times s$)F_1 sexual hybrid; **3:** ($v \times s$)F_1 sexual hybrid polyploid; **4:** *v* (from control). **Middle, 1:** *s* nonfused from a fusion trial; **2:** ($v + s$) fusion hybrid; **3:** *v* nonfused from a fusion trial. **Bottom, 1** and **2:** *s* plants; **3:** ($v + s$) a smaller plant from the same callus used next above row; **4** and **5:** *v* plants.

a 20% plating efficiency for amphidiploid hybrids (no results were reported for *N. langsdorffii*) are not relevant to this discussion since the protoplasts were cultivated on a different (liquid) medium (46).

Carlson (7) stated in response to a question that he considers it "good fortune and not part of our intended selective system" that 33 calluses appearing after treatment with sodium nitrate were shown to be auxin-autotrophic upon transfer to an auxin-free medium. Three of these calluses regenerated to form amphidiploid hybrid plants, which is to be expected if somatic cells were fused. Progeny were apparently obtained from only one of the three plants. Using the same $NaNO_3$-containing medium, other investigators have not been able to obtain fusion of protoplasts of vacuolated mature cells (*see above*). The survival of 33 calluses out of 2×10^7 protoplasts on a nonselective medium is good luck. Whether the calluses not regenerating new plants (but which were undoubtedly auxin autotrophs) were also hybrids is not known; habituation or a mutation to auxin autotrophy could have occurred.

Until recently, I have considered the above to be one of three examples

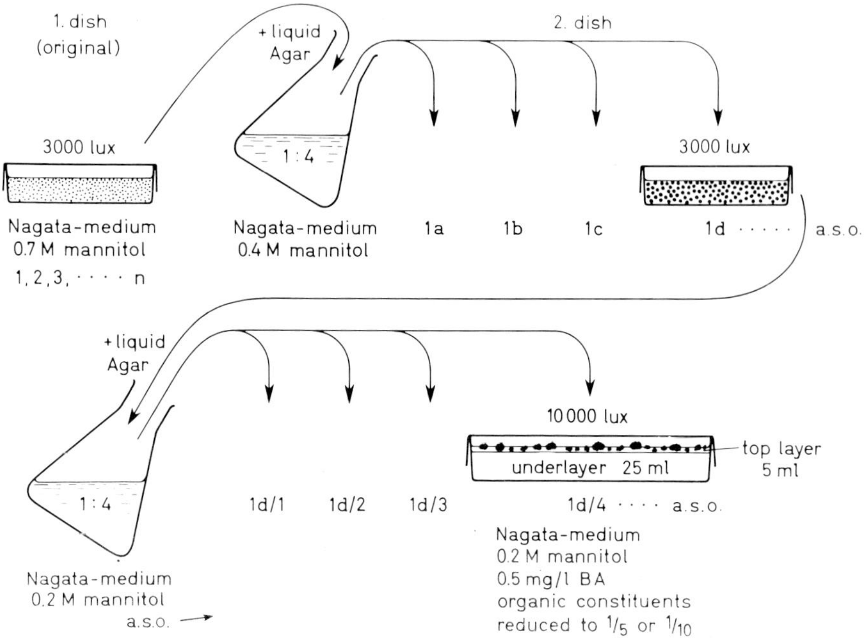

FIG. 4. A scheme of the culture of protoplasts and calluses derived from protoplast fusion. Experimental protocol as follows: morning—preparation of protoplasts; early afternoon—fusion and immediate plating in 60-mm dishes (1, 2, 3, etc.) and illuminating at 300 lux for 72 hr followed by 3,000 lux for approximately 2 weeks. Dishes then diluted approximately 1:4 with liquid agar and replated (1a, 2b, etc.). Approximately 2 weeks later a third dilution was made with replating onto underlayer in 90-mm dishes and maintained under about 10,000 lux. All incubations performed at 28°C.

of somatic hybridization and the first of such in plants (40,43). But from the contributions of Kung et al. (33), it has now become known that the plastids carry the genetic determinants of the peptide pattern of the large subunit of the "Fraction 1 protein." The peptide pattern of the descendants of the single plant obtained by Carlson et al. (8) appears to be that of the *N. glauca* and that of the sexual hybrid, *N. glauca* ♀ × *N. langsdorffii* ♂. One would expect that a mixture of peptides of the large subunits of *N. glauca* and *N. langsdorffii* would be present in the hybrids resulting from somatic fusion. In Carlson's hybrid, however, the single characteristic enabling one to distinguish a sexual from a somatic hybrid strongly suggests that the plant is a sexual hybrid. I consider it a justifiable hypothesis that it is a sexual hybrid, particularly in view of the history of its origin. Kung et al. (33) think that the genetic information of the *N. langsdorffii* plastids was "not expressed," but that only describes the fact that the small subunit of Fraction 1 (coded for by nuclear genes) contains peptides from both parents, whereas those of the large subunit (coded for by plastids) contain genetic information from only one parent.

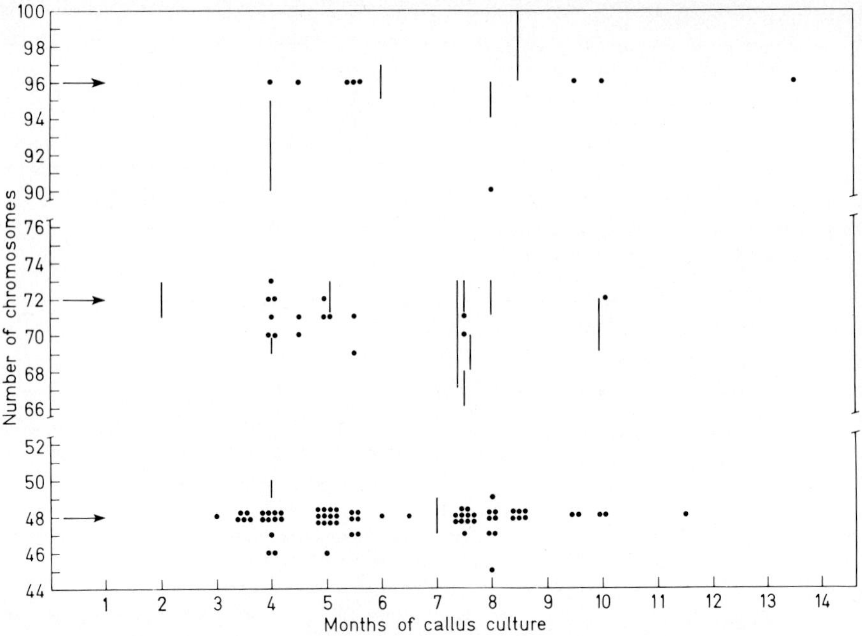

FIG. 5. Chromosome number of somatic hybrids at various callus ages. Hybrids created by fusion of "haploid" (24 chromosomes) protoplasts from two different varieties of tobacco. Additional details in ref. 44.

In 1974 I saw calluses of somatic hybrids between *N. glauca* and *N. langsdorffii* which Kao had made using the PEG technique. Since that time, plants must have been regenerated from these calluses. It should not be difficult to determine just how the peptide pattern of the large subunit of the Fraction 1 protein is constituted, using a somewhat greater number of plants. One can ascertain whether the large subunit of the Fraction 1 protein is only that of *N. glauca* (or sometimes that of *N. langsdorffii*) or whether a mixture is present. My hypothesis can easily be refuted and that of Kung et al. (33) confirmed if a mixture of these peptide patterns is not found in any plants derived from the somatic hybrids, and especially if no peptide patterns of *N. langsdorffii* are found.[1] Including the *N. glauca* + *N. langsdorffii* hybrids, there will be only three known cases in which plants have been generated from somatic cell hybrids. But see also the Petunia hybrids described by Cocking (*this volume*). All four of these types

[1] Note added in proof—At the recent (July, 1976) UNESCO-UNDP-IRCO Conference "Cell Genetics in Higher Plants," H. H. Smith briefly reported on these hybrid plants. None had the morphology and chromosome number of the amphidiploid and the "parasexual hybrid" (8). Thus, the question of what was the nature of the first published "parasexual hybrid" remains unanswered.

of plants can be produced sexually, at least in principle. This is also true for *Sphaerocarpos* gametophytes (from which the sexual hybrids would be made) as they would have exactly the same genetic constitution as the protoplasts prepared from haploids of differing sexes. In this case, one could also make somatic hybrids from the same sex, thereby providing an interesting possibility of varying the sexual behavior of other dioecious plants. Even if the grandiose plans to somatically hybridize legumes with the grasses are not soon realized, there still remain modest, but nevertheless quite interesting, applications for these new methods.

We have made our *v* + *s* (fusion) hybrids from vegetative haploid material and obtained somatic hybrids that were identical in ploidy to the sexually produced hybrids. The situation is different if one begins by using the normal diploid protoplasts, or if one chooses as one partner a diploid and, as the other, a haploid. In the latter case, one obtains triploids and by use of colchicine one can thus proceed to hexaploids. In many cases, however, one can also obtain the hexaploid in the conventional manner by crossing diploids with tetraploids.

Figure 6 represents the comparison of conventional sexual crosses (×) with the somatic hybridization using fusion (+). This scheme clearly demonstrates that in the case of plants such as *Nicotiana* wherein genetic information of the plastids is not carried by the pollen, the phenomenon of maternal inheritance should occur in sexual crosses. Somatic hybridization would be expected to provide the possible transfer of plastid-borne information from both parents into hybrids. In certain cases this phenomenon can be of some importance for plant breeding.

Kao and co-workers (27,28) have extensively employed cells of widely separated parents to produce cell hybrids which underwent further division (soybeans + peas, soybeans + barley, soybeans + *N. glauca*). It is improbable that one will ever regenerate plants out of such combinations. In order to produce a somatic hybrid that could not be produced sexually, we think one should choose as donors plants as closely related as possible and in which the cycle: plant → protoplast → callus → plant is possible. *Nicotiana tabacum* and *Petunia hybrida* are obvious candidates; Pogliaga (54) has successfully crossed *N. tabacum* with *P. parodii*. We have observed that a cross of *Nicotiana tabacum* (*ss* ♀) with a petunia mutant (Mu_1 ♂) does not give fertile seeds, and we have failed in numerous attempts to obtain hybrid plants by fusion of *Petunia hybrida* Mu_1 and *N. tabacum ss*, *s*, and *v*. From some 7×10^7 protoplasts, we have obtained thousands of calluses and hundreds of regenerated plants. Unfortunately, we do not know if one obtains complementation in putative hybrids from crosses of *s* or *v* with the yellow-green, light-insensitive petunia mutant Mu_1. Indeed, under the conditions in which *v* (*vv*) or *s* (*ss*) calluses are light green to yellow, the petunia calluses are green. During the search for possible

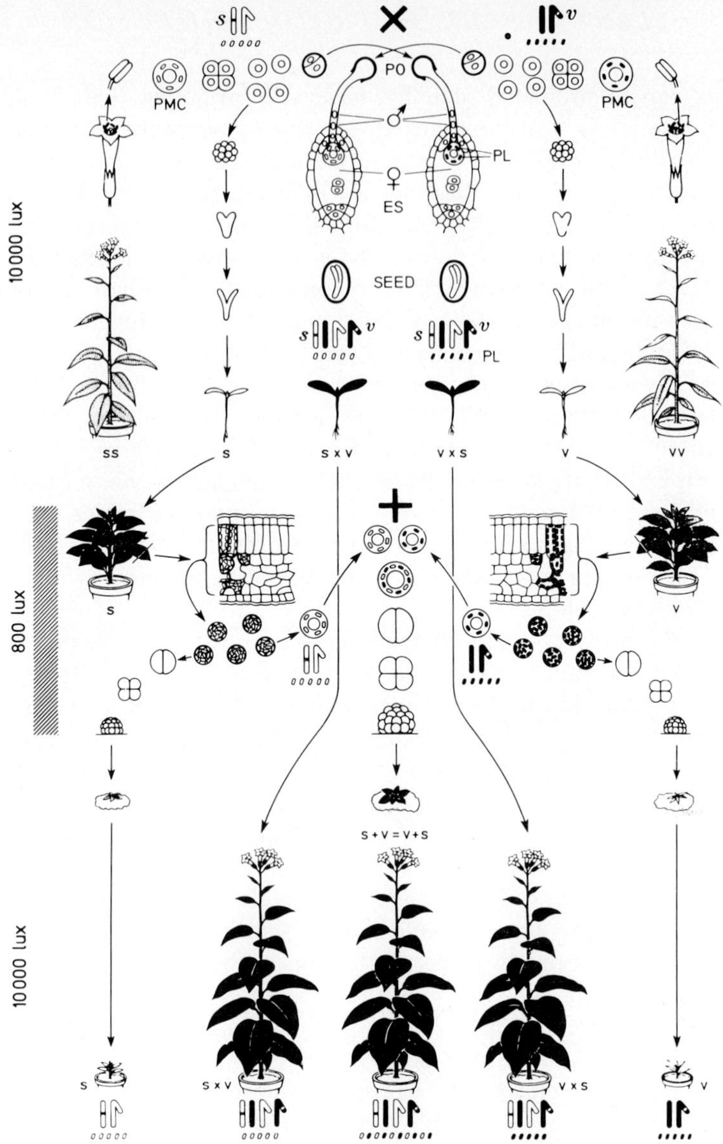

FIG. 6. Summary of conventional, sexual (X), and somatic (+) hybridization, depicting the selection of somatic hybrid cells, calluses, and plants complemented to normal chlorophyll color and light sensitivity. Cells originated from two light-sensitive, nonallelic recessive mutants of tobacco; *ss* and *vv* indicate the "diploid," and *s* and *v* the "haploid" state. Two pairs of chromosomes symbolize the genome. The plastom (see Fig. 1 legend) is depicted by *0* and ●, there being no difference between (*s* × *v*) and (*v* × *s*). This scheme also shows (a) production of "haploids" according to Guha and Maheswari (23); (b) conversion of "haploid embryo" plants, in faint light (800 lux), to more or less normal green plants; (c) preparation of protoplasts from these plants; (d) fusion of *s* and *v* protoplasts and the selection, under high light intensity, of the "hybrid callus" and plants therefrom or the slow growth and pale color of plants, under 10,000 lux, derived from the nonhybridized protoplasts. PMC, pollen mother cells; PO, pollen with tube (microgametophyte); ES, embryo sack; PL, plastids or proplastids transmitted only by the egg cell (not by the pollen tube) and mixed after fusion. (From ref. 42.)

dark greens, we have nearly always found regenerated petunias, several of which had varying chromosome numbers. A few calluses contained mosaics of petunia and tobacco, and some do not regenerate to plants.

During a visit to our laboratory, Dr. Zenkteler attempted to obtain sexual hybridization *"in vitro"* between *Petunia hybrida* and *Nicotiana tabacum.* He frequently obtained fertilization, but he always found proembryos that contained only a few cells and never found full embryonic development. This shows clearly that the incompatibility between the genomes does not lie in fertilization but can originate in postfertilization processes. Naturally, we cannot be certain that a similar event has occurred in our fusion experiments, that is, failure of embryonic development following a successful fusion of somatic cells. If one carries out a fusion such as *s* or *v* + petunia (24 + 14 = 38 chromosomes) or *ss* or *vv* + petunia (48 + 14 = 62 chromosomes) and does not find the correct number of chromosomes, one cannot unequivocally state that no hybridization has occurred. It is known that cell hybrids produced from human and mouse cells continuously lose their human chromosomes. If the culture medium contains a selective factor(s) for one or more of the human chromosomes, then the genome which contains mouse and a few human chromosomes can be maintained for a long time. If one has been able to establish such a hybrid plant cell culture, then one should make every effort to generate plants from it.

SUMMARY

1. Today one can greatly accelerate plant breeding studies by the use of plants generated from the products of meiosis. These techniques have thus far made only sparse contributions toward improvement of plants that are important for feeding mankind.

2. Mutation and selection, especially in haploid cells or protoplast cultures, have applicability in plant breeding at the present time.

3. Protoplasts of higher plants frequently are resistant to unnatural conditions, for example, high pH. This enables them to take up RNA from solutions of comparatively low concentrations; the use of low concentrations apparently diminishes enzymic destruction of RNA taken up. The use of these unnatural conditions frequently increases the probability of taking foreign genetic information into the plant cells.

4. At the present time, it is possible with relatively few types of plants to demonstrate the cycle: plant → protoplast → callus → plant. The success of completing this cycle is a prerequisite for obtaining hybrid plants developed from the fusion of protoplasts prepared from the two different plants.

5. A further prerequisite is an effective fusion medium. Fusions occur at high pH in the presence of calcium ions; other parameters of the medium

favoring fusion are a low osmotic pressure and an elevated temperature (up to 37°C).

6. Tobacco protoplasts have a considerable negative surface charge; therefore, spontaneous fusion, except by neighboring cells, is prevented. Calcium ions were shown to neutralize the surface charge.

7. High rates of fusion are obtained using polyethylene glycol; this rate is enhanced by increased alkalinity and the addition of calcium ions.

8. Selected phospholipids are able to convert the surface charge of the tobacco protoplasts to a weak positive from a strong negative charge, suggesting a means of creating an artificial sexuality in somatic cells.

9. Numerous somatic *cell* hybrids can be formulated from donors having only a slight taxonomic relationship, whereas only four somatic *plant* hybrids have been described; the latter can also be prepared by sexual crosses.

10. Incompatibility is not only confined to the process of fertilization, but can also be observed at subsequent stages of plant development, as demonstrated by incomplete proembryo maturation.

11. In general, successful somatic hybridizations following fusion of protoplasts are strongly dependent on the use, as soon as possible, of an appropriate selection system. Recessive nonallelic genes restricting growth and development (but complementing one another in the hybrid, enabling normal growth) are especially useful.

ACKNOWLEDGMENT

The assistance of Drs. Milton Gordon and Mary-Dell Chilton in translating the manuscript is gratefully acknowledged.

REFERENCES

1. Aoki, S., and Takebe, I. (1969): Infection of tobacco mesophyll protoplasts by tobacco mosaic virus ribonucleic acid. *Virology,* 39:439–448.
2. Bergmann, L. (1959*a*): Über die Kultur von Zellsuspensionen von Daucus carota. *Naturwissenschaften,* 46:20–21.
3. Bergmann, L. (1959*b*): A new technique for isolating and cloning cells of higher plants. *Nature,* 184:648–649.
4. Bergmann, L. (1960): Growth and division of single cells of higher plants in vitro. *J. Gen. Physiol.,* 43:841–851.
5. Binding, H., Binding, K., and Straub, J. (1970): Selektion in Gewebekulturen mit haploiden Zellen. *Naturwissenschaften,* 57:136–139.
6. Carlson, P. S. (1973*a*): Methionine sulfoximine-resistant mutants of tobacco. *Science,* 180:1366–1368.
7. Carlson, P. S. (1973*b*): Towards a parasexual cycle in higher plants. *Colloque Int. CNRS,* 212:497–505 and 548–549.
8. Carlson, P. S., Smith, H. H., and Dearing, R. (1972): Parasexual interspecific plant hybridization. *Proc. Natl. Acad. Sci. U.S.A.,* 69:2292–2294.
9. Chaleff, R. S., and Carlson, P. S. (1974): Somatic cell genetics of higher plants. *Annu. Rev. Genet.,* 8:267–278.

10. Chupeau, Y., Bourgin, J. P., Missioner, C., Dorion, N., and Morel, G. (1974): Preparation et culture de protoplastes de divers Nicotiana. *C.R. Acad. Sci.* [*D*] (*Paris*), 278:1565–1568.
11. Cocking, E. C. (1965): Ferritin and tobacco mosaic virus uptake, and nuclear cytoplasmic relationships in isolated tomato fruit protoplasts. *Biochem. J.*, 95:28–29.
12. Cocking, E. C. (1966*a*): Electron microscope studies on isolated plant protoplasts. *Z. Naturforsch.* [C], 21b:581–584.
13. Cocking, E. C. (1966*b*): An electron microscopic study of the initial stages of infection of isolated tomato fruit protoplasts by tobacco mosaic virus. *Planta*, 68:206–214.
14. Cocking, E. C. (1970): Virus uptake, cell wall regeneration and virus multiplication in isolated plant protoplasts. *Int. Rev. Cytol.*, 28:89–124.
15. Cocking, E. C. (1972): Plant cell protoplasts—isolation and development. *Annu. Rev. Plant Physiol.*, 23:29–50.
16. Cocking, E. C., and Pojnar, E. (1969): An electron microscopic study of the infection of isolated tomato fruit protoplasts by tobacco mosaic virus. *J. Gen. Virol.*, 4:305–312.
17. Cooperation Group of Haploid Breeding of Tobacco of Shangtung Institute of Tobacco and Peking Institute of Botany, Academia Sinica (1974): Success of breeding the new tobacco cultivar Tan-Yuh Nr. 1". *Acta Botanica Sinica*, 16:300–303 (English summary).
18. Deka, P. C., and Sen, S. K. (1976): Differentiation in calli originated from isolated protoplasts of rice (Oryza sativa L.) through plating technique. *Mol. Gen. Genet.*, 145:239–244.
19. Devor, K. A., Teather, R. M., Brenner, M., Schwarz, H., Würz, H., and Overath, R. (1976): Membrane hybridization by centrifugation analyzed by lipid phase transitions and reconstitutions of NADH-oxidase activity. *Eur. J. Biochem.*, 63:459–467.
20. Gautheret, R. (1942): Hétéroauxines et cultures de tissues végétaux. *Bull. Soc. Chim. Biol.*, 24:13–47.
21. Gleba, Y. Y., Butenko, R. G., and Sytnik, K. M. (1975): Protoplast fusion and parasexual hybridization in Nicotiana tabacum. *XII International Botanical Congress, Leningrad*, Abstr., p. 290.
22. Grout, B. M. W., Willison, J. H. M., and Cocking, E. C. (1973): Interactions at the surface of plant cell protoplasts; an electrophoretic and freeze-etch study. *Bioenergetics*, 4:311–328.
23. Guha, S., and Maheswari, S. C. (1966): Cell division and differentiation of embryos in the pollen grains of Datura in vitro. *Nature*, 213:97–98.
24. Hollingshead, C. (1930*a*): A lethal factor in Crepis effective only in interspecific hybrid. *Genetics*, 15:114.
25. Hollingshead, C. (1930*b*): Cytological investigations of hybrids and hybrid derivatives of Crepis capillaris and Crepis tectorum. *Univ. California Publ. Agr. Sci.*, 6:55.
26. Illmensee, K., and Mintz, B. (1976): Totipotency and normal differentiation of single teratocarcinoma cells cloned by injection into blastocysts. *Proc. Natl. Acad. Sci. U.S.A.*, 73:549–553.
27. Kao, K. N., Constabel, F., Michayluk, M. R., and Gamborg, O. L. (1974): Plant protoplast fusion and growth of intergeneric hybrid cells. *Planta*, 120:215–227.
28. Kao, K. N., and Michayluk, M. R. (1974): A method for high frequency intergeneric fusion of plant protoplasts. *Planta*, 115:355–367.
29. Kao, K. N., and Michayluk, M. R. (1975): Nutritional requirements for growth of Vicia hajastana cells and protoplasts at a very low population density in liquid media. *Planta*, 126:105–110.
30. Kao, K. N., Miller, R. A., Gamborg, O. L., and Harvey, B. L. (1970): Variation in chromosome number and structure in plant cells grown in suspension cultures. *Am. J. Genet. Cytol.*, 12:297–301.
31. Kasha, K. J. (Ed.) (1974): *Haploids in Higher Plants—Advances and Potential*. University of Guelph Press, Guelph.
32. Keller, W. A., and Melchers, G. (1973): The effect of high pH and calcium on tobacco leaf protoplast fusion. *Z. Naturforsch.* [C], 28c:737–741.
33. Kung, S. D., Gray, J. C., Wildman, S. G., and Carlson, P. S. (1975): Polypeptide composition of fraction 1-protein from parasexual hybrid plants in the genus Nicotiana. *Science*, 187:353–355.
34. Laboratory of Genetics, Kwangtung Institute of Botany, Academia Sinica (1975): Studies on another culture in vitro in Oryza sativa subsp. Shien I. The role of basic medium and

supplemental constituents in callus induced from another and in differentiation of root and shoot. *Acta Genetica Sinica,* 2:81–89 (English summary).
35. Laboratory of Plant Cell and Tissue Culture, Institute of Genetics, Academia Sinica (1975): Primary study on induction of pollen plants of Zea mays. *Acta Genetica Sinica,* 2:138–143 (English summary).
36. Maliga, P., Breznovits, A., and Matton, L. (1973): Streptomycin resistant plants from callus culture of haploid tobacco. *Nature* [*New Biol.*], 244:29–30.
37. Melchers, G. (1939): Genetik und Evolution. Bericht eines Botanikers. *Z. Verebungslehre,* 76:229–259.
38. Melchers, G. (1965): Einige genetische Gesichtspunkte zu sogenannten Gewebekulturen. *Ber. Dtsch. Bot. Ges.,* 78:21–29.
39. Melchers, G. (1968): Genetical aspects in callus culture work. In: *Proceedings of International Symposium on Plant Growth Substances, Jan. 1967,* edited by S. M. Sircar, pp. 89–90. Calcutta University.
40. Melchers, G. (1975): Genetik und Pflanzenzüchtung mit mikrobiologischen Methoden. *Planta Med.* [*Suppl.*], pp. 5–34.
41. Melchers, G., and Bergmann, L. (1958): Untersuchungen an Kulturen von haploiden Geweben von Antirrhinum majus. *Ber. Dtsch. Bot. Ges.,* 71:459–473.
42. Melchers, G., and Labib, G. (1973): Plants from protoplasts, significance for genetics and breeding. *Colloque Int. CNRS,* 212:367–372.
43. Melchers, G., and Labib, G. (1974): Somatic hybridization of plants by fusion of protoplasts. I. Selection of eight resistant hybrids of "haploid" light sensitive varieties of tobacco. *Mol. Gen. Genet.,* 135:277–294.
44. Melchers, G., and Sacristán, M. D. (1976): Somatic hybridization of plants by fusion of protoplasts. II. The chromosome numbers of somatic hybrid plants of 4 different fusion experiments. *Recueil de Travaux Original Dédiés à la Memoire de Georges Morel.* Masson et Cie, Paris (*in press*).
45. Mintz, B., and Illmensee, K. (1975): Normal genetically mosaic mice produced from malignant teratocarcinoma cells. *Proc. Natl. Acad. Sci. U.S.A.,* 72:3585–3589.
46. Murashige, T., and Skoog, F. (1962): A revised medium for rapid growth and bioassays with tobacco tissue cultures. *Physiol. Plant.,* 15:479–497.
47. Nagata, T., and Takebe, I. (1971): Plating of isolated tobacco mesophyll protoplasts on agar medium. *Planta,* 99:12–20.
48. Nakamura, A., Yamada, T., Kadotani, N., and Itagaki, R. (1974): Improvement of flue-cured tobacco variety MC 1610 by means of haploid breeding method and investigations on some problems of this method. In: *Haploids in Higher Plants—Advances and Potential,* edited by K. J. Kasha, pp. 277–278. University of Guelph Press, Guelph.
49. Otsuki, Y., Shimomura, T., and Takebe, I. (1972): Tobacco mosaic virus multiplication and expression of the N gene in necrotic responding tobacco varieties. *Virology,* 50:45–50.
50. Otsuki, Y., and Takebe, I. (1973): Multiple infection of tobacco protoplasts by TMV, CMV and PVX. *Ann. Phytopathol. Soc. Jpn.* (Abstr.), 39:224.
51. Otsuki, Y., and Takebe, I. (1974): Double infection of tobacco protoplasts by ordinary and tomato strains of TMV. *Ann. Phytopathol. Soc. Jpn.* (Abstr.), 40:210–211.
52. Otsuki, Y., and Takebe, I. (1976): Double infection of isolated tobacco mesophyll protoplasts by an unrelated plant virus. *J. Gen. Virol.,* 30:309–316.
53. Ouyang, T. W., Hu, H., Chuang, C. C., and Tseng, C. C. (1973): Induction of pollen plants from anthers of Triticum aestivum L. cultured in vitro. *Scientia Sinica,* 16:79–95.
54. Pogliaga, H. H. (1952): Hibridio intergenerico "Nicotiana × Petunia." *Rev. Argentina Agronomia,* 19:171–178.
55. Power, J. B., and Cocking, E. C. (1970): Isolation of leaf protoplasts: macromolecule uptake and growth substance response. *J. Exp. Bot.,* 21:64–70.
56. Power, J. B., and Cocking, E. C. (1971): Fusion of plant protoplasts. *Sci. Prog.,* 59:181–198.
57. Power, J. B., Cummins, S. E., and Cocking, E. C. (1970): Fusion of isolated plant protoplasts. *Nature,* 225:1016–1018.
58. Power, J. B., Frearson, E. M., and Cocking, E. C. (1971): The preparation and culture of spontaneously fused tobacco leaf spongy-mesophyll protoplasts. *Biochem. J.,* 123:29–30.

59. Sacristán, M. D. (1967): Auxin-Autotrophie und Chromosomenzahl. *Mol. Gen. Genet.,* 99:311–321.
60. Sacristán, M. D. (1975): Clonal development in tumorous cultures of Crepis capillaris. *Naturwissenschaften,* 62:139–140.
61. Sacristán, M. D., and Melchers, G. (1969): The caryological analysis of plants regenerated from tumorous and other callus cultures of tobacco. *Mol. Gen. Genet.,* 105:317–333.
62. Sacristán, M. D., and Wendt-Gallitelli, M. F. (1973): Tumorous cultures of Crepis capillaris: chromosomes and growth. *Chromosoma,* 43:279–288.
63. Sakano, K., Kung, S. D., and Wildman, S. G. (1974): Identification of several chloroplast DNA genes which code for the large sub-unit of Nicotiana fraction 1-proteins. *Mol. Gen. Genet.,* 130:91–97.
64. Sarkar, S. (1976): Use of protoplasts for plant virus studies. In: *Methods in Virology, Vol. 6,* edited by K. Maramorosch and H. Koprowski. Academic Press, New York.
65. Sarkar, S., Upadhya, M. D., and Melchers, G. (1974): A highly efficient method of inoculation of tobacco mesophyll protoplasts with ribonucleic acid of tobacco mosaic virus. *Mol. Gen. Genet.,* 135:1–9.
66. Schieder, O. (1974): Selektion einer somatischen Hybriden nach Fusion von Protoplasten auxotropher Mutanten von Sphaerocarpos donnellii Aust. *Z. Pflanzenphysiol.,* 74:357–365.
67. Stebbins, L. (1950): *Variation and Evolution in Plants.* Columbia University Press, New York.
68. Takebe, I., Aoki, S., and Sakai, F. (1975): Replication and expression of tobacco mosaic virus genome in isolated tobacco leaf protoplasts. In: *Proceedings of the Second John Innes Symposium,* edited by R. Markham, D. R. Davis, D. A. Hopwood, and R. W. Horne, pp. 115–124. North Holland Publishing Co., Amsterdam.
69. Takebe, I., Labib, G., and Melchers, G. (1971): Regeneration of whole plants from isolated mesophyll protoplasts of tobacco. *Naturwissenschaften,* 58:318–320.
70. Takebe, I., and Otsuki, Y. (1969): Infection of tobacco mesophyll protoplasts by tobacco mosaic virus. *Proc. Natl. Acad. Sci. U.S.A.,* 64:843–848.
71. Takebe, I., Otsuki, Y., and Aoki, S. (1968): Isolation of tobacco mesophyll cells in intact and active state. *Plant Cell Physiol.,* 9:115–124.
72. Thomas, E., Hoffmann, F., and Wenzel, G. (1975): Haploid plantlets from microspore of rye. *Z. Pflanzenzuchtung,* 75:106–113.
73. Thomas, E., Hoffmann, F., Potrykus, I., and Wenzel, G. (1976): Protoplast regeneration and stem embryogenesis of haploid androgenetic rape. *Mol. Gen. Genet.,* 145:245–247.
74. Thomas, E., and Wenzel, G. (1975): Embryogenesis from microspores of Brassica napus. *Z. Pflanzenzüchtung,* 74:77–81.
75. Wallin, A., Glimelius, K., and Eriksson, T. (1974): Effects of polyethylene glycol on aggregation and fusion of Daucus carota protoplasts. In: *Third International Congress of Plant Tissue and Cell Culture,* Abstr. No. 207.
76. Wang, C. C., Chu, C. C., Sun, C. S., Wu, S. H., Yin, K. C., and Hsu, C. (1973): The androgenesis in wheat (Triticum aestivum) anthers cultured in vitro. *Scientia Sinica,* 16:218–222.
77. Wang, C. C., Sun, C. S., and Chu, Z. C. (1974): On the conditions for the induction of rice pollen plantlets and certain factors affecting the frequency of induction. *Acta Botanica Sinica,* 16:43–54 (English summary).
78. Wang, Y. C., Sun, C. S., Wang, C. C., and Chien, N. F. (1973): The induction of the pollen plantlets of Triticale and Capsicum annum from another culture. *Scientia Sinica,* 16:147–151 (English summary).
79. Wenzel, G., Hoffmann, F., Potrykus, I., and Thomas, E. (1975): The separation of viable rye microspores from mixed populations and their development in culture. *Mol. Gen. Genet.,* 138:293–297.
80. Wenzel, G., and Thomas, E. (1974): Observations on the growth in culture of anthers of Secale cereale. *Z. Pflanzenzüchtung,* 72:89–94.
81. White, P. R. (1939): Potentially unlimited growth of excised plant cell callus in an artificial nutrient. *Am. J. Bot.,* 26:59–64.

Recombinant Molecules: Impact on Science and Society, edited by R. F. Beers, Jr. and E. G. Bassett. Raven Press, New York © 1977.

22. Genetic Engineering and Crop Improvement

Peter S. Carlson and *Thomas B. Rice

Department of Crop and Soil Sciences, Michigan State University, East Lansing, Michigan 48824

INTRODUCTION

Most yield advances in crop species have been achieved by individuals whose training and focus have been directly concerned with agricultural production. The past several decades have seen the development of new knowledge and techniques in the basic sciences, genetics in particular, which hold promise for crop improvement but which have yet to be applied toward that end. If the new approaches can be made routinely available to plant breeders and agronomists, then progress in plant improvement will be enhanced. Restructuring of a cell's genetic organization, the process currently termed "genetic engineering," has always played a central role in crop improvement. This chapter considers briefly some genetic aspects of crop improvement and suggests ways that recently developed, novel approaches to genetic manipulation may complement the more classic means of genetic engineering.

CLASSIC PLANT BREEDING

Plant breeding is an ancient science. The origins of our current crop species are buried in prehistory; all evidence indicates that crop species were domesticated during the Stone Age. Crop species arose from weed species that underwent natural hybridization resulting in increased genetic variability and subsequent selection for desirable phenotypes by prehistoric peoples. Methods of reaping and sowing in the field or methods for storage or preparation can be effective selection screens for plants just as growth on a Petri plate is for a bacterial colony. These early plant breeders searched for, recovered, and propagated genetic variants or recombinants that displayed agronomically desirable traits. The transformation of weed

* Present address: Plant Genetics Group, Pfizer Central Research, Groton, Conn. 06340.

species into crop plants was accomplished in the absence of modern science or of any knowledge of mendelian genetics.

Contemporary plant breeders employ essentially the same strategy with success. Their approach involves the production of populations with a broad genetic base followed by selection at the whole plant level for recombinants with positive agronomic alterations. Genetic manipulation is practiced without knowing the basis of the separate components that comprise the character being modified. Selection for traits such as yield is practiced at the endpoint of the complex biological processes which produce a whole plant. Mendelism and a knowledge of transmission genetics provide a conceptual basis for what is occurring during the breeder's genetic manipulations, but this knowledge is not necessary for crop improvement.

In most current breeding programs, the availability of genetic variability is not the limiting factor in crop and variety improvement. There is a wide range of genetic diversity in the surviving natural population of most crop species. The focus of breeding efforts is centered on selecting desirable recombinant types which emerge from any particular cross or segregating population. To select desirable recombinants, it is first necessary to define which alterations or recombinants to select. Currently the assays of agronomic utility and subsequent selection are based on observations of whole plant phenotypes. Consequently, only major alterations can be recognized. These alterations appear as statistically significant changes in characteristics of bulk populations. Assaying at the endpoint of a number of complex biochemical, physiological, and developmental processes hides many potentially useful recombinants in the complexity of the buffered processes producing a whole plant.

The complexity of plant biology and crop productivity yield is also expressed in the genetics of agriculturally important traits. The majority of these traits appear to be controlled by "polygenes," and their transmission is analyzed by quantitative methods. The quantitative inheritance of these traits is a reflection of the complex biological processes that underlie their expression and of the lack of well-defined mutants with which to analyze these processes. Quantitative inheritance is a phenomenon associated with crosses involving naturally occurring genetic variability and complex biological end products. There is no reason to expect that mutants affecting these processes could not be produced once the individual components are known, nor that the genetics and biochemistry of such traits would be any different from those found in other organisms (e.g., metabolic pathways). For the time being, however, the plant breeder has no choice but to use the phenotype of the endpoint as the basis for selection. Significant progress could be made in the improvement of breeding techniques if we could establish reliable and meaningful assays at critical points in a number of the component processes of agronomic traits. Examples of such processes are nitrogen metabolism, photosynthesis, water relations, mineral nutrition, transport within the plant, and grain development. With these critical processes individually analyzed and assayed, one could combine genotypes demonstrating optimal performance at different steps in a process to produce a new, highly productive variety.

THE ROLE OF MOLECULAR BIOLOGY

Recent advances in molecular biology have provided methods of genetic manipulation that should be applicable to agricultural plant species. This is certainly an exciting prospect. Despite the rapid progress in expanding our knowledge of basic genetic and biochemical mechanisms in lower organisms, this knowledge has had no direct impact on crop improvement. This lack of impact may be ascribed in large part to conceptual and experimental differences in the disciplines of molecular biology and plant breeding.

Molecular biology is composed of two basic elements: the *reductionistic* world view of basic science and a powerful set of analytical experimental tools. One example of this approach was the use of defined genetic variants combined with precise biochemical methods to elucidate the mechanisms that regulate metabolic pathways in a variety of organisms. In contrast, plant breeding as currently practiced has, of necessity, a more *holistic* approach. Plant breeders have to operate within difficult constraints. They have little choice in either their experimental materials or the problems that confront them. No strong correlations have been established between yield and any of the individual processes that contribute to the final product. The experimental and technological requirements of plant breeding and the constraints of the plant system are different from those imposed by molecular biology. The question is: can the novel methods of "genetic engineering" defined in microbial systems really be applied to crop improvement?

Two different approaches presently being developed attempt to extend the methods and techniques of molecular biology from microbial investigation to application for crop improvement: one involves cellular manipulations and the other involves DNA manipulations. Cellular manipulations hold the potential for developing an experimental system for crop species suitable for more refined analytical techniques. Using single somatic cells as experimental organisms, it is possible to achieve mutant production, analysis, and hybridizations not possible using whole plants. Such techniques may permit important cellular processes to be characterized to the extent that useful directed modification is possible. Engineering at the level of the genetic material itself (i.e., manipulation of DNA) has great potential for the manipulation of defined genetic elements, although much work needs to be done before such sophisticated techniques can be applied to specific agricultural problems.

Although many tools of the molecular biologist are available to the plant geneticist at the present time, some limitations prevent their application to breeding problems. The first problem with these approaches is a technical one. Regeneration of whole plants from single cells is essential for application of the technology of *in vitro* genetic manipulation to higher plants. However, this step has not yet been routinely accomplished with any major food crop. The second problem arises from the real needs of the plant breeder. In most instances, the availability of genetic variability is not the limiting factor in crop improvement; the ability to recognize and recover useful recombinants sets the limit. Hence the production of genetic variability via cellular mutation or hybridization provides no uniquely useful tool at the present time. The third problem results from the developmental biology of agronomic characters. Many agronomic traits are tissue specific; their expression is found in only one or a few tissues within the plant and often is not found in cells cultured *in vitro*. If a particular trait is not expressed in culture, there is no reason to expect that the trait can be altered and screened for by *in vitro* methods. The fourth problem involves the genetics of agricultural traits. Mutant selection systems and DNA manipulations allow modification of single gene traits. Most agronomic traits as they are now defined are polygenic in inheritance. Small additive, stepwise modifications would be difficult to recognize. Currently, genetic modification of crop plants using cellular or DNA manipulations should prove appropriate when the alteration involves single gene traits which are not tissue specific and for which there are good selective techniques. The technology involved in these approaches will almost certainly be improved to overcome the limitations discussed above. However, at present, single gene traits which are not tissue specific are rare, as are appropriate selective systems. Possible examples of such traits include disease resistance or tolerance to ion toxicity, but the range is limited.

It would appear that molecular biology is not yet relevant to crop improvement. The difficulty is that current efforts have attempted to transfer the experimental results (defined genetic manipulation) directly without also extending the reductionistic approach of molecular biology. The immediate need is not to find new ways to generate genetic variability but to find new ways to screen critically the variability already provided by nature and to identify the biochemical, physiological, and developmental components of agronomic traits. Once individual components and the rate-limiting steps of important traits are identified, designing methods of selection for altered traits is possible. We identify the first task as the extension of the world view of molecular biology to agriculturally important characters.

Perhaps an appropriate approach to this task would be to choose a character defined from the breeder's perspective and attempt to apply molecular biological tools to its analysis (i.e., attempt a genetic-biochemical study of the character). Agronomic traits are vastly more complex than bacterial metabolic pathways (e.g., lactose catabolism). A trait such as yield, for example, which represents the integrated product of innumerable specific pathways, would have to be broken down into a large number of subcomponents. At first glance the question of how to begin to dissect such complex traits appears difficult at best. One apparent approach to identifying rate-limiting processes that are important in crop production is by analysis of heterosis.

HETEROSIS AND THE BIOLOGY OF CROP IMPROVEMENT

Effective genetic manipulation of agronomic traits, especially if using *in vitro* selection techniques, requires identification of the relevant metabolic processes and specific rate-limiting steps. Many such traits, including yield, drought tolerance, time of maturity, protein content, and tillering ability are complex quantitative traits under the control of multiple genes (polygenes) (1,3). The final phenotype is separated from the basic biochemical steps, the units of selection, by several levels of biological organization and environment-genotype interactions. Unfortunately, the definition of polygene and statistical methods for its analysis are not compatible with the analytical approaches of molecular genetics. Likewise, biochemical approaches have been frustrated by the complexity of quantitative traits even when they include the analysis of divergent genotypes. For example, despite considerable effort, no strong correlation has yet been found between final yield and the productivity of any distinct biochemical pathway (2).

Characterization of heterosis will provide information about the processes that limit expression of quantitative traits. Heterosis (hybrid vigor) refers to the beneficial effects (increased size and productivity) observed in most F_1 hybrids derived from crosses between unrelated parental lines (1,3). This classically defined genetic phenomenon is described almost entirely by statistical methods and remains essentially uncharacterized in molecular terms. Known genetic and molecular mechanisms can account for the phenomenon of heterosis (6). However, there is no direct evidence that any of these mechanisms are in fact involved in heterosis. Several cases have been described where the activity of a specific enzyme (e.g., nitrate reductase, mitochondrial oxidative phosphorylation) was higher in the hybrid than in either parental line (4,7). However, no causal relationship has been established between such increases in enzyme activity and overall increases observed for growth or yield. Despite uncertainty concerning the mechanisms involved, it is clear that constraints in one or more of the systems that limit yield in the parents have been relieved in a superior hybrid. Such hybrids offer an opportunity to identify physiological and developmental traits that are of direct importance to yield and the rate-limiting steps contributing to these traits. This information would eventually result in development of techniques to predict the combining ability of prospective parental lines.

We need an experimental system dealing with a specific character that satisfies both the classic definition of a heterotic effect and the requirements for effective manipulation and analysis of the discrete genetic and biochemical components involved. One approach would be to use specific stress conditions to modulate the magnitude of the hybrid advantage. For example, if the superior performance of a hybrid genotype is due in part to more efficient conversion of a particular input (e.g., light, CO_2, nitrate, phosphate) to a useful product, then (a) the advantage should diminish in the absence of any essential input because this condition will now limit the final phenotype; (b) the hybrid advantage should be magnified under limiting conditions where growth is directly related to the level of that specific input; and (c) the relative response of the hybrid and parental genotypes should be unaffected if the heterotic effect is not related to the stress conditions. It should therefore be possible to take a hybrid(s) clearly exhibiting heterosis under field conditions, raise it under controlled, specified stress conditions in the laboratory, and thereby identify parameters which alter the relative performance of the hybrid and parental genotypes. Using this approach, it should be possible to pinpoint specific biochemical pathways that contribute to hybrid vigor. Extended study can focus on individual steps within those pathways, elucidate the mechanisms responsible for heterosis, and identify the rate-limiting steps underlying yield.

METHODS AND OBSERVATIONS

The experimental approach cited above was tested by growing eight inbred strains of corn and their F_1 hybrids in a controlled environment with nutrient solutions containing different levels of nitrate. We sought to discover whether increased efficiency in the processes of nitrate uptake and utilization could account for heterosis in any of these hybrids. Preliminary results and the experimental procedures are presented in Table 1. Each set of inbred strains and their hybrids were fertilized either with a nutrient solution containing a standard level (6 mM) of nitrate or a modified nutrient solution containing either no nitrate or reduced levels of nitrate. The height, fresh weight, and dry weight of the aerial portion of the plants were measured after approximately 60 days of growth; the ratios between the values for the hybrid and the values for the better of the two parental strains were used as a measure of heterosis. All four hybrids surpassed the better parent for each of the traits measured (except height, OH43 × B14) (Table 1). In three of four cases, the heterotic response disappeared as predicted when nitrate was removed from the medium. Continued superiority of the M017 × W28 hybrid at no nitrate presumably reflects higher levels of nitrogen in the hybrid seed. One of the four hybrids (CO113 × CO109) showed increased

TABLE 1. *Relative growth response of F_1 hybrids and inbred parental varieties to different levels of nitrate*

Hybrid genotype	Trait	Relative growth (%, hybrid/better parent)			
		Nitrate concentration			
		6 mM	None	6 μM	60 μM
CO113 × CO109	Dry weight	125[a]	107[b]	187	157[c]
	Fresh weight	120[a]	98[b]	187	173[c]
	Height	118[a]	102[b]	118	126[c]
OH43 × B14	Dry weight	144[c]	69[c]	64	109
	Fresh weight	144[c]	75[c]	61	111
	Height	100[c]	88[c]	79	110
MO17 × W28	Dry weight	134	163		196
	Fresh weight	230	192		204
	Height	142	126		135
H93 × VA26	Dry weight	202	104		78
	Fresh weight	172	130		97
	Height	130	93		88

[a] Average of 4 experiments.
[b] Average of 2 experiments.
[c] Average of 3 experiments.

The values in the table represent an average of 1 to 4 experiments. In each experiment the relative growth is indicated as a percentage in which the performance of the hybrid is compared to the better of the two inbred parental lines. The performance of each genotype is based on the average of measurements of 9 to 12 plants grown in 12 × 18 in flats. Flats containing 5 to 6 in washed, sterile sand over 4 in of Perlite were sown with 24 seeds (2 genotypes) at a uniform spacing. Flats were maintained at 27°C in growth chambers with 20-hr days (3,000 foot candles at 4 feet from lights, 4 A.M. to 12 P.M.; 5,000 foot candles 11 A.M. to 7 P.M.; and 7,200 foot candles 2 to 5 P.M. and 4-hr nights. Flats were watered twice daily with deionized water. Starting 10 days after planting, the seedlings were watered thrice weekly with a nutrient solution. One liter of solution per flat was applied and allowed to drain through. Control flats received a Long Ashton nutrient solution composed of the following mineral salts (concentration in milligrams per liter):

Macronutrients: KNO_3, 202; $Ca(NO_3)_2 \cdot 4H_2O$, 472; $MgSO_4 \cdot 7H_2O$, 184; $NaH_2PO_4 \cdot H_2O$, 184.

Micronutrients: $MnSO_4 \cdot H_2O$, 1.69; $CuSO_4 \cdot 5H_2O$, 0.125; $ZnSO_4 \cdot 7H_2O$, 0.29; H_3BO_3, 3.1; NaCl, 5.85; $(NH_4)_6Mo_7O_{24} \cdot 4H_2O$, 0.0088.

Iron: Ferric tartrate, 5.0.

The experimental flats received a modified nutrient solution which contained either low levels of nitrite or no nitrite. The macronutrient composition of these modified nutrient solutions is listed in Table 3. The concentration of the micronutrients and iron remained unchanged. The aerial portion of each plant was measured, harvested, and weighed 56 to 65 days after planting. Dry weights were determined after 6 days of drying at 95°C.

heterosis at low levels of nitrate indicating, according to our hypothesis, that increased efficiency of uptake and/or assimilation of nitrate contributes in part to this heterosis. The H93 × VA26 hybrid does not show a response at low levels of nitrate and the OH43 × B14 hybrid responds slightly at 60 μM nitrate compared with 6 μM nitrate, but there is no stimulation of heterotic effect above control levels despite the higher levels of nitrate reductase reported in the hybrid (8). An attempt was made with CO113 × CO109 to localize further the heterotic effect within the nitrate assimilation pathway by using the intermediates—nitrite, ammonium, and L-glutamine—as nitrogen sources (Table 2); the macronutrient composition of solutions used in these experiments is shown in Table 3. The magnitude of the heterotic effect decreases when later intermediates are used; this suggests that the overall advantage results from the cumulative effect of

TABLE 2. *Relative growth response of CO11 × CO109 hybrid and inbred parental varieties to different nitrogen supplements*[a]

	Relative growth (%, hybrid/better parent)					
	Nitrogen supplement					
	6 mM NO_3	0 μM NO_3	60 μM NO_3	60 μM NO_2	60 μM NH_4	60 μM Gln
Dry weight	125[b]	107	157[c]	115	97	148
Fresh weight	120[b]	98	173[c]	149	105	159
Height	118[b]	102	126[c]	118	110	124

[a] Values are an average of 2 or more experiments.
[b] Average of 4 experiments.
[c] Average of 3 experiments.

The values in the first three columns are included from Table 1 for comparison with the results obtained when the CO113 × CO109 hybrid received nutrient solution supplemented with alternate nitrogen sources. These nitrogen compounds were added to the modified nutrient solution containing no nitrate as indicated in Table 3. Plants were grown following the procedures described in Table 1.

TABLE 3. *Macronutrient composition of modified Long Ashton nutrient solutions with different nitrogen supplements*

	Concentration (mg/l)					
	Nitrogen supplement					
Macronutrients	60 μM NO_3	6 μM NO_3	0 μM NO_3	60 μM NO_2	60 μM NH_4	60 μM Gln
KNO_3	2.02	0.202	—	—	—	—
$Ca(NO_3)_2 \cdot 4H_2O$	4.72	0.472	—	—	—	—
KNO_2	—	—	—	5.1	—	—
$(NH_4)_2SO_4$	—	—	—	—	7.92	—
L-Glutamine	—	—	—	—	—	8.76
KH_2PO_4	269.3	272	272	272	272	272
$Ca(H_2PO_4)_2 \cdot H_2O$	499	504	504	504	504	504
$MgSO_4 \cdot 7H_2O$	184	184	184	184	184	184
$NaH_2PO_4 \cdot H_2O$	184	184	184	184	184	184

small increments at several sites rather than a large advantage at a single enzymatic step. The results are not strictly comparable, however, since alternate uptake pathways are important for these different nitrogen sources (5,9). The enzymes that catalyze these steps are being characterized in current studies.

DISCUSSION

The results are not surprising: (a) heterosis can result from increased efficiency for nitrate uptake and assimilation, and (b) the metabolic processes responsible for heterosis are different in different hybrid genotypes. These preliminary experiments do show, however, that this is a productive approach for the identification of specific processes (including the rate-limiting step of yield) that contribute to hybrid vigor. Once these steps have been identified, it is then appropriate to use the experimental system of molecular biology to undertake genetic modification in the service of crop improvement.

ACKNOWLEDGMENTS

We thank Ms. B. Floyd for excellent technical assistance and Drs. D. Parke, M. Christianson, R. Locy, and R. Malmberg, Mr. T. Jacobs and Ms. A. Klein for helpful comments. The work described in this chapter was supported by Contract No. E(11–1)–2528 of the U.S.E.R.D.A.

REFERENCES

1. Allard, R. W. (1960): *Principles of Plant Breeding*. John Wiley & Sons, New York.
2. Brown, A. W. A., Byerly, T. C., Gibbs, M., and San Pietro, A. (Eds.) (1975): *Crop Productivity—Research Imperatives*. Michigan Agricultural Experiment Station, East Lansing.
3. Gowen, J. W. (Ed.) (1962): *Heterosis*. Iowa State College Press, Iowa City.
4. Hageman, R. H., Leng, E. R., and Dudley, J. W. (1967): A biochemical approach to corn breeding. *Adv. Agronomy,* 19:45–86.
5. Heimer, Y. M., and Filner, P. (1971): Regulation of the nitrate assimilation pathway in cultured tobacco cells. III. The nitrate uptake system. *Biochem. Biophys. Acta,* 230:362–372.
6. Rice, T. B., and Carlson, P. S. (1975): Genetic analysis and plant improvement. *Annu. Rev. Plant Physiol.,* 26:279–308.
7. Sarkissian, I. V., and Srivasta, H. K. (1973): Some molecular aspects of mitochondrial complementation and heterosis. In: *Genes, Enzymes and Populations,* edited by A. M. Srb, pp. 53–60. Plenum Press, New York.
8. Schrader, L. E., Peterson, D. M., Leng, E. R., and Hageman, R. H. (1966): Nitrate reductase activity of maize hybrids and their parental inbred. *Crop Sci.,* 6:169–173.
9. Sims, A. P., Folkes, B. F., and Bussey, A. H. (1968): Mechanisms involved in the regulation of nitrogen assimilation in microorganisms and plants. In: *Recent Aspects of Nitrogen Metabolism in Plants,* edited by E. J. Hewitt and C. V. Cutting, pp. 91–114. Academic Press, New York.

Recombinant Molecules: Impact on Science and Society, edited by R. F. Beers, Jr. and E. G. Bassett. Raven Press, New York © 1977.

23. Discussion

Moderator: E. W. Nester

Dr. H. Z. Liu: Dr. Brill, which organism codes for the leghemoglobin?

Dr. W. J. Brill: It is clear that the plant codes for the globin part of leghemoglobin. There is some debate as to which organism codes for the heme. In fact, people have been able to isolate the messenger RNA from polyribosomes of soybean nodules; that messenger RNA makes the globin part of leghcmoglobin.

Dr. T. M. Powledge: Dr. Brill, we have not heard much about risk this afternoon, which is a nice change. Do you foresee any?

Dr. W. J. Brill: Do I foresee any risks?

Dr. T. M. Powledge: Either with the work itself or with the success of the work?

Dr. W. J. Brill: Well, there is a fair chance nothing will come of all of this. There are always risks. One might envision that ammonia-excreting mutants could fill the lakes with ammonia. I do not think that is going to be a problem because the biggest hurdle is to get ammonia-excreting, depressed mutants to compete in nature.

As you saw in my second slide, 12 to 24 ATP molecules are consumed for every nitrogen fixed. Thus, there is a tremendous selection against fixing nitrogen under conditions where it is not really needed.

Mr. R. Ridell: Dr. Brill, on what basis did you map the *nif*B gene to the left of *his*D?

Dr. W. J. Brill: That was by transduction, and it seems to be confirmed by preliminary deletion analysis.

Mr. Howard: P-1 transduction?

Dr. W. J. Brill: Yes.

Dr. C. A. Thomas, Jr.: Dr. Brill, it seems to me that in all speculations regarding transfer of *nif* genes into strange hosts, one supposes there is available a strong reducing potential equivalent to, for example, that of reduced rubridoxin. Is that a reasonable expectation? Is such a potential available in most of these putative hosts?

Dr. W. J. Brill: Yes, I think that potential exists in just about any cell. You can use many artificial electron donors in nitrogen-fixing systems *in vitro,* including chemicals.

Dr. F. R. Blattner: Dr. Nester, I would like to know what the role of the inverted repeat sequence in the plasmid in the crown gall system is, and whether there is any evidence that it is involved in the transfer.

Dr. E. W. Nester: I think it is much too early—as far as our group is concerned—to discuss that now. We have no information on whether inverted repeat sequences are in the plasmid.

Dr. F. R. Blattner: Are there any plasmids in which there are deletions?

Dr. E. W. Nester: Some plasmids have suffered deletions and, as far as I know, all those have been avirulent. The size of plasmids that can be isolated from virulent strains ranges from approximately 96 to 160 million. So perhaps there are some natural deletions. But we have not analyzed in detail any sequence homology between these plasmids.

Dr. Melchers: May I add a comment to this? In our laboratory we have a mutant of *Agrobacterium* that does not induce tumors but often backmutates and induces tumors. This mutant has the plasmid, as shown by investigators from the Netherlands.

It is possible to find mutants in the plasmids, but, for bacterial geneticists, there is a smaller certainty of that.

Dr. M. P. Gordon: Dr. Brill, I understand that attempts to seed fields with different strains of *Rhizobium* do not work well because there is a more complicated relationship between the bacteria and the host plant. Particular bacterial strains seem to be good for a particular field.

Do mutants actually do the work under field conditions, or do resident bacteria take over and crowd them out?

Dr. Brill: That is probably the most important point to consider in improving *Rhizobium*. There have been numerous studies involving inoculation of fields with commercial cultures of *Rhizobium*. Often there is some kind of marker on the cells with which the field was inoculated. Frequently, if soybeans have been grown on that field with another *Rhizobium* inoculant, perhaps the commercial strain will not take over even though the field inoculated at high numbers. There are many fields, however, in which a given plant has not been introduced previously. These particular mutant strains are being tested in those fields.

I truly predict that the strains that I mentioned, SM31 and SM35, will be unsuccessful in field experiments because of that very problem. The wild type was chosen only because of its superiority for growth in the laboratory, which probably works against it when it is "fighting" in the field. However, there are reports and techniques in which one can inoculate with a high concentration of cells, even a field filled with different kinds of *Rhizobium*. Under certain conditions the newly added inoculum is the first one to inoculate the plants. But this is a very serious problem.

What are we going to do now? We are trying two things: One, to take *Rhizobium* that is successful in a certain field, bring it into the laboratory, and see if we can obtain mutants similar to the ones I showed you. The other approach is to see if we can transform in and screen for the mutant phenotype using these mutants as a donor and a "successful" strain as a recipient.

Dr. B. J. Luberoff: I am a layman in this field and I would like to combine what I heard in this morning's session with what I heard this afternoon.

Dr. Carlson made the point that there apparently is a discourse, although it may be a synthetic one that he just created, between plant breeders on one side and another group called scientific molecular biologists. There is a discourse going on. Some gaps are being bridged; some jargon is being mutually understood. I believe I once heard Dr. Young say that this kind of discourse was apparently not happening between those with clinical concerns and molecular biologists concerned with bacteria.

I have heard much discussion that the lay public should be involved in this whole thing, going right down to dishwashers in the laboratories—blowing the whistle, as it were.

I would like to ask Dr. Carlson to begin the discussion addressing a specific question. If there is a discourse between the plant breeder and the laboratory scientists working on molecular biology, can you tell us if you think similar discourses are occurring elsewhere? If they are not, how might they be encouraged and stimulated, and, particularly, what could a journalist do about it?

Dr. P. S. Carlson: That is a difficult question. What I said was really a distillation of my own experience for the past 2 years. I was a naive molecular biologist. My plant breeding friends are really tied to the earth. The plant breeder is a renaissance biologist. He has to know genetics, pathology, physiology, soil science, weather, and everything else. So he is concretely welded to the earth.

A molecular biologist can play with ideas, and the plant breeder, I think, has the right to expect that when ideas are played with, they should at least be relevant to his concerns.

I do not know of any other individuals who are attempting to do the same kind of resolution that I have attempted, and maybe that is pretension on my part. I think this kind of thing is needed, but it may be a long time coming because we speak different languages. We operate within different contexts. The relevance is not there yet.

Trying to define an experiment that really turns me on as a molecular biologist and is relevant to a plant breeder is quite difficult. A few experiments appeal to both groups. I think the burden of proof lies with the molecular biologist or genetic engineer. Until the molecular biologist or genetic engineer can generate a variety of plants that make the farmers some money, we are just dealing with an idea.

Dr. G. Melchers: May I comment on this? My experience is that good plant breeders are not interested in molecular biology. They have had big results in the last 10 or 20 years and they found we are playing around, and I think that is right. They are not interested in it. They see the successes of Borlaug, Chase, and others. They believe that we are speaking nonsense.

Dr. P. S. Carlson: That is because genetic variability is not the limiting thing here.

Dr. G. Melchers: That is not right.

Dr. P. S. Carlson: It is the putting together–

Dr. G. Melchers: Different cases, however. In some cases you need more variability, in other cases you need no variability. One cannot generalize.

Dr. E. C. Cocking: I would like to comment on the situation in the United Kingdom. As you know, the Agricultural Research Council is attempting to handle this subject systematically. Agricultural research, of course, stems and flows down through the plant breeding institutes in the United Kingdom. What we have in the United Kingdom–and I was at a typical meeting about a month ago–is an ongoing dialogue among plant breeders and workers in this general field and people from the plant breeding institutes. The plant breeding institutes act as a buffer zone between this innovative type of work, which often goes on in the universities, and the plant breeder work. Of course, this is not unusual because there are often many things in plant breeding in which the innovative events have occurred in universities. These have filtered through the plant breeding institutes to the plant breeder in the United Kingdom.

On the broader issue of the dialogue with the public at large, I think the important thing is to recognize that this is an ongoing exercise. It is not sufficient to have this sort of meeting or some other sort of meeting as a one-time thing. If it is not this subject it is going to be another. You want a forum for such topics, certainly as an annual event.

Our counterpart of your AAAS is the British Association for the Advancement of Science, which could be regarded as the establishment of British science. Because it is representative of the whole spectrum of plant science, it has taken on the role of affording the opportunity in its annual meeting for such a dialogue and also an ongoing dialogue throughout the year.

So I think one would have the framework in varying degrees to have a continuing dialogue on the agricultural scene. The Agricultural Research Council has already set up this ongoing dialogue with the plant breeder.

Dr. T. A. Kass: Could Professor Melchers ask his questions of Professor Carlson that he had at the end of his presentation, and could we hear the answers to those?

Dr. P. S. Carlson: Why don't you pose them?

Dr. G. Melchers: In your first publication on fusion, you stated that you had 25% fusion events.

Dr. P. S. Carlson: No, we said we had 33 regenerated calli and felt we had 25% fusion. I think now that what we had was clumping and not fusion.

Dr. Melchers: That is a good answer. Not fusion but clumping. Thank you very much.

Dr. Carlson: I would say that when polyethylene glycol is used to effect higher levels of fusion, the hybrids fall all over.

The second question was with regard to the maternal protoplasm and particularly Fraction 1 protein. Many new hybrids have been made by Kao and Smith; they have been analyzed by Wildman, and, interestingly enough, when plants are regenerated only one of two chloroplast types seems to predominate: either *glauca* or *langsdorffii.*

Dr. Melchers: One or the other?

Dr. Carlson: One or the other. There are some possibilities that they occur; sometimes they do not. I think the answer is that we do not know yet.

Dr. Melchers: There are no mixers?

Dr. Carlson: There are some mixers but I am not sure of the data because I am not involved with the experiment. By and large, they are only of one kind.

Dr. Melchers: They say they could find 10% mixtures. They say so in published work. If there were 10% of one kind, they certainly could find them by their methods.

Dr. Carlson: No, you misunderstand. In our publication it was noted that the resolution of the technique was such that if 10% of the protein was of one parental type and 90% of the others, we would still recognize it. What we found, in fact, was that it was from zero to less than 10% of one parental type. Now, in subsequent hybrids that have been made—and a number of them have been analyzed—you would expect that when we find both *glauca* and *langsdorffii,* we really have a chimeral mixture of chloroplasts. What the data actually suggest is that one of the two chloroplast types predominates and the other drops out.

There are some mixtures but there seems to be some incompatibility within the cells having mixtures of chloroplasts in tobacco.

Dr. Melchers: It must not be incompatibility. It can be genetic drift.

Dr. Carlson: Could be on the nucleus, too.

Dr. Melchers: No, the nuclei are identical. That is no reason.

Dr. E. C. Cocking: Having been involved in the introduction of sodium nitrate as a fusing agent, I think it would now be interesting, as more systems are producing somatic hybrids with various selection procedures, for somebody to compare the various fusion methods in the same system.

In fact, we have experiments under way between *Petunia parodii* and *Petunia hybrida* using the selection scheme which I outlined to you. We used sodium nitrate fusion, calcium, high pH, polyethylene glycol with and without calcium or alkali washes, and the absence of fusion-inducing agents as treatments.

One of the things that has become evident to us is that although with certain of the fusion-inducing agents high-percentage fusion may be achieved initially, often few of the resultant heterokaryons survive. This may differ between the various fusion-inducing agents. My guess, for what it is worth, is that the final analysis will reveal that the actual frequencies of somatic hybrid formation effected by the various fusion-inducing methods will not be significantly different.

Dr. W. Szybalski: I have two questions for Drs. Brill or Eaton Signer concerning nitrogen fixation. What is the rate-limiting step from the practical point of view? Would it help to add additional genes to *Rhizobium* by using plasmids so there would be 10 or 20 or 300 copies?

The second question is: what is the mechanism of nitrogen supply in rice plant in the Orient which has been grown for ages with little fertilizer? One way or another they have to provide for the nitrogen deficiency.

Dr. W. J. Brill: I think energy is limiting: ATP once again.

I do not predict that one can increase nitrogen fixation by adding more genes or more enzymes to the system. We have evidence from studies not mentioned in my chapter. If the *Rhizobium* level is kept constant, plants grown under a big canopy where they have large leaf area can get more energy. We feel they seem to fix more nitrogen. So I do not think adding more enzyme to the system will permit more nitrogen fixation. It is a very inefficient enzyme. Analysis of the genera *Azotobacter, Clostridium,* and *Klebsiella,* when grown in adequate nitrogen, indicates that 2% of their protein is nitrogenase—a very sluggish enzyme. Nitrate reductase is two orders of magnitude more active, and you need, therefore, two orders of magnitude less enzyme, so you have to add extra molybdenum. There are approximately 30 iron atoms per molecule of active nitrogenase, thereby creating a need for extra iron. This has to be brought into the cell and placed into the enzyme, each step an energy-requiring one.

In practice, I think it is best to take advantage of systems that are already working, such as the legume system. I feel that this will probably yield faster agronomic results if anything will yield them. We have a system that is working. There are things that we can manipulate easily without getting mutants. One can use the effectiveness assay for screening plants. I think it is important to recognize that such methods will help the plant breeder to identify quickly what plants have a potential for high levels of nitrogen fixation.

The second question is: how can one grow, for generation after generation, rice—which is not a nitrogen-fixing plant—without adding bulk fertilizer to these paddies? The answer is that nitrogen-fixing bacteria are very active in these paddies. At different stages, different organisms are involved. One also observes an abundance of blue-green algae. They are fascinating biological systems that use light as an energy source and furnish nitrogen to the paddy.

In an interesting symbiosis, a tiny aquatic plant called *Azolla* is commonly found floating on paddy waters. At the base of the fern-like plant are foci where blue-green algae aggregate and fix nitrogen for the *Azolla*. This situation is commonly found in rice paddies. In fact, in North Viet Nam the *Azolla* plants are harvested from lakes and rivers and added to the soil as fertilizer.

So without becoming too deeply involved in molecular biology, there are some untried, or perhaps only moderately tried, systems that will probably increase the use of nitrogen-fixing systems.

No one has yet isolated the *Azolla* separate from the blue-green algae and reinfected them. Nevertheless, many labs are working on these systems and I am sure breakthroughs will occur.

Dr. F. E. Young: I did not intend to make a speech but since Dr. Luberoff commented on my earlier comments, I would like to elaborate on exactly what I was trying to transmit. I get the impression at times that we are caught on the horns of two dilemmas. One is apathy, the other is indignation. Apathy takes the form of our unwillingness to read broadly enough and to look widely enough to understand the horizons around us.

What we are specifically talking about, I guess, is a course on medical molecular biology for molecular biologists. As I listened—and I do not wish to construe this as a snide remark because I am deficient in many areas of scientific expertise—I became impressed with the fact that the existence of multiply-resistant antibiotic strains had not been appreciated. Yet those of us who worry about illness, disease, and patients harboring microorganisms acknowledge that the frequency of multiple antibiotic resistance is very high. We made a field trip to Iraq during a time of the mercury poisoning there to try to see whether the heavy metals had any influence on plasmids. We were chagrined to find in this country, where antibiotics were freely available over the counter, that over 90% of the individuals in the most rural environment or in cities had isolable multiply-resistant antibiotic forms of *Staphylococci.* To hear about Dr. Falkow's epidemic in the nursery may be a surprise, but it is also disconcerting because these events are happening.

What I was trying to devise was some forum where we could best communicate to those who are so involved in elegant molecular techniques, considering the problems that exist in medicine, infectious disease, and medical microbiology.

I guess it is almost a tragedy and possibly a travesty of education that many of our microbiology departments in medical schools are not equipped to teach this discipline.

Dr. H. Z. Liu: I would like to comment on Dr. Melcher's question on somatic fusion between *N. glauca* and *langsdorffii.* I do not know how Kao produced somatic hybrids under his conditions of incubation. Reading your publications, I recall that most of your trials were illuminated with 800 lux.

In Dr. H. H. Smith's laboratory, we produced somatic hybrids with polyethylene glycol incubated under 5 to 10 lux until cell division occurred when all the chloroplasts degenerated. Thus, the Fraction 1 protein that you expected might not exist under those conditions.

Dr. G. Melchers: I did not understand his question.

Dr. H. Z. Liu: Chloroplasts degenerated when coming from the mixture of two sources because all colorless new shoots were induced at least 30 days after the initial fusion.

Dr. G. Melchers: But protoplasts are organized through genomes and plastomes. One chloroplast comes from the other.

Dr. H. Z. Liu: Not from that mature chloroplast but from a protoleukoplast.

Dr. Carlson: There is still a genetic continuity between the leukoplasts and the developing chloroplasts, so one would have to assume, on the basis of classic genetic evidence, that there would be transmission.

Dr. Liu: One may overtake the other. I did not say it had to be 50%.

One last comment about genetic engineering. The view about genetic engineering having to be defined at the molecular level seems to be personal interpretation. It seems to me that plant breeders and animal breeders are genetic engineers, and this has been going on since the Stone Age.

Recombinant Molecules: Impact on Science and Society, edited by R. F. Beers, Jr. and E. G. Bassett. Raven Press, New York © 1977.

24. Introduction to Section D: Virus Vectors

Daniel Nathans

Department of Microbiology, Johns Hopkins University School of Medicine, Baltimore, Maryland 21205

To introduce this section of the volume, I want to make some brief general comments about viruses as vehicles for recombinant DNA. The use of viruses for cloning and propagating DNA segments is based on the same principles as the use of bacterial plasmids, namely, insertion of DNA into a preexisting replicon. Viruses have an additional feature: the recombinant DNA may be encapsidated, resulting in a transducing particle that can be efficiently delivered into other cells, as determined by the host range of the virus. In theory, any viral DNA that is infectious can be used as a vehicle. However, the size of the inserted DNA segment will be limited by the capacity of the virion capsid, except possibly for those filamentous bacteriophages whose length appears to be determined by the length of encapsidated DNA. Alternatively, if the viral DNA is propagated under conditions in which encapsidation does not occur (e.g., with a viral segment missing late genes), one might expect that even larger fragments of DNA could be inserted into the viral replicon.

So far, the two viruses that have been used as vehicles are coliphage λ and the primate virus simian virus 40 (SV40), each of which is probably the best understood virus of its type. In the case of λ, viable recombinants can be constructed by replacing the extensive nonessential region of λ DNA with foreign DNA (see Murray and Blattner et al., *this volume*). Moreover, in the λ vehicles developed by the Murrays and by Davis, formation of infectious particles is dependent on the insertion of DNA, since without such insertion the vehicular DNA is too short to be encapsidated. This last feature provides a powerful selection for recombinants with DNA inserts.

In the case of SV40 a rather detailed physical map of the viral DNA is available which allows rational construction of recombinant molecules, using SV40 genes and signals (see Fareed et al., Goff and Berg, and Hamer, *this volume*). The segment of DNA needed for viral DNA replication includes the early or "A" gene and a specific origin of replication. Unlike λ, the small SV40 DNA molecule has no extensive nonessential region, and therefore to obtain encapsidated recombinant DNA, one needs to substitute a segment of viral DNA by the segment to be inserted, thus forming a defective genome similar to naturally occurring defective variants of SV40. And as with natural variants, two general types of constructed recombinants can be visualized: those that retain one or more functional genes and the origin of viral DNA replication, and those that retain only the origin of replication, the single *cis* element required for viral DNA replication. The first class can be cloned by complementation in the presence of an appropriate SV40 mutant, and the second class can be cloned by using wild-type SV40 as a helper virus to supply all *trans* functions. If the recombinant DNA is of a different size than the helper virus, recombinant DNA can be readily purified from helper virus DNA by electrophoresis, or recombinant virions can be separated from helper virions by their buoyant density difference.

Since SV40 DNA integrates into cellular DNA, one may also be able to use viral DNA as a vehicle to insert specific DNA segments into chromosomes of mammalian cells. One approach to this end is to use cloned recombinants. Another approach is based on the observations of van der Eb and his colleagues that a segment of SV40 DNA containing only the early gene plus contiguous sequences can transform cells about as efficiently as whole molecules of SV40 DNA. Therefore, it should be possible to construct recombinant molecules containing the transforming segment, transform cells directly with this DNA, and thereby select cells which have integrated nonviral genes. Both integrated recombinant DNA and that present in cells infected with recombinant virus should be useful in studying gene regulation in mammalian cells.

Recombinant Molecules: Impact on Science and Society, edited by R. F. Beers, Jr. and E. G. Bassett. Raven Press, New York © 1977.

25. Making Use of Coliphage Lambda

Kenneth Murray

Department of Molecular Biology, University of Edinburgh, Edinburgh EH9 3JR, Scotland, U.K.

INTRODUCTORY REMARKS

Two of the chapters in the first section of this volume (N. E. Murray and W. Szybalski) were concerned with the use of bacteriophage λ as a vehicle for cloning fragments of DNA. The generation and rejoining of DNA fragments were also discussed (R. J. Roberts), so here I wish to describe some of the features of lambda I think are especially useful in research on recombinant DNA.

ESSENTIAL FEATURES OF LAMBDA

Figure 1 summarizes in a simplified form some of the essential features of lambda (6). The mature phage contains a single linear, duplex DNA molecule (3.1×10^7 MW). At each end are short, single-stranded projections of complementary sequence by which, through normal Watson-Crick base pairing, the DNA forms a circular molecule when it is injected by the phage into its host cell. Genes on the left of the linear map code for head and tail proteins of the phage. Much of the central region is inessential and can be deleted without seriously impairing phage growth. *Red* represents the phage recombination system, *O* and *P* are concerned with replication of the phage DNA, and *S* and *R* code for proteins that lyse the host cell. The *c*I gene codes for a repressor protein which can interact at the sites (*dotted arrows* in the lower part of the figure) to regulate expression of the phage genes in both directions from P_L and P_R (*broken arrows*). *N* and *Q* are positive regulatory genes, the products of which interact at the positions shown by dotted arrows. *Q* is necessary for the expression of genes *S, R,* and genes to the left of them (*long broken arrow inside the circle*). Thus, after circularization of the chromosome, gene *Q* activates the expression of genes *A, C,* etc., (i.e., those on the left of the linear map) as well as genes *R* and *S*. Also shown in the figure is the attachment site, X, by which the phage chromosome may be inserted by site-specific recombination into its host chromosome (where it may be stably replicated along with the host), and

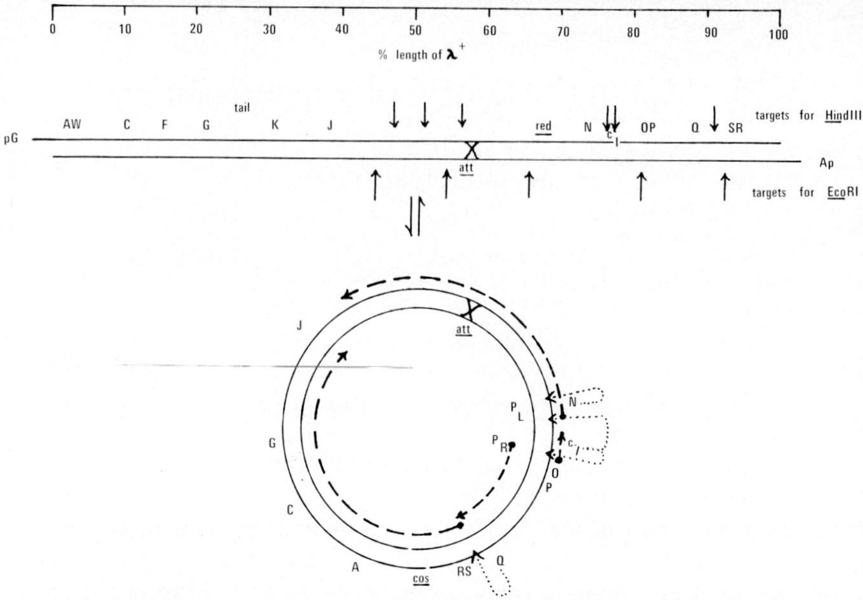

FIG. 1. A simplified version of the lambda chromosome showing some of its more important functions which are discussed in the text. (Reproduced with permission from *Endeavour*, XXXV (126), 1976.)

the positions of targets for the restriction enzymes *Hin*dIII and *Eco*RI in the wild-type chromosome. Expression to the left from the phage promoter, P_L, is extremely efficient, and is regulated by the phage gene *N* product. I will describe later the way in which some of these functions can be exploited, but first I will describe some of the methods for recovery of recombinant DNA molecules.

RECOVERY OF *in vitro* RECOMBINANTS

The wild-type phage contains several targets for most of the restriction enzymes commonly used in this area of research. It is therefore necessary to manipulate the phage to produce derivatives that have either a single target for a particular restriction enzyme and have some space made in the chromosome so that another fragment of DNA can be accommodated in this chromosome, or have two targets spanning a relatively large segment of DNA that can be replaced by a DNA fragment from another source (15,16,18,20,24). Having inserted new fragments of DNA between the two arms of the lambda chromosome, these new molecules must be recovered and propagated. There are three ways in which this can be done. One is the transfection process analogous to bacterial transformation in which the

DNA preparation is absorbed by competent bacterial cells which are then plated out and infected cells give rise to a phage plaque (13).

Two other procedures, however, can also be useful and are potentially more efficient. One of these is the well-known helper-mediated transformation process of Kaiser and Hogness (11), which is illustrated here with an example from some unpublished work by D. Ward and N. Murray. The chromosome represented at the top of Fig. 2 is one from which we were attempting to develop a vector for use with fragments of DNA produced by digestion with the restriction enzyme from *Bacillus amyloliquefaciens* (*Bam*I) (25), for which the wild-type λ chromosome has five targets (5,19). Four of these are readily manipulated, but site 1, on the left, is probably in an essential gene and so far we have not succeeded in removing this site. In the example in Fig. 2, targets 3 and 5 have been removed by deletions and targets 2 and 4 remain. The fragment of DNA between targets 2 and 4 is probably about 25% of the phage chromosome. The objective is to replace the segment of phage DNA between these two targets by a DNA fragment from some other source produced by digestion with the same enzyme. After that fragment is inserted between the two phage segments B and D, the left-hand terminal fragment, A, must be attached in order to form a viable DNA molecule, and the efficiency of joining four fragments together is normally rather low. The proportion of such molecules in the population may be greatly improved by use of a helper phage. Host cells are infected with a phage that carries genetic markers (e.g., *amber* mutations in genes *J* and *R*)

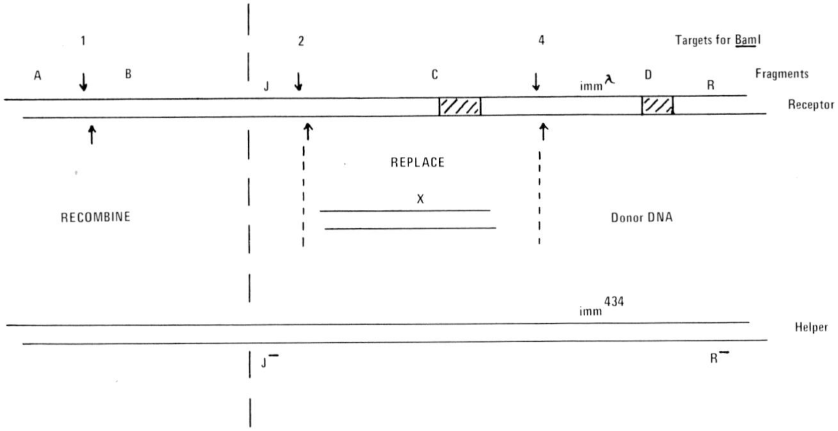

FIG. 2. The use of recombination with a helper phage to enhance the recovery of the *in vitro* recombinant DNA. The genome of the receptor phage (upper) retains targets 1, 2 and 4 for *Bam*I. Correct joining of fragments A, B, and D gives a molecule too small for packaging, so that fragment C or a replacement, X, of suitable size is essential for viability. Viable phage genomes of the type A, B, X, and D are formed by *in vitro* joining of the four fragments or by recombination left of the broken line between the partial genome B.X.D and the helper phage and are recognized as J^+ *spi*$^-$ *imm*$^\lambda$ R^+ phages.

to distinguish it from the DNA to be recovered and the products of the ligase reaction are then added. The helper phage also has an immunity different from that of the phage to be prepared. Selection against the helper can therefore be made by plating the infected cells on indicator bacteria that are immune to it or are unable to suppress the amber mutations. A left cohesive end may be incorporated into the *in vitro* recombinant molecules (B-X-D) by a recombinant event, mediated by phage or bacterial systems, occurring anywhere to the left of the broken vertical line in Fig. 2. In appropriate cells this process proves to be quite efficient. The desired recombinants can be recognized readily because replacement of the central segment of DNA results in phage that have the so-called Spi^- phenotype and will plate on a P2 lysogen (26). Douglas Ward has used this selection quite successfully, and, of course, the Spi^- phage can then be screened directly for specific genes. In the case of incorporated segments of *E. coli* DNA, for example, this is achieved readily by complementation of auxotrophs in the usual way.

The *in vitro* packaging procedures developed by Barbara and Thomas Hohn and co-workers (7,8) are also very useful for recovery of the recombinant DNA molecules. Two lysogens are used, one of which is defective in genes for one essential head protein and the other in genes for another (Fig. 3). If λ DNA is present in a reaction mixture to which the gene products indicated in Fig. 3 are progressively added, mature phage are formed and these can then be propagated in bacterial cells in the normal way. Of the two lysogens used, one is defective in gene D and the other in gene E,

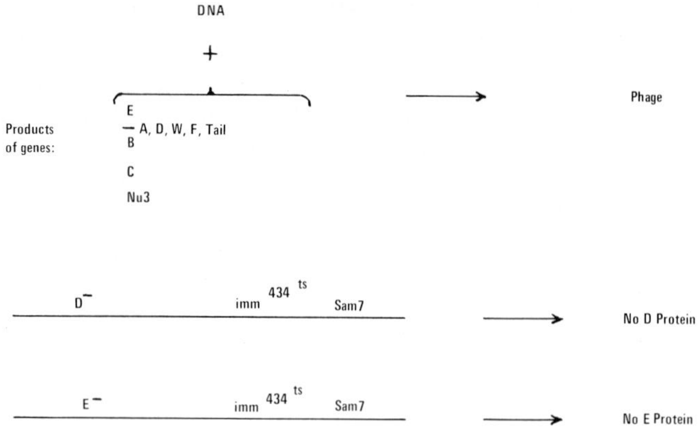

FIG. 3. Packaging of λ DNA *in vitro.* Mature phage particles are formed on incubation under appropriate conditions of phage DNA with the products of the genes shown in the upper part of the figure. Cells lysogenized with either of the phages whose genotypes are described in the lower part of the figure cannot, upon induction, furnish viable phage because each is defective in an essential component. A mixture of the two induced lysogens, however, affords complementation in the usual way, and mature phage particles can be formed that contain either the endogenous phage DNA or a phage DNA added to the mixture. (From refs. 7 and 8.)

so that neither can form a mature phage itself, but on mixing extracts of both cells normal E protein and normal D protein are present so that all of the components necessary for normal packaging of the genome are available. If the recombinant DNA molecules are added to such extracts of cells they will be packaged into phage, but the endogenous phage DNA (from the lysogens) is, of course, also packaged. The endogenous phage can be distinguished from the new (*in vitro* recombinant) ones by use of an appropriate immunity marker. In the case illustrated in Fig. 3, the two strains used to provide the components for phage assembly have the immunity of phage 434, so that *in vitro* recombinant molecules made in a vector with a different immunity can be recovered selectively by plating the complex reaction mixture on *E. coli* lysogenized with λ imm^{434}. This reaction is very efficient, giving 10^{-3} to 10^{-4} phages per DNA molecule (7), which is at least an order of magnitude higher than is normally obtained by the transfection method (B. Hohn and K. Murray, *unpublished observations*).

SELECTION AND SCREENING OF RECOMBINANTS

I would like to discuss the selection and screening of recombinant phages that have been recovered by any of the procedures just described. Of course, many simple methods are available, but here I shall select just two examples for more detailed description. The first uses what we call replacement receptors (16). If a phage DNA molecule has two targets for a particular restriction enzyme, and removal of the DNA fragment between these targets is such that a molecule formed by joining the two outer fragments is too small to be packaged into a mature phage, then it is obviously necessary that additional DNA be inserted into the space between the two outer fragments of the phage DNA (Fig. 4). If *E. coli* DNA is digested with the appropriate restriction enzyme, mixed with the restricted vector, treated with ligase, and the mixture used to transfect competent cells, the resulting phage population can then be screened for any particular *E. coli* gene by plating the phage on a host strain mutant in that gene. For example, phages that have acquired a fragment of *E. coli* DNA containing a suppressor gene would be detected by their ability to complement a suitable amber mutant. A convenient strain for this purpose is an *E. coli lacZ* amber mutant because, although this mutation prevents the cell from making β-galactosidase, this is not an essential function and there are convenient color tests for the production of this enzyme. On the so-called MacConkey indicator plates, *E. coli* cells (or turbid phage plaques) are red if they produce β-galactosidase and so hydrolyze lactose, whereas they remain colorless if they do not; thus, phage that can supply the suppressor function to relieve the amber mutation in *lac* are manifest as red plaques. Another color test is based on the use of indicator plates containing 5-bromo-4-chloro-3-indolyl-β-D-galactoside, hydrolysis of which by β-galactosidase releases an indole

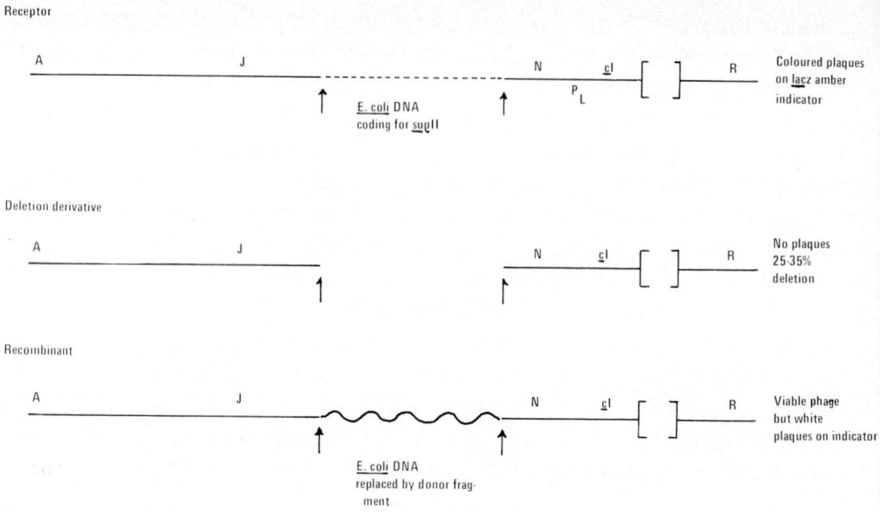

FIG. 4. Replacement receptors. The phage genome has two targets for the restriction endonuclease which span a segment of *E. coli* DNA carrying a suppressor gene. Removal of the central segment and joining of the two flanking fragments gives a DNA molecule too small for packaging. Replacement of the central segment by a fragment of DNA from another source permits the formation of viable phage, but these are no longer able to suppress an *amber* mutation (in a nonessential function) in the host cell. (From refs. 16 and 17.)

derivative which gives blue plaques (2); the advantage of this indicator is that it can also be used with clear plaques, whereas the MacConkey indicator requires the establishment of repression to produce turbid plaques. Clearly, phage containing the suppressor gene (or the *lac* gene) can also be used as a vector because, instead of selecting phage that have acquired the suppressor fragment, the experiment can be inverted, as it were, to detect those that have lost it. Replacement of the suppressor fragment by DNA from another source now yields a phage that cannot suppress the *lac* amber host, so the color test can be used to select recombinants as white (or colorless) plaques.

Another particularly useful screening method involves insertion into the immunity region of the vector (16). Such vectors have a considerable amount of their chromosome deleted and retain one target for a particular restriction enzyme in the immunity region (Fig. 5). Insertion of any piece of DNA at this target destroys the ability of the phage to produce a functional repressor so that the recombinants give clear plaques which are readily distinguished from the turbid plaques of parental phages formed by rejoining of the two fragments from the parent phage DNA molecule.

These two systems have been used extensively in our work during the past year (1,16,17). The replacement systems and their attendant screening

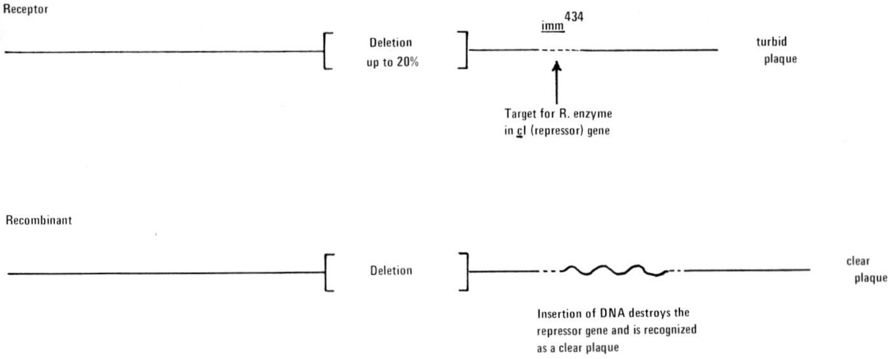

FIG. 5. Immunity insertion receptors. The genome of the receptor phage has only one target for the restriction enzyme, and this target is located within the repressor gene. Insertion of any DNA fragment at this position leads to the formation of a defective repressor so that the recombinant phage is readily detected by a morphological change. (From refs. 16 and 17.)

methods depend on the correct expression of the inserted sequences. The phage regulatory systems can be exploited to enhance the level of expression of inserted genes as was described earlier in this volume (N. E. Murray). Phage defective in genes *S* or *R* are unable to lyse their host cell so that phage and phage gene products accumulate within the cells. An *amber* mutation in gene *Q* also prevents cell lysis (Fig. 1), but, in addition, synthesis of the phage's late gene products (genes on the left of the linear map) is curtailed so that the proportion of the cell's synthetic activity directed from the phage promoter, P_L, is greatly increased. The level of expression of genes inserted into the phage so that they are under the influence of P_L may thus be elevated, often to a high degree by suitable further genetic manipulation of the phage, as has been illustrated in the favorable case of the *trp* operon of *E. coli* (14; N. E. Murray, *this volume*). At present, similar procedures are being used to improve the production of several enzymes of interest to nucleic biochemists – DNA ligase, DNA polymerase, and DNA modification and restriction enzymes, all of which are fairly obvious choices. This work is proceeding, but the high levels of expression obtained with the *E. coli trp* operon have yet to be attained.

HYBRIDS OF LAMBDA WITH EUKARYOTIC DNA

Screening populations of recombinant phages for those carrying a particular gene is often straightforward if the inserted DNA sequence is of prokaryotic origin (1,16), since complementation assays can be used. With the simpler eukaryotes, however, the genome is itself much more complex, metabolic pathways often differ appreciably from those of *E. coli,* and, although some preliminary results (22; J. Carbon, *this volume*) have now been obtained which encourage the use of complementation tests to detect expression of incorporated genes, these are certainly less readily applicable than with incorporated prokaryotic DNA segments. Alternative methods for screening the recombinant populations are therefore necessary, and when a suitable probe is available, molecular hybridization techniques offer an obvious approach.

The *in situ* hybridization methods used so successfully in cytological analyses (9) were shown to be applicable to individual phage plaques in test experiments with a phage having two targets for the *Eco*RI enzyme (10). Digestion of the phage DNA with this enzyme released the fragment between these two sites; from it a complementary RNA copy was made by the use of RNA polymerase and radioactively labeled nucleoside triphosphates. This RNA was shown to serve as a probe for phages that carry the complementary sequence (Fig. 6). With the RNA probe used in these experiments, hybridization against single plaques of the phage represented in Fig. 6 about 2,700 cpm were incorporated into a plaque, whereas in controls with phage that had the corresponding section deleted from the chromosome

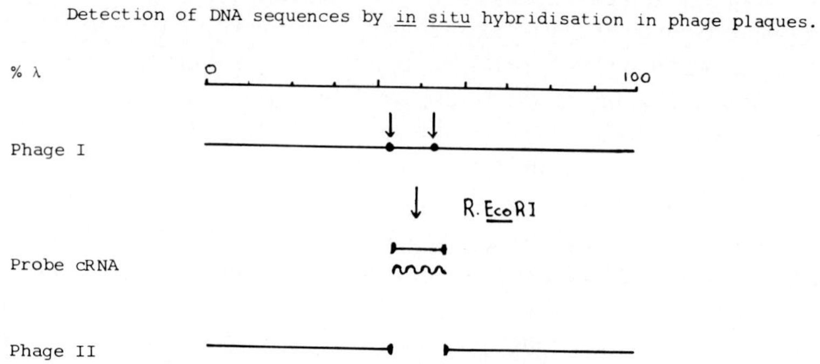

Cts/min retained after hybridisation with labelled cRNA probe:

In plaques		Control sample
Phage I	Phage II	from lawn
2698	36	21

FIG. 6. *In situ* hybridization on individual phage plaques. The genome of the phage represented in the upper part of the figure contains two targets for the *Eco*RI enzyme. A transcript of the fragment contained within these two targets was made *in vitro* with RNA polymerase and radioactively labeled nucleoside triphosphates. This RNA was used as a probe to test the efficiency of hybridization on individual plaques of phage containing these sequences, the phage with this segment deleted, as well as T4 phage and a section of the bacterial lawn serving as controls. (From ref. 10.)

(or with T4 phage or a sample of the bacterial lawn), essentially no hybridization was observed. Similar experiments have been done with bacterial colonies carrying plasmids (4).

A modified version of the plaque hybridization method has been used to screen phages carrying sea urchin DNA for those that contain the histone genes (P. Mounts, E. M. Southern, and K. Murray, *unpublished observations*). In these experiments, sea urchin DNA was digested with the *Eco*RI enzyme and fractionated by electrophoresis on agarose gel. DNA fragments containing histone genes were located by hybridization with histone messenger RNA after elution from the gel (21) and purified by equilibrium centrifugation in CsCl before being inserted into an immunity receptor phage of the type described earlier (16). The recombinant phages, which were detected as clear plaques, were then screened by hybridization against histone mRNA; 3 of 200 plaques examined gave a positive result. (These experiments were carried out in a laboratory of P3 category under the Asilomar guidelines.) Although the sea urchin histone genes have already been cloned in a plasmid (12), their insertion into a suitable phage should facilitate studies of expression of these genes in bacteria. Obviously the histone gene fragment can be inserted into the phage genome in either of two orientations, and analysis of DNA from the three phages by digestion with other restriction enzymes showed that two had the fragment in one orientation and one in the other (P. Mounts, E. M. Southern, and K. Murray, *unpublished observations*). This is illustrated schematically at the top of Fig. 7. The lambda system permits one to ensure that the inserted fragment is in the correct orientation for expression from the phage promoter, P_L, because the two DNA strands can be separated readily by equilibrium centrifugation in CsCl in the presence of poly(G,U) (23) and the strand carrying a coding sequence identified by hybridization against histone mRNA.

The experiment illustrated in the lower part of Fig. 7 was done with two recombinants in which the fragment of sea urchin DNA had been inserted into the immunity region of the phage, but in opposite orientations. The results show that the histone mRNA hybridized exclusively against the heavy strand of the DNA from one of the phages and exclusively against the light strand of DNA from the other. Since the mRNA used was a mixture of all the histone mRNA, this demonstrates that all of the histone genes are present on the same strand of the DNA. This conclusion was reached also by Gross et al. (3) from similar experiments with sea urchin histone genes inserted into one of our immunity receptors (16).

In order to facilitate expression of these inserted genes, it is necessary to remove the fragment from the immunity region and transfer it to a position where expression will be mediated by P_L (Fig. 7). After this transposition, one selects those phages in which the fragment has been inserted in the orientation such that expression of its genes will be in the leftward direc-

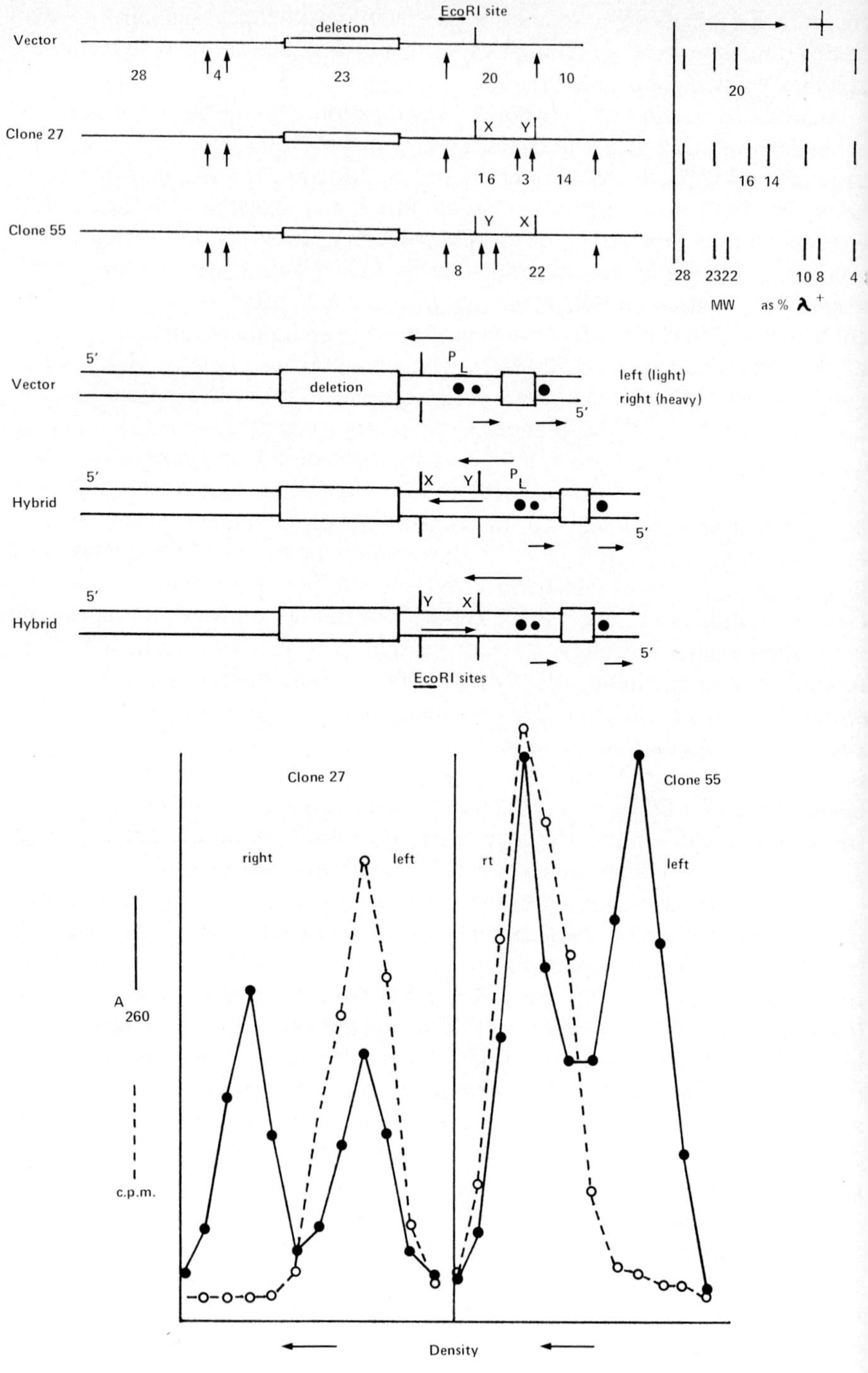

EcoRI site
deletion
Vector
28
4
23
20
10
Clone 27
X
Y
16
3
14
Clone 55
8
22
20
16 14
28 2322
10 8
4
MW as % λ+
5′
P
L
left (light)
right (heavy)
Hybrid
EcoRI sites
Clone 27
Clone 55
right
left
rt
A
260
c.p.m.
Density

tion (i.e., histone mRNA will hybridize against the light, but not the heavy, strand of the phage DNA). This phase of the study is at present being pursued by Keith Peden in my laboratory.

I hope that this brief account provides an indication of some of the ways in which the sophisticated genetic background of phage lambda can be exploited in research on recombinant DNA molecules.

ACKNOWLEDGMENT

I am greatly indebted to my collaborators mentioned in the text, particularly to my wife Noreen who designed and executed the genetic manipulations involved, to my assistant Sandra Bruce, to Douglas Ward, Barbara Hohn, Phoebe Mounts, and Edwin Southern for permission to cite their unpublished work, and to Sydney Brenner for drawing my attention to the use of *in vitro* packaging experiments for recovery of recombinant DNA molecules as phage. The research described has been supported in part by the Science Research Council and the Medical Research Council.

REFERENCES

1. Borck, K., Beggs, J. D., Brammar, W. J., Hopkins, A. S., and Murray, N. E. (1976): The construction in vitro of transducing derivatives of phage lambda. *Mol. Gen. Genet.*, 146:199–208.
2. Davies, J., and Jacob, F. (1968): Genetic mapping of the regulator and operator genes of the lac operon. *J. Mol. Biol.*, 36:413–417.
3. Gross, K., Schaffner, W., Telford, J., and Birnstiel, M. (1976): Polarity and asymmetry of the histone coding sequences. *Cell* (*in press*).
4. Grunstein, M., and Hogness, D. S. (1975): Colony hybridization: a method for the isolation of cloned DNAs that contain a specific gene. *Proc. Natl. Acad. Sci. U.S.A.*, 72:3961–3965.

FIG. 7. Analysis of DNA from hybrids of lambda and sea urchin DNA. The upper part of the figure illustrates schematically the distinction between two recombinants with the DNA fragment in opposite orientations by electrophoretic analysis on agarose gels of a digest of the DNA from the two phages (and their parent phage) with a second restriction enzyme. The vertical arrows show targets for this restriction enzyme, and the numbers between them (and beneath the bands separated by gel electrophoresis as shown on the upper right of the figure) give the size of the DNA fragments as percentage of λ^+ DNA. An immunity insertion receptor was used for cloning the sea urchin DNA, but for expression of the inserted genes it is desirable (perhaps necessary) to transfer the DNA fragment to a different receptor where expression is regulated by the phage promoter, P_L. Two orientations are again possible, the desirable one having the direction of transcription from right to left, as shown in the central part of the figure: the horizontal arrows indicate directions of transcription from the more important promoters which are located by the dots. This may be determined by separation of the two strands of the phage DNA by equilibrium centrifugation followed by hybridization of fractions from the CsCl gradient with radioactive mRNA, as illustrated in the lower part of the figure for DNA, from two of the original clones in the immunity insertion receptor. (From P. Mounts, E. M. Southern, and K. Murray, *unpublished observations.*)

5. Haggerty, D. M., and Schleif, R. F. (1976): Location in bacteriophage lambda DNA of cleavage sites of the site-specific endonuclease from Bacillus amyloliquefaciens H. *J. Virol.*, 18:659–663.
6. Hershey, A. D. (Ed.) (1971): *The Bacteriophage Lambda.* Cold Spring Harbor Laboratory, Cold Spring Harbor, New York.
7. Hohn, B. (1975): DNA as substrate for packaging into bacteriophage lambda in vitro. *J. Mol. Biol.*, 98:93–106.
8. Hohn, B., Wurtz, M., Klein, B., Lustig, A., and Hohn, T. (1974): Phage lambda DNA packaging in vitro. *J. Supramolec. Structure,* 2:302–317.
9. John, H. A., Birnstiel, M. L., and Jones, K. W. (1969): RNA-DNA hybrids at the cytological level. *Nature,* 233:582–587.
10. Jones, K. W., and Murray, K. (1975): A procedure for detection of heterologous DNA sequences in lambdoid phage by in situ hybridization. *J. Mol. Biol.*, 96:455–460.
11. Kaiser, A. D., and Hogness, D. S. (1960): The transformation of E. coli with DNA isolated from bacteriophage λdg. *J. Mol. Biol.*, 2:392–415.
12. Kedes, L. H., Chang, A. C. Y., Houseman, D., and Cohen, S. N. (1975): Isolation of histone genes from unfractionated sea urchin DNA by subculture cloning in E. coli. *Nature,* 255:533–538.
13. Mandel, M., and Higa, A. J. (1970): Calcium-dependent bacteriophage DNA infection. *J. Mol. Biol.*, 53:159–162.
14. Moir, A., and Brammar, W. J. (1976): The use of specialized transducing phage in the amplification of enzyme production. *Mol. Gen. Genet.* (*in press*).
15. Murray, K., and Murray, N. E. (1975): Phage lambda receptor chromosomes for DNA fragments made with restriction endonuclease III of Haemophilus influenzae and restriction endonuclease I of E. coli. *J. Mol. Biol.*, 98:551–564.
16. Murray, K., Murray, N. E., and Brammar, W. J. (1975): Restriction enzymes and the cloning of eukaryotic DNA. In: *Proceedings Xth Federation of European Biochemical Societies Meeting,* Vol. 38, pp. 193–207. North-Holland, Amsterdam.
17. Murray, N. E., Brammar, W. J., and Murray, K. (1976): Lambdoid phages that simplify the recovery of in vitro recombinants. *Mol. Gen. Genet.* (*in press*).
18. Murray, N. E., and Murray, K. (1974): Manipulation of restriction targets in phage λ to form receptor chromosomes for DNA fragments. *Nature,* 251:476–481.
19. Perricaudet, M., and Tiollais, P. (1975): Defective phage lambda chromosome, potential vector for DNA fragments obtained after cleavage by endonuclease BamI. *FEBS Lett.*, 56:7–11.
20. Rambach, A., and Tiollais, P. (1974): Bacteriophage lambda having EcoRI endonuclease sites only in the nonessential region of the genome. *Proc. Natl. Acad. Sci. U.S.A.*, 71:3927–3930.
21. Southern, E. M. (1975): Fractionation of DNA fragments by preparative gel electrophoresis (*unpublished*).
22. Struhl, K., Cameron, J. R., and Davis, R. W. (1976): Functional genetic expression of eukaryotic DNA in E. coli. *Proc. Natl. Acad. Sci. U.S.A.*, 73:1471–1475.
23. Szybalski, W., Kubinski, H., Hradecna, Z., and Summers, W. C. (1971): Analytical and preparative separation of the complementary DNA strands. *Methods Enzymol.*, 21:383–413.
24. Thomas, M., Cameron, J. R., and Davis, R. W. (1974): Viable molecular hybrids of bacteriophage lambda and eukaryotic DNA. *Proc. Natl. Acad. Sci. U.S.A.*, 71:4579–4583.
25. Wilson, G. A., and Young, F. E. (1975): Isolation of a sequence-specific endonuclease (BamI) from Bacillus amyloliquefaciens. *J. Mol. Biol.*, 97:123–125.
26. Zissler, J., Signer, E., and Schaefer, F. (1971): The role of recombination in growth of bacteriophage lambda. II. Inhibition of growth by prophage P2. In: *The Bacteriophage Lambda,* edited by A. D. Hershey. Cold Spring Harbor Laboratory, Cold Spring Harbor, New York.

Recombinant Molecules: Impact on Science and Society, edited by R. F. Beers, Jr. and E. G. Bassett. Raven Press, New York © 1977.

26. Construction and Testing of Safer Phage Vectors for DNA Cloning*

Bill G. Williams, David D. Moore, James W. Schumm, David J. Grunwald, Ann E. Blechl, and Frederick R. Blattner

Laboratory of Genetics, University of Wisconsin, Madison, Wisconsin 53706

CHARON PHAGES

Several laboratories have reported construction of λ phages for DNA cloning (1,2,4,6). Over the last 3 years in our laboratory, 12 derivatives of phage λ have been constructed as vectors for *Hin*dIII and/or *Eco*RI fragments of foreign DNA. They have been designated Charon phages after the mythological boatsman of the river Styx. Their structures and genotype designations are shown in Fig. 1, which also shows the disposition of *Eco*RI and *Hin*dIII sites relative to the replaceable region of the genome. Clones are constructed by replacing the DNA between the restriction sites in each vector or by inserting into the single site. The various Charon phages are adapted to serve several purposes including the cloning of various sized DNA fragments, as well as those terminated with the two different types of cohesive sequences produced by *Hin*dIII and *Eco*RI. There are also several genetic indicator functions including the *lac, red,* and *cI* genes, which are inactivated by fragment insertion in one or more of the vectors (see Fig. 1). Two of the vectors, Charons 3 and 4, have been chosen for further development and testing as safe vectors. Amber mutations in capsid genes A and B were crossed into these phages, and the suffix "A" (for amber) was appended to the Charon designation. Charon 3A is suitable for cloning DNA fragments ranging up to 8.4 kilobase pairs (KBP), and Charon 4A is suitable for fragments from 7.9 to 19 KBP. Thus, each phage is useful for cloning a different size range of DNA fragments.

These structures have been extensively documented by genetic tests, heteroduplex mapping in the electron microscope, and gel electrophoresis of fragments obtained by digestion with restriction endonucleases.

Both these vectors have been used successfully for cloning.

---

* This is paper no. 2031 from the Laboratory of Genetics, University of Wisconsin.

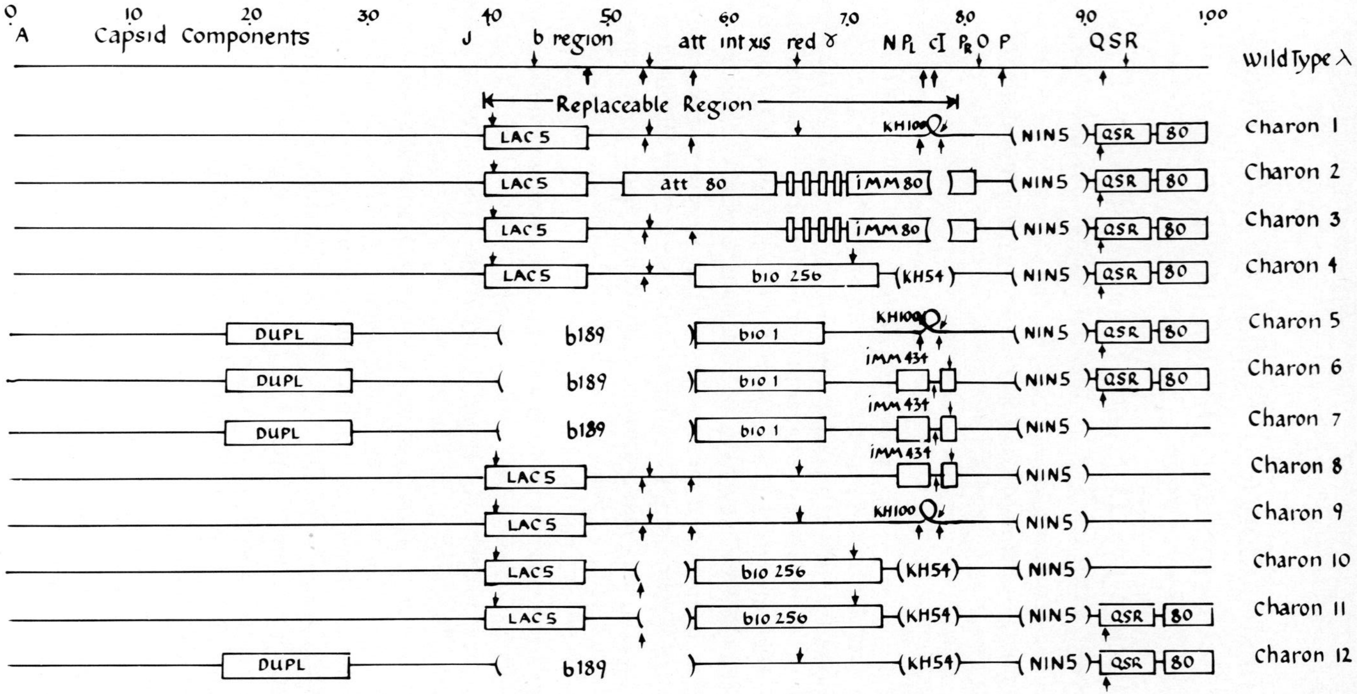

FIG. 1. Charon phages for DNA cloning. The DNA molecule of phage λ is shown with naturally occurring *Eco*RI sites indicated by downward arrows and *Hin*dIII sites shown by upward arrows. In each vector, the genome has been manipulated to remove sites in regions essential for phage growth, and to position sites appropriately for DNA insertion. Charons 3A and 4A were constructed by introducing amber mutations into genes A and B at the left end of the map.

Strategy and Tactics for Biological Containment with Phage λ

Phage λ can replicate either lytically or as a permanent resident of a bacterium as a plasmid or lysogen. As far as lytic growth in a natural environment is concerned, phage λ probably has limited capabilities. For efficient growth, λ requires exacting laboratory conditions. Moreover, the phage has a limited host range. As reported at the Torrey Pines meeting by R. W. Davis, none of over 1,800 *E. coli* strains isolated from streams and hospital patients was able to plate λvir. On the other hand, after establishment in a bacterium as a plasmid or lysogen, the cloned segment could be perpetuated in a natural environment at least as well as the bacterium.

Our principal strategy, therefore, has been to develop phages in which the lytic mechanism is preserved while the plasmid or lysogen forming capabilities are blocked. Secondly, the amber mutations reduce the host range even further so that the phage vector will be unable to propagate even on other laboratory strains of *E. coli*. Thirdly, we have endeavored, wherever possible, to enhance the growth of the vector on the laboratory host normally used for propagation so that small culture volumes can be employed.

Safety Features Incorporated into the Charon Phages

To supplement the inherent properties of the phage λ *E. coli* K-12 system which are advantageous in terms of biological containment, the following are present in both vectors:

1. Immunity region deletion: Each of the Charon phages contains a deletion which removes the DNA from the region that codes for the phage repressor. Without this important gene, the lysogenic state cannot be maintained and infected cells are efficiently killed. In addition, clones derived from Charon 4A are missing the *int* gene and the *att* site.
2. N^+ genotype: Clones made from Charons 3A and 4A are wild-type for phage gene *N*, which causes a major barrier to plasmid formation. It is believed that this is due to uncontrolled transcription past the *t*r1 and *t*r2 termination sites, which promotes active replication of the phage and increases the expression of lytic functions *S* and *R* which kill the cell.
3. *nin* deletion: Each of the Charon phages contains the *nin*5 mutation which deletes the *t*r2 termination site. This ensures that whether gene *N* functions, cell lytic functions *S* and *R* will be strongly expressed. This additionally reduces probability of stable plasmid formation because of consequent cell lysis.
4. Amber mutations: Due to the presence of amber mutations, the phages have restricted host range in that they require a suppressor

tRNA to make two capsid proteins. This must be supplied by the *su*$^+$ bacterial host used for propagation in the laboratory. Suppressor strains are probably rare in nature. (S. Brenner reported that of 34 isolates from nature that could be tested with phage T4, none contained a suppressor.)

5. Restriction-modification: Charon 3A and Charon 4A can be propagated on a nonmodifying bacterium which provides an additional barrier to lytic propagation on bacterial strains containing a restriction system.

Propagation on the Laboratory Host

The Charon phages and clones derived from them are efficiently propagated by the "preadsorb-dilute-shake" (PDS) method developed in our laboratory. In this method, 6×10^5 plaque-forming units of phage are mixed with 0.3 ml of a stationary culture of cells, usually strains KH802 or KH803, and 0.3 ml 0.01 M $MgCl_2$, 0.01 M $CaCl_2$. Preadsorption is for 10 min at 37°C; the mix is diluted into 1 liter NZ broth + 0.01 M $MgCl_2$ and shaken overnight at 37°C. After lysis, 3 ml chloroform is added which immediately kills all bacteria without affecting the phage. Starting with 10^5 input phage, the method generally yields between 2×10^{13} and 2×10^{14} phage in a 1 liter culture. This corresponds to 1 to 10 mg phage DNA not counting purification losses. The usual host used for propagation is strain KH803, which carries the SuII amber suppressor and lacks the K restriction and modification systems (hsr$^-$ hsm$^-$).

An important feature of this propagation method is that it does not require the induction of a lysogen, and thus allows a vector that is lacking the phage's repressor gene to be used.

TESTS VALIDATING THE EFFECTIVENESS OF BIOLOGICAL CONTAINMENT

Test System

To evaluate the containment of Charon 3A and Charon 4A, we made tests using the β-galactosidase gene on the *lac*5 substitution of the Charon phages to stand for the cloned segment. This gene is readily assayable both as a phage and as a plasmid by use of the chromogenic indicator dye, 5-bromo-4-chloro-3-indolyl-β-D-galactoside (XG), which turns blue upon cleavage by the β-galactosidase enzyme. When a DNA segment is cloned in either of our phages, the *lac*5 segment is replaced by the cloned DNA, and thus the *lac* segment is analogous in position to a cloned fragment.

Phages carrying *lac*5 can be readily identified by the intense blue color of the plaques on dye indicator plates. This holds true whether the bacterial lawn is *lac*$^+$ or *lac*$^-$.

For determination of *lac*$^+$ bacterial phenotype, it is necessary to provide the permease function. Thus, bacterial strains were constructed for this study which are chromosomally deleted for the *lac* operon, but which contain an F′ episome that is constitutive for the *lac* permease function (y^+z^-). We constructed both *hsm*$^+$ *su*° and *hsm*$^+$ *su*II *lac*z$^-$ y$^+$ indicators as well as their λ *imm*21 and λ att80 *imm*21 QSR80 lysogens. These bacterial strains allow the determination of *lac* z function independent of y function. To prove this, we lysogenized the *lac* z$^-$ y$^+$ indicator strains with *lac*5. Whereas the original indicator strains produced white colonies on nutrient plates containing dye (NZXG plates) and failed to grow on minimal plates with lactose as the sole carbon source (*lac*-min XG plates), the lysogens produced blue colonies on both indicator plates. Thus, we have a reliable measure of the presence of the cloned fragment in a bacterium. Moreover, using these bacteria in conjunction with K modified and unmodified preparations of Charon 3A and Charon 4A, we were able to assess the effects of the amber mutations, K restriction, and presence of the prophages separately and in various combinations. Figure 2 shows an imaginative rendition of the genotypes of the bacterial test strains employed.

SPECIFIC TESTS

Phage Survival Under Various Treatments

In Table 1 are data used to examine the sensitivity of Charon phages to a number of conditions that might be expected during dissemination. We found the phage to be extremely sensitive to acid. At pH 3, phages were inactivated with a half-time much less than 1 min and were undetectable at the 10^{-10} level after 2.75 hr. This makes it extremely unlikely that a phage could traverse the stomach. We found that the Charon phages are also very sensitive to desiccation on the laboratory bench as well as to detergents.

Shaking in Raw Sewage

When Charon phages were shaken in raw sewage, we found a reasonably rapid inactivation rate (approximately 10^{-1}/day). This means that the criterion of 10^{-8} survival would be met in 8 days. The rate of inactivation in fresh tap water was even greater.

Effect of Dilution

The adsorption of phages to sensitive bacteria is a diffusion-limited, bimolecular process that becomes less rapid at increasing dilutions. The

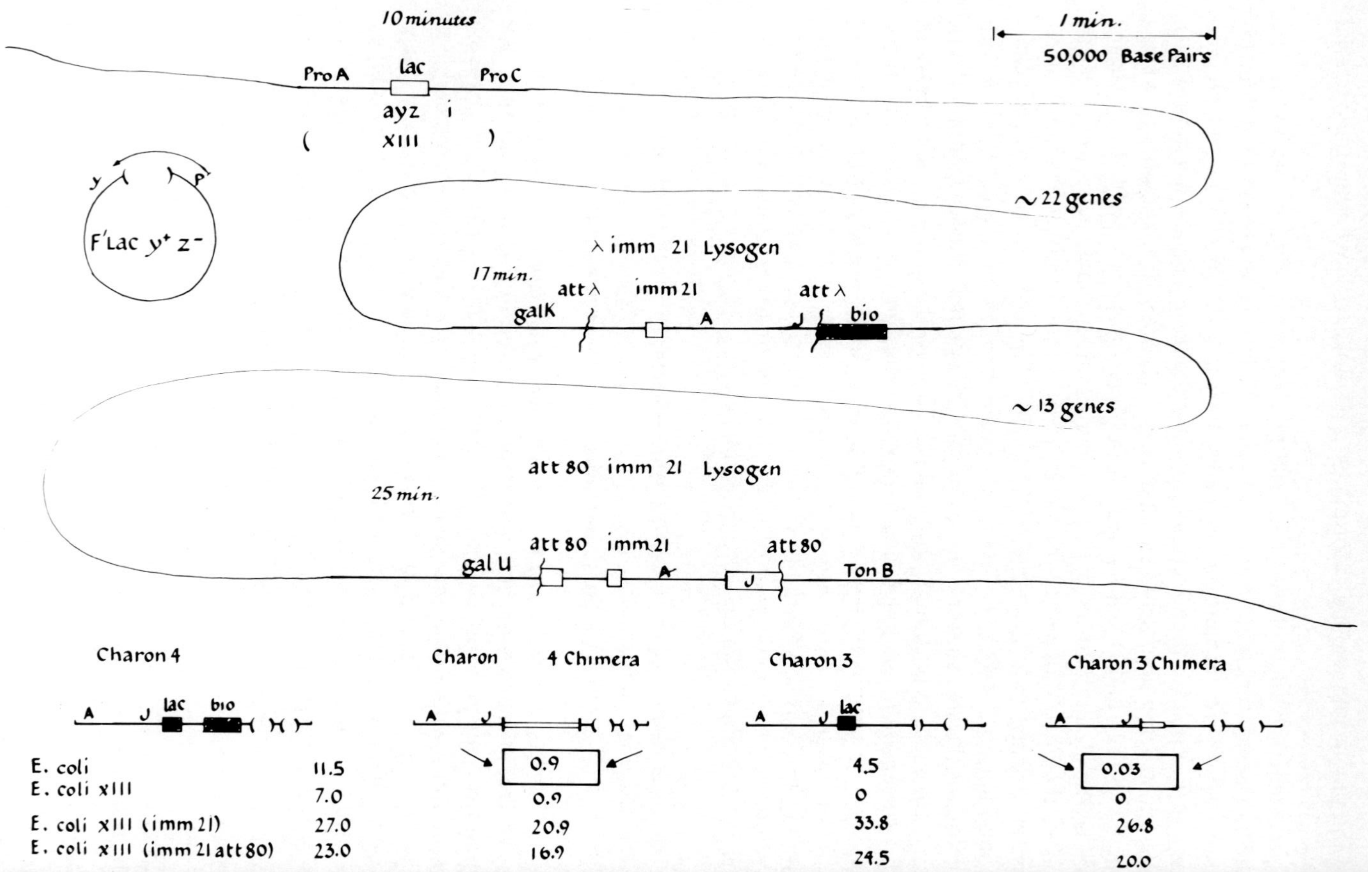
10 minutes
1 min.
50,000 Base Pairs
Pro A
lac
Pro C
ayz
i
x111
F'Lac y+ z−
~22 genes
λ imm 21 Lysogen
17 min.
att λ
imm 21
att λ
galK
A
J
bio
~13 genes
att 80 imm 21 Lysogen
25 min.
att 80
imm 21
att 80
gal U
A
J
Ton B
Charon 4
Charon 4 Chimera
Charon 3
Charon 3 Chimera
E. coli
E. coli x111
E. coli x111 (imm 21)
E. coli x111 (imm 21 att 80)
11.5
7.0
27.0
23.0
0.9
0.9
20.9
16.9
4.5
0
33.8
24.5
0.03
0
26.8
20.0

time for 1/e of a dilute population of phages to adsorb depends linearly on the concentration of sensitive bacteria according to the following relationship:

$$t = k/c \quad (5)$$

where t = time in minutes, c = the number of sensitive bacteria per milliliter, and k is a constant. The value of k for λ was recently determined to be 2×10^9 by J. Salstrom (*personal communication*). Thus, in a culture at 2×10^9 bacteria/ml, it would take 1 min for 37% of the phages to find bacteria, whereas at 10^7 bacteria/ml (which was found to be the titer of our sample of raw sewage), it would take 200 min if all bacteria in sewage were sensitive to the phage.

Since these bacteria were not all *E. coli* and fewer than 1 in 10^3 *E. coli* found in the aqueous environment are sensitive to λ, we estimate that 2×10^5 min (140 days) would be needed for adsorption of 37% of the phage population. This is certainly a substantial barrier to lytic propagation of Charon phages, considering that the phages themselves are inactivated at a rate of 10^{-1}/day.

Lack of Propagation on Nonpermissive Strains

We have measured several criteria of propagation of Charon phages on nonpermissive laboratory strains (*data not shown*). We tested both su° strains which fail to suppress the amber mutations, and hsr^+ strains which restrict the unmodified phages. These strains might be considered representative of the rare, naturally occurring strains able to adsorb λ.

We found that modified Charons 3A and 4A plated with efficiencies lower than 10^{-8} on su° strains. Moreover, K restriction produced more than a thousand-fold reduction in plating efficiency. The combined effect was too low to be measured. We also found that Charon 4A kills infected cells even though the infection is nonproductive.

We also attempted to measure the average burst size of Charon 4A in single cycle growth experiments. Since low plating efficiencies do not neces-

---

FIG. 2. Genetic elements in bacterial strains used for testing of safer phage vectors. The *E. coli* chromosome is shown as a wavy line with the sites of genes relevant to this study positioned approximately to scale. In all strains the chromosomal deletion × 111 at 10 min was used to remove the *lac* operon, and the F′*lac* $y^+ z^-$ episome was introduced to provide the permease function coded by gene *y* under the constitutive control of an episomal promoter. Over half of the *z* gene is missing from the episome. Strain M96 C is su°, whereas CSH18 is suII and can therefore suppress the amber mutation in Charons 3A and 4A. These bacteria were also lysogenized with λ*imm*21 prophages either at the λ attachment site at 17 min or at the φ80 attachment site at 25 min to test the effect of phage homology in the bacterial chromosome. Charon 4 also has extensive homology at the *bio* region at 18 min. The extent of homology in kilobase pairs between the vectors tested and the various bacteria used as host is shown in the table at the bottom.

TABLE 1. *Survival of Charon 4 phage after various treatments*

Treatment	Tempera-ture, C	Treatment period	Initial titer	Final titer	Survival[a]
Incubation in 0.2M formate buffer, pH 3.0 in NZ broth[b]	25°	2.75 hr	7×10^{9}	$<1 \times 10^{0}$	$<1.3 \times 10^{-10}$
"		5 min	7×10^{9}	1.1×10^{5}	1.4×10^{-5}
"		30 sec	7.9×10^{9}	1.4×10^{7}	1.8×10^{-3}
Incubation in NZ broth adjusted to pH 3.0 w/HCl[b]	25°	10 min	1.5×10^{7}	2×10^{2}	1.2×10^{-5}
"	"	30 sec	1.9×10^{7}	2.0×10^{5}	1.1×10^{-2}
Incubation in 1% SDS[c]	25°	30 min	9.6×10^{9}	2.0×10^{5}	2.1×10^{-7}
Incubation in 0.25% SDS[c]	"	"	9.6×10^{9}	4.5×10^{5}	4.7×10^{-5}
Desiccation on lab bench[d]	25°	6 hr	1.0×10^{10}	1.99×10^{6}	2.0×10^{-4}
Desiccation in Petri dish[d]	"	"	1.0×10^{10}	8.7×10^{5}	8.7×10^{-5}
Shaking in raw sewage[e]	37°	120 hr	3.4×10^{6}	1.0×10^{1}	2.9×10^{-6}
"	"	60 hr	3.4×10^{6}	$<1.0 \times 10^{3}$	$<2.9 \times 10^{-4}$
"	"	12 hr	3.4×10^{6}	2.1×10^{5}	6.2×10^{-2}
Shaking in NZ broth[e]	37°	120 hr	2.9×10^{6}	3.0×10^{2}	1.0×10^{-4}
"	"	60 hr	2.9×10^{6}	4.0×10^{3}	1.4×10^{-3}
"	"	12 hr	2.9×10^{6}	3.0×10^{5}	1.0×10^{-1}
Shaking in tap water[e]	37°	12 hr	3.4×10^{9}	3.0×10^{7}	8.8×10^{-3}

[a] Calculated from final/initial titers.

[b] The pH of a suspension of Charon 4 in NZ broth was lowered to 3.0 from 7.0 with 2 M formic acid or HCl.

[c] Prepared by mixing equal volumes of a suspension of Charon 4A in broth and 2% SDS or 0.5% SDS in φ80 buffer (0.1 M NaCl, 0.01 M Tris, pH 7.4).

[d] One milliliter of a suspension of Charon 4A in NZ was allowed to dry for 6 hr on a laboratory bench or in a Petri dish and rehydrated with NZ and titered.

[e] Aliquots of Charon 4A unmodified phage with a CsCl purified preparation were added to 10 ml of the indicated substance. City of Madison untreated raw sewage (including University of Wisconsin effluent) was obtained immediately before treatment. Its bacterial titer was 1×10^{7}/ml; the naturally occurring phage titer was low and no blue plaques were noted. Therefore, Charon titers could be determined as blue plaques on XG plates.

sarily mean correspondingly poor growth (a burst of 5 or greater is needed to produce a plaque), we felt this should be done to supplement the data on plating efficiency. These experiments posed considerable difficulties in removal or inactivation of unadsorbed phages, and it was difficult to sample a sufficient volume of the dilute lysates required to prevent readsorption. However, in no nonpermissive case did we observe any growth. In various experiments, we were able to set upper limits on burst size ranging from 10^{-6} to 10^{-3} of that observed in a productive infection.

Lysogen or Plasmid Formation by Charon Phages in Permissive Bacteria

In the experiments summarized in Table 2, we determined the number of bacteria surviving in lysates that might perpetuate a cloned segment in a

TABLE 2. *Stable* lac+ *transductants in lysates*

Phage	Bacteria	Phage in,/ml	Phage out,/ml	Blue colony-forming units/ml lysate	Blue colonies out/phage out
Charon 3	CSH18	6×10^{2}	3.2×10^{10}	3×10^{2}	9.4×10^{-9}
Charon 3A	CSH18	6×10^{2}	9.7×10^{10}	5×10^{2}	5.2×10^{-9}
Charon 4	CSH18	6×10^{2}	2.2×10^{10}	$<1 \times 10^{1}$	$<4.5 \times 10^{-10}$
Charon 4A	CSH18	6×10^{2}	3.7×10^{10}	$<1 \times 10^{1}$	$<2.7 \times 10^{-10}$
Charon 4A	CSH18(λi^{21})	6×10^{2}	3.9×10^{10}	$<1 \times 10^{1}$	$<2.6 \times 10^{-10}$

Lysates of Charon phages were grown by the PDS protocol on *lac* $z^- y^+$ cells (CSH 18). Some 50-ml aliquots of the lysates were centrifuged to pellet cells. The pellets were washed twice in 100 ml cold ϕ80 buffer and resuspended in 5 ml ϕ80 buffer. A 1-ml aliquot was treated with rabbit anti-λ serum (5% by volume) and incubated 30 min at 37°C, after which it was diluted into 1 liter of prewarmed NZ broth and grown for 2 hr shaking at 37°C. This procedure was repeated and the resulting washed and resuspended cells were essentially devoid of free phage. Cell counts were taken at each stage of the procedure on nonselective plates to monitor cell growth. The washed cells were plated on *lac*-min XG plates. Cell counts were corrected for growth during the workup to determine the number of *lac*+ cells originally present in the lysates. Reconstruction experiments were done by adding a λ-resistant λ p*lac* 5 lysogen of CSH 18 and a Charon 3 transductant of CSH 18 to lysates, which demonstrated that such cells were not lost in the above protocol and that their growth rates were the same as those of CSH 18.

stable form. Charon phages were grown by the PDS method on *lac* $z^- y^+$ cells. After lysis, surviving cells were separated from free phage and plated on XG dye indicator plates with lactose as the sole carbon source. Reconstruction experiments were done to verify our ability to demonstrate extremely low levels of *lac*+ bacteria in lysates. The results (Table 2) showed *lac*+ bacteria present in Charon 3 and 3A lysates at 9.4×10^{-9} and 5.2×10^{-9} of the level of phages produced. These bacteria were stable in their *lac*+ phenotype and did not plate Charon 3. We conclude that they represent a low level of plasmid or lysogen formation with Charon 3. In Charon 4 and Charon 4A lysates, no *lac*+ colonies were observed, although they could be detected in these experiments at less than 4.5×10^{-10} and 2.7×10^{-10} of the levels of phages produced. The same experiment was also done with *lac*− bacteria containing the λ*imm*21 prophage. The λ*imm*21 prophage made no difference in the results.

Chloroform Sensitivity of *E. coli* and Transformants

The sensitivity to chloroform of the putative Charon 3 and 3A plasmids was tested. As with other *E. coli*, fewer than 3 in 10^{11} cells survived a 4-min treatment with chloroform. Based on these experiments, we conclude that the bacteria in Charon 3A and Charon 4A lysates do not represent a major route for dissemination of cloned DNA segments, especially when chloroform is used after lysis.

Lysogen or Plasmid Formation by Charon Phages in Nonpermissive Bacteria

A final series of experiments was undertaken to assess the possibility of lysogen or plasmid formation in nonpermissive strains of *E. coli.* These experiments examined the likelihood that a recombinant phage that had escaped into nature might subsequently be stably integrated into a naturally occurring bacterium. Although the probability of an encounter under dilute conditions with a nonpermissive bacterium that could adsorb λ would be low (as discussed above), it would be cause for concern if the integration step turned out to be highly efficient. Our results (Table 3) show that this is not the case. When su° *lac* z^- y^+ bacteria were challenged with Charon 3A or Charon 4A under conditions where adsorption was substantially complete, the observed frequency of stable lac^+ bacteria ranged from 10^{-10} to 10^{-8} per input phage with 8.5×10^{-9} as an average of 12 experiments. There was little difference between Charon 3A or Charon 4A, nor did the K modification appear to affect the results.

Lysogen or Plasmid Formation by Charon Phages in Nonpermissive Bacteria Carrying Lambdoid Prophages

We repeated the preceding experiments in nonpermissive strains carrying the prophages λ *att*λ *imm*21 *QSR*80 and λ *att*80 *imm*21 *QSR*80 in an at-

TABLE 3. *Gene transfer to nonpermissive bacteria by Charon phages*

Phage[a]	Lysogen contained in bacterium	Average frequency of stable transductants/phage input	No. of experiments averaged
3A m	None	8.4×10^{-9}	2
3A u	"	6.4×10^{-9}	4
4A m	"	1.8×10^{-8}	2
4A u	"	1.3×10	4
3A m	λ*imm* 21	3.2×10^{-6}	2
3A u	"	5.7×10^{-7}	4
4A m	"	1.0×10^{-6}	2
4A u	"	2.5×10^{-7}	4
3A m	λ*imm* 21 *att*80 *h*80	1.5×10^{-6}	1
3A u	"	1.2×10^{-8}	4
4A m	"	2.5×10^{-7}	2
4A u	"	1.2×10^{-7}	4

[a] m designates previous growth on a hsm^+ host and u on a hsm^- host. The indicated phages were preadsorbed to fresh, stationary phase *lac* z^- y^+ bacteria. One milliliter of the mixture was diluted into 200 ml prewarmed NZ broth and shaken at 37°C for 2–3 hr. Periodically aliquots were plated on NZ plates to determine growth during this incubation; cell growth range was 1.5–78 (average = 26.1)-fold. The culture was centrifuged and the pellet washed with 100 ml cold ϕ80 buffer, recentrifuged, and the resuspended bacteria spread on *lac*-min XG plates. In each experiment, blue colonies appearing after 24 hr at 37°C were counted, corrected for growth, and the frequency of stable transductants/input phage was computed. Reconstruction experiments showed that the transformed cells grew at the same rate as the *lac* parents on this medium.

tempt to learn whether increased homology between bacterial DNA and phage DNA would lead to increased levels of transfer of the *lac* genes to the nonpermissive recipient. These prophages share at least 20 KBP of homology with Charons 3A and 4A and are integrated at different sites in the host genome.

Although this test system embodies an extremely unlikely degree of homology, the increases observed were only on the order of 100-fold (Table 3). Considering that all but a few base pairs of bacterial homology are replaced in the cloning process, and that *E. coli* sensitive to λ and carrying prophages with such a high degree of homology to λ must be rare, we are satisfied that far fewer than one in 10^8 phage particles released into the natural environment would be perpetuated in this way.

SUMMARY

Two phage λ vectors have been constructed for cloning foreign DNA and tested for parameters of growth, survival, and gene transfer in simulations of the natural environment. We have identified five substantial barriers to the propagation of Charon A phages in a natural environment such as a sewer.

1. The phages are sensitive to inactivation by low pH, desiccation, and detergents which might be encountered during dissemination.
2. Under dilute conditions, it would be difficult for a phage to find a bacterium.
3. Once found, the bacterium would likely be lacking an appropriate exposed adsorption site.
4. Even if the adsorption and injection took place, it would be likely that the bacterium would destroy the injected DNA by restriction.
5. If the DNA survived restriction, the lack of a suppressor tRNA would prevent a burst and the bacterium would be killed; in any event, the bacterium, if it survived, would not be likely to perpetuate the cloned fragment as a plasmid or lysogen, even if it carried a prophage with substantial homology to the Charon vehicle.

ESTIMATED OVERALL CONTAINMENT—A COMMENT

The proposed guidelines for Recombinant DNA Research (3) specify that an EK2 host-vector system should not permit the survival of the cloned DNA fragment in other than specially designed and carefully regulated laboratory environments at a frequency greater than 10^{-8}. Although such numerical estimates are somewhat arbitrary, we shall attempt to evaluate the containment potential of the phages in numerical terms.

In a sewer, perpetuation requires that phage survive long enough to find a receptive bacterium, and then either successfully infect it by lysis, or convert it to a carrier plasmid or lysogen. Since the estimated time for inactivation to a 10^{-8} level in raw sewage was much shorter than the estimated time to find a sensitive bacterium (8 versus 140 days), we conclude that there is less than an extremely small chance that a released bacteriophage would ever find a sensitive bacterium. Moreover, in tests on pure receptive bacteria chosen to represent the nonpermissive natural host, the efficiency of plating and of plasmid or lysogen formulation were both lower than 10^{-8}. Even on lambdoid lysogens of these simulated natural hosts, the frequency of perpetuation was less than 3×10^{-7}, and such lysogens probably do not represent one-third of the natural bacterial population. Thus, the overall frequency of perpetuation in a sewer would be less than $10^{-8} \times 10^{-8}$ considering the problems of both finding a receptive bacterium and perpetuating the cloned segment in it.

In the case of dissemination of accidental ingestion, the low survival of phages in acid conditions (10^{-10} in 2.75 hr) would also limit the probability of perpetuation to levels within the EK2 criterion.

We also measured the production of transformed bacteria in lysates used for phage production and found fewer than 10^{-8} such bacteria per phage produced. Moreover, these bacteria were killed effectively by the routine addition of chloroform to the lysate (survival less than 3×10^{-11} in 4 min).

Although we would hesitate to multiply the survival numbers to obtain estimates in the 10^{-16} range, we feel that the presence of multiple barriers to dissemination which individually afford protection in the 10^{-8} range provides a strong argument that the overall margin of containment afforded by the Charon phages is conservatively within the EK2 requirement.

REFERENCES

1. Murray, K., and Murray, N. E. (1975): Phage lambda receptor chromosomes for DNA fragments made with restriction endonuclease III of Haemophilus influenzae and restriction endonuclease I of Escherichia coli. *J. Mol. Biol.,* 98:551–564.
2. Murray, N. E., and Murray, K. (1974): Manipulation of restriction targets in phage λ to form receptor chromosomes for DNA fragments. *Nature,* 251:476–481.
3. National Institutes of Health (1976): Guidelines for recombinant DNA research. *Federal Register,* 41(No. 431):27,902–27,943.
4. Rambach, A., and Tiollais, P. (1974): Bacteriophage having EcoRI endonuclease sites only in the nonessential region of the genome. *Proc. Natl. Acad. Sci. U.S.A.,* 71:3927–3930.
5. Schlesinger, M. (1932): The size of the particle and the specific gravity of the bacteriophage determined by centrifugation. *Z. Hyg. Infektionskrankh.,* 114:161–176.
6. Thomas, M., Cameron, J. R., and Davis, R. W. (1974): Viable molecular hybrids of bacteriophage lambda and eukaryotic DNA. *Proc. Natl. Acad. Sci. U.S.A.,* 71:4579–4583.

Recombinant Molecules: Impact on Science and Society, edited by R. F. Beers, Jr. and E. G. Bassett. Raven Press, New York © 1977.

27. Propagation of a Fragment of Adenovirus DNA in *Escherichia coli* after Covalent Linkage to a Lambda Vector

Pierre Tiollais, Michel Perricaudet, *Ulf Pettersson, and *Lennart Philipson

*Department of Molecular Biology, Institut Pasteur, 75015 Paris, France; and *Department of Microbiology, The Wallenberg Laboratory, Uppsala University, 75237 Uppsala, Sweden*

INTRODUCTION

Techniques have recently been introduced which make it possible to join genes from unrelated species. Hybrid DNA molecules can be constructed *in vitro* with DNA ligase between fragments with cohesive ends, generated either by restriction endonucleases (16) or by terminal addition of poly(dA) and poly(dT) sequences with terminal transferase (9). When one of the moieties in the hybrid DNA is a prokaryotic self-replicating entity, such as a plasmid (4,8) or a derivative of the bacteriophage lambda (20,26,32), the recombinant DNA can be propagated in *E. coli.* A eukaryotic vector which replicates prokaryotic DNA in eukaryotic cells has also been described (21).

Although this new technology may entail potential hazards (1), the nature of which is hard to predict at present, it may have several important applications in molecular biology, medicine, agriculture, and industry. In order to assess the potential risks and benefits, it is necessary to establish whether genes which are inserted into a heterologous vector can be accurately expressed. Hybrids containing bacterial DNA (2,3,20,30,32), DNA from *Drosophila melanogaster* (6,31,32,35), ribosomal genes of *Xenopus laevis* (18), histone genes from sea urchins (10), and short complementary DNA copies of globin messenger RNA (25,27) have been propagated in *E. coli.* Small segments of lambda DNA containing the leftward operator have been replicated in monkey kidney cells (21). So far, there is no evidence for correct transcription and translation of eukaryotic genes after propagation of hybrids in prokaryotic hosts.

Because of the difficulties in isolating DNA fragments which correspond to single defined eukaryotic genes and the relative ease of isolating fragments of animal virus DNA which contain the gene(s) for one or a few well-characterized polypeptides, we have constructed and studied hybrids between lambda DNA and fragment *Eco*RI-B of Ad2 DNA. This fragment contains sequences which code for the adenovirus DNA binding protein (13,14). The experimental details of this study have been presented elsewhere (34).

Because of the potential biohazards involved in the construction of hybrids between DNA from bacteriophage lambda and Ad2, the experiments were conducted according to the recommendations (1) of the Asilomar meeting. We will also discuss the possibility of using adenovirus genome as a vector in eukaryotic cells.

METHODS AND RESULTS

Selection of Hybrids Between Lambda DNA and Fragments of Ad2 DNA

The DNA of wild-type lambda cannot be used as a vector for construction of recombinant DNA since it contains five cleavage sites for endonuclease *Eco*RI, some in regions which are essential for phage replication. Instead, a vector phage was constructed by first isolating a mutant (which had lost all cleavage sites for endonuclease *Eco*RI) followed by an *in vivo* genetic recombination between this phage and λ-plac5, giving rise to a recombinant λ-plac3 with only three cleavage sites for endonuclease *Eco*RI (26) (Fig. 1). All cleavage sites are in the region of the genome which is nonessential for replication. A derivative of λ-plac3 was constructed *in vitro* by cleavage with *Eco*RI and subsequent ligation, giving rise to λ-2 (Fig. 1), which has two cleavage sites for endonuclease *Eco*RI and a 16% deletion which corresponds to one of the internal *Eco*RI fragments of λ-plac3. An amber mutation was introduced in the *P* gene in the final vector, λ-2Pam. The DNA of this phage yields approximately 1,000 transfectants per nanogram of DNA when uncleaved, but less than 0.1/ng of DNA after cleavage with *Eco*RI. DNA from λ-2Pam was cleaved with *Eco*RI and the two largest fragments, which contain the essential genes for lambda replication, were purified by sucrose gradient centrifugation and used as vector fragments. Ligation of a mixture of the two vector fragments and fragment λ-*Eco*RI-C gave approximately 10 plaques per nanogram, which is at least 100 times greater than the plating efficiency for the vector fragments alone (Table 1). Hybrid genomes which have an integrated DNA fragment between the left and the right arm of the vector can thus be selected (Fig. 1).

Hybrids between the lambda DNA and fragments of Ad2 DNA were constructed and propagated according to the rules defined by the Asilomar conference (1) in a moderate-risk laboratory. The Ad2 DNA fragments were purified by polyacrylamide gel electrophoresis in order to eliminate the Ad2-*Eco*RI-A fragment, which contains the genes necessary for induction

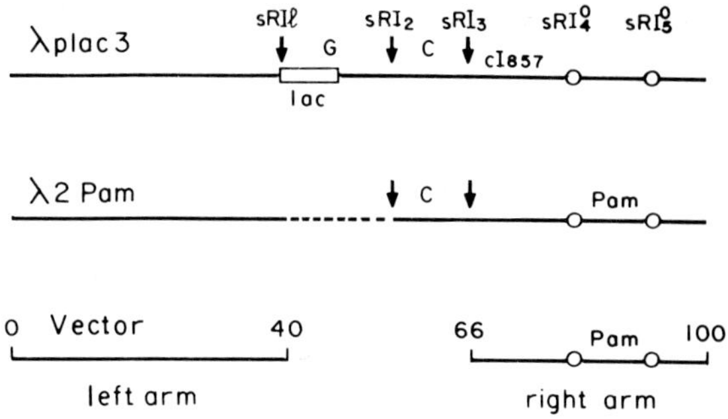

FIG. 1. Lambda vector for *Eco*RI DNA fragments. After hydrolysis of λplac3 or λ2Pam DNA with endonuclease *Eco*RI, the two largest fragments were separated by sucrose gradient centrifugation and concentrated by hydroxylapatite chromatography (34). The vector which consists of a left arm (0 to 40%) and a right arm (66 to 100%) contains the sRI_4^0 and sRI_5^0 mutations and a Pam mutation.

and maintenance of transformation (5,7). Furthermore, since fragment *Eco*RI-A of Ad2 DNA is terminal and thus contains only one cohesive end, the risks for accidental construction of hybrids with transforming sequences should be remote. In order to decrease the potential risk for dissemination of the hybrid phage, investigators introduced an amber mutation into the *P* gene of the vector, and the hybrid phage was propagated in a suppressor strain of *E. coli* ($C600r_k^-m_k^-$ rec B^-C^-). After growth in this strain, the hybrid DNA is not modified and will therefore be restricted in other hosts. The plating efficiency of λ-2Pam on different strains of *E. coli* is shown in Table 2. The plating efficiency of the vector phage is approximately 2×10^{-8} in wild-type *E. coli* relative to the host system used for propagation.

The positive selection technique described above was then used to isolate hybrids between the fragments *Eco*RI-B or *Eco*RI-F of Ad2 DNA and vector fragments derived from λ-2Pam. As shown in Table 1, approximately the same plating efficiency was observed for a mixture containing the vector

TABLE 1. *Transfection efficiency of vector and hybrid DNA*

DNA[a]	Transfectants/ng DNA
λplac3 or λ2P$^+$	800–1,000
λ2Pam	500–750
λ2Pam + *Eco*RI endonuclease	0.05
Vector[b] *Pam* + DNA ligase	0.1
Vector *Pam* + λ-*Eco*RI-C fragment + DNA ligase	12
Vector *Pam* + adenovirus 2-*Eco*RI-B fragment + DNA ligase	11
Vector *Pam* + adenovirus 2-*Eco*RI-F fragment + DNA ligase	5

[a] Transfection was carried out on *E. coli* C600 $r_k^-m_k^-$ rec B^-C^-.
[b] The two terminal fragments of λ2*Pam* were purified by sucrose gradient centrifugation and used as vector.

TABLE 2. *Plating efficiency of bacteriophage λ2Pam on different hosts*

Phage	Host bacterium	PFU[a] on $C600r_K^+m_K^+SuII$ / PFU on $C600r_K^-m_K^-SuII$	PFU on $W3350r_K^+m_K^+Su^-$ / PFU on $C600r_K^-m_K^-SuII$
λ2Pam	$C600r_K^+m_K^+SuII$	1	1.5×10^{-6}
λ2Pam	$C600r_K^-m_K^-SuII$	5×10^{-4}	2×10^{-8}

[a] Plaque-forming unit.

fragments and fragment Ad2 *Eco*RI-B as for the same mixture and fragment λ-*Eco*RI-C. A mixture of vector fragments and fragment Ad2-*Eco*RI-F had a lower plating efficiency, probably because this fragment (1.1×10^6 daltons) is much smaller than fragments Ad2-*Eco*RI-B (2.8×10^6 daltons) and λ-*Eco*RI-C (3.1×10^6 daltons). Different hybrid clones were selected and propagated for several passages. They were stable and the DNA was readily encapsidated into phage particles. It should, however, be emphasized that this system always selects for recombinant genomes, since genomes consisting of the vector fragments alone are too small to be encapsidated.

Phages Containing Hybrid DNA

DNA extracted from ^{32}P-labeled hybrid phages or λ-2Pam was analyzed by electrophoresis in agarose gels after cleavage with endonuclease *Eco*RI. As shown in Fig. 2, DNA from λ-2Pam contained fragment λ-*Eco*RI-C, whereas the hybrid clones contained the vector fragments and a fragment which was of the same size as the corresponding *Eco*RI-B fragment of Ad2 DNA. When the relative distribution of radioactivity in the fragments was determined for several clones, it was apparent that some clones contained multiple copies of the inserted fragment. A maximum of three copies of fragment Ad2-*Eco*RI-B has been observed in hybrid DNA. In host cells which allow recombination, hybrids containing trimers of fragment Ad2-*Eco*RI-B are rapidly converted to hybrids with dimers and monomers of the inserted fragment, probably due to recombination. No evidence for amplification of the inserted fragments has been obtained. All hybrid phages used for our studies on transcription contained single copies of the inserted fragment.

Since the cleavage site for endonuclease *Eco*RI is a palindrome, each fragment can be inserted in two directions as schematically shown in Fig. 3. The two possible types of hybrids containing fragment *Eco*RI-B can, however, be distinguished since there exists one cleavage site for endonuclease *Bam*HI at position 0.60 on the Ad2 genome (28), which is close to one end of fragment Ad2-*Eco*RI-B. On the DNA of λ-2Pam, there exists one cleavage site for endonuclease *Bam*HI at position 0.71, next to the left end of the right vector arm (22). Cleavage of hybrid DNA with endonuclease

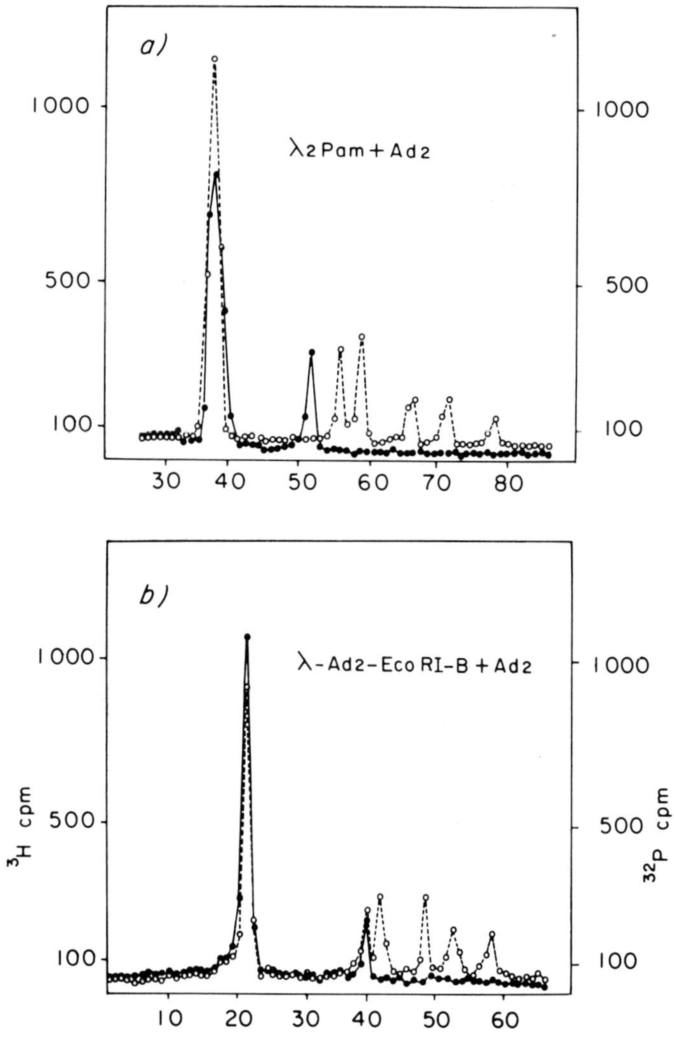

FIG. 2. Characterization of the lambda-adenovirus 2 hybrids. ^{32}P-labeled DNA of λ2Pam and λ-Ad2-*Eco*RI-B was extracted and mixed with ^{3}H-thymidine labeled Ad2 DNA. After hydrolysis with *Eco*RI, the DNA fragments were analyzed by agarose gel electrophoresis (33). **a:** λ2Pam DNA mixed with Ad2 DNA; **b:** λ-Ad2-*Eco*RI-B DNA mixed with Ad2 DNA. (○- - - -○) ^{3}H-labeled Ad2 DNA; (●——●) ^{32}P-labeled λ or hybrid DNA.

*Bam*HI will therefore generate a fragment of 1.8×10^6 daltons if the l-strand of the Ad2-*Eco*RI-B fragment is continuous with the l-strand of lambda, whereas a fragment of 4.0×10^6 daltons will be generated if the l-strand of the Ad2 fragment is continous with the r-strand of the lambda genome (Fig. 3). Two clones, λ-Ad2-*Eco*RI-B4 and λ-Ad2-*Eco*RI-B7 respectively,

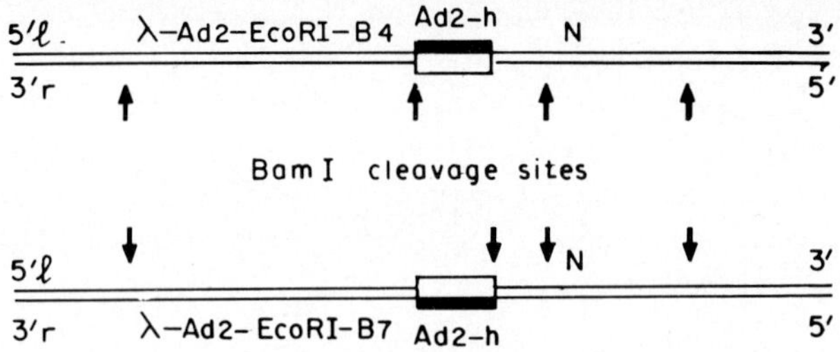

FIG. 3. Cleavage maps of the λ-Ad2-*Eco*RI-B hybrids. The hybrids are Pam, red⁻, att⁻, and int⁻. They possess two *Eco*RI cleavage sites and four *Bam*HI cleavage sites. In the case of λ-Ad2-*Eco*RI-B4, the h-strand of fragment Ad2-*Eco*RI-B corresponds to the l-strand of lambda. In the case of λ-Ad2-*Eco*RI-B7, the h-strand of fragment Ad2-*Eco*RI-B corresponds to the r-strand of lambda.

were isolated in which the Ad2-*Eco*RI-B fragment was inserted in opposite directions as shown in Fig. 3.

Adenovirus RNA Sequences Expressed in *E. coli* Infected with Hybrid Phage

Adenovirus-specific RNA sequences were detected by filter hybridization (23) within 3 min after infection of *E. coli* with hybrid phage, but no hybridization was observed when RNA from mock-infected *E. coli* was analyzed. RNA, extracted early and late after infection with the two hybrid phages containing fragment Ad2-*Eco*RI-B inserted in opposite directions (Fig. 3), was also hybridized to the separated strands of Ad2 DNA. In both cases, "early" RNA hybridized preferentially to the strand which is continuous with the l-strand of the lambda genome. However, "late" RNA showed an increased hybridization to the strand continuous with the r-strand of lambda, as would be expected if transcription of the inserted fragments was controlled by the lambda promoters (15,29).

In order to determine the fraction of the two strands of fragment Ad2-*Eco*RI-B which could be saturated with RNA from *E. coli* infected with hybrid phage, we prepared ^{32}P-labeled separated strands of this fragment (24). The separated strands were incubated under annealing conditions with differing amounts of "late" RNA from *E. coli* infected with the hybrid clone λ-Ad2-*Eco*RI-B7, and the fraction of each fragment strand in the hybrid was determined by chromatography on hydroxylapatite (24). RNA extracted late after infection with this hybrid phage was found to saturate 37% of the l-strand and 60% of the h-strand of the fragment. Cytoplasmic RNA extracted late after infection of HeLa cells with Ad2 saturates 50 to 55% of the l-strand and 40 to 45% of the h-strand of this fragment (24).

In order to demonstrate that transcription of the inserted fragment of hybrid phage is controlled by promoters on the lambda genome, we extracted RNA after infection with the hybrid clone λ-Ad2-*Eco*RI-B7 [derived from a λ-lysogenic strain of *E. coli* (W3350Su$^-$) which lacks suppressor tRNA]. Hybridization experiments showed that the transcription of the inserted fragment was reduced by 97% compared to RNA from non-lysogenic strains. These results suggest that the lambda repressor is able to prevent the transcription of the inserted adenovirus sequences in cells infected with hybrid phage.

Failure to Detect Adenovirus-Specific Gene Products after Infection with Hybrid Phage

Fragment Ad2-*Eco*RI-B contains sequences which code for an adenovirus-specific DNA binding protein (14). Hence, we looked for the presence of polypeptides derived from the adenovirus-specific DNA binding protein after infection of *E. coli* with λ-Ad2-*Eco*RI-B hybrids. So far we have been unable to detect, by SDS-polyacrylamide gel electrophoresis, any new polypeptides after infection with hybrid phages λ-Ad2-*Eco*RI-B4 and λ-Ad2-*Eco*RI-B7. In addition, new polypeptides could not be detected by immunoprecipitation with an antiserum prepared against the DNA binding protein. Our negative results may indicate that some of the sequences coding for the DNA binding protein are not transcribed or that eukaryotic messenger RNA sequences cannot be translated in prokaryotes. Experiments are in progress to resolve this problem.

The Ad2 Genome Contains Sequences which Probably are Nonessential for Replication

It might be possible to use adenovirus DNA as a vector for insertion and propagation of unrelated DNA fragments, provided that the genome retains sequences which are essential for replication. Several groups of defective and nondefective Ad2-SV40 hybrid viruses have been isolated. They all contain significant deletions of Ad2 sequences which have been mapped by electron microscopy using heteroduplex techniques (11,12,17). The deletions that exist in the hybrid viruses, Ad2$^+$ND$_{1-5}$, Ad2^{++}HEY, and Ad^{++}LEY are shown in Fig. 4; also depicted is a relevant portion of a map showing RNA sequences that are expressed early and late after lytic infection (24). Considering the variable and consistently smaller amount of SV40 DNA inserted in place of the deleted Ad2 DNA in the nondefective Ad2$^+$ND$_{1-5}$ hybrids (11,12), it appears that at least part of these deleted sequences are unnecessary for Ad2 replication. Another nondefective variant of Ad2 lacking SV40 DNA sequences has recently been isolated; it also has a deletion which includes the *Eco*RI-D/E junction (Newell and Kelly, *personal communication*), again suggesting that there is a region not essential for replication of Ad2 in human cells. The hybrid viruses

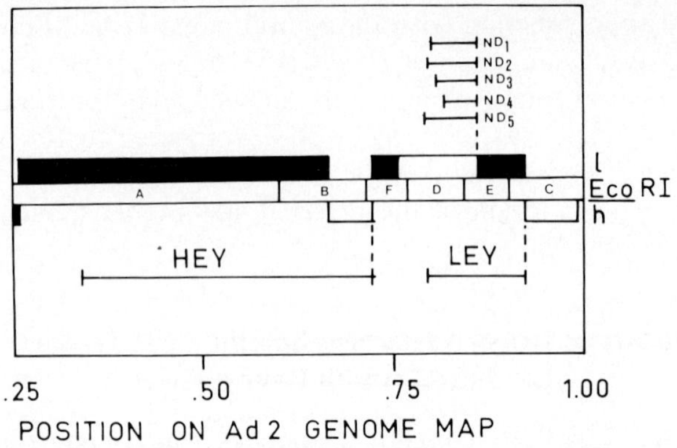

FIG. 4. Relationship between "early" and "late" genes on the Ad2 genome and deletions in DNAs from SV40-Ad2 hybrids. The bars indicate the size and location of deletions in the nondefective hybrid viruses (ND_{1-5}) and in strains $Ad2^{++}$LEY and HEY. (Data from refs. 11 and 12.) Solid bars represent late regions and open bars correspond to early regions for adenovirus messenger RNA in lytic infection of HeLa cells (24).

$Ad2^{++}$HEY and $Ad2^{++}$LEY contain deletions of Ad2 DNA, which include regions of late genes on the l-strand of Ad2-*Eco*RI-A-B and *Eco*RI-E-C, respectively. Deletion of these regions coding for major virus capsid proteins can explain the fact that these two sets of viruses are defective (12). The fact that the deletions in $Ad2^{+}ND_{1-5}$ extend over the *Eco*RI-D/E junction and that this region probably is nonessential suggests that it would be possible to use adenovirus DNA as a vector provided that the appropriate *Eco*RI sites can be eliminated. The DNA of Ad5 has only two cleavage sites for *Eco*RI at positions 0.76 and 0.83 (28). The latter is located in the presumed nonessential region, which suggests that it could be used for insertion of *Eco*RI fragments of unrelated DNA. Furthermore, Ad3 DNA appears to have two *Eco*RI cleavage sites within the presumed nonessential region (19) and may thus be used directly as a vector, provided that the Ad3 genome has a topography similar to that of Ad2 and Ad5.

DISCUSSION AND SUMMARY

Hybrids between lambda DNA and fragments of Ad2 DNA have been constructed. Phages containing hybrid genomes could be selected because a minimal length of the lambda DNA is required to generate plaques after transfection. The presence of integrated *Eco*RI fragments of Ad2 DNA in the lambda genome was demonstrated by electrophoresis of recombinant DNA after cleavage with endonucleases *Eco*RI and *Bam*HI. The latter enzyme was also used to establish the orientation of fragment Ad2-*Eco*RI-B in the hybrid DNA. Adenovirus RNA sequences which obviously originate from both strands of the inserted fragment were detected after infection. The observed transcription pattern suggests that the leftward and rightward promoters of the lambda genome control the transcription of the inserted fragment. This conclusion was corroborated by the finding that transcription of adenovirus sequences was repressed in lysogenic bacteria after superinfection with hybrid phage. The adenovirus RNA transcribed in bacteria infected with hybrid phage does not appear to be translated. These results may suggest that eukaryotic genes inserted into prokaryotic vectors are not accurately transcribed nor translated therein. If these findings are confirmed in other systems, they will have great implications for the application of the recombinant DNA technology and for the assessment of the potential risks. The hybrids can, however, be used to produce and clone large quantities of Ad2 fragments. Bacterial RNA containing Ad2 sequences can also be used for *in vitro* translation, which may help to understand the presumed restricted translation of eukaryotic messenger RNA in prokaryotic systems.

ACKNOWLEDGMENTS

We are grateful to Professor H. Lundbäck and Dr. Alf Lindberg at the National Bacteriological Laboratory for allowing access to a containment laboratory in the early phase of this work. We also want to express our gratitude to Drs. J. Monod, L. Montagnier, and H. Buc for encouragement and helpful discussions. Finally, we want to thank Miss Gun-Inger Lindh and Mr. Hans-Jürg Monstein for technical assistance. This study was supported by grants from the C.N.R.S., Collège de France, and D.G.R.S.T. in France, and from the Medical Research Council and the Cancer Society in Sweden.

REFERENCES

1. Berg, P., Baltimore, D., Brenner, S., Roblin, R. O., and Singer, M. F. (1975): Asilomar conference on recombinant DNA molecules. *Science,* 188:991–994.
2. Cameron, J. R., Panasenko, S. M., Lehman, I. R., and Davis, R. W. (1975): In vitro construction of bacteriophage λ carrying segments of the Escherichia coli chromosome: Selection of hybrids containing the gene for DNA ligase. *Proc. Natl. Acad. Sci. U.S.A.,* 72:3416–3420.
3. Chang, A. C. Y., and Cohen, S. N. (1974): Genome construction between bacterial species in vitro: Replication and expression of Staphylococcus plasmid genes in Escherichia coli. *Proc. Natl. Acad. Sci. U.S.A.,* 71:1030–1034.
4. Cohen, S. N., Chang, A. C. Y., Boyer, H. W., and Helling, R. B. (1973): Construction of biologically functional bacterial plasmids in vitro. *Proc. Natl. Acad. Sci. U.S.A.,* 70:3240–3244.
5. Gallimore, P. H., Sharp, P. A., and Sambrook, J. (1974): Viral DNA in transformed cells. II. A study of the sequences of adenovirus 2 DNA in nine lines of transformed rat cells using specific fragments of the viral genome. *J. Mol. Biol.,* 89:49–72.
6. Glover, D. M., White, R. L., Finnegan, D. J., and Hogness, D. S. (1975): Characterization of six cloned DNAs from Drosophila melanogaster, including one that contains genes for ribosomal RNA. *Cell,* 5:149–155.
7. Graham, F. L., Abrahams, P. J., Waarner, S. O., Mulder, C., de Vries, F. A. J., Fiers, W., and van der Eb, A. J. (1974): Studies on in vitro transformation by DNA and DNA frag-

ments of human adenovirus and simian virus 40. *Cold Spring Harbor Symp. Quant. Biol.*, 39:637–650.

8. Hershfield, V., Boyer, H. W., Yanofsky, C., Lowett, M. A., and Helinski, D. R. (1974): Plasmid ColEl as a molecular vehicle for cloning and amplification of DNA. *Proc. Natl. Acad. Sci. U.S.A.*, 71:3455–3459.
9. Jackson, D. A., Symons, R. H., and Berg, P. (1972): Biochemical method for inserting new genetic information into DNA of simian virus 40: Circular SV40 DNA molecules containing lambda phage genes and the galactose operon of Escherichia coli. *Proc. Natl. Acad. Sci. U.S.A.*, 69:2904–2909.
10. Kedes, L. H., Chang, A. C. Y., Houseman, D., and Cohen, S. N. (1975): Isolation of histone genes from unfractionated sea urchin DNA by subculture cloning in E. coli. *Nature*, 255:533–536.
11. Kelly, T., and Lewis, A. M., Jr. (1973): Use of nondefective adenovirus-simian virus 40 hybrids for mapping the simian virus 40 genome. *J. Virol.*, 12:643–652.
12. Kelly, T., Lewis, A. M., Levine, A. S., and Siegel, S. (1974): Structure of two adenovirus-simian virus 40 hybrids which contain the entire SV40 genome. *J. Mol. Biol.*, 89:113–126.
13. Levine, A. J., van der Vliet, P. C., Rosenwirth, B., Rabete, J., Frenkel, G., and Ensinger, M. (1974): Adenovirus-infected, cell-specific, DNA-binding proteins. *Cold Spring Harbor Symp. Quant. Biol.*, 39:559–566.
14. Lewis, J. B., Atkins, J. F., Baum, P. R., Solem, R., Gesteland, R. F., and Anderson, C. W. (1976): Location and identification of the genes for adenovirus type 2 early polypeptides. *Cell*, 7:141–151.
15. Lozeron, H. A., Dahlberg, J. E., and Szybalski, W. (1976): Processing of the major leftward mRNA of coliphage lambda. *Virology*, 71:262–277.
16. Mertz, J. E., and Davis, R. W. (1972): Cleavage of DNA by R_1 restriction endonuclease generates cohesive ends. *Proc. Natl. Acad. Sci. U.S.A.*, 69:3370–3374.
17. Morrow, J., and Berg, P. (1972): Cleavage of simian virus 40 DNA at a unique site by a bacterial restriction enzyme. *Proc. Natl. Acad. Sci. U.S.A.*, 69:3365–3369.
18. Morrow, J. F., Cohen, S. N., Chang, A. C. Y., Boyer, H. W., Goodman, H. W., and Helling, R. B. (1974): Replication and transcription of eukaryotic DNA in Escherichia coli. *Proc. Natl. Acad. Sci. U.S.A.*, 71:1743–1747.
19. Mulder, D., Sharp, P. A., Delius, H., and Pęttersson, U. (1974): Specific fragmentation of DNA of adenovirus serotypes 3, 5, 7, and 12, and adeno-simian virus 40 hybrid virus $Ad2^+ND1$ by restriction endonuclease R·EcoRI. *J. Virol.*, 14:68–77.
20. Murray, N. E., and Murray, K. (1974): Manipulation of restriction targets in phage λ to form receptor chromosomes for DNA fragments. *Nature*, 251:476–481.
21. Nussbaum, A. L., Davoli, D., Ganem, D., and Fareed, G. C. (1976): Construction and propagation of a defective simian virus 40 genome bearing an operator from bacteriophage lambda. *Proc. Natl. Acad. Sci. U.S.A.*, 73:1068–1072.
22. Perricaudet, M., and Tiollais, P. (1975): Defective bacteriophage lambda chromosome, potential vector for DNA fragments obtained after cleavage by Bacillus amyloliquefaciens endonuclease (Bam I). *FEBS Lett.*, 56:7–11.
23. Pettersson, U., and Philipson, L. (1975): Location of sequences on adenovirus genome coding for 5.5S RNA. *Cell*, 6:1–4.
24. Pettersson, U., Tibbetts, C., and Philipson, L. (1976): Hybridization maps of early and late messenger RNA sequences on the adenovirus type 2 genome. *J. Mol. Biol.*, 101:479–501.
25. Rabbitts, T. H. (1976): Bacterial cloning of plasmids carrying copies of rabbit globin messenger RNA. *Nature*, 260:221–225.
26. Rambach, A., and Tiollais, P. (1974): Bacteriophage λ having EcoRI endonuclease sites only in the nonessential region of the genome. *Proc. Natl. Acad. Sci. U.S.A.*, 71:3927–3930.
27. Rougeon, F., Kourilsky, P., and Mach, B. (1975): Insertion of a rabbit beta-globin gene sequence into an Escherichia coli plasmid. *Nucleic Acids Res.*, 2:2365–2378.
28. Sambrook, J., Williams, J., Sharp, P. A., and Grodzicker, T. (1975): Physical mapping of temperature-sensitive mutations of adenovirus. *J. Mol. Biol.*, 97:369–390.
29. Szybalski, W., Bovre, K., Fiandt, M., Hayes, S., Hradecna, Z., Kumar, S., Lozeron, H. A., Nijkamp, H. J. J., and Stevens, W. F. (1970): Transcriptional units and their controls in

Escherichia coli phage λ: Operons and scriptons. *Cold Spring Harbor Symp. Quant. Biol.*, 35:341–353.
30. Tanaka, T., and Weisblum, B. (1975): Construction of a colicin El-R factor composite plasmid in vitro: Means for amplification of desoxyribonucleic acid. *J. Bacteriol.*, 121:354–362.
31. Tanaka, T., Weisblum, B., Ichnoss, M., and Inman, R. (1975): Construction and characterization of a chimeric plasmid composed of DNA from Escherichia coli and Drosophila melanogaster. *Biochemistry*, 14:2064–2072.
32. Thomas, M., Cameron, J. R., and Davis, R. W. (1974): Viable molecular hybrids of bacteriophage lambda and eukaryotic DNA. *Proc. Natl. Acad. Sci. U.S.A.*, 71:4579–4583.
33. Tiollais, P., Gallibert, F., Auger, M. A., and Lepetit, A. (1972): L'électrophorèse des acides ribonucléiques en gel de polyacrylamide. *Biochimie*, 54:339–354.
34. Tiollais, P., Perricaudet, M., Pettersson, U., and Philipson, L. (1976): Propagation in *E. coli* of bacteriophage lambda with integrated fragments of adenovirus 2 DNA. *Gene*, 1:49–63.
35. Wensink, P., Finnegan, D. J., Donelson, J. E., and Hogness, D. S. (1975): System for mapping DNA sequences in chromosomes of Drosophila melanogaster. *Cell*, 3:315–325.

Recombinant Molecules: Impact on Science and Society, edited by R. F. Beers, Jr. and E. G. Bassett. Raven Press, New York © 1977.

28. Construction of Hybrid Viruses Containing SV40 and Lambda Phage DNA Segments and Their Propagation in Cultured Monkey Cells

Stephen P. Goff and Paul Berg

Department of Biochemistry, Stanford University School of Medicine, Stanford, California 94305

INTRODUCTION

Recently, techniques have been developed which permit the biochemical construction and propagation of novel transducing phages (2,16,18) and plasmids (4,15,25) carrying specific genetic markers. The generality of these procedures permits the introduction, into bacteriophage genomes and plasmids, of DNA segments from organisms that do not ordinarily interact genetically with such phages and plasmids.

We wish to describe the successful construction and propagation of a transducing animal virus. Approximately 2 kilobases (kb) of DNA were removed from the late region of the SV40 genome by sequential cleavages with *Hpa*II and *Bam*HI endonucleases (at 0.735 and 0.13, respectively, on the SV40 DNA map) and a segment of about 1.5 kb of λ phage DNA was inserted in its place. The resulting hybrid DNA was cloned and propagated in CV-1 monkey kidney cells by mixed infections (14) with *tsA*58, an early mutant of SV40 (22). In this chapter we describe the genetic manipulations, propagation, physical characterization, and expression of the resulting transducing virus. A more detailed description of these experiments has been published elsewhere (7). While this paper was being prepared, the construction of other hybrid DNAs containing contributions from SV40 and λ phage DNAs was reported (6,17); in these cases, however, the absence of any functional SV40 genes in the viral vector makes cloning and propagation of the hybrid virus more cumbersome in that wild-type SV40 is required as the helper.

RESULTS

Preparation of SV40 DNA Vector (SVGT-1)

Earlier experience had established that naturally arising mutants of SV40, lacking nearly the entire late region (SV40 map coordinates 0.70 to 0.15), can be propagated in mixed infections with a helper virus that can provide the missing functions (14); a suitable helper virus for complementing such late deletion mutants is *tsA*30 or *tsA*58, thermosensitive early mutants of SV40 (22). Accordingly, a vector suitable for cloning and propagating foreign DNA segments was prepared by excising virtually the entire late region of the viral DNA by two successive cleavages with *Hpa*II and *Bam*HI restriction endonucleases; these enzymes cleave SV40 DNA at map positions 0.735 (20) and 0.13 (F. E. Young, *personal communication*), respectively (Fig. 1A). The large segment of the viral DNA (0.6 SV40 genome length), hereafter referred to as SVGT-1, contains the origin of SV40 DNA replication [at 0.67 (5,19)]; it was separated from the 0.4 SV40 genome length segment by two sequential electrophoreses in agarose (Fig. 1A).

SVGT-1 was modified for joining to the λ phage DNA segment by digestion with an excess of λ-exonuclease (12) to remove approximately 50

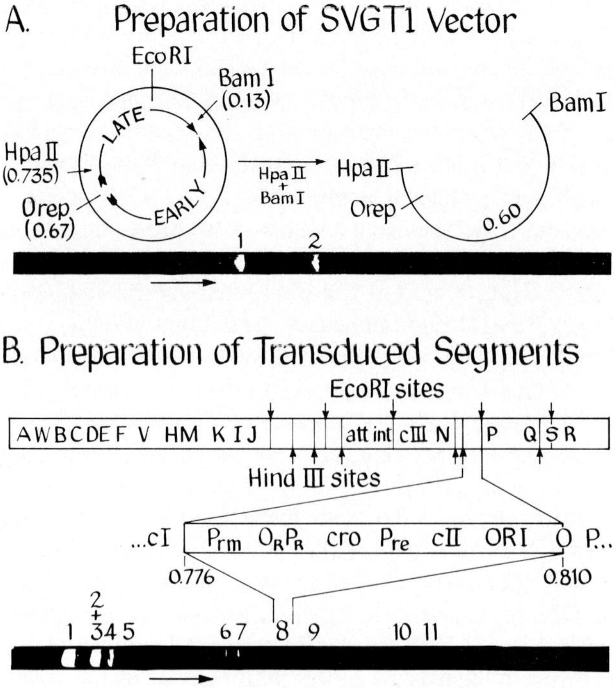

FIG. 1. A: SV40 form I DNA was cleaved with *Hpa*II and *Bam*HI restriction endonucleases at the sites shown. The two resulting DNA fragments were separated by electrophoresis in agarose gels (direction of migration: left to right), stained, and photographed as mentioned in the experimental procedures.

B: Linear λ DNA was cleaved with *Eco*RI and *Hind*III restriction endonucleases at the sites shown. The resulting fragments were electrophoresed on agarose gels (migration left to right), stained, and photographed as mentioned above. (From ref. 7.)

nucleotides from the 5′ end of each strand (10,13) followed by incubation with deoxynucleotidyl terminal transferase (11) and ^{3}H-dATP to add about 200 deoxyadenylate residues per exposed 3′ end (10,13). The modified fragment recovered after these two reactions was uncontaminated with intact circular or full-length linear SV40, as judged by its low infectivity. Whereas circular and full-length linear SV40 DNA produce 2×10^6 and 2 to 4×10^5 plaque-forming units (PFU)/μg DNA, respectively, in the standard plaque assay (14), the purified, modified fragment produced less than 20 PFU/μg DNA (no plaques were detected after infections using up to 0.05 μg of the DNA fragment).

Preparation of λ Phage DNA Insert

Lambda phage DNA contains five *Eco*RI endonuclease cleavage sites (24) and six *Hin*dIII endonuclease cleavage sites (1); digestion with both enzymes generated 12 fragments ranging in size from 0.3 to 23 kb. The longer and intermediate size fragments were separated readily by electrophoresis in agarose (Fig. 1B). By comparison with SV40 DNA marker fragments of known length, the sizes of the intermediate λ fragments 6, 7, 9, and 10 were found to be 1.91, 1.81, 1.28, and 0.91 kb (± 0.05 kb), respectively.

Fragment 8 (Fig. 1B) is 1.48 kb in length and contains ORI, the origin of λ DNA replication, and two structural genes, *CII* and *cro,* as well as four transcriptional promoters: P_{re} and P_{rm} promoters regulate *CI* gene transcription, a promoter near ORI controls transcription of the short OOP RNA transcript, and P_r, the repressor-regulated promoter, controls late gene transcription (8). After elution of fragment 8 from the gel, it was treated with λ phage exonuclease and then with deoxynucleotidyl terminal transferase and ^{32}P-dTTP as mentioned above.

Construction and Propagation of λ-SVGT-1 Hybrid Viruses

SVGT-1 containing ^{3}H-poly(dA) termini and the λ DNA insert with ^{32}P-poly(dT) termini were mixed and annealed as described in the experimental procedures. The annealed DNA was used without further purification or treatment to infect monolayers of CV-1P in the presence or absence of *tsA*58. At 41°C, *tsA*58 DNA alone produced no plaques, the annealed DNA alone gave no plaques (<200 PFU/μg DNA), but infection with the two DNAs together produced approximately 2.5×10^3 PFU/μg annealed DNA.

Thirty-four plaques from the mixedly infected cultures were extracted, and after more *tsA*58 virus was added to each of the extracts, monolayers of CV-1 cells were infected and incubated for 14 days at 41°C. Viral DNA was extracted from each of the infected cultures (9), and the presence of the λ phage DNA segment was detected by measuring the reassociation kinetics of ^{3}H-labeled λ fragment 8 DNA in the presence and absence of the isolated viral DNA. There was either no or only a slight increase in the rate of annealing of the labeled λ DNA with the viral DNA obtained from 22 of the plaques (Fig. 2B); hence, these do not contain λ DNA sequences. Three of the plaques yielded viral DNA which increased the reassociation rate of the λ DNA probe only moderately (Fig. 2C); these probably contain low levels of λ-SVGT-1 hybrid but were not examined further. However, DNA from infections with 9 of the 34 plaques caused a striking increase in the reannealing rate of the labeled λ DNA fragment (Fig. 2A). As is shown below, each of these nine plaques contained, in addition to the helper virus, particles with a recombinant virus genome in which the λ phage DNA segment was covalently joined by short poly(dA-dT) segments to SVGT-1.

Structure of the Putative λ-SVGT-1 Hybrid DNAs

Each of the putative λ-SVGT-1 hybrid DNA preparations contains *tsA*58 DNA; consequently, when such DNA samples are digested with *Hin*dIII endonuclease, electrophoresis in agarose should reveal all of the fragments expected from SV40 DNA. In addition, a new fragment (about 0.6 SV40 fractional length) should be generated containing the λ DNA and poly dA:dT segments with short SV40 DNA "tails" (see Fig. 3).

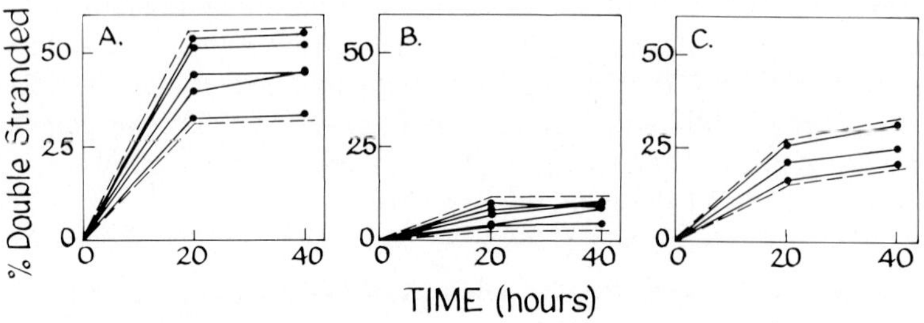

FIG. 2. Annealing kinetics of DNA obtained from plaques resulting from transfection with the hybrid and helper DNA. ^{3}H-labeled λ fragment 8 DNA was denatured and allowed to reassociate in the presence of viral DNA that had been obtained from infections with each of 34 plaques.

A: Nine of the DNA stocks greatly accelerated the reassociation of the labeled λ fragment 8 DNA: only 5 representative clones are shown, but the reassociation kinetics for each of the DNA samples yielded curves that fell within the limits marked by the dotted lines.

B: Twenty-two of the DNA stocks did not affect the rate of annealing of the labeled λ fragment 8 DNA into double-stranded form. The curves are for the annealing of the labeled probe alone and with 4 of the 22 clones. The other 18 curves fell within the limits of the dotted lines.

C: Three of the DNA samples promoted the annealing of λ fragment 8 DNA only slowly. This would occur if the amount of hybrid DNA in these preparations was very low. (From ref. 7.)

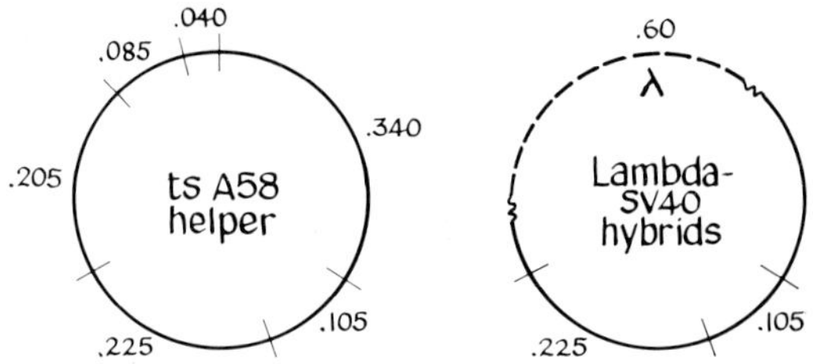

FIG. 3. Fragments produced by *Hin*dIII endonuclease digestion of helper and hybrid virus genones. The vertical slashes indicate the sites of cleavage and the numbers the length of the fragments expressed in SV40 fractional lengths. *tsA*58 DNA yields all the fragments of wild-type SV40; λ-SVGT-1 hybrid DNA yields two of these fragments plus a novel fragment of 0.60 SV40 fractional length. (From ref. 7.)

Figure 4 shows that this expectation was fulfilled; *Hin*dIII endonuclease digests of each of the nine putative λ-SVGT-1 hybrid DNA preparations gave an electrophoretic pattern qualitatively similar to the ones shown in Fig. 4A.

Using a procedure developed by Southern (21), which permits transfer of denatured DNA from such gels to nitrocellulose sheets without disturbing their relative positions, one can detect the λ DNA-containing band by *in situ* hybridization with a radioactively labeled probe. ^{32}P-labeled λ cRNA, prepared with whole λ DNA and *E. coli* RNA polymerase, was hybridized to such nitrocellulose "imprints," and, after suitable washing, autoradiograms of the nitrocellulose sheets were prepared (Fig. 4B). Only the large *Hin*dIII endonuclease-generated fragments contained λ DNA as judged by their hybridization to the labeled λ cRNA.

There was some variability in the size and number of the large fragments produced by the *Hin*dIII endonuclease digestion (track 3 in Fig. 4A). Plaque purification of each λ-SVGT-1 hybrid eliminated the multiplicity of large fragments within a given hybrid DNA preparation (Fig. 5); therefore, this heterogeneity can be ascribed to the occurrence of mixtures of hybrid DNAs within the initial plaque. Since DNA obtained from cloned λ-SVGT-1 virus stocks yields only a single large band after *Hin*dIII endonuclease digestion, the structure of the hybrid DNA is probably stable during subsequent multiplication. The slight variability in the size of the large fragment among clones is probably due to differences in the length of the poly(dA-dT) sequence that joins the λ and SV40 DNA segments of the various hybrid DNAs.

Heteroduplex Analysis of the Putative λ-SVGT-1 Hybrid DNA

A novel and characteristic heteroduplex structure should be formed when full-length single strands from *Eco*RI endonuclease-cleaved, wild-type SV40 DNA are annealed to the strands of the large fragment formed

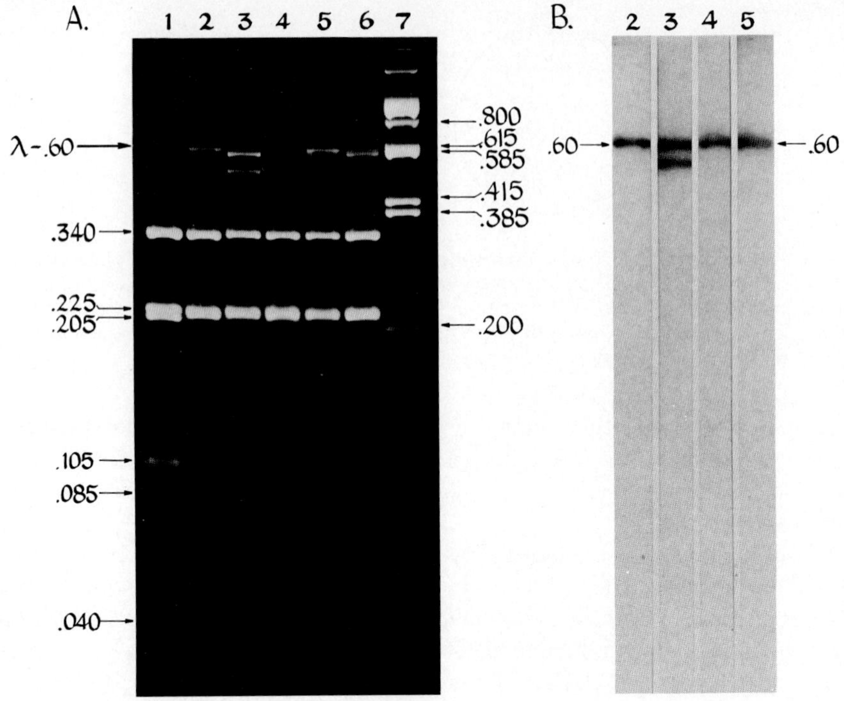

FIG. 4. Agarose gel electrophoresis of DNA fragments produced by *Hin*dIII endonuclease digestion of hybrid DNAs.

A: DNA bands were visualized by staining the gel with ethidium bromide. Track 1: Wild-type SV40 DNA. Tracks 2 through 6: λ-SVGT-1 DNA prepared from plaque isolates 7, 9, 10, 18, and 32. The DNA used in these digestions was obtained from infections with virus from the plaques. The new fragments produced in the digestions of the hybrid DNAs are 0.60 fractional SV40 units in length. Each of the nine isolates containing λ DNA sequences yielded similar digestion products. Track 7: Length standards prepared by partial cleavage of SV40 DNA with *Hpa*I endonuclease.

B: Autoradiogram after hybridization of a replica of the gel in Fig. 4A with ^{32}P-labeled λ cRNA. The DNA in each gel track was transferred to nitrocellulose strips, hybridized *in situ* to the labeled cRNA, and then autoradiographed. DNA from plaque isolates 7, 9, 10, and 18 are shown. (From ref. 7.)

by cleavage of λ-SVGT-1 hybrid DNA with *Hin*dIII endonuclease; in that fragment, the λ DNA segment is joined by a poly(dA-dT) bridge to SV40 DNA (Fig. 6). The "tails" of SV40 DNA anneal to a full-length linear SV40 DNA strand generating a circular structure that is part single-stranded and part duplex DNA. Because the λ DNA sequence is flanked by poly(dA) at one join and by poly(dT) at the other, that segment occurs as a loop joined by a short "neck" of dA-dT duplex to the SV40 DNA heteroduplex region (Fig. 7). The size of the inserted segment as determined by length measurement of the single-stranded heteroduplex loop is 1.46 kb (± 0.06 kb), identical to the size of λ fragment 8. The existence of these

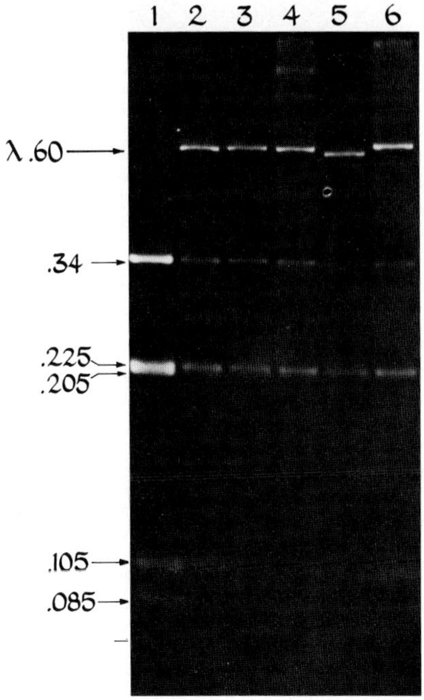

FIG. 5. Homogeneity and stability of the λ DNA segment during propagation of λ-SVGT-1 hybrid virus in CV-1 cells. The original plaque isolates 18, 9, 10, 16, and 32 were cloned twice by plaque purification using *tsA*58 as helper. Then DNA stocks were prepared from each cloned sample, digested with *Hin*dIII endonuclease, and electrophoresed as described in Fig. 4A. Track 1 is wild-type SV40 DNA and tracks 2 through 6 are isolates 18, 9, 10, 16, and 32 after cloning. (From ref. 7.)

heteroduplex structures is thus quantitatively consistent with the predicted structure of the large *Hin*dIII endonuclease-cleavage product and, therefore, of the λ-SVGT-1 hybrid DNA.

Orientation of the Insert

The inserted λ DNA fragment can, in principle, occur in either of two orientations relative to the SVGT-1 DNA. Hybrids of both orientations

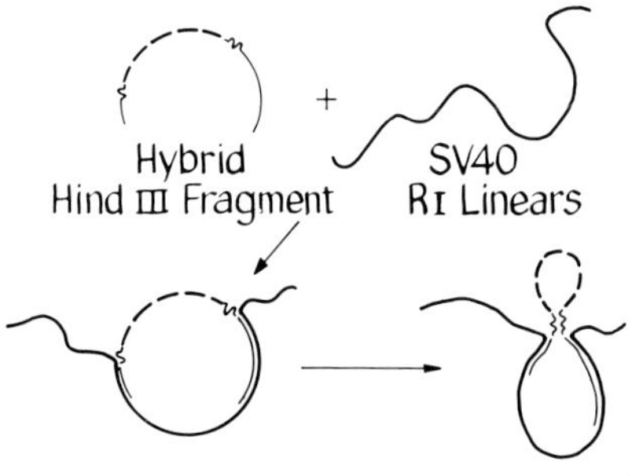

FIG. 6. Schematic representation of the heteroduplexes expected from *Eco*RI endonuclease-cleaved SV40 linear DNA and the λ DNA-containing segment produced by *Hin*dIII endonuclease cleavage of λ-SVGT-1 hybrid DNA. Heteroduplexes are formed between the linear SV40 DNA and the SV40 DNA tails of the *Hin*dIII-generated fragment. When the poly(dA) and poly(dT) sequences on the same strand anneal to form a short "snapback" region, the λ DNA segment appears as a single-stranded loop; with higher formamide concentrations, the λ DNA, poly(dA), and poly(dT) segments are extended. (From ref. 7.)

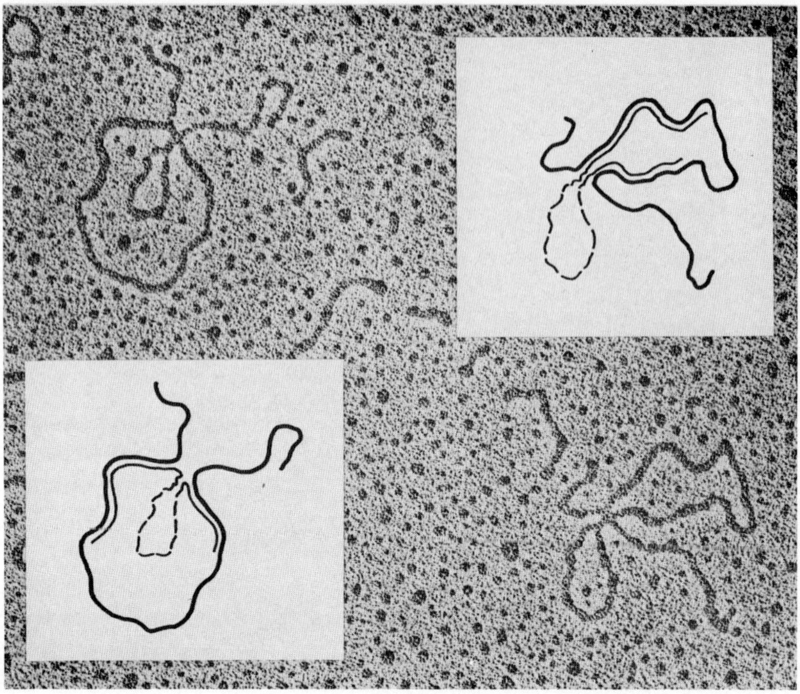

FIG. 7. Electron micrographs of heteroduplexes diagrammed in Fig. 6. In the tracings, heavy lines represent the linear SV40 DNA, the thin lines represent the SV40 tails of the fragment derived from hybrid DNA, and the dashed lines represent λ DNA segments. (From ref. 7.)

were, in fact, isolated. The orientation of the insert was determined by cleavage of the hybrid DNA with *Hin*cII endonuclease which cleaves the λ DNA insert at two sites and the SVGT-1 at two sites (Fig. 8A). Four fragments are thus produced: one containing only λ DNA (0.118 SV40 fractional length), one containing only SV40 DNA (0.20 SV40 fractional length), and two fragments containing λ and SV40 DNA as well as poly (dA-dT) segments. The lengths of these hybrid fragments should differ, in a predictable way, for the two possible orientations.

Figure 8A shows the fragment lengths (in SV40 fractional length) that should be produced by *Hin*cII endonuclease digestion of the hybrids having each orientation, assuming that the dA-dT regions are about 50 base pairs long. Agarose gel electrophoresis of *Hin*cII endonuclease-digested DNA from λ-SVGT-1 clones 9 and 18 shows that the fragment pattern characteristic of the two possible orientations does occur in two isolates (Fig. 8B). All cloned isolates of λ-SVGT-1 yielded fragment patterns similar to either clone 9 or 18. Thus, all the hybrids tested must contain λ fragment 8; no

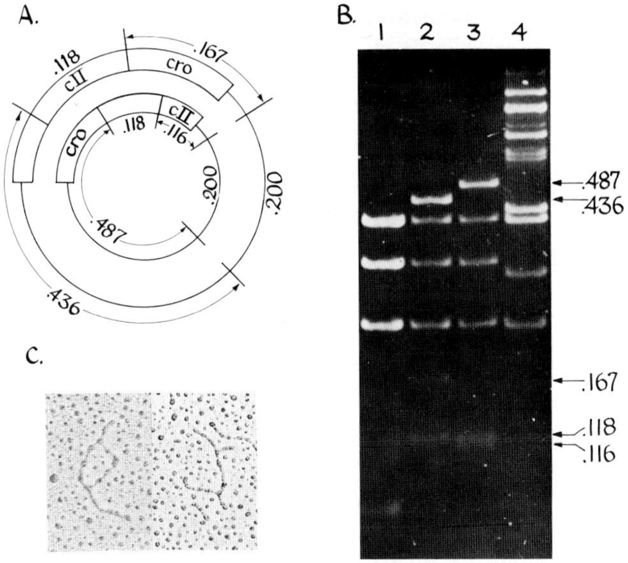

FIG. 8. The orientation of the inserted λ DNA sequences relative to the SVGT-1 vector DNA. **A:** The two possible orientations of the inserted λ DNA. Single lines represent SV40 DNA; bars depict λ DNA sequences. Radial lines are the cleavage sites for the *Hin*cII restriction endonuclease; numbers are the predicted lengths of the fragments in SV40 fractional lengths, assuming that the poly(dA-dT) join segments are 50 bp long.

B: Agarose gel electrophoresis of fragments produced by *Hin*cII endonuclease cleavage of λ-SVGT-1 hybrid DNAs. Track 1: Wild-type SV40 DNA. Track 4: A set of DNA length markers. Tracks 2 and 3: DNA from cloned hybrids 18 and 9, respectively, after cleavage with *Hin*cII endonuclease. Cleavage of the λ-SVGT-1 *tsA*58 helper DNA present in these stocks yields all the bands expected from SV40 DNA; in addition, several new bands, notably those at 0.436 and 0.487 SV40 fractional length, are observed at the predicted positions for the two possible orientations.

C: Heteroduplexes between the λ DNA-containing fragments produced by *Hin*dIII endonuclease cleavage of λ-SVGT-1 (clone 9) and λ-SVGT-1 (clone 18) hybrid DNAs. The two hybrid DNAs were treated with *Hin*dIII endonuclease to produce fragments containing the entire λ DNA segment with short SV40 DNA tails, denatured and renatured as described in the experimental procedures. In the left heteroduplex, the SV40 DNA tails have annealed while the inverted λ DNA sequences remain unpaired. In the other heteroduplex, the λ DNA sequences have annealed, leaving single strands of SV40 DNA. (From ref. 7.)

other fragment of λ DNA produced by *Hin*dIII and *Eco*RI digestion contains the characteristic *Hin*cII cleavage sites.

The existence of opposite orientations of the inserted sequence can also be shown by heteroduplex analysis. The large *Hin*dIII endonuclease-generated fragments of clones 9 and 18 (which contain the entire λ DNA insert) were mixed, denatured, and reannealed to generate heteroduplex molecules. Although both *Hin*dIII endonuclease cleavage fragments are the same length and contain the λ DNA insert, they do not form perfectly matched homoduplexes. Instead, when the SV40 segments are paired (Fig.

8C), the opposite orientation of the λ DNA segments prevents pairing of that region and creates a single-stranded bubble; when the λ DNA segments base pair, the SV40 tails are single-stranded (Fig. 8C). Structures of these types are diagnostic of an inversion.

Expression of the λ DNA Segment During Propagation of the λ-SVGT-1 Hybrid DNA

Our experiments establish that the λ-SVGT-1 hybrid DNA replicates during infection. Of interest is whether the λ DNA sequence is transcribed and whether the resulting mRNA sequence is translated into one or more polypeptides. To answer this question, we examined the RNA and proteins produced following infection of a monkey cell with a mixture (approximately 1:1) of cloned λ-SVGT-1 DNA (either clone 9 or 18) and *tsA*58 DNA. Fifteen to twenty days after infection (about 50% cytopathic effect was observed), CV-1 cultures were labeled for 1 hr with 5 mCi of ^{32}P-orthophosphate and the total cellular RNA was isolated; viral DNA was isolated from similarly infected but unlabeled cultures. To examine the proteins, we labeled primary AGMK cells approximately 20 days after infection for 4 hr with ^{35}S-methionine (100 μCi) in methionine-free culture medium; total cell and viral proteins were then isolated and examined by polyacrylamide gel electrophoresis and autoradiography.

During the infection, the λ-SVGT-1 hybrid DNA replicated nearly as well as the *tsA*58 helper DNA; the ratio of hybrid to helper DNA, as estimated by agarose gel electrophoresis, was between 0.3 and 1.0. But, at the same time, when substantial amounts of SV40-specific RNA were being synthesized, there was no formation of λ-specific RNA (as judged by filter hybridization) and and virtually none by measurement of reassociation kinetics (Table 1). These results indicate that although the λ-SVGT-1 hybrid DNA replicates nearly as well as SV40 DNA itself, transcripts of the λ DNA sequence either are not synthesized or do not persist in the infected cells; substantially the same results were obtained with hybrids having either of the two possible orientations of the inserted λ DNA sequence.

The search for new polypeptides in the λ-SVGT-1 hybrid infected cells was also fruitless. Regardless of which λ-SVGT-1 hybrid was used for infection, only host proteins and the characteristic SV40-induced proteins (VPI, VP2, and VP3) were observed after autoradiography of the polyacrylamide gels (*unpublished results*).

DISCUSSION

Our cloning vector, SVGT-1, lacks the region of the SV40 genome that codes for the viral capsid proteins; but unlike the vector developed in

TABLE 1. *Synthesis of SV40- and λ-specific RNA following infection of CV-1 cells with λ-SVGT-1 and SV40 virus*

RNA from cells infected with:	% of ^{32}P-RNA hybridized to: SV40 DNA	λ DNA	% of total RNA homologous to: SV40 DNA	λ fragment 8 DNA
Nothing	<0.05	<0.02	<0.02	<0.0001
SV40 alone	1.97	<0.05	0.47	<0.0001
*tsA*58 plus hybrid 9	2.21	<0.02	0.49	0.0003
*tsA*58 plus hybrid 18	1.20	<0.03	0.10	<0.0001
λ cRNA	—	35.2	—	0.86

Analysis of RNA isolated from infected CV-1 cells. Cells were infected as indicated, labeled with ^{32}P-inorganic phosphate, and total RNA isolated. The first pair of columns shows the percentage of the input RNA that hybridized to filters containing SV40 or λ DNA. No λ-specific RNA was labeled in the hybrid-infected cells above the background of 0.05%. The second pair of columns shows the fraction of the total RNA homologous to SV40 or λ fragment 8 DNA as measured by the acceleration of the annealing of the labeled DNA probe by the added RNA. The values for SV40 RNA have been calculated taking into account that the predominant SV40 RNA (the 16S late mRNA) can increase the annealing rate of only 24% of one strand of the labeled DNA. (From ref. 7.)

Fareed's laboratory (6,17), SVGT-1 still contains an intact, functional *A* gene, which is essential for viral DNA replication (23). Since *tsA*58 DNA or virus cannot produce plaques on a monkey kidney cell monolayer at 41°C, and hybrid genomes constructed from SVGT-1 cannot direct the synthesis of viral capsid proteins under any circumstances, no infectious virus or plaques are produced except in cells infected with both genomes. This obligatory reciprocal complementation simplifies the task of detecting and cloning the sought-after recombinant genomes, as their plaques need to be discriminated from only the relatively few thermoresistant revertants of the helper virus and from SVGT-1 that became cyclized without the inserted segment. By contrast, complementation with wild-type virus, as required with Fareed's vector (6,17) demands more complicated procedures for detecting and cloning the hybrid regions.

There are several limitations to the use of SVGT-1 for cloning foreign DNA segments. One stems from the inability to encapsidate SV40 DNA molecules larger than about 5 kb; since SVGT-1 is itself about 3 kb, the insert must be 2 kb or smaller to be propagated as a virion. To clone larger DNA segments requires development of additional SVGT vectors. Presently, we are exploring smaller SVGT vectors that contain only the origin of DNA replication or that region plus a single gene that could complement an appropriate *ts* or defective helper virus. An alternative way to reduce, or possibly eliminate, the size limitation of the cloned DNA segments is to construct vectors that can be propagated as plasmid-like elements in animal cells.

Although our analysis of the fine structure of the hybrid DNA is preliminary, no gross changes in the structure of either SVGT-1 or λ DNA segments seem to have occurred during multiplication of the virus during several plaque purification steps and the preparation of viral DNA. Although this agrees with the experience in Fareed's laboratory (6,17), this question will have to be examined periodically during propagation of the hybrid viruses, inasmuch as there is no selection for maintenance of the original sequence during viral multiplication.

A puzzling outcome of our attempts to assess λ DNA expression during multiplication of λ-SVGT-1 genomes was the failure to detect more than a trace of λ-specific RNA sequences, irrespective of the orientation of the inserted sequence. Several explanations may be considered: first, transcription of the λ DNA segment did not occur because the poly(dA-dT) segment proximal to the λ DNA sequence interrupted transcription at that point; second, some not yet understood feature of the λ DNA sequence precludes its transcription by the mammalian RNA polymerase; third, the λ DNA segment is transcribed, but the lifetime of the transcript is too brief to permit its detection; or fourth, the major late SV40 16S RNA transcript is initiated from an SV40 DNA sequence that is missing from the SVGT-1 vector. Each of these possibilities needs to be explored further, but our present belief is that the poly(dA-dT) join closest to the origin of replication probably does not prevent transcription of the L-strand past that point because viable SV40 mutants with poly(dA-dT) inserts at that location exist (3).

SUMMARY

This chapter describes the successful construction and propagation of a transducing animal virus. A segment of DNA, approximately 2 kb in length, was removed from the late region of the SV40 genome by sequential cleavages with *Hpa*II and *Bam*HI endonucleases (at 0.735 and 0.13, respectively, on the SV40 DNA map). A segment of about 1.5 kb of λ phage DNA containing ORI (the origin of λ DNA replication), the two structural genes *CII* and *cro,* and four transcriptional promoters was inserted into the late region of SV40 by the poly(dA-dT) joining procedure. The resulting hybrid DNAs were cloned and propagated as virions in CV-1 monkey kidney cells by mixed infections at 41°C with *tsA*58, an early mutant of SV40. The location, size, and orientation of the inserted λ DNA segment was verified by restriction endonuclease digestions and by heteroduplex analysis. Clones with each of the two possible orientations of the λ DNA segment were isolated. CV-1 cells infected with λ-SV40 hybrid virus contain little or no λ-specific RNA or proteins, even though the hybrid virus replicates nearly as well as the helper virus.

ACKNOWLEDGMENTS

These experiments were supported by research grants from the U.S. Public Health Service and the American Cancer Society. S. G. is a Smith

Kline & French predoctoral scholar as well as a U.S. Public Health Service Trainee.

REFERENCES

1. Allet, B., and Bukhari, A. I. (1975): Analysis of bacteriophage mu and lambda-mu hybrid DNAs by specific endonucleases. *J. Mol. Biol.*, 92:529–540.
2. Cameron, J. R., Panasenko, S. M., Lehman, I. R., and Davis, R. W. (1975): In vitro construction of bacteriophage λ carrying segments of the Escherichia coli chromosome: selection of hybrids containing the gene for DNA ligase. *Proc. Natl. Acad. Sci. U.S.A.*, 72:3416–3420.
3. Carbon, J., Shenk, T. E., and Berg, P. (1975): Biochemical procedure for production of small deletions in simian virus 40 DNA. *J. Mol. Biol.*, 98:1–15.
4. Clarke, L., and Carbon, J. (1975): Biochemical construction and selection of hybrid plasmids containing specific segments of the Escherichia coli genome. *Proc. Natl. Acad. Sci. U.S.A.*, 72:4361–4365.
5. Danna, K. J., and Nathans, D. (1972): Bidirectional replication of simian virus 40 DNA. *Proc. Natl. Acad. Sci. U.S.A.*, 69:3097–3100.
6. Ganem, D., Nussbaum, A. L., Davoli, D., and Fareed, G. C. (1976): Propagation of a segment of bacteriophage λ DNA in monkey cells after covalent linkage to a defective simian virus 40 genome. *Cell*, 7:349–359.
7. Goff, S. P., and Berg, P. (1976): Construction of hybrid viruses containing SV40 and λ phage DNA segments and their propagation in cultured monkey cells. *Cell*, 9:695–705.
8. Hershey, A. D. (Ed.) (1971): *The Bacteriophage Lambda.* Cold Spring Harbor Laboratory, Cold Spring Harbor, New York.
9. Hirt, B. (1967): Selective extraction of polyoma DNA from infected mouse cell cultures. *J. Mol. Biol.*, 26:365–369.
10. Jackson, D., Symons; R., Berg, P. (1972): Biochemical method for inserting new genetic information into DNA of simian virus 40: circular SV40 DNA molecules containing lambda phage genes and the galactose operon of Escherichia coli. *Proc. Natl. Acad. Sci. U.S.A.*, 69:2904–2909.
11. Kato, K. I., Goncalves, J. M., Houts, G. E., and Bollum, F. S. (1967): Deoxynucleotide-polymerizing enzymes of calf thymus gland. *J. Biol. Chem.*, 242:2780–2789.
12. Little, J. W., Lehman, I. R., and Kaiser, A. D., (1967): An exonuclease induced by bacteriophage λ. *J. Biol. Chem.*, 242:672–678.
13. Lobban, P. E., and Kaiser, A. D. (1973): Enzymatic end-to-end joining of DNA molecules. *J. Mol. Biol.*, 78:453–471.
14. Mertz, J. E., and Berg, P. (1974): Defective simian virus 40 genomes: isolation and growth of individual clones. *Virology*, 62:112–124.
15. Morrow, J. F., Cohen, S. N., Chang, A. C. Y., Boyer, H. W., Goodman, H. M., and Helling, R. B. (1974): Replication and transcription of eukaryotic DNA in Escherichia coli. *Proc. Natl. Acad. Sci. U.S.A.*, 71:1743–1747.
16. Murray, N. E., and Murray, K. (1974): Manipulation of restriction targets in phage λ to form receptor chromosomes for DNA fragments. *Nature*, 251:476–481.
17. Nussbaum, A. L., Davoli, D., Ganem, D., and Fareed, G. C. (1976): Construction and propagation of a defective simian virus 40 genome bearing an operator from bacteriophage λ. *Proc. Natl. Acad. Sci. U.S.A.*, 73:1068–1072.
18. Rambach, A., and Tiollais, P. (1974): Bacteriophage λ having EcoRI endonuclease sites only in the nonessential region of the genome. *Proc. Natl. Acad. Sci. U.S.A.*, 71:3927–3930.
19. Salzmann, M. P., Fareed, G. C., Seebring, E. D., and Thoren, M. M. (1974): The mechanism of SV40 DNA replication. *Cold Spring Harbor Symp. Quant. Biol.*, 38:257–265.
20. Sharp, P. A., Sugden, B., and Sambrook, J. (1975): Detection of two restriction endonuclease activities in Haemophilus parainfluenzae using analytical agarose-ethidium bromide electrophoresis. *Biochemistry*, 12:3055–3063.
21. Southern, E. M. (1975): Detection of specific sequences among DNA fragments separated by gel electrophoresis. *J. Mol. Biol.*, 98:503–517.

22. Tegtmeyer, P. (1972): Simian virus 40 deoxyribonucleic acid synthesis: the viral replicon. *J. Virol.,* 10:591–598.
23. Tegtmeyer, P. (1974): Altered patterns of protein synthesis in infection by SV40 mutants. *Cold Spring Harbor Symp. Quant. Biol.,* 39:3–15.
24. Thomas, M., and Davis, R. W. (1974): Studies on the cleavage of bacteriophage lambda DNA with EcoRI restriction endonuclease. *J. Mol. Biol.,* 91:315–328.
25. Wensink, P. C., Finnegan, D., Donelson, J. C., and Hogness, D. S. (1974): A system for mapping DNA sequences in the chromosomes of Drosophila melanogaster. *Cell,* 3:315–325.

Recombinant Molecules: Impact on Science and Society, edited by R. F. Beers, Jr. and E. G. Bassett. Raven Press, New York © 1977.

29. Cloning of a Segment from the Immunity Region of Bacteriophage λ DNA in Monkey Cells

*George C. Fareed, Dana Davoli, and
**Alexander L. Nussbaum

Department of Biological Chemistry, Harvard Medical School, Boston, Massachusetts 02115

INTRODUCTION

We have been concerned with the fate of prokaryotic DNA segments in mammalian cells and with the development of methods for cloning both prokaryotic and eukaryotic DNA in such cells. The ability to propagate foreign DNA segments and specific host cell genes in appropriate eukaryotic host cells should facilitate the analysis of the mechanism and control of eukaryotic gene expression. These studies have been initiated by an investigation of the fate, in monkey kidney cells, of recombinant DNA molecules containing specific segments of bacteriophage λ DNA and eukaryotic vector fragments from SV40 DNA.

Our approach for obtaining mammalian transducing viruses arose from studies of reiteration mutants of papova viruses (8,11–13,22). These mutants, derived from serial undiluted passage of SV40 or polyoma in permissive cells, characteristically have deleted a large part of the wild-type (WT) genome and reiterated in tandem a small part bearing the origin for viral DNA replication. Structural analysis (11,19,20,22) has indicated that this origin is the only required *cis* function for replication. This was further supported by the finding of Lai and Nathans (21) that the termination site in SV40 DNA replication was nonspecific. Thus, in analogy to the cloning of foreign DNA segments in *Escherichia coli* by enzymatic insertion into a suitable plasmid or phage λ replicon (4,6,7,14,16–18,25,31,33), one could use a segment of SV40 DNA containing its initiation site as the vector for a eukaryotic or prokaryotic DNA segment of appropriate size for encapsidation.

Present address: * Department of Microbiology and Immunology and Molecular Biology Institute, University of California, Los Angeles, California 90024; and ** Boston Biomedical Research Institute, Boston, Massachusetts 02114.

We found in a previous study (13) that the short, monomeric segments from SV40 reiteration mutants can serve as vectors for propagating specific segments of λ DNA in monkey kidney cells. In the presence of helper SV40 DNA to supply *trans* functions, the hybrid genomes are replicated and those whose sizes are near that of WT SV40 DNA [5,000 base pairs (bp)] are encapsidated in progeny virions. In our initial experiments, a defective hybrid bearing a specific 520 bp segment from the λ immunity region was propagated but not cloned in monkey cells (13).

In a subsequent report (27) we described the partial cloning in monkey cells of a 2,300 bp segment neighboring that 520 bp segment. The cloning of this hybrid in mammalian cells not only showed the efficacy of defective SV40 vectors but also provided an opportunity for examining a well-characterized prokaryotic genetic regulatory site, the leftward operator (O_L), in a eukaryotic environment. This region of λ seemed appropriate for the construction of novel recombinant DNA molecules since it lacked the replication functions of λ, and the O_L that it contained could be specifically identified in a hybrid molecule by λ repressor binding. We shall review here the construction, propagation, and physical structure of this SV40-λ hybrid.

MATERIALS AND METHODS

Cells, Virus, and Decontamination

Both primary African green monkey kidney cells (Flow Laboratories) and the CV-1 line of monkey kidney cells were subcultured in Eagle's minimal essential medium supplemented with 5% tryptose phosphate broth (Difco), 0.5% glucose, and 10% calf serum. The small plaque strain 776 of SV40 and the triplication mutant a_3 were propagated as described before (12). Virus and DNA infections were performed in a vertical laminar flow hood (Bioscience Inc.) in a limited access tissue culture room equipped with an exhaust air filter. All media and washings from virus and DNA infections were aspirated into Wescodyne® (West Chemical Products, Inc.) for disposal. All flasks, culture dishes, or pipettes which came into contact with infected cultures were soaked in Wescodyne® prior to disposal.

Cleavage of DNAs with Bacterial Restriction Endonucleases

Restriction endonucleases from *E. coli* RY-13 (endo R.*Eco*RI), *Hemophilus influenzae* d (endo R.*Hin*dII+III and R.*Hin*dIII), *H. parainfluenzae* (endo R.*Hpa*II), and *H. aegyptius* (endo R.*Hae*III) were employed as previously described (12,13). The restriction endonuclease from *Bacillus amyloliquefaciens* H (endo R.*Bam*I) was the phosphocellulose fraction described by Wilson and Young (34). A 1-μl sample of the purified enzyme completely converted 1 μg of superhelical SV40 DNA I to unit length linear molecules in a 100-μl reaction mixture consisting of 6 mM Tris (pH 7.5),

5 mM $MgCl_2$, and 6 mM 2-mercaptoethanol during a 30-min incubation at 37°C.

For preparation of the vector DNA harboring an SV40 DNA replication origin, the DNA from a triplication mutant, a_3, was prepared from infected monkey cells and the monomer segment, *a*, was isolated after endo R.*Eco*RI cleavage (9,12). An 80-μg sample of fragment *a* was incubated in a 500-μl reaction mixture containing 60 μl endo R.*Bam*I for 90 min. The reaction was terminated by addition of EDTA (20 mM) and Sarkosyl® (0.1%), and the two cleavage products (*aBam*-A and *aBam*-B) were purified by preparative gel electrophoresis in a 3.0% polyacrylamide–0.5% agarose slab gel (13). The larger of the two fragments (*aBam*-A) contains the SV40 replication origin based on previous physical structure of fragment *a* (19) and the known location for endo R.*Bam*I cleavage in SV40 DNA (C. Mulder and M. Green, *unpublished observations*). A 2-μg sample of *aBam*-A was further cleaved with endo R.*Hin*dIII (13).

A 115-μg sample of the λ *Eco*RI-B fragment (2,26,30) was cleaved using endo R.*Hin*dIII (13) and endo R.*Bam*I (100 μl in a 2.7-ml reaction mixture incubated at 37°C for 6 hr). The products of cleavage were separated by electrophoresis through 2% polyacrylamide–0.5% agarose slab gels, and the 2,400 bp segment was excised and eluted (12,13).

In Vitro Recombination of DNA Segments

For the formation of hybrid DNA molecules, the following were added to a 130-μg reaction mixture (12,13) of bacteriophage T4 polynucleotide ligase (32): 0.25 μg fragment *aBam*-A, 0.5 μg fragment *aBam*-A further cleaved by endo R.*Hin*dIII, and 1.5 μg 2,400 bp segment from phage λ DNA. The reaction with T4 polynucleotide ligase and subsequent DNA infection of monkey cells in the presence of WT (helper) SV40 DNA were performed as described before (12,13).

Cloning and Characterization of the Hybrid

A modification of the infectious center method of Fried (10) was employed for cloning hybrid genomes as described by Nussbaum et al. (27). The procedures for analysis of the purified hybrid DNA using DNA-DNA hybridization, λ repressor binding, restriction endonucleases, and electron microscopy have been detailed before (5,12,13).

RESULTS

Linkage of a DNA Fragment from the Immunity Region of Phage λ to a Defective SV40 Replicon

Figure 1A shows the coordinates of the pertinent restriction endonuclease cleavage sites within the second largest fragment (7,160 bp) resulting from

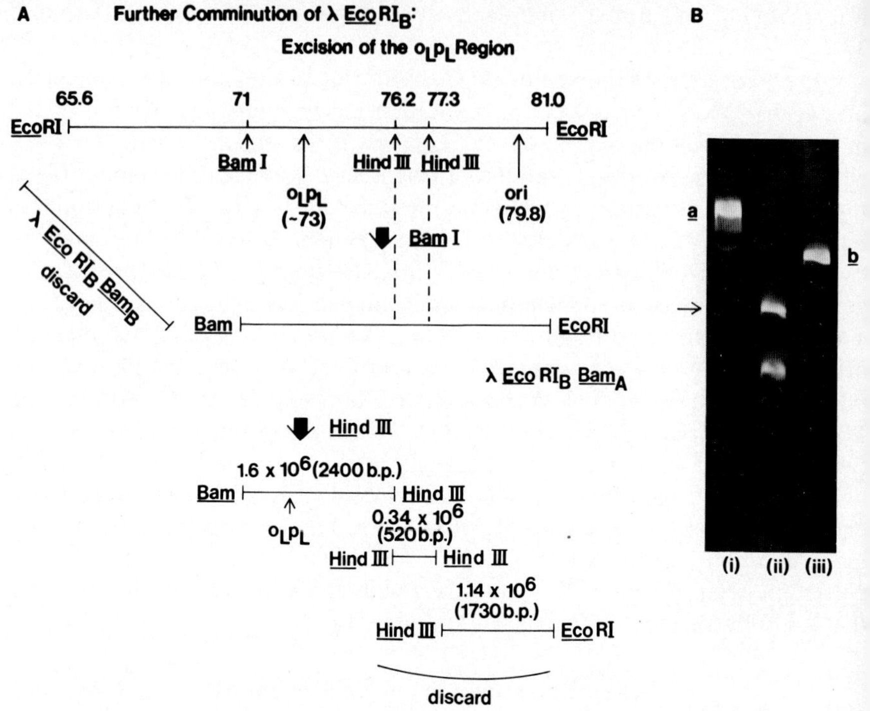

FIG. 1. Preparation of DNA fragments used for *in vitro* recombination. **A:** Schematic representation of the excision of the λ DNA fragment carrying the leftward operator, O_LP_L, from the immunity region located in λ endo R.*Eco*RI fragment B. See text for details. **B:** SV40 triplication mutant (a_3) was cleaved with endo R.*Eco*RI (9,12) and analyzed by 3.0% polyacrylamide–0.5% agarose disc gel electrophoresis (i) as described before (12). Gels (ii) and (iii) were run in parallel. An equivalent sample of a_3 (1.5 μg) was cleaved simultaneously by endo R.*Eco*RI and *Bam*I (ii). The arrow locates the *aBam*-A fragment which bears the SV40 origin. Gel (iii) provides a M.W. standard (8×10^5) generated by endo R.*Eco*RI cleavage of reiteration mutant b_4 (3).

endo R.*Eco*RI cleavage of phage λ DNA. This *Eco*RI-B fragment (26,30) contains a *Bam*I site at 71 map units (29) and two *Hin*dIII sites at 76.2 and 77.3 (3). After treatment with endo R.*Bam*I and endo R.*Hin*dIII, the second largest resulting fragment (2,400 bp) was isolated. This fragment contains O_L (3,23), one *Bam*I cohesive end (34), and one *Hin*dIII cohesive end (28).

A defective triplication mutant (a_3) of SV40 (Fig. 2) bearing the initiation site for SV40 DNA replication in each monomeric segment has been described previously (8,9,12,19). The monomeric segment (*a*, 1,670 bp) was isolated after incubation with endo R.*Eco*RI [Fig. 1B (i)] and further cleaved with endo R.*Bam*I [Fig. 1B (ii)] to yield two subfragments (940 bp and 730 bp). These structures are illustrated schematically in Fig. 2. A portion of the larger of the two (*aBam*-A), which contains the SV40 origin, was treated with endo R.*Hin*dIII to give a family of molecules sharing at

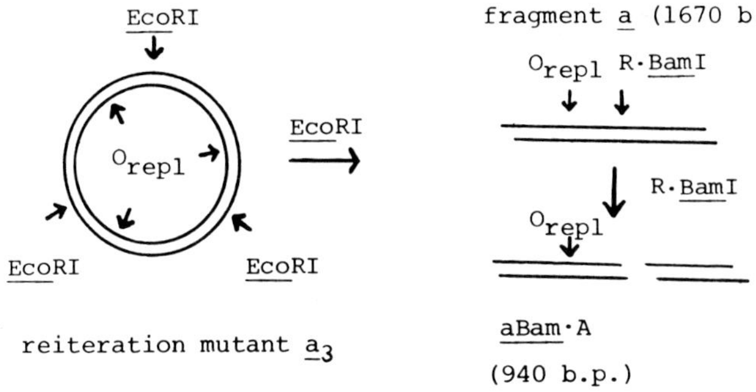

Formation of SV40 vector aBam·A by cleavage of SV40 reiteration mutant a_3 with R·EcoRI and R·BamI. (O_{repl}, origin of replication)

FIG. 2. Schematic diagram of the SV40 vector.

least one cohesive terminus with the λ 2,400 bp segment and also bearing an origin for SV40 DNA replication.

The three DNA preparations (*aBam*-A, *aBam*-A further cleaved by *Hin*dIII, and the λ 2,400 bp segment) were mixed and subjected to the action of T4 polynucleotide ligase as described in the section "Materials and Methods." The resulting complex mixture of ligation products was added to WT SV40 DNA and used to infect monkey kidney cells. The infection was permitted to go to completion, and the viral lysate so formed was used to infect new cultures of monkey cells (27). A small level of binding of the denatured DNA to filters containing single-stranded λ DNA (1 to 2% of the amount obtained with SV40 DNA filters) was detected. Moreover, shortened genomes in addition to those of WT size were demonstrated by 1.4% agarose gel electrophoresis (*data not shown*).

Partial Cloning and Analysis of the SV40-λ Hybrid

In order to obtain a viral stock greatly enriched for the hybrid, we employed an infectious center cloning procedure as described in Materials and Methods. Of 38 infectious center plaques surveyed, one (*i.c.*219) demonstrated homology with λ DNA. A viral stock was prepared from this plaque, and ^{32}P-labeled viral DNA was purified from cells infected with this stock. Examination of this viral DNA by electrophoresis through 1.4% agarose revealed two species migrating more rapidly than WT superhelical SV40 DNA (Fig. 3A). The faster-migrating species (*i.c.*219–f2) constituted 56% of the superhelical DNA, and the slower (*i.c.*219–f1) amounted to 28%, as judged by densitometry. The f2 DNA was resistant to both endo R.*Eco*RI and R.*Bam*I, whereas f1 DNA was resistant only to endo R.*Eco*RI (Fig.

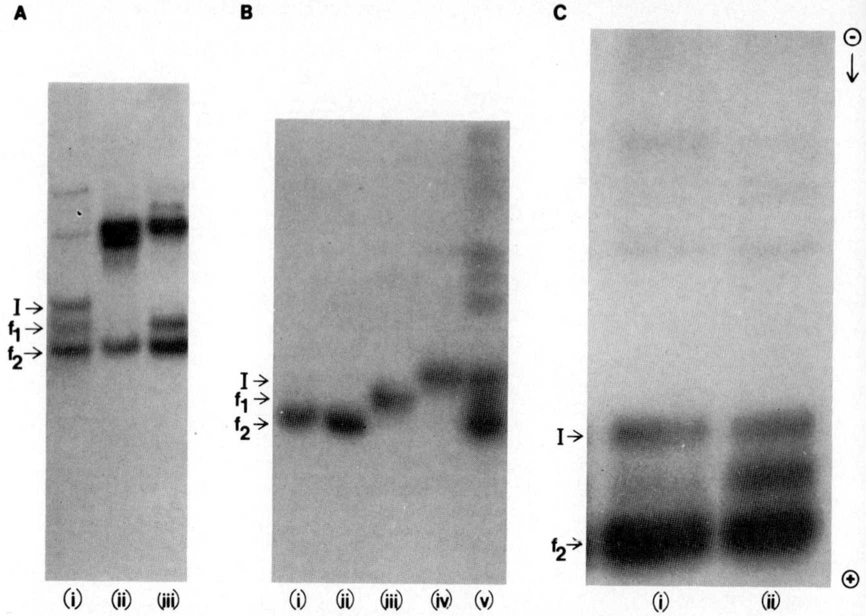

FIG. 3. Properties of the DNA derived from plaque *i.c.*219. The infectious center method of Fried (10) using DNA infection rather than virus infection was employed to clone the SV40-λ hybrid. Thirty-eight infectious center plaques were aspirated (0.2 ml) and 0.1 ml of each was used to infect a fresh culture of CV-1 cells in 35-mm dishes. After labeling with ^{3}H-thymidine (12), viral DNA was selectively extracted and tested for hybridization to λ DNA on filters (13). One plaque aspirate, *i.c.*219, proved to be positive and the remainder of the aspirate from this plaque was used to infect monkey cells in a 150-mm dish. The lysate resulting from this infection provided the viral stock for preparing intracellular viral DNA analyzed below (12,13). **A:** Autoradiogram of *i.c.*219 Form I (closed circular, superhelical) ^{32}P-labeled DNA which had been subjected to electrophoresis in a 1.4% agarose slab gel before (i) and after treatment with endo R.*Bam*I (ii) or endo R.*Eco*RI (iii). **B:** Autoradiogram under the same conditions as above of the purified *i.c.*219 species. (i): DNA Form I surviving after cleavage with combined endo R.*Eco*RI and endo R.*Bam*I [purified by neutral sucrose sedimentation (12)]; (ii–iv): the f2, f1, and WT (DNA I) species were subjected to electrophoresis after prior separation on a preparative slab gel (13); (v): unfractionated *i.c.*219 DNA. **C:** Enrichment of the *i.c.*219–f2 content of the mixture after passaging. Purified *i.c.*219–f2 DNA was used in a DNA infection in the presence of helper (WT) DNA (12). ^{32}P-labeled DNA Form I was prepared as before (12,13), and electrophoresis was carried out as above. (i): Composition after one additional passage via purified DNA infection. (ii): Composition of the original mixture. (From ref. 27).

3A). Thus, it was possible to purify the predominant, shortened DNA species either by preparative agarose gel electrophoresis and excision of individual DNA bands or by isolation of the *Eco*RI- and *Bam*I-resistant DNA after neutral sucrose sedimentation. These purified species are shown in Fig. 3B. Monkey cells were mixedly infected with purified f2 DNA and WT SV40 DNA, thereby generating a viral stock containing 74% f2 DNA and approximately 4% f1 DNA (Fig. 3C). The size of the f2 DNA, based on its mobility in 1.4% agarose, was determined to be 80% that of WT SV40

DNA (Fig. 4A); its superhelical density was identical to that of SV40 DNA I, based on isopycnic analysis in an ethidium bromide–CsCl density gradient. An evaluation of the homology of *i.c.*219 DNA with λ DNA was performed (27). Of the three species isolated by preparative agarose gel electrophoresis, f2 DNA was strongly bound to the λ filter.

From the structure of the λ component linked to the SV40 replicon in the original infecting chimeric DNA, the partially cloned chimera was expected to harbor the O_L site to which λ repressor binds (5). As Fig. 5 illustrates, the binding of λ repressor to f2 DNA was easily detected at repressor levels which bound SV40 DNA to an insignificant extent.

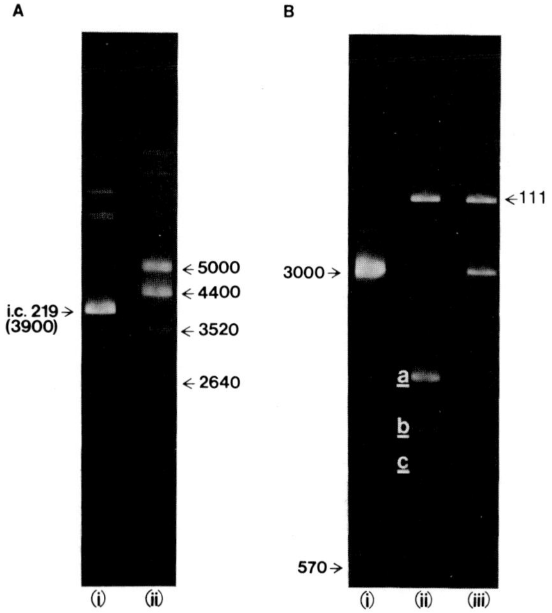

FIG. 4. Agarose gel electrophoretic analysis of the partially purified SV40-λ hybrid DNA. The superhelical SV40-λ hybrid DNA was purified as described in Fig. 3. **A**(i): A 1-μg sample of the hybrid (*i.c.*219) was examined after electrophoresis at 50 V through a 10 × 0.6 cm, 1.4% agarose disc gel. The direction of migration is from top to bottom, and the conditions for electrophoresis, subsequent staining with ethidium bromide, and photography are as described before (12). The size of this DNA was determined to be 3,900 bp based on the relative electrophoretic mobility of superhelical viral DNAs in a 1.5-μg sample [**A**(ii)] containing WT SV40 DNA (5,000 bp) and reiteration mutants d_5 (4,400 bp) and d_4 (3,520 bp). This sample was prepared as described by Ganem et al. (12) and subjected to identical conditions for electrophoresis as in **A**(i). **B:** A 2-μg sample of the hybrid DNA was cleaved with endo R.*Hpa*II and the largest of the three fragments (3,000 bp) was examined by 1.4% agarose gel electrophoresis (i) as described above. The gels (ii) and (iii), run in parallel to (i), contained linear DNA fragments for size calibration. The sample in (ii) consisted of 0.7 μg SV40 DNA III (unit length linears) and 0.7 μg of a mixture of fragments *a* (1,670 bp), *b* (1,210 bp), and *c* (1,030 bp) from different reiteration mutants of SV40 (8). In gel (iii) were 0.5 μg of SV40 DNA III and 0.6 μg of a 3,000 bp fragment from SV40 DNA produced by cleavage with endo R.*Hpa*II and R.*Bam*I.

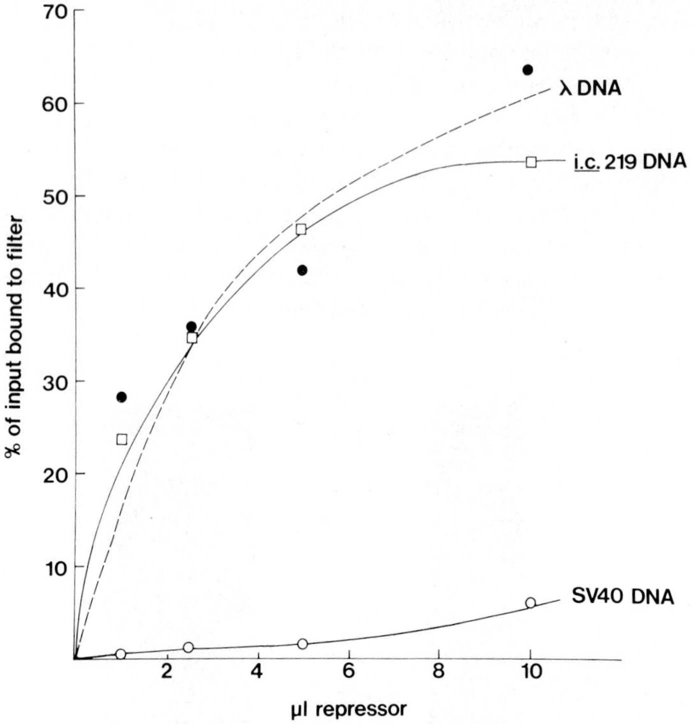

FIG. 5. Affinity of repressor for *i.c.*219 DNA. Viral DNA (*i.c.*219-f2, SV40 or λ) was exposed to increasing amounts of λ repressor protein and passed through nitrocellulose filters (5). Materials retained by filters are expressed as percent of input radioactivity. The binding assay (total volume, 0.6 ml) mixture contained 0.02 μg of f2 ^{32}P-DNA (10,000 cpm) and either λ ^{3}H-DNA (0.25 μg; 5,000 cpm) or SV40 ^{3}H-DNA (0.03 μg; 5,000 cpm) and the indicated volumes of λ repressor (3×10^{-8} M). Binding of f2 by λ repressor was abolished by addition of nonradioactive λ DNA in excess. (From ref. 27.)

Several restriction endonucleases known to produce specific fragments from either the vector DNA or the λ immunity region were used to examine the physical structure of the hybrid DNA. Although the endo R.*Hae*III cleavage pattern of λ DNA [Fig. 6A (i)] was too complex to permit comparison with f2 [Fig. 6A (v)], this cleavage of the λ *Eco*RI-B fragment appeared to yield several fragments with identical migration patterns to those from f2, suggesting the presence of λ DNA sequences in the hybrid. Interestingly, cleavage of both the λ *Eco*RI-B [Fig. 6A (iv)] and f2 [Fig. 6 (vi)] with endo R.*Hpa*II revealed two identical small fragments corresponding to 570 and 350 bp. These fragments are known to be located in the region of the leftward operator of λ (1,3). The larger *Hpa*II fragment from the hybrid DNA was determined to be 3,000 bp by agarose gel electrophoresis (Fig. 4B).

It has been shown (3,23) that two *Hpa*II cleavage sites in the region of

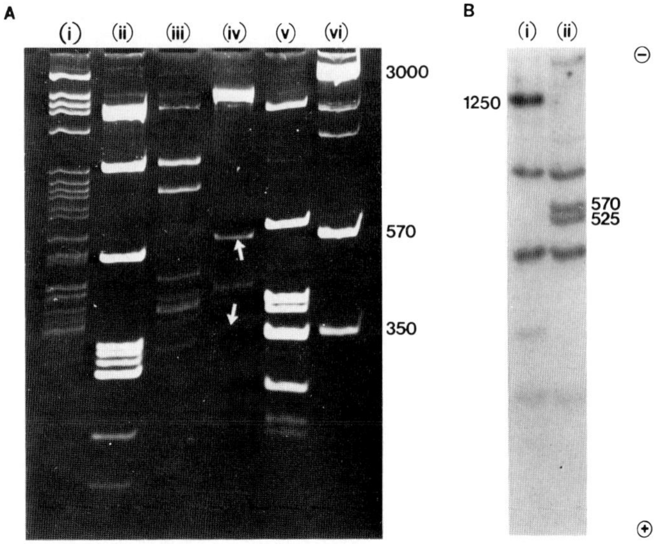

FIG. 6. Structural studies with restriction endonucleases. Indicated DNA quantities are subjected to enzymatic cleavage and applied to a 3.0% polyacrylamide–0.5% agarose slab gel (12,13). **A:** Slot (i): λ DNA (4.7 μg) after treatment with endo R.*Hae*III. (ii): SV40 DNA (4.3 μg) after treatment with endo R.*Hae*III. (iii): λ-endo R.*Eco*RI fragment B (1.7 μg) after endo R.*Hae*III treatment. (iv): λ*Eco*RI-fragment B after endo R.*Hpa*II treatment. (v): *i.c.*219-f2 DNA (2 μg) after endo R.*Hae*III treatment. (vi): f2 after endo R.*Hpa*II treatment. **B:** Autoradiogram of *i.c.*219-f2 ^{32}P-DNA after cleavage with endo R.*Hin*dII + III (i) or with both endo R.*Hin*II + III and endo R.*Hpa*II (ii). (From ref. 27.)

O_L are within a *Hin*dII+III fragment of 1,125 bp. Cleavage of f2 with endo R.*Hin*dII+III produced a 1,250 bp fragment instead of the 1,125 bp fragment which contained *Hpa*II cleavage sites (Fig. 6B). Endo R.*Hpa*II and R.*Hin*dII+III cleavage of the hybrid DNA generated the expected 570 bp segment from within the 1,250 bp fragment. The increased size of this *Hin* fragment in the hybrid DNA suggests that the original *Hin*dIII site at 76.2 λ map units is not retained and that a new *Hin* site is provided most likely from the SV40 segment.

The intensity of fluorescence of endo R.*Hae*III fragments of the hybrid DNA [Fig. 6A (v)] suggested that certain larger fragments were present in molar excess. This was corroborated by a densitometric analysis of an autoradiogram of the *Hae*III fragments from ^{32}P-labeled f2 DNA (Fig. 7). The presence of reiterated sequences will be further discussed in the section after next.

Confirmation of the Structure of the λ Segment in the Chimeric DNA

The above results suggest that the f2 DNA contains a region of λ DNA from which *Hpa*II yields two fragments of 350 and 570 bp. Maniatis and

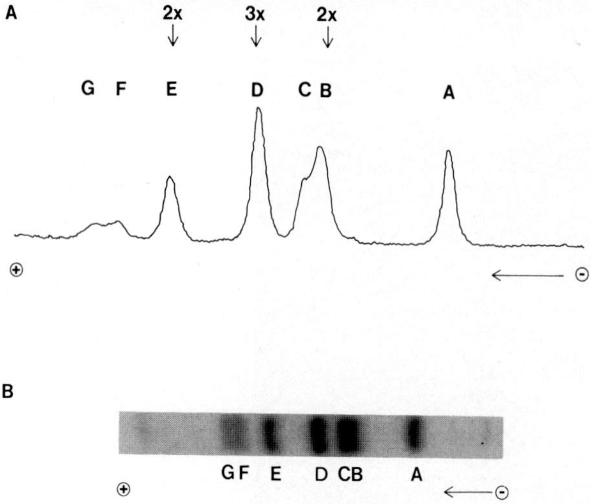

FIG. 7. Autoradiogram of *Hae*III fragments from *i.c.*219 f2 ^{32}P-DNA. The upper **A** densitometer tracing of the *Hae*III electrophoresed fragments (see Fig. 6) shown in **B** indicated that two fragments (*Hae*-B and -E) were in twofold excess and one (*Hae*-D) was in threefold molar excess.

Ptashne (23) have previously shown that a repressor binding site of λ is present in the 350 bp segment. To verify that the site is present on the 350 bp segment of the chimera, unfractionated *i.c.*219 DNA was treated with endo R.*Hpa*II, incubated with λ repressor, and filtered through a nitrocellulose disc. As expected, of the two small fragments only the 350 bp segment was bound (Fig. 8).

The conclusions reached above were fully confirmed by electron microscopic heteroduplex analysis. The chimeric circular DNA was nicked, denatured, and allowed to anneal to the denatured λ *Eco*RI-B fragment. The homology between the two DNAs as seen in Fig. 9 represents approximately 32% of the λ fragment or 2,300 bp. An additional heteroduplex analysis was performed to map more precisely the λ segment in the hybrid. The λ *Eco*RI-B DNA fragment was cleaved by endo R.*Hin*dIII. As can be seen in Fig. 10A, this produces three fragments, one of which (that spanning 65.6 units to 76.2 units) should form a heteroduplex with the hybrid. Such a heteroduplex is seen in Fig. 10B. Fifteen heteroduplexes of this type were measured, and the linear single-stranded DNA extending from the circular molecule was calculated to be 36% of the λ *Eco*RI-B DNA fragment. The duplex region of the heteroduplex was difficult to measure since the double-stranded/single-stranded boundary was not well defined, but, since only one linear tail was seen, the result proved that the hybrid preserved that part of the inserted λ segment up to the *Hin*dIII site at 76.2 λ map units. The *Hin*dIII cleavage of the λ *Eco*RI-B fragment was incomplete as shown by

λ Repressor Binds 350 bp Fragment

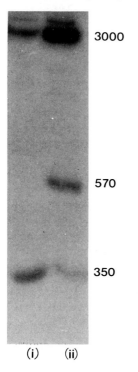

FIG. 8. Preparative isolation of the 350 bp fragment from the endo R.*Hpa*II digest of *i.c.*219-f2 ^{32}P-DNA. Unfractionated *i.c.*219 DNA (2.3 × 10^6 cpm, 12.4 μg) was cleaved with endo R.*Hpa*II and concentrated by alcohol precipitation in the presence of 200 μg of calf thymus DNA. It was then exposed to repressor protein (10^{-10} moles) in a total volume of 2 ml. The resulting complex was retained on a filter and eluted with an SDS-containing buffer (23). A 10% aliquot of the eluate (7,000 cpm) was applied to a 3.0% polyacrylamide–0.5% agarose slab gel and subjected to electrophoresis and autoradiography (i) as described in Fig. 6. (ii): *i.c.*219-f2 DNA cleaved with endo R.*Hpa*II as in Fig. 6A. The 3,000 bp segment in (i) was retained probably due to nonspecific binding to the filter, whereas of the two small fragments only the 350 bp fragment was retained. (From ref. 27.)

the formation of a few heteroduplexes of the type seen in Fig. 10C and diagrammed in Fig. 10D. This heteroduplex resulted from the annealing of the hybrid single-stranded DNA circle with a fragment of λ *Eco*RI-B spanning the λ genome from 65.6 to 77.3 λ map units. The results from these experiments show that the region spanning 71 to 76 λ map units is present in the SV40-λ hybrid. This segment contains the *N* gene and its transcriptional regulatory site, the leftward operator-promoter (O_L/P_L), but does not include the *Hin*dIII and *Bam*I cleavage sites.

Structure of the Vector Portion of the Hybrid

The size of the hybrid on 1.4% agarose gels is 3,900 bp (Fig. 4A). The λ segment found in this hybrid is approximately 2,300 bp and the vector segment used (*aBam*-A) was 940 bp. If one λ and one vector segment were present in the hybrid, such a molecule would be expected to have a total size of 3,240 bp. The extra 660 bp found in the 3,900 bp genome could be accounted for if a large portion of the *aBam*-A vector had been duplicated *in vivo* to form a molecule bearing two origins for SV40 DNA replication.

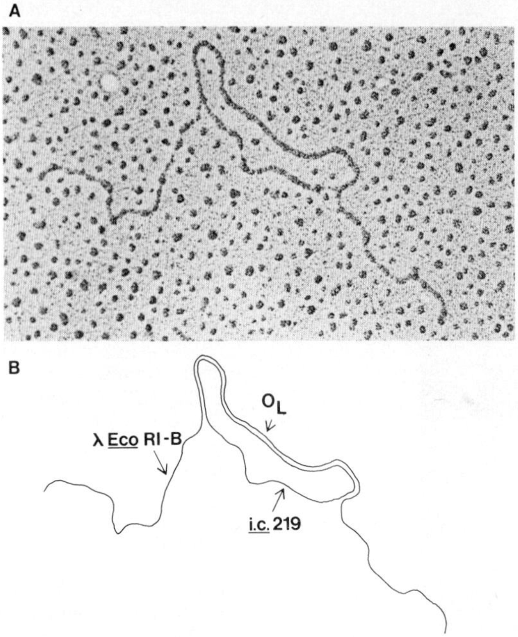

FIG. 9. Electron microscopic heteroduplex analysis of *i.c.*219-f2 DNA. The hybrid superhelical DNA, prepared as described in Fig. 3, was completely converted to relaxed circles containing one phosphodiester bond interruption (15). These molecules were then denatured and renatured in the presence of an equimolar quantity of λ *Eco*RI fragment B, spread for electron microscopy in 50% formamide over a 20% formamide hypophase (13), and visualized **(A)** at a magnification of 55,000 in a Zeiss EM10 electron microscope. As shown in the diagrammatic insert **(B)**, the region of homology between λ *Eco*RI-B and *i.c.*219-f2 is 2,300 bp and includes the leftward operator (O_L) of phage λ. Data from measurements of 17 heteroduplexes. (From ref. 27.)

Further studies of the SV40 sequences in the hybrid were performed to test this hypothesis.

It was shown before that cleavage of the hybrid with endo R.*Hpa*II produces three fragments. The largest fragment, which is 3,000 bp (Fig. 4B), contains all of the SV40 vector segment within the hybrid in addition to small amounts of the λ DNA at both ends, since the three *Hpa*II sites are within the λ portion of the hybrid. This *Hpa*II-3,000 bp segment was analyzed by cleavage with endo R.*Hin*dII+III or R.*Hae*III (Fig. 11a and 11d, respectively). Also shown is the cleavage pattern of the vector *aBam*-A (Fig. 11b and 11e) and of SV40 DNA (Fig. 11c and 11f). Cleavage of *aBam*-A with endo R.*Hind* yields two fragments D_1 and I, which co-migrate with two of the fragments produced by endo R.*Hind* cleavage of the 3,000 bp fragment. Comparison of the *Hae* cleavage pattern of *aBam*-A and the 3,000 bp fragment also reveals two fragments of identical size. Densito-

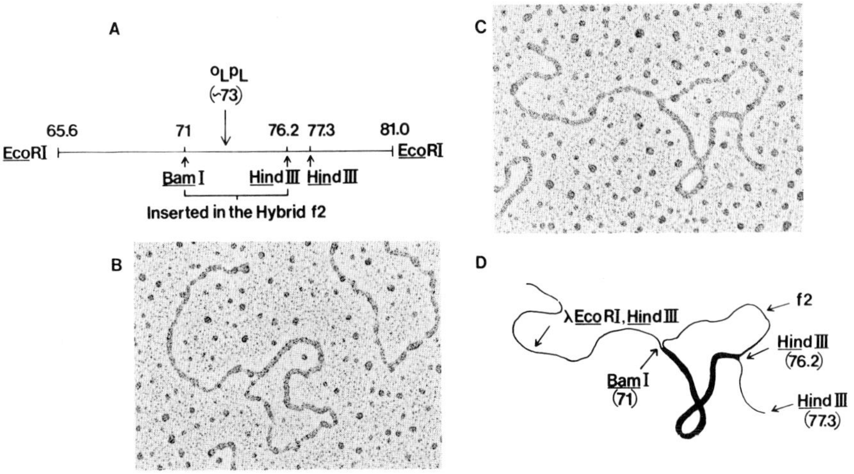

FIG. 10. A: Schematic representation of the λ *Eco*RI-B fragment. The segment of λ *Eco*RI-B DNA extending from 71 λ map units (cleavage site for endo R.*Bam*I) to 76.2 λ map units (one of the two sites for endo R.*Hin*dIII) is present in the SV40-λ hybrid as indicated. **B:** Heteroduplex between the SV40-λ hybrid and λ *Eco*RI-B, *Hin*dIII fragment. The hybrid DNA was purified, then nicked according to the method of Greenfield et al. (15), and allowed to anneal to denatured λ *Eco*RI-B (further cleaved with endo R.*Hin*dIII) as previously described (13). The conditions used for formation of heteroduplexes and for mounting for electron microscopy were the same as in Fig. 9. A partially duplex region is located in the circular portion of this representative heteroduplex molecule. The presence of one linear single-stranded segment extending from the circle proves that the hybrid DNA contains that portion of λ *Eco*RI-B up to the *Hin*dIII cleavage site at 76.2 map units. The single-stranded, linear tail measures 0.359 ± 0.02 λ *Eco*RI-B lengths as expected for that part of *Eco*RI-B, *Hin*III fragment between 65.6 and 71 λ map units, which is not present in the hybrid genome. **C:** Heteroduplex between the hybrid and λ *Eco*RI-B, *Hin*dIII (incomplete cleavage). A few heteroduplexes of this type were seen which probably were created from annealing of a hybrid molecule with the λ *Eco*RI-B fragment cleaved only once by *Hin*dIII at 77.3 map units. **D:** Diagrammatic representation of the heteroduplex shown in **C** with the duplex region of homology drawn as a thick line.

metric tracings of these gels revealed that those two *Hin* fragments from the hybrid were present in a twofold molar excess as compared to the remaining *Hin* fragments. This suggested that the vector DNA, *aBam*-A, was duplicated in the hybrid genome.

The 3,000 bp fragment was denatured and annealed in the presence of single-stranded circular molecules of the triplication mutant a_3 to corroborate these conclusions. The heteroduplex formed between a linear 3,000 bp fragment and a circular a_3 molecule is depicted in Fig. 12. Two regions of homology are separated by a deletion loop. The regions of homology represent a tandem duplication of a portion of *aBam*-A in the hybrid. The deletion loop is that portion of monomer *a* present in the a_3 DNA but not included in the *aBam*-A vector. The single-stranded tails extending from the circle are the segments of λ DNA still remaining in the 3,000 bp *Hpa*II

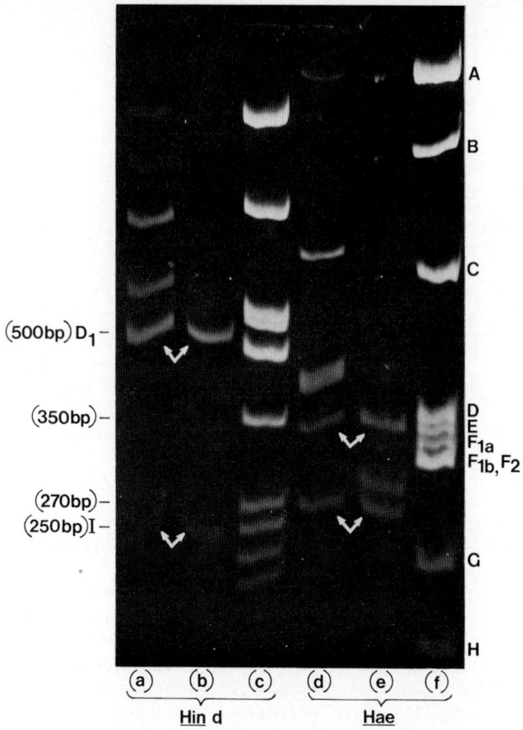

FIG. 11. Analysis of the SV40 vector sequences in the hybrid DNA using restriction endonucleases. Reaction mixtures for endo R.*Hin*dII + III (a, b, and c) or endo R.*Hae*III (d, e, and f) were prepared containing the *Hpa*II-3,000 bp fragment from the hybrid (0.8 μg, a and d), fragment *aBam*-A (1.0 μg, b and e), and WT SV40 DNA I (2.0 μg, c and f). The *Hpa*II-3,000 bp fragment had been purified by neutral sucrose sedimentation (12) after endo R.*Hpa*II cleavage of the hybrid DNA (Fig. 4B), and the *aBam*-A fragment produced by endo R.*Eco*RI and R.*Bam*I cleavage of a_3 DNA was prepared as described in "Materials and Methods." The conditions for nuclease cleavage and gel electrophoretic analysis in a 3.0% polyacrylamide–0.5% agarose slab gel have been previously detailed (13). The *Hin*-D_1 (500 bp) and *Hin*-I (250 bp) fragments (a and b) have previously been mapped in the triplication mutant, a_3, genome. The *Hin*-D_1 is known to harbor the SV40 DNA replication origin (19). Small white arrows point to pairs of *Hin* or *Hae* fragments of identical electrophoretic mobility from the *Hpa*II-3000 bp fragment and *aBam*-A. Densitometric determinations using the negative of this photograph and from autoradiographs of ^{32}P-labeled hybrid DNA revealed that *Hin*-D_1 (500 bp) and -I (250 bp) were in twofold molar excess in the *Hpa*-II 3,000 bp fragment (a).

fragment. This heteroduplex substantiates the presence of a tandem duplication of most of the vector segment *aBam*-A in the chimerical molecule. Based on the known physical structure of fragment *a,* these sequences include the origin region of SV40 DNA near the junction of *Hin*-A and *Hin*-C and a portion of *Hin*-K (19).

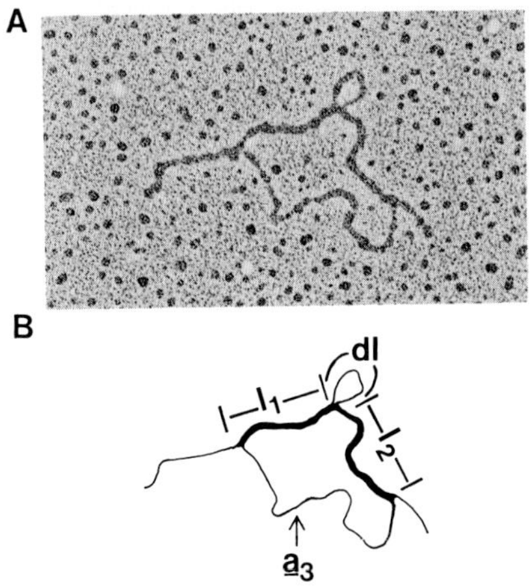

FIG. 12. Heteroduplex between reiteration mutant a_3 and the *Hpa*II-3,000 bp fragment. The a_3 DNA I (12) was nicked (15), then denatured and renatured in the presence of an equimolar quantity of the *Hpa*II-3,000 bp fragment from the SV40-λ hybrid. This fragment had been purified after endo R.*Hpa*II cleavage as described in Fig. 11, and electron microscopic analysis was carried out. Measurements were made of 15 heteroduplexes using single-stranded and duplex molecules of a_3 as standards. The two duplex regions (shown as a heavy line in diagram **B** and lettered l_1 and l_2), which are 0.184 ± 0.006 and 0.182 ± 0.006 of the a_3 genome, respectively, represent a tandem duplication of the majority of the vector segment, *aBam*-A, preserved in the hybrid. The single-stranded tails (thin lines in **B**) are the remaining segments of λ DNA included in the *Hpa*II-3,000 bp fragment.

DISCUSSION

In this chapter we have described the partial cloning of a 2,300 bp λ segment in monkey cells and physical mapping of the resulting SV40-λ hybrid genome. The heterogeneous progeny viral DNA from the original infection with the *in vitro* ligation products was inadequate for detailed structural studies of the prokaryotic segment, and only after enrichment for the hybrid with the cloning procedure were we able to perform these studies. The heteroduplex analysis of the λ sequences in the chimera showed no deletion or substitution loops and indicated that its primary structure was preserved, at least to a resolution of 100 bp. The endo R.*Hpa*II cleavage profile revealed that the 920 bp region bearing O_L in λ DNA was preserved intact, and this was further substantiated by the preparative λ repressor binding of the *Hpa*II-350 bp segment. However, loss of small regions was detected in the chimera as judged by loss of both the *Bam*I and *Hin*dIII cleavage sites. The SV40 and λ portions in the original recombinant molecule were most probably linked through their *Bam*I ends during the *in vitro* ligation. Therefore, the *Hin*dIII site on the λ segment and the *Eco*RI site on the SV40 vector, which are noncomplementary, could have been lost during the *in vivo* closure of the linear hybrid genome to form a circular molecule (24).

A subsequent *in vivo* duplication of the SV40 replication origin region in the hybrid could have been associated with the deletion of the *Bam*I site. Such a duplication could not have occurred *in vitro* since an *in vitro* joining of two *aBam*-A segments would have resulted in an inversion of one fragment relative to the other. Therefore, the structure of the SV40 vector in the hybrid very likely represents another instance in which the origin of replication of SV40 has been reiterated upon passage *in vivo* (11,12,19,22, 24). This reiteration of the origin has provided a selective advantage in replication over the helper SV40 DNA since greater than 70% of the progeny viral DNA molecules generated by infecting monkey cells with equivalent amounts of hybrid and WT SV40 DNAs are SV40-λ hybrid molecules (Fig. 3C). For this SV40-λ hybrid, the reiteration of the SV40 vector sequences has also increased its size to allow for encapsidation in progeny virus.

There are several potential applications of this method of propagating foreign DNA in mammalian cells. Transcriptional and translational control mechanisms could be studied by examining the fate in different mammalian cell lines of a variety of prokaryotic genetic sequences (e.g., a λ operator, rRNA or tRNA genes, or genes coding for specific enzymes) covalently linked to SV40 DNA. The chimeric molecule described in this chapter can be isolated from infected cells as a nucleoprotein complex (SV40 minichromatin) whose characterization should aid in analyzing the arrangement of histones around a well-defined regulatory site, and their effects on the interaction of a repressor with its specific site of action.

Hybrid genomes constructed with a noncomplementing SV40 replicon such as that used in this work must be propagated and cloned with wild-type helper genomes. This constraint offers no simple means for selection for the hybrid genomes in a mixed population. To overcome this deficit, investigators could employ vector segments containing intact early or late gene regions. For example, excision of a suitable fragment from the late gene region of SV40 and insertion of a prokaryotic segment of similar size would create a defective hybrid capable of complementing early (*tsA*) mutants. We have recently employed this approach to clone a bacterial tRNA suppressor gene in monkey cells (D. H. Hamer, G. C. Fareed, D. Davoli, and C. A. Thomas, *manuscript in preparation*).

ACKNOWLEDGMENTS

This investigation was supported by research grant No. CA-14885 from the National Institutes of Health, U.S. Public Health Service. One of the authors (D.D.) is a predoctoral trainee supported by an Institutional National Research Service Award in Viral Oncology (U.S.P.H.S.), and another (G.C.F.) is the recipient of Public Health Service Research Career

Program Award No. CA-00057. We are grateful to M. Ptashne and R. Sauer for supplying us with λ repressor.

REFERENCES

1. Allet, B. (1973): Fragments produced by cleavage of λ deoxyribonucleic acid with the Hemophilus parainfluenzae restriction enzyme HpaII. *Biochemistry,* 12:3972–3977.
2. Allet, B., Jeppersen, P. G. N., Katagiri, K. J., and Delius, H. (1973): Mapping the DNA fragments produced by cleavage of λ DNA with endonuclease RI. *Nature,* 241:120–123.
3. Allet, B., and Solem, R. (1974): Separation and analysis of promoter sites in bacteriophage lambda DNA by specific endonucleases. *J. Mol. Biol.,* 85:475–484.
4. Cameron, P., Pirrotta, V., Steinberg, R., Hopkins, N., and Ptashne, M. (1970): In vitro construction of bacteriophage λ carrying segments of the Escherichia coli chromosome: Selection of hybrids containing the gene for DNA ligase. *Proc. Natl. Acad. Sci. U.S.A.,* 72:3416–3420.
5. Chadwick, P., Pirrotta, V., Steinberg, R., Hopkins, N., and Ptashne, M. (1970): The λ and 434 phage repressors. *Cold Spring Harbor Symp. Quant. Biol.,* 35:283–294.
6. Chang, A. C. Y., Lansman, R. A., Clayton, D. A., and Cohen, S. N. (1975): Studies of mouse mitochondrial DNA in Escherichia coli: Structure and function of eukaryotic-prokaryotic chimeric plasmids. *Cell,* 6:241–244.
7. Cohen, S. N., Chang, A. C. Y., Boyer, H. W., and Helling, R. B. (1973): Construction of biologically functional bacterial plasmids in vitro. *Proc. Natl. Acad. Sci. U.S.A.,* 70: 3240–3244.
8. Davoli, D., and Fareed, G. C. (1975): Formation of reiterated simian virus 40 DNA. *Cold Spring Harbor Symp. Quant. Biol.,* 39:137–146.
9. Fareed, G. C., Byrne, J., and Martin, M. A. (1974): Triplication of a unique genetic segment in a simian virus 40-like virus of human origin and evolution of new viral genomes. *J. Mol. Biol.,* 87:275–288.
10. Fried, M. (1974): Isolation and partial characterization of different defective DNA molecules derived from polyoma virus. *J. Virol.,* 13:939–946.
11. Fried, M., Griffin, B. E., Lund, E., and Robberson, D. L. (1975): Polyoma virus—a study of wild-type mutant and defective DNAs. *Cold Spring Harbor Symp. Quant. Biol.,* 39:45–52.
12. Ganem, D., Nussbaum, A. L., Davoli, D., and Fareed, G. C. (1976*a*): Isolation, propagation and characterization of replication requirements of reiteration mutants of simian virus 40. *J. Mol. Biol.,* 101:57–83.
13. Ganem, D., Nussbaum, A. L., Davoli, D., and Fareed, G. C. (1976*b*): Propagation of a segment of bacteriophage λ DNA in monkey cells after covalent linkage to a defective simian virus 40 genome. *Cell,* 7:349–359.
14. Glover, D. M., White, R. L., Finnegan, D. J., and Hogness, D. S. (1975): Characterization of 6 cloned DNAs from Drosophila melanogaster including one that contains genes for ribosomal RNA. *Cell,* 5:149–157.
15. Greenfield, L., Simpson, L., and Kaplan, D. (1975): Conversion of closed circular DNA molecules to single-nicked molecules by digestion with DNAase I in the presence of ethidium bromide. *Biochim. Biophys. Acta,* 407:365–375.
16. Hamer, D. H., and Thomas, C. A., Jr. (1976): Molecular cloning of DNA fragments produced by restriction endonucleases. *Proc. Natl. Acad. Sci. U.S.A.,* 73:1537–1541.
17. Hershfield, V., Boyer, H. W., Yanofsky, C., Lovett, M. A., and Helinski, D. R. (1974): Plasmid Co1E1 as a molecular vehicle for cloning and amplification of DNA. *Proc. Natl. Acad. Sci. U.S.A.,* 71:3455–3459.
18. Kedes, L. H., Chang, A. C. Y., Houseman, D., and Cohen, S. N. (1975): Isolation of histone genes from unfractionated sea urchin DNA by subculture cloning in E. coli. *Nature,* 255:533–538.
19. Khoury, G., Fareed, G. C., Berry, K., Martin, M. A., Lee, T. N. H., and Nathans, D. (1974): Characterization of a rearrangement in viral DNA: Mapping of the circular simian virus 40-like DNA containing a triplication of a specific one-third of the viral genome. *J. Mol. Biol.,* 87:289–301.

20. Lai, C., and Nathans, D. (1974*a*): Deletion mutants of simian virus 40 generated by enzymatic excision of DNA segments from the viral genome. *J. Mol. Biol.*, 89:179–193.
21. Lai, C., and Nathans, D. (1974*b*): Non-specific termination of simian virus 40 DNA replication. *J. Mol. Biol.*, 97:113–118.
22. Lee, T. N. H., Brockman, W. W., and Nathans, D. (1975): Evolutionary variants of simian virus 40: Cloned substituted variants containing multiple initiation sites for DNA replication. *Virology*, 66:53–69.
23. Maniatis, T., and Ptashne, M. (1973): Structure of lambdal operators. *Nature*, 246:133–136.
24. Mertz, J. E., Carbon, J., Herzberg, M., Davis, R. W., and Berg, P. (1975): Isolation and characterization of individual clones of simian virus 40 mutants containing deletions, duplications and insertions in their DNA. *Cold Spring Harbor Symp. Quant. Biol.*, 39:69–84.
25. Morrow, J. F., Cohen, S. N., Chang, A. C. Y., Boyer, H. W. Goodman, H. M., and Helling, R. B. (1974): Replication and transcription of eukaryotic DNA in Escherichia coli. *Proc. Natl. Acad. Sci. U.S.A.*, 71:1743–1747.
26. Murray, N. E., and Murray, K. (1974): Manipulation of restriction targets in phage λ to form receptor chromosomes for DNA fragments. *Nature*, 251:476–481.
27. Nussbaum, A. L., Davoli, D., Ganem, D., and Fareed, G. C. (1976): Construction and propagation of a defective simian virus 40 genome bearing an operator from bacteriophage λ. *Proc. Natl. Acad. Sci. U.S.A.*, 73:1068–1072.
28. Old, R., Murray, K., and Roizes, G. (1975): Recognition sequence of restriction endonuclease III from Hemophilus influenzae. *J. Mol. Biol.*, 92:331–339.
29. Perricaudet, M., and Tiollais, P. (1975): Defective bacteriophage lambda chromosome, potential vector for DNA fragments obtained after cleavage by Bacillus amyloliquefaciens endonuclease (BamI). *FEBS Lett.*, 56:7–11.
30. Thomas, M., and Davis, R. W. (1975): Studies on the cleavage of bacteriophage lambda DNA with EcoRI restriction endonuclease. *J. Mol. Biol.*, 91:315–328.
31. Timmis, K., Cabello, F., and Cohen, S. N. (1975): Cloning, isolation and characterization of replication regions of complex plasmid genomes. *Proc. Natl. Acad. Sci. U.S.A.*, 72:2242–2246.
32. Weiss, B., Jacquemin-Sablon, A., Live, T. R., Fareed, G. C., and Richardson, C. C. (1968): Enzymatic breakage and joining of deoxyribonucleic acid. VI. Further purification and properties of polynucleotide ligase from Escherichia coli infected with bacteriophage T4. *J. Biol. Chem.*, 243:4543–4555.
33. Wensink, P. C., Finnegan, D. J., Donelson, J. C., and Hogness, D. S. (1974): System for mapping DNA sequences in chromosomes of Drosophila melanogaster. *Cell*, 3:315–325.
34. Wilson, G. A., and Young, E. A. (1975): Isolation of a sequence-specific endonuclease (BamI) from Bacillus amyloliquefaciens H. *J. Mol. Biol.*, 97:123–125.

Recombinant Molecules: Impact on Science and Society, edited by R. F. Beers, Jr. and E. G. Bassett. Raven Press, New York © 1977.

30. SV40 Carrying an *Escherichia coli* Suppressor Gene

* Dean H. Hamer

Department of Biological Chemistry, Harvard Medical School, Boston, Massachusetts 02115

INTRODUCTION

This chapter reports studies of a recombinant between simian virus 40 (SV40) and a fragment of *E. coli* DNA carrying the suppressor tRNA gene $su^{+}III$. Evidence is presented that this recombinant virus can be propagated in monkey cells for many generations with no observable sequence alterations in the bacterial DNA fragment. It is also shown that the bacterial DNA is transcribed in productively infected monkey cells, and some preliminary results on the origin and fate of these transcripts are described.

These experiments were initiated with two goals in mind. The first was to develop a general technique for propagating and studying the expression of defined DNA fragments in eukaryotic cells. I chose to work with SV40 because its genetic and physical maps are well-defined, and its replication and transcription have been extensively studied. This allows one to know which portion of the viral DNA will serve as an effective vehicle molecule and to place the inserted DNA under the transcriptional control of the virus. Furthermore, SV40 can transform a wide variety of species, and all transformants have at least part of the viral genome inserted into one or more of their chromosomes (20). Thus, SV40 transducing viruses might eventually be used to obtain cell lines with stable alterations in their chromosomal DNA.

A second motive, perhaps more ambitious, was the hope that the SV40-$su^{+}III$ recombinant might be used to construct suppressor lines of mammalian cells. The *E. coli* $su^{+}III$ gene specifies a tRNA molecule which translates the amber codon UAG as tyrosine (14). Recently Capecchi, Hughes, and Wahl (5) reported that UAG is recognized as a signal for polypeptide chain termination in a mammalian cell-free protein synthesizing system, and that termination can be suppressed by the addition of *E. coli* suppressor

* Present address: Laboratory of Molecular Genetics, National Institute of Child Health and Human Development, National Institutes of Health, Bethesda, Maryland 20014.

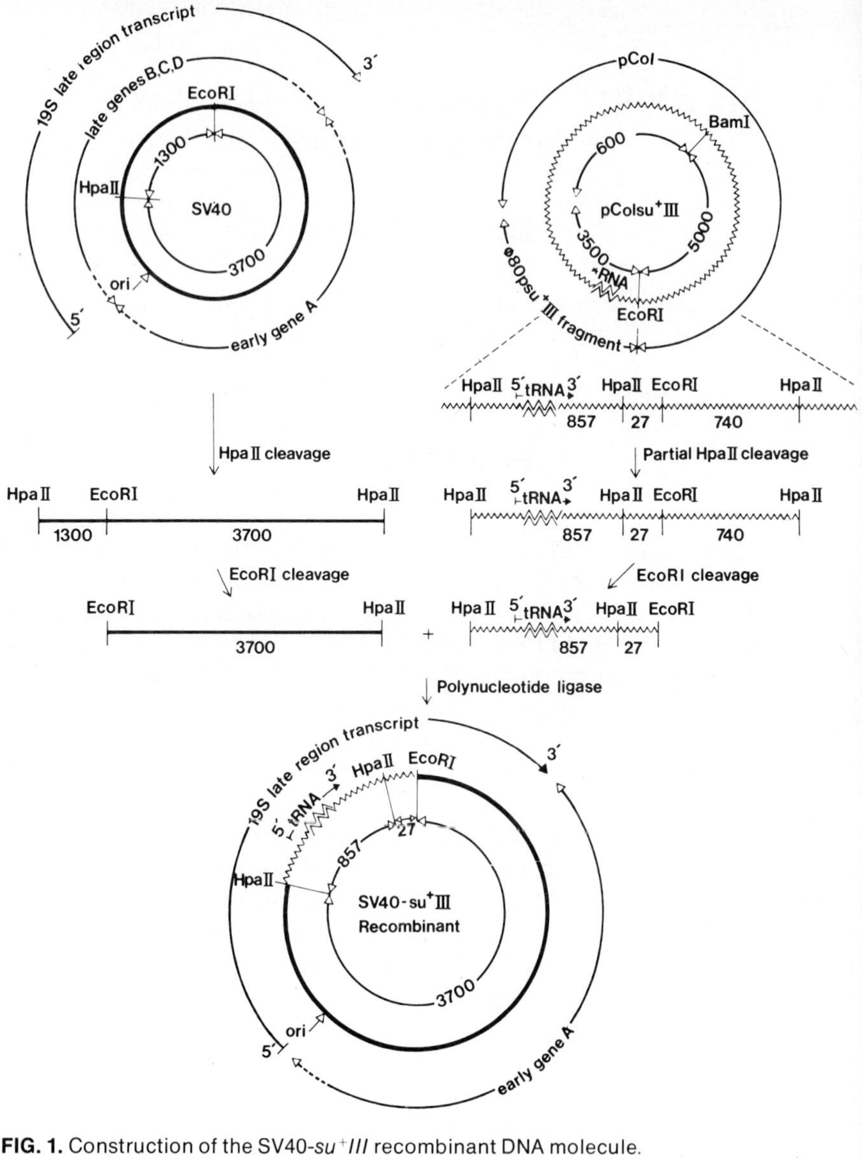

FIG. 1. Construction of the SV40-*su*$^{+}$*III* recombinant DNA molecule.

The SV40 vehicle fragment: Superhelical SV40 ^{3}H-DNA was prepared from cells infected at low multiplicity with a plaque-purified virus stock. The DNA was cleaved with *Hpa*II, and full-length linear molecules were isolated by electrophoresis through 1% agarose/Tris-acetate-EDTA gels. This preparation was digested with *Eco*RI, and the 3,700-bp vehicle fragment was recovered by electrophoresis through 2% agarose/Tris-borate-EDTA gels. As a final purification step, this material was treated with *Hae*II and *Sal*I, for which it has no sites, and electrophoresed through 1.4% agarose/Tris-phosphate-EDTA gels. The final preparation appeared homogeneous on two types of gels, under conditions where a 0.1% (by weight) contaminant would have been seen. No nondefective molecules could be detected by a plaque assay on monkey tissue culture cells.

The *su*$^{+}$*III* fragment: Superhelical pCol*su*$^{+}$*III* ^{32}P-DNA was partially digested with *Hpa*II, and a 1,624-bp fragment containing the *su*$^{+}$*III* gene was purified by electrophoresis tRNA. These observations led to the speculation that monkey cells infected with the SV40-*su*$^{+}$*III* recombinant virus might transcribe the bacterial suppressor gene; that these transcripts might be appropriately processed, modified, and aminoacylated; and that the resulting functional suppressor tRNA might be sufficiently abundant to compete with the cells' termination factors. This would have provided a solution to the familiar "chicken and egg" dilemma: amber mutants cannot be recognized without suppressor strains, but suppressor strains cannot be recognized without amber mutants. Unfortunately, I have found that monkey cells infected with the SV40-*su*$^{+}$*III* recombinant virus do not synthesize detectable levels of mature suppressor tRNA.

CONSTRUCTION AND CHARACTERIZATION OF THE SV40-*su*$^{+}$*III* RECOMBINANT VIRUS

The SV40 Vehicle Fragment

The vehicle used in these experiments was a 3,700-base pair (bp) fragment of wild type SV40 DNA obtained by cleavage with restriction endonucleases *Eco*RI and *Hpa*II, both of which have a single site on the viral genome (25,27,32). As shown in Fig. 1, this fragment contains the origin of viral DNA replication (7,9) and the early gene *A*, but it lacks sequences necessary for the expression of the late genes *B, C*, and *D* (6,21,33,35). Therefore, recombinants made with this vehicle are defective; they can be propagated as virus only if the essential late gene products are supplied, in *trans*, by a helper virus.

The *su*$^{+}$*III* Fragment

The source of the bacterial suppressor gene was an *E. coli* plasmid called pCol*su*$^{+}$*III*. This plasmid was isolated by ligating *Eco*RI fragments of the transducing phage ϕ80 p*su*$^{+}$*III* (30) to the ColE1-ampicillin resistance factor pCol (34), transforming a Lac^{-}_{amber} Trp^{-}_{amber} strain of *E. coli*, and selecting for Lac^{+} Trp^{+} ampicillin-resistant clones (*manuscript in prepara-*

---

through 0.9% agarose/Tris-acetate-EDTA gels. This partial product was cleaved with *Eco*RI and the desired 884-bp *su*$^{+}$*III* fragment with one *Hpa*II terminus and one *Eco*RI terminus was isolated by electrophoresis through 2% agarose/Tris-phosphate-EDTA gels. This fragment was further purified by treatment with a mixture of *Bam*I and *Sal*I, for which it has no sites, and electrophoresis through 4% polyacrylamide/Tris-borate-EDTA gels. This procedure yielded a fragment preparation which appeared homogeneous on two types of electrophoresis gels, under conditions where a 0.1% (by weight) contaminant would have been detected. A transformation assay showed that the *su*$^{+}$*III* fragment was not capable of autonomous replication in *E. coli*.

The SV40-*su*$^{+}$*III* recombinant: A 2-ml reaction mixture containing 2.8 μg of the ^{3}H-labeled SV40 fragment and 112 μg of the ^{32}P-labeled *su*$^{+}$*III* fragment was treated with bacteriophage T4 polynucleotide ligase. Covalently closed circular molecules were then purified by two centrifugations in CsCl-ethidium bromide gradients. The final product contained 0.056 μg of the SV40 fragment and 4.5 μg of the *su*$^{+}$*III* fragment.

tion). The plasmid DNA was partially cleaved with *Hpa*II to give a 1,624-bp fragment which, upon treatment with *Eco*RI, yielded an 884-bp fragment with one *Hpa*II terminus, one *Eco*RI terminus, and an internal *Hpa*II site close to the *Eco*RI terminus (Fig. 1). This fragment contains a single copy of the $tRNA^{Tyr}$*$su^{+}III$* structural sequence, as well as the presumed promoter and transcription termination regions.

Ligation of DNA Fragments

Since both *Eco*RI (17) and *Hpa*II (12) produce unique cohesive termini, the SV40 and $su^{+}III$ fragments can anneal to one another in a single orientation and can then be covalently closed by treatment with polynucleotide ligase. This procedure yields circular recombinant molecules with two important features. First, they have a length 92% that of wild-type SV40 DNA and can therefore be incorporated into virus particles (10). Second, the $su^{+}III$ fragment is inserted such that the 5′ end of the tRNA structural sequence is proximal to the 5′ end of the 19S SV40 late region transcript (2).

To isolate recombinant molecules, I incubated a solution containing 0.57 pmoles/ml of ^{3}H-labeled SV40 fragment and 95 pmoles/ml of ^{32}P-labeled $su^{+}III$ fragment with bacteriophage T4 polynucleotide ligase, then centrifuged it to equilibrium in a CsCl-ethidium bromide gradient. Covalently closed circular molecules, representing about 2% of the SV40 DNA and 4% of the $su^{+}III$ fragment DNA, were collected and purified by a second banding. A rough calculation suggests that greater than 99% of the SV40 DNA in this preparation should be in circular SV40-$su^{+}III$ hybrids, as opposed to SV40-SV40 dimers or more complex forms. However, such hybrid molecules constitute only about 1% of the total DNA, most of which is in $su^{+}III$-$su^{+}III$ dimers.

Propagation of SV40-$su^{+}III$ Recombinant Genomes

To propagate the SV40-$su^{+}III$ recombinant, I infected a flask of monkey cells with 1.6 μg of the recombinant DNA preparation described above, together with 0.05 μg of SV40 tsA_{239} DNA as the conditional early gene mutant helper. The cells were incubated at 41°C, which is nonpermissive for the *tsA* helper. Under these conditions, virus should be produced only by those cells infected with *both* the recombinant, providing early gene product, and the helper, providing late gene products. Control flasks were infected with recombinant DNA alone or with helper DNA alone. Twelve days later, the flasks were freeze-thawed and the resulting lysates were plaque-assayed on monkey cells at 41°C. As expected, the cells infected with recombinant plus helper DNA gave a high yield of virus [8×10^{5} plaque-forming units (PFUs)/ml], whereas the cells infected with recombinant DNA alone or with helper DNA alone gave no detectable virus (0 plaques/0.4 ml). Thus, the recombinant and helper genomes complemented one another.

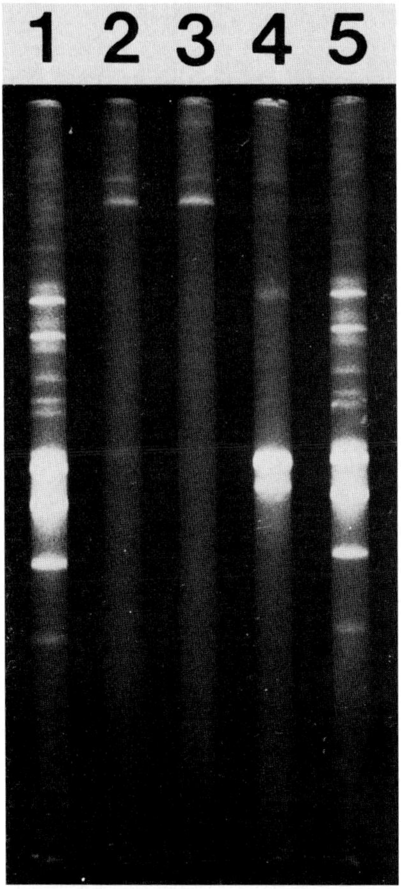

FIG. 2. Electrophoretic analysis of intracellular viral DNA.

Confluent monolayers of about 10^7 monkey cells (line TC7) were infected (10) with 1.6 μg SV40-*su*$^+$*III* recombinant DNA (isolated as described in the legend to Fig. 1), 0.05 μg SV40 tsA_{239} helper DNA, or a mixture of 1.6 μg recombinant DNA plus 0.05 μg helper DNA. Appropriate physical containment and decontamination procedures were employed for this infection and all subsequent experiments with the recombinant virus (4,28). After 12 days' incubation at 41°C, the cultures were freeze-thawed, and 1 ml of each lysate was used to infect about 10^7 confluent monkey cells. After 2 hr absorption at 37°C, 9 ml medium containing 2% fetal calf serum was added, and the cultures were shifted to 41°C. At 24 hr post-infection, 250 μCi ^{3}H-thymidine (50 Ci/mmole) was added. At 60 hr post-infection, the cultures were extracted by the method of Hirt (18), and the supernatant fluids were centrifuged in CsCl-ethidium bromide gradients. Yields of superhelical DNA were 2.4 $\times$ 10^4 cpm for cells infected with the recombinant alone lysate, 2.6 $\times$ 10^4 cpm for cells infected with helper alone lysate, and 2.2$\times$ 10^6 cpm for cells infected with the recombinant plus helper lysate.

The superhelical DNA preparations were extracted with isoamyl alcohol, dialyzed, and analyzed by electrophoresis through 1% agarose gels (16). **1** and **5:** Superhelical marker DNAs include (from the bottom of the gel) SV40 reiteration mutant *d* trimers (2,636 bp), tetramers (3,514 bp), and pentamers (4,393 bp) (10) and wild-type SV40 (5,000 bp). **2:** Superhelical DNA from cells infected with the recombinant alone lysate. **3:** Superhelical DNA from cells infected with the helper alone lysate. **4:** Superhelical DNA from cells infected with the recombinant plus helper lysate.

Has the SV40-*su*$^+$*III* recombinant DNA been incorporated into virus particles? To answer this, I took advantage of the fact that only those genomes which were encapsidated in the original DNA infection will be transferred and replicated in a subsequent infection (10). Accordingly, monkey cells were infected with the recombinant plus helper lysate, and intracellular viral DNA was prepared 3 days later by Hirt extraction (18) and CsCl-ethidium bromide gradient centrifugation. Analysis of this superhelical DNA by agarose gel electrophoresis (Fig. 2) revealed two species of viral DNA. About 70% of the viral DNA migrated at the position of helper (5,000 bp), while the remainder migrated at the position expected for the SV40-*su*$^+$*III* recombinant (4,584 bp). Figure 2 also shows the gel profiles of the superhelical DNA recovered from cells infected with the recombinant alone and helper alone lysates. This material contained no viral DNA; the faint bands near the tops of the gels are mitochondrial species plus some chromosomal DNA. These results provide further evidence that the recombinant and helper complemented one another, and they indicate that the recombinant DNA was packaged into virions. The fact that *only* the expected recombinant and helper species were observed in the progeny viral DNA from the mixed infection suggests that no gross sequence rearrangements or recombination occurred during the initial DNA infection.

The SV40-*su*$^+$*III* Recombinant Has the Expected Structure

The 4,584-bp recombinant species was purified by gel electrophoresis and characterized by DNA-RNA hybridization, restriction endonuclease cleavage, and heteroduplex analysis.

Table 1 presents the results which were obtained when the SV40-*su*$^+$*III* recombinant virus DNA, along with various control DNAs, were immobilized on nitrocellulose filters and challenged with a preparation of *E. coli* ^{32}P-tRNA containing about 40% tRNA^{Tyr}*su*$^+$*III*. Hybridization was observed to DNA from the SV40-*su*$^+$*III* recombinant and from the suppressor plasmid pCol*su*$^+$*III*, but not to DNA from wild-type SV40 or from the parental nonsuppressor plasmid pCol. Thus, the recombinant virus contains the suppressor tRNA structural sequence.

The *Hpa*II cleavage profile of the SV40-*su*$^+$*III* recombinant virus DNA is shown in Fig. 3. As expected, two equimolar fragments were observed. One of these was indistinguishable from the SV40 vehicle fragment, whereas the other co-migrated with the 857-bp *Hpa*II fragment of pCol*su*$^+$*III* known to contain the suppressor gene.

Are the *su*$^+$*III* fragments grown in monkey cells and *E. coli* identical? To answer this, I mixed ^{3}H-labeled SV40-*su*$^+$*III* recombinant virus DNA with ^{32}P-labeled pCol*su*$^+$*III* DNA, digested the mixture with *Hpa*II, isolated the *su*$^+$*III* fragments by preparative agarose gel electrophoresis, cleaved them with *Hae*III, and analyzed the products on a polyacrylamide

TABLE 1. *Hybridization of suppressor tRNA to SV40-*su⁺III *recombinant virus DNA*

DNA on filter (1.2 μg/filter)	^{32}P-tRNA hybridized (cpm)
None	3
SV40	16
SV40-*su⁺III* recombinant	443
pCol	14
pCol*su⁺III*	211

E. coli CA275 *su⁻* was infected with φ80h am_1 p*su⁺III* and labeled with 200 μCi/ml $^{32}PO_4$. Total ^{32}P-tRNA was isolated by the method of Abelson et al. (1) through the DEAE-cellulose step. About 40% of this preparation migrated at the position of $tRNA^{Tyr}$ when analyzed by 8% polyacrylamide gel electrophoresis and autoradiography. Superhelical DNA samples were nicked and denatured by boiling for 20 min, cooled to 0°C, adjusted to 6 × SSC (SSC is 0.15 M NaCl, 0.015 M Na citrate, pH 7), and immobilized on 25-mm nitrocellulose filters as described by Gillespie and Spiegelman (13). The filters were incubated with 19,800 cpm of ^{32}P-tRNA in 4 ml of 2 × SSC containing 0.2% SDS at 65°C for 16 hr. The filters were then washed with 2 × SSC, treated with 25 μg/ml of boiled RNAase A, re-washed with 2 × SSC, dried, and counted.

gel. Figure 4 shows that the ^{3}H-viral and ^{32}P-plasmid DNAs yielded exactly the same set of five fragments, ranging in length from 390 to 25 bp.

The structure of the recombinant virus DNA was further examined by electron microscopic heteroduplex analysis. Figure 5 shows a heteroduplex between the SV40-*su⁺III* recombinant and pCol*su⁺III*. In this experiment, the plasmid DNA was converted to a linear form by cleavage with *Bam*I at a single site about 5,000 bp distant from the suppressor gene. The two DNAs show a short region of homology, which had a length equal to that of the *su⁺III* fragment and was located at the expected position on the linear plasmid molecule. No deletion or substitution loops were observed in the double-stranded region of the heteroduplex.

Taken together, these results show that the sequence of the *su⁺III* fragment has been faithfully preserved, at least to a resolution of about 50 bp. Similar results were obtained with DNA from virus stocks submitted to one additional passage. Thus, the recombinant can be stably propagated for many generations.

TRANSCRIPTION OF THE BACTERIAL DNA IN MONKEY CELLS

Detection of Bacterial Transcripts

The filter hybridization method of Gillespie and Spiegelman (13) was used to detect bacterial RNA synthesized in monkey cells. Cells were infected, at 41°C, with the SV40-*su⁺III* recombinant plus SV40 tsA_{239} helper

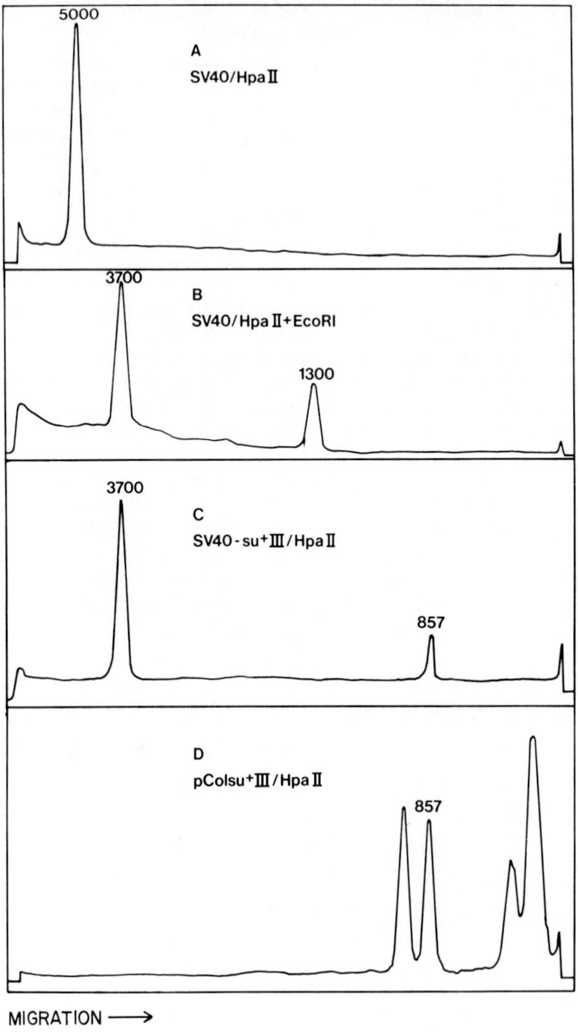

FIG. 3. Analysis of SV40-*su*$^+$*III* recombinant virus DNA by *Hpa*II cleavage.

The DNAs were cleaved with *Hpa*II or *Hpa*II plus *Eco*RI and electrophoresed through 2% agarose gels, which were stained with ethidium bromide and scanned in a fluorimeter (16,23). **A:** SV40 cleaved with *Hpa*II. **B:** SV40 cleaved with *Hpa*II plus *Eco*RI. **C:** SV40-*su*$^+$*III* recombinant virus DNA cleaved with *Hpa*II. **D:** pCol*su*$^+$*III* cleaved with *Hpa*II.

virus lysate, pulse-labeled with ^{3}H-uridine at 63 to 64 hr after infection, then extracted by the hot phenol–sodium dodecyl sulfate (SDS) procedure (32). As controls, mock-infected and wild-type SV40-infected cells were labeled and extracted in parallel. The total cell ^{3}H-RNA preparations were then assayed for viral and bacterial transcripts by hybridization to appropriate denatured DNAs immobilized on filters. In these experiments, each of

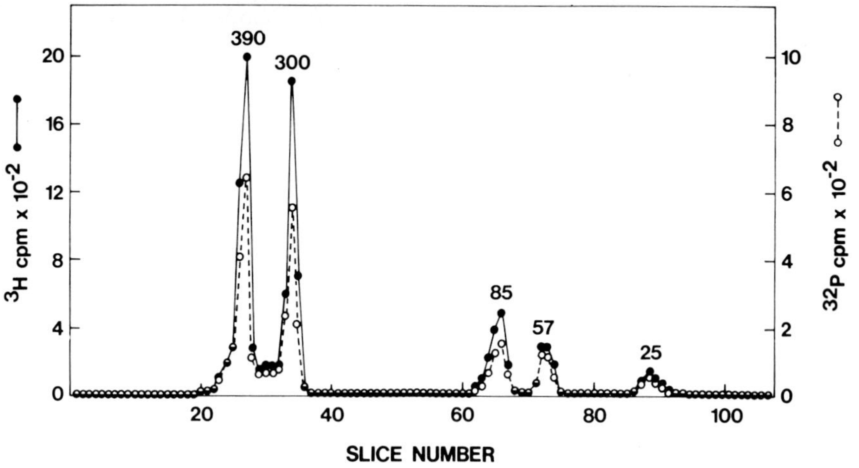

FIG. 4. The *su*$^+$*III* fragments grown in monkey cells and *E. coli* have identical *Hae*III cleavage patterns.

A mixture of SV40-*su*$^+$*III* recombinant virus ^{3}H-DNA and pCol*su*$^+$*III* ^{32}P-DNA was treated with *Hpa*II, and the 884-bp *su*$^+$*III* fragments were isolated by electrophoresis through a 2% agarose gel. This preparation was digested with *Hae*III, and the resulting fragments were separated on an 8% polyacrylamide gel. The gel was fractionated into 1-mm slices, which were solubilized and counted for both isotopes (16).
●——●, SV40-*su*$^+$*III* recombinant virus ^{3}H-DNA, cpm × 10^{-2}/slice;
○——○, pCol*su*$^+$*III* ^{32}P-DNA, cpm × 10^{-2}/slice.

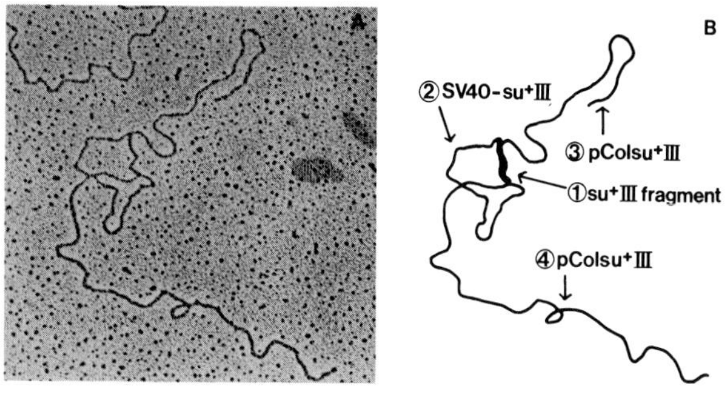

CONTOUR LENGTHS (% OF SV40)

SEGMENT	EXPECTED	FOUND (mean±SD)
①	17.7	16.9±1.0
②	74.0	71.6±1.7
③	100	101
④	172	170

FIG. 5. Heteroduplex analysis of SV40-*su*$^+$*III* × pCol*su*$^+$*III*.

Nicked circular SV40-*su*$^+$*III* recombinant virus DNA was heteroduplexed with pCol*su*$^+$*III* DNA which had been converted to a linear form by cleavage with *Bam*I. Samples were prepared for electron microscopy as described previously (11). Length measurements of segments 1 and 2 were made on 7 molecules. The preparation of *Bam*I used in this experiment had significant nicking activity, which resulted in an obvious shortening of one or both single-stranded arms on most of the heteroduplexes observed. Therefore, the length measurements of segments 3 and 4 were made on the single molecule shown. The internal standard was SV40-*su*$^+$*III* circular DNA, which was found by gel electrophoresis to have a length 0.92 that of wild-type SV40.

the filter-bound DNAs was present in excess over its complementary RNA in solution. Table 2 shows that the RNA from cells infected with the SV40-*su$^+$III* recombinant virus stock hybridized to both SV40 and pCol*su$^+$III* DNAs, but not to pCol DNA. In contrast, the RNA from cells infected with wild-type SV40 hybridized only to SV40 DNA, whereas the RNA from mock-infected cells showed no significant binding to any of the DNAs tested.

In the experiment presented in Table 2, the ratio of bacterial to viral transcripts was about 0.1. What ratio would be expected if the bacterial and viral RNAs were being synthesized and degraded at exactly the same rates? Assuming that all of the cells are infected with both recombinant and helper virus, and that the entire bacterial fragment is transcribed, the ratio would be

$$\frac{F_b L_b}{(1 - F_b)L_s + F_b(L_s - L_b)} = 0.12$$

where F_b = fraction of the virus stock which is recombinant virus = 0.3
L_b = length of the inserted bacterial DNA = 884 bp
L_s = length of the transcribed viral DNA = 2,500 bp for late region transcripts.

The finding that the observed ratio is 0.1, or 83% of the expected ratio, suggests that a substantial fraction of the bacterial RNA is stable relative to SV40 RNA.

TABLE 2. *Hybridization analysis of monkey cell ^{3}H-RNA*

Virus used to infect monkey cells	Input ^{3}H-cpm (10^3)	^{3}H-cpm hybridized Blank	SV40	pCol	pCol*su$^+$III*
None	390	2	11	2	8
SV40	378	3	2,230	8	9
SV40-*su$^+$III* + SV40 *tsA$_{239}$*	389	9	2,868	10	314

Confluent monolayers of about 2×10^7 monkey cells were infected with 2 ml medium, wild-type SV40, or SV40-*su$^+$III* recombinant plus SV40 *tsA$_{239}$* helper virus lysate as indicated. Following 2 hr absorption at 37°C, 9 ml medium containing 2% fetal calf serum was added, and the cultures were shifted to 41°C. At 62 hr post-infection, 1 mCi of ^{3}H-uridine (38 Ci/mmole) was added. One hour later, the cultures were lysed with SDS, and total cell RNA was prepared by the hot phenol method (31). Aliquots of the ^{3}H-RNA in 0.5 ml of 2 × SSC, 0.2% SDS were incubated for 20 hr at 65°C with filters containing no DNA, 1.0 μg SV40 DNA, 2.9 μg pCol DNA, and 2.9 μg pCol*su$^+$III* DNA. The filters were washed with 2 × SSC, treated with 25 μg/ml boiled RNAase A (25 μg/ml), re-washed with 2 × SSC, dried, and counted (13). When the amount of DNA was increased by 250%, the amount of hybridization varied by less than 10%. The ^{3}H-labeled material bound to the pCol*su$^+$III* filter was rendered 99% acid soluble by treatment with boiled RNAase A but was completely resistant to treatment with DNAase I.

Size Distribution of the Bacterial Transcripts

Figure 6 shows the pattern obtained when total 1-hr pulse-labeled ^{3}H-RNA from cells infected with the SV40-*su*$^+$*III* recombinant plus SV40 *tsA*$_{239}$ helper virus stock was centrifuged through a sucrose gradient and each fraction was assayed for viral and bacterial transcripts by hybridization to SV40 and pCol*su*$^+$*III* DNA filters. The total RNA had the usual pattern of high molecular weight rRNA precursors plus mature 28S, 18S, and 4S species. The SV40-specific RNA showed a peak at 19S, together with some material at 16S, as expected for late region transcripts (2). Two size classes of bacterial transcripts were observed. Approximately half of the bacterial RNA sedimented as a sharp peak at 19S, the same position

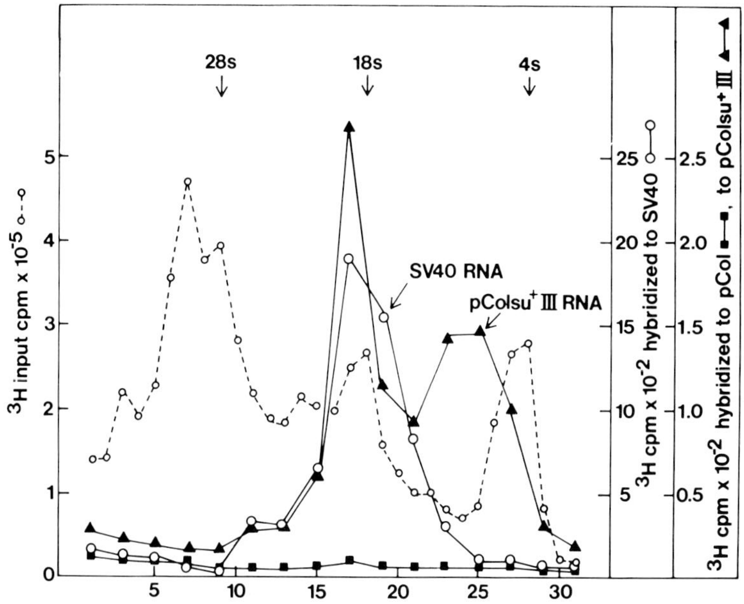

FIG. 6. Size distribution of the bacterial RNA synthesized in monkey cells.

One-hour pulse-labeled total cell ^{3}H-RNA was isolated from monkey cells infected with the SV40-*su*$^+$*III* recombinant plus SV40 *tsA*$_{239}$ helper virus lysate as described in the legend to Table 2. A 360-μg sample of this RNA was mixed with 16,000 cpm of monkey cell ^{32}P-RNA as a marker, heated at 65°C for 5 min, cooled to 0°C, layered on an 11-ml 15 to 30% sucrose gradient, and centrifuged 14 hr at 30,000 rpm and 20°C in the Beckman SW41 rotor. Fractions of 0.37 ml were collected and counted for Cherenkov radiation to determine the positions of the 28S, 18S, and 4S ^{32}P-RNA markers. Every other fraction was diluted to 1 ml of 2 × SSC, 0.2% SDS, and assayed for bacterial and viral ^{3}H-RNA by filter hybridization as described in the legend to Table 2.

○----○, Input ^{3}H-cpm × 10^{-5}; ○——○, ^{3}H-cpm × 10^{-2} hybridized to SV40; ■——■, ^{3}H-cpm × 10^{-2} hybridized to pCol; ▲——▲, ^{3}H-cpm × 10^{-2} hybridized to pCol*su*$^+$*III*.

as the viral transcripts. Presumably, this material consists of hybrid molecules arising from initiation at the SV40 late gene promoter, read through of the bacterial DNA fragment, and termination at the normal viral site. Due to the polarity with which the SV40 and $su^{+}III$ fragments were joined, such molecules would be expected to contain the suppressor tRNA structural sequence. The remainder of the bacterial transcripts sedimented as a broad peak around 7S, which overlapped the 4S region of the gradient. The origin of these low molecular weight species is not known.

Monkey Cells Do Not Synthesize Suppressor tRNA

Two approaches were used to determine if monkey cells infected with SV40-$su^{+}III$ recombinant plus SV40 tsA_{239} helper virus synthesize suppressor tRNA. In the first experiment, infected monkey cells were pulse-labeled with ^{3}H-tyrosine, and tRNA was prepared under conditions where the aminoacyl bond is stable. To test this monkey cell ^{3}H-tyrosyl-tRNA for suppressor tRNA, I took advantage of an unusual feature of the *E. coli* $tRNA^{Tyr}$ sequence: namely, there are no guanines in the 19-nucleotide sequence at the 3′ amino acid acceptor end of the molecule (14). Thus, when *E. coli* ^{3}H-tyrosyl-$tRNA^{Tyr}$ is digested with RNAase T1, which cleaves at guanine, the ^{3}H-tyrosine should be found linked to an unusually long oligonucleotide. Accordingly, the infected monkey cell ^{3}H-tyrosyl-tRNA was digested with RNAase T1, and the resulting oligonucleotides were separated according to length on a DEAE-cellulose column. *E. coli* ^{3}H-tyrosyl tRNA was digested in parallel and run on a separate column. The chromatographs are presented in Fig. 7. As expected, the ^{3}H-tyrosine in the *E. coli* sample eluted at the high salt concentration (0.3 M NaCl) characteristic of long oligonucleotides. In contrast, all of the ^{3}H-tyrosine in the monkey cell preparation eluted at a low salt concentration (0.1 M NaCl). Thus, monkey cells do not synthesize appreciable quantities of aminoacylated suppressor tRNA.

In the second experiment, I prepared tRNA from monkey cells infected with the SV40-$su^{+}III$ recombinant plus SV40 tsA_{239} helper virus stock and attempted to charge it with ^{3}H-tyrosine using *E. coli* synthetase. This experiment would detect mature suppressor tRNA molecules even if they were not aminoacylated by the monkey cells. The assay is specific because *E. coli* synthetase shows little reactivity toward mammalian $tRNA^{Tyr}$ (8). The results of this experiment, presented in Table 3, were completely negative; reactions containing the infected monkey cell tRNA accepted no more ^{3}H-tyrosine than did reactions containing no RNA. Control experiments demonstrated that this negative result was not due to a synthetase inhibitor in the monkey cell preparation, nor to degradation of the 3′ amino acid acceptor termini of the monkey cell tRNA.

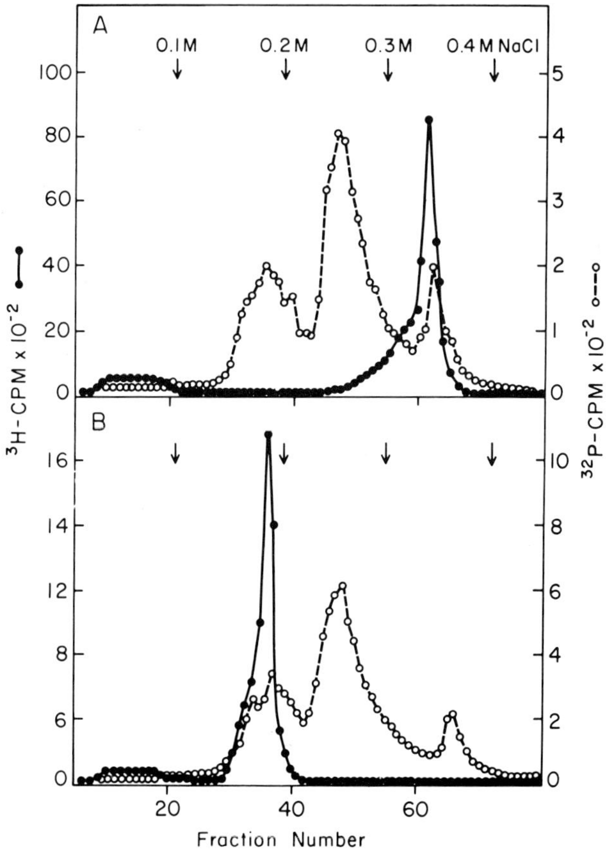

FIG. 7. DEAE-cellulose chromatography of RNAase T1 digests of ^{3}H-tyrosyl-tRNAs. *E. coli* ^{3}H-tyrosyl-tRNA was prepared by charging total *E. coli* tRNA *in vitro* (19,26). ($tRNA_I^{Tyr}$ and $tRNA_{II}^{Tyr}$ have the same 3′ sequence.) To prepare the monkey cell ^{3}H-tyrosyl-tRNA, I infected a monolayer of about 2×10^7 confluent cells, at 41°C, with 2 ml of the SV40-*su*$^+$*III* recombinant plus SV40 tsA_{239} helper virus lysate. At 60 hr post-infection, the culture was labeled with 50 μCi/ml ^{3}H-tyrosine (54 Ci/mmole) for 10 min. ^{3}H-tyrosyl-tRNA was prepared by phenol extraction and DEAE-cellulose chromatography at pH 4.5 (36). The ^{3}H-tyrosyl-tRNAs, together with *E. coli* ^{32}P-tRNA, were digested for 4 hr at 37°C with 125 units of RNAase T1 in 0.5 ml 0.1 M Na acetate, 0.002 M EDTA, 7 M urea, pH 4.5. The digests were diluted with 4.5 ml of 7 M urea and loaded onto DEAE-cellulose columns equilibrated with 0.01 M Na acetate, 7 M urea, pH 4.5. The columns were eluted with 10-ml portions of equilibration buffer containing 0.1, 0.2, 0.3, and 0.4 M NaCl. Fractions of 0.5 ml were collected and counted for both isotopes. **A:** *E. coli* ^{3}H-tyrosyl-tRNA. **B:** Monkey cell ^{3}H-tyrosyl-tRNA.

●——●, ^{3}H-cpm $\times 10^{-2}$/fraction; ○----○, ^{32}P-cpm $\times 10^{-2}$/fraction.

TABLE 3. *Monkey cell tRNA cannot be charged with tyrosine by* E. coli *synthetase*

Experiment	tRNA	A_{260} units	^{3}H-amino acid	pmoles incorporated
1.	None	0	Tyrosine	0.045
	E. coli	0.85		6.1
	Mock-infected monkey cells	0.64		0.054
	SV40-infected monkey cells	0.60		0.054
	SV40-*su*$^{+}$*III* + SV40 *tsA*$_{239}$-infected monkey cells	1.11		0.058
2.	None	0	Tyrosine	0.069
	E. coli	0.85		6.7
	SV40-*su*$^{+}$*III* + SV40 *tsA*$_{239}$-infected monkey cells	0.96		0.066
	E. coli plus SV40-*su*$^{+}$*III* + SV40 *tsA*$_{239}$-infected monkey cells	0.85 plus 0.96		6.9
3.	None	0	Arginine	0.008
	E. coli	0.85		8.1
	SV40-*su*$^{+}$*III* + SV40 *tsA*$_{239}$-infected monkey cells	1.11		8.0

Each 0.1-ml reaction mixture contained 100 mM Tris (pH 7.4), 10 mM magnesium acetate, 5 mM KCl, 2 mM ATP, 0.29 nmoles ^{3}H-L-tyrosine (2.4×10^{7} cpm/nmole) or 0.43 nmoles ^{3}H-L-arginine (1.0×10^{7} cpm/nmole), the indicated amount of deacylated tRNA, and excess *E. coli* mixed synthetases (19,26). After 15 min incubation at 37°C, the amount of ^{3}H-amino acid incorporated into 2 N HCl-insoluble material was determined. The deacylated tRNAs were prepared by phenol extraction, DEAE-cellulose chromatography, incubation in 1 M Tris (pH 9.1) for 30 min at 37°C, and ethanol precipitation (1). Experiment 1 shows that none of the monkey cell tRNAs tested could be charged with tyrosine. Similar results, but with lower backgrounds, were obtained when the amount of incorporation was determined by DEAE-cellulose chromatography. Experiment 2 demonstrates that monkey cell tRNA does not inhibit the charging of *E. coli* tRNA with tyrosine. Experiment 3 shows that monkey cell tRNA is nearly as effective an acceptor for arginine as is *E. coli* tRNA. This result was expected, since *E. coli* synthetase shows good reactivity with mammalian $tRNA^{Arg}$ (3), and indicates that the monkey cell tRNA was not degraded during the purification procedures. The monkey cell tRNA could also be charged with alanine and tryptophan.

DISCUSSION

I have described the construction of an SV40 genome in which a portion of the late gene region is replaced by a fragment of *E. coli* DNA carrying the suppressor tRNA gene *su*$^{+}$*III*. This SV40-*su*$^{+}$*III* recombinant was replicated and encapsidated in monkey cells when late gene products were supplied by coinfecting SV40 *tsA* helper. The structure of the recombinant genome was studied by DNA-RNA hybridization, restriction endonuclease cleavage, and electron microscopic heteroduplex analysis. The results showed that the recombinant had exactly the structure expected from its manner of construction, at least to a resolution of about 50 bp.

The use of SV40 in propagating foreign DNA in mammalian cells has been demonstrated by two contributors to this volume. G. C. Fareed has described the construction of hybrids between fragments of bacteriophage λ DNA and vehicles derived from SV40 reiteration mutants. These hybrids contain the SV40 origin of replication but no other viral genes, and hence they require wild-type SV40 as helper. S. P. Goff has made recombinants with λ DNA using a vehicle-helper system similar to the one I have described. However, in his experiments the vehicle and foreign DNA fragments were joined by poly(dA-dT) joints, rather than by ligation of the cohesive termini produced by restriction endonucleases.

Useful Features of the Vehicle-Helper System

Several advantages of the late region deletion vehicle plus conditional early gene mutant helper system deserve comment.

First, virus is produced, under nonpermissive conditions, only by those cells mixedly infected with both recombinant and helper DNA. This provides a strong selection for recombinant viruses, even though it is not possible to select directly for the functioning of the inserted DNA fragment.

A second, more technical point relates to the use of vehicle and foreign DNA fragments both of which have one terminus made by *Eco*RI and the other made by *Hpa*II. This ensures that the foreign DNA fragment is joined to the vehicle fragment in only one orientation. Moreover, since the vehicle cannot undergo intramolecular joining, circular recombinant molecules can easily be purified *before* the initial DNA infection. This turns out to be much simpler than purifying recombinant molecules from the complex mixture of species arising after infection with an unfractionated ligation reaction mixture.

A third advantage, at this point only a theoretical one, is that the vehicle contains all the information necessary for cell transformation (15). We have infected rat embryo cells with the SV40-*su*$^{+}$*III* recombinant virus DNA and have obtained typical SV40 t-antigen-positive transformants (G. C. Fareed and D. H. Hamer, *unpublished observations*). Experiments to determine if these transformants contain integrated bacterial DNA are in progress.

Do Mammalian Cells Possess Restriction Systems?

I have shown that the bacterial *su⁺III* fragment propagated in monkey cells is indistinguishable, at least at a resolution of about 50 bp, from the authentic plasmid fragment. Likewise, Fareed and his co-workers (11,29) found no major sequence alterations in the λ DNA fragments which they propagated in monkey cells using SV40 reiteration mutant vehicles. In contrast to the latter authors, however, I do not believe these results provide any evidence against the existence of restriction enzymes in mammalian cells. If monkey cells did possess a restriction system, the expected result would be a *quantitative* decrease in the infectivity of the recombinant DNA in the initial DNA infection, not a *qualitative* change in the structure of the progeny viral molecules. The mass lysate technique used to propagate both the SV40-*su⁺III* and SV40-λ recombinants does not provide the necessary quantitative data. Furthermore, the fragments which have been studied are all quite small and might have no sites for the hypothetical restriction endonucleases simply by chance. For example, the *E. coli* RI restriction endonuclease finds one site per 4,000 bp; the chance that an 884-bp fragment would have no sites for this enzyme is 0.8.

Synthesis of Bacterial RNA in Monkey Cells

Monkey cells infected with the SV40-*su⁺III* recombinant virus synthesize RNA complementary to the inserted bacterial DNA fragment. These bacterial transcripts are quite abundant—in fact, they are present in quantities close to those which would be anticipated if the bacterial RNA were synthesized and degraded at the same rates as SV40 RNA. About half of the bacterial transcripts labeled in 1 hr had the same length as SV40 late region transcripts. Presumably, these molecules are initiated at the normal SV40 promoter and contain the suppressor tRNA structural sequence. The rest of the bacterial transcripts were short and included tRNA-size molecules. However, attempts to detect either free or charged suppressor tRNA were unsuccessful. This suggests that monkey cells are deficient in at least one of the several steps involved in tRNA biosynthesis in *E. coli*. Clearly the SV40-*su⁺III* recombinant will not provide a useful tool for recognizing amber mutations in mammalian genes. Perhaps a more fruitful approach would be to purify a yeast suppressor tRNA gene and insert it into SV40.

S. P. Goff has found that monkey cells infected with his SV40-λ recombinant do not synthesize detectable amounts of λ-specific RNA. The discrepancy between his results and mine might be attributed to either of two reasons. First, it is possible that transcription initiated at the SV40 late region promoter does not continue through the λ DNA, perhaps due to the presence of the poly(dA-dT) joint. However, this seems unlikely, as SV40 derivatives with a short stretch of poly(dA-dT) at the *Hpa*II site are viable (24). Second, perhaps λ RNA is synthesized but is unstable. One could speculate that the $su^{+}III$ transcripts are stabilized by secondary structural features not possessed by the λ transcripts. Alternatively, poly(A) or poly(U) sequences in the λ RNA might serve as recognition sites for cellular RNAases.

Are These Experiments Biohazardous?

The experiments described in this chapter follow the guidelines established in the "Summary Statement of the Asilomar Conference on Recombinant DNA Molecules" (4) and the National Institutes of Health "Guidelines for Research Involving Recombinant DNA Molecules" (28). Three points merit discussion. First, the bacterial DNA fragment which was inserted into SV40 does not contain toxigenic genes and is not capable of autonomous replication in *E. coli.* This fragment was purified to better than 99% homogeneity, as shown by both a biophysical assay (gel electrophoresis) and a biological assay (transformation). Secondly, the SV40-$su^{+}III$ recombinant is defective. It can be propagated as virus only if late gene products are supplied by a helper virus. This limits the ability of the recombinant to be propagated outside of controlled laboratory conditions. Thirdly, appropriate physical containment and decontamination procedures were employed to further minimize the probability of accidental dissemination of the recombinant virus.

Possible Uses of Mammalian Transducing Viruses

This chapter has described an SV40 recombinant carrying a well-defined fragment of bacterial DNA. Recently, techniques have been developed for the cloning, in *E. coli,* of mammalian genes, such as the rabbit globin gene (22). Using the method described here, it should be possible to insert such purified mammalian genes into the SV40 late region. From the results obtained with the SV40-$su^{+}III$ recombinant, it seems likely that monkey cells infected with such viruses would synthesize substantial quantities of RNA complementary to the inserted gene. Such mammalian transducing viruses might provide useful tools for studying RNA processing, distinguishing between transcriptional and translational control mechanisms, and mapping *cis*-acting regulatory regions.

ACKNOWLEDGMENTS

These experiments were performed in collaboration with G. C. Fareed, C. A. Thomas, Jr., and D. Davoli. I thank D. Ganem for sparking my interest in this project and collaborating on preliminary experiments; A. Landy for bacterial strains, phage, and data on restriction sites; S. Falkow for pCol; J. Morrow for useful discussions; and R. Klodner, M. DePamphilis, D. Tapper, and C. Richardson for enzymes. This investigation was supported by grants from the National Institutes of Health to G. C. Fareed and C. A. Thomas, Jr. D. H. H. is a predoctoral trainee supported by the National Institute of General Medical Sciences.

REFERENCES

1. Abelson, J. N., Gefter, M. L., Barnett, L., Landy, A., Russell, R. C., and Smith, J. O. (1970): Mutant tyrosine transfer ribonucleic acids. *J. Mol. Biol.*, 47:15–28.
2. Acheson, N. (1976): Transcription during productive infection with polyoma virus and simian virus 40. *Cell,* 8:1–12.
3. Anderson, W. F. (1969): Evolutionary conservation of the synthetase recognition site of alanine transfer ribonucleic acid. *Biochemistry,* 8:3687–3691.
4. Berg, P., Baltimore, D., Brenner, S., Roblin, R. O., III, and Singer, M. F. (1975): Summary statement of the Asilomar conference on recombinant DNA molecules. *Proc. Natl. Acad. Sci. U.S.A.*, 72:1981–1984.
5. Capecchi, M. R., Hughes, S. H., and Wahl, G. M. (1975): Yeast super-suppressors are altered tRNAs capable of translating a nonsense codon in vitro. *Cell,* 6:269–278.
6. Chou, J. Y., and Martin, R. G. (1974): Complementation analysis of simian virus 40 mutants. *J. Virol.*, 13:1101–1109.
7. Danna, K. J., and Nathans, D. (1972): Bidirectional replication of simian virus 40 DNA. *Proc. Natl. Acad. Sci. U.S.A.*, 69:3097–3100.
8. Doctor, B. P., and Mudd, J. A. (1963): Species specificity of amino acid acceptor ribonucleic acid and aminoacyl soluble ribonucleic acid synthetases. *J. Biol. Chem.*, 238: 3677–3681.
9. Fareed, G. C., Garon, C. F., and Salzman, N. P. (1972): Origin and direction of simian virus 40 deoxyribonucleic acid replication. *J. Virol.*, 10:484–491.
10. Ganem, D., Nussbaum, A. L., Davoli, D., and Fareed, G. C. (1976*a*): Isolation, propagation and characterization of replication requirements of reiteration mutants of simian virus 40. *J. Mol. Biol.*, 101:57–83.
11. Ganem, D., Nussbaum, A. L., Davoli, D., and Fareed, G. C. (1976*b*): Propagation of a segment of bacteriophage λ-DNA in monkey cells after covalent linkage to a defective simian virus 40 genome. *Cell,* 7:349–360.
12. Garfin, D. E., and Goodman, H. M. (1974): Nucleotide sequences at the cleavage sites of two restriction endonucleases from Hemophilus parainfluenzae. *Biochem. Biophys. Res. Commun.*, 59:108–116.
13. Gillespie, D., and Spiegelman, S. (1965): A quantitative assay for DNA-RNA hybrids with DNA immobilized on a membrane. *J. Mol. Biol.*, 12:829–842.
14. Goodman, H. M., Abelson, J., Landy, A., Brenner, S., and Smith, J. D. (1968): Amber suppression: a nucleotide change in the anticodon of a tyrosine transfer RNA. *Nature,* 217:1019–1024.
15. Graham, F. L., Abrahams, P. J., Mulder, C., Heijneker, H. L., Warnaar, S. O., de Vries, F. A. J., Fiers, W., and Van der Eb, A. J. (1974): Studies on in vitro transformation by DNA and DNA fragments of human adenoviruses and simian virus 40. *Cold Spring Harbor Symp. Quant. Biol.*, 39:637–650.
16. Hamer, D. H., and Thomas, C. A., Jr. (1975): The cleavage of Drosophila melanogaster DNA by restriction endonucleases. *Chromosoma,* 49:243–268.
17. Hedgpeth, J., Goodman, H. M., and Boyer, H. W. (1972): DNA nucleotide sequence restricted by the RI endonuclease. *Proc. Natl. Acad. Sci. U.S.A.*, 69:3448–3452.
18. Hirt, B. (1967): Selective extraction of polyoma DNA from infected mouse cell cultures. *J. Mol. Biol.*, 26:365–369.
19. Kelmers, A. D., Novelli, G. D., and Stulberg, M. P. (1965): Separation of transfer ribonucleic acids by reverse phase chromatography. *J. Biol. Chem.*, 240:3979–3983.

20. Ketner, G., and Kelly, T. J., Jr. (1976): Integrated simian virus 40 sequences in transformed cell DNA: analysis using restriction endonucleases. *Proc. Natl. Acad. Sci. U.S.A.*, 73:1102–1106.
21. Lai, C. J., and Nathans, D. (1974): Mapping the genes of simian virus 40. *Cold Spring Harbor Symp. Quant. Biol.*, 39:53–60.
22. Maniatis, T., Kee, S. G., Efstratiadis, A., and Kafatos, F., (1976): *Cell* (*in press*).
23. Manteuil, S., Hamer, D. H., and Thomas, C. A., Jr. (1975): Regular arrangement of restriction sites in Drosophila DNA. *Cell*, 5:413–422.
24. Mertz, J. E., Carbon, J., Herzberg, M., Davis, R. W., and Berg, P. (1974): Isolation and characterization of individual clones of simian virus 40 mutants containing deletions, duplications and insertions in their DNA. *Cold Spring Harbor Symp. Quant. Biol.*, 39:69–84.
25. Morrow, J. F., and Berg, P. (1972): Cleavage of simian virus 40 DNA at a unique site by a bacterial restriction enzyme. *Proc. Natl. Acad. Sci. U.S.A.*, 69:3365–3369.
26. Muench, K. H., and Berg, P. (1966): Preparation of aminoacyl ribonucleic acid synthetases from Escherichia coli. In: *Procedures in Nucleic Acid Research*, edited by G. L. Cantoni and D. R. Davies, p. 375. New York, Harper & Row.
27. Mulder, D., and Delius, H. (1972): Specificity of the break produced by restricting endonuclease R_1 in simian virus 40 DNA, as revealed by partial denaturation mapping. *Proc. Natl. Acad. Sci. U.S.A.*, 69:3215–3219.
28. National Institutes of Health (1976): Guidelines for research involving recombinant DNA molecules. *Federal Register*, 41 (431):27,902–27,943.
29. Nussbaum, A. L., Davoli, D., Ganem, D., and Fareed, G. C. (1976): Construction and propagation of a defective simian virus 40 genome bearing an operator from bacteriophage λ. *Proc. Natl. Acad. Sci. U.S.A.*, 73:1068–1072.
30. Russell, R. L., Abelson, J. N., Landy, A., Gefter, M. L., Brenner, S., and Smith, J. D. (1970): Duplicate genes for tyrosine transfer RNA in Escherichia coli. *J. Mol. Biol.*, 47:1–13.
31. Scherrer, K. (1969): Isolation and sucrose gradient analysis of RNA. In: *Fundamental Techniques in Virology*, edited by K. Habel and N. P. Salzman, p. 413. New York, Academic Press.
32. Sharp, P. A., Sugden, B., and Sambrook, J. (1973): Detection of two restriction endonuclease activities in Haemophilus parainfluenzae using analytical agarose-ethidum bromide electrophoresis. *Biochemistry*, 12:3055–3063.
33. Shenk, T. E., Rhodes, C., Rigby, P. W. J., and Berg, P. (1974): Mapping of mutational alterations in DNA with S_1 nuclease: the location of deletions, insertions and temperature-sensitive mutations in SV40. *Cold Spring Harbor Symp. Quant. Biol.*, 39:61–67.
34. So, M., Gill, R., and Falkow, S. (1975): The generation of an ApR cloning vehicle which allows detection of inserted DNA. *Mol. Gen. Genet.*, 142:239–249.
35. Tegtmeyer, P. (1972): Simian virus 40 deoxyribonucleic acid synthesis: the viral replicon. *J. Virol.*, 10:591–598.
36. Yang, W. K., and Novelli, G. D. (1971): Analysis of isoaccepting tRNA's in mammalian tissues and cells. *Methods Enzymol.*, 20(C):44–55.

Recombinant Molecules: Impact on Science and Society, edited by R. F. Beers, Jr. and E. G. Bassett. Raven Press, New York © 1977.

31. Discussion

Moderator: David Baltimore

Dr. D. Baltimore: I would like to open the discussion by considering questions of the technical aspects of the use of lambda as a cloning vehicle as presented in the first chapters of the section by Dr. Murray and Dr. Blattner.

I would first like to ask if any of the panelists have questions for each other and then turn it over to the audience.

Dr. L. Philipson: Dr. Hamer, you referred to the fact that you purified your DNA fragments. I think that this is an important point and you skipped over it lightly. How do you purify them?

Dr. D. H. Hamer: The first step in the purification I mentioned was the biological purification, putting the fragment on a plasmid, a small replicon. What I did not mention was that the particular plasmids used in these experiments had a sustained deletion of about 2,500 base pairs of the phage DNA (thereby reducing the number of restriction endonuclease sites) and made the purification of the fragment much simpler.

In order to make the fragment, we did a three-step procedure. The first step was to take supercoiled plasmid DNA, make partial digest with *Hpa*II, and isolate that overlapping 1,600-base pair fragment on an agarose gel in a Tris-acetate-EDTA buffer system.

We cleaved this fragment with *Eco*RI and isolated the desired fragment on agarose gel, this time using a Tris-phosphate-EDTA system. In the final purification step, we treated that fragment extensively with two different restriction enzymes for which it has no sites. These are the endonucleases *Sal*I and *Bam*I. Then we repurified the fragment by electrophoresis on a polyacrylamide gel in a Tris-borate-EDTA system.

We characterized that fragment before using it to infect cells. Two assays were done. The first was by a physical assay and simply was a gel electrophoresis, an analysis on acrylamide gels and agarose gels in different buffer systems. In each experiment we ran two parallel gels. One had approximately 2 μg of fragment; the other had about 2 ng. We stained the gels and scanned them in a very sensitive Zeiss fluorimeter. The gels with 2 μg of fragment show a single big peak and some background noise; whereas the gels with 2 ng of fragment show a small peak which is evident over the background by severalfold. We concluded that these fragments are electrophoretically homogeneous under conditions where 0.1% contamination would have been detected.

We also performed a biological assay. To ensure that the fragment would not be capable of autonomous replication in *E. coli,* we treated about 30 μg of fragment with DNA ligase under the same conditions eventually used to make recombinant molecules and then used that preparation to transform *E. coli* with double amber mutation. We next looked for suppressor plus transformants by plating on appropriate minimal media. After ligation 30 μg of that fragment gave no such transformants, whereas 100 ng of plasmid DNA gave 10^4 transformants.

So we concluded that it is incapable of replicating in *E. coli* and that the preparation is not contaminated with replicating fragments which can be rescued by ligation.

Dr. S. D. Ehrlich: Dr. Murray, in your lambda phages containing tRNA inserts can you recognize hybrid phages which have the insert in either of the two directions?

Dr. K. Murray: Yes, the suppressor is expressed in either orientation.

Dr. D. H. Hamer: May I make a brief comment? The same is true with the suppressor fragment in either orientation in plasmids also. They are both expressed.

Dr. S. D. Ehrlich: Dr. Hamer, you had a very detailed description of the verification of the purity of the inserted segment. Nevertheless, if I recall correctly, you mentioned in your chapter that you prepared something like 10^6 hybrid phages, or the lysate titer was 10^6, but you could detect purity to a level of 10^3. Would you comment on that?

Dr. D. H. Hamer: These are not direct plaquing experiments. In these experiments cells are infected with a small amount of DNA and then allowed to elaborate virus over the period of 2 weeks so that the final titer is many orders of magnitude greater than the initial number of input molecules.

In these experiments, the amount of SV40 vehicle fragment used to infect was about 0.02 μg, corresponding to less than 2×10^4 plaque-forming units.

Furthermore, we showed that the vehicle fragment we have used gave absolutely no virus at all under these conditions even when a 100-fold higher amount was used to infect.

I guess the proof of the pudding is that in our experiments only two species of DNA were recovered, the appropriate recombinant and the helper. We have been unable to detect any contaminants in the preparation.

Dr. P. K. Reddy: Dr. Hamer, in your experiments you described the insertion of the suppressor tRNA gene into the SV40 DNA, and then you looked for functional tRNA in the cells. Have you considered that in the case of *E. coli,* the tRNA is required to have modified nucleotides in order to function?

Dr. D. H. Hamer: Obviously the synthesis of functional suppressor tRNA in *E. coli* involves a large number of steps. It involves at least two specific ribonucleases, various enzymes which can modify the RNA, and, in some cases, modification is necessary for functional suppressor tRNA. At least it affects its ability to be charged and its ability to direct protein synthesis. So perhaps it is not too surprising that not all of these activities are present in mammalian cells.

Dr. D. Baltimore: Have you done any experiments on whether the processing of precursors taken from *E. coli* can be carried out by eukaryotic extracts? Have you checked if the tRNA is charged by charging enzymes of mammalian origin?

Dr. D. H. Hamer: With regard to the first point, I have done no such experiments. I do not know of any real data available. Dr. Arthur Landy might have some information about this. With regard to the second question, it was shown over 10 years ago that tyrosine tRNA from *E. coli* is poorly charged by mammalian cells. It is not clear whether that is a result of the sequence of the tRNA or failure of the appropriate modifications.

Dr. D. Baltimore: But if you put the tRNA that you were trying to make into a mammalian extract and use the assay that you had, would it be charged?

Dr. D. H. Hamer: I do not think the tRNA made in *E. coli* would be charged by the mammalian extract.

Mr. D. A. Konkel: I have a question for Dr. Blattner. Your vectors can carry no amber mutation in the host lysis function, the *S* gene. Dr. Leder's vectors do carry such a mutation, the idea being that by suppressing the host lysis, one can generate many more copies of the DNA because the cell does not lyse.

The question is: Do you think that the safety considerations gained by not having that amber mutation overcome the increased usefulness of having the mutation?

Dr. F. R. Blattner: An *S*-7 mutation is normally used for overproduction of DNA by starting with a lysogen with a temperature-sensitive repressor. The phages are induced by heating the lysate and then the replication takes place in the cell.

A typical yield for an experiment like that would be around 5×10^{11} phage particles per milliliter after lysis with chloroform.

The safety disadvantage is that one has a repressor gene which is active at 30°C as a part of the vector. Also, one of the important genes that would be involved in killing has been mutated away.

We have attempted to get around this by devising a phage that will grow lytically to 5×10^{11}. We have succeeded in pushing it to one-fifth of that value. Therefore, we are satisfied that the gain in safety outweighs the decrease in yield.

Mr. D. A. Konkel: I think the Leder vector is nonlysogenic, also, isn't it?

Dr. F. R. Blattner: It has C-1857.

Dr. D. Baltimore: I think that was an answer.

Dr. Nathans would like to interpolate a short comment related to some of the earlier presentations.

Dr. D. Nathans: I want to mention some experiments carried out in my lab, also using SV40 as a cloning vehicle. In particular, a series of experiments that have not been covered in this volume so far involves the demonstration of a lambda DNA fragment that is hooked to the early region of SV40 and used to transform mouse cells. It produces transformants which contain covalently linked SV40 DNA and lambda DNA integrated in the cellular genome.

These experiments were carried out as follows. We linked the large RI-A fragment of lambda from the left end—containing genes for the structural proteins—to an SV40 fragment which was the RI, *Hpa*II 74% piece. This contains the entire early region and has been shown by Van der Eb and his colleagues and by Walter Scott and Bill Brockman to be sufficient to transform cells.

We used this linear fragment to transform 3T3 mouse cells and selected transformants therefrom; the transforming efficiency of this fragment compared to that of the SV40 fragment alone was on the order of approximately one-fifth. We then selected a number of transformants. I will talk about one of those in particular.

The critical experiment was to analyze the cellular DNA of these transformants in a way that Kelly and Kettner did for SV40 transformed cells—and Bochan did at Cold Spring Harbor—that is, to analyze a small amount of cellular DNA cleaved with a restriction enzyme. In this case we chose the *Hin*d III because there are a number of restriction sites in the SV40 portion but no restriction sites in the lambda portion. Clearing the cellular DNA of transformed cells with *Hin*d III and electrophoresing the fragments on agarose result in a series of large numbers of fragments. These were then transferred by Southern's technique to Millipore sheets and hybridized with complementary RNA made with SV40 DNA as template and complementary RNA made with the RI-A fragment of lambda. We found that with SV40, there were two large fragments of cellular DNA that contain SV40 sequences, plus the expected fragments from the early region of SV40 DNA. We presume, then, that these large fragments contain SV40 sequences linked to cellular DNA. When we did the same experiment in a parallel manner with transcript from lambda DNA, two fragments with mobility identical to that of the SV40-containing sequences were found. Of course, we did not then find the labeled SV40 DNA fragment, from which we infer that there is, within the cellular DNA, a segment of DNA containing an unknown amount of lambda sequences and the input SV40 sequences.

It is clear from the size of these fragments that this entire fragment is about 15 million molecular weight; the lambda fragment had not been included and we are now mapping this, using *Hin*d III restriction enzyme to see how much of the lambda sequences are present. But I think this does indicate that one can make this kind of adduct, get it incorporated within the cellular genome, and then go on to see whether it is expressed. Obviously, of more interest would be to put in other kinds of genes.

Dr. D. Baltimore: Do you think the SV40 is necessary in there except as a signal that the integration has occurred?

Dr. D. Nathans: We have no idea whether any viral functions are required for integration. But it does serve, as Dr. Baltimore indicated, as a way to select those cells that have incorporated the SV40 part—namely, they are transformed and they are easily recognized and can be picked.

Mr. J. J. Sninsky: Several years ago, before the inception of *in vitro* genetic engineering, Dr. Carl Merril at NIH had some evidence which seemed to indicate that λpgal was able to incorporate into galactosemic fibroblasts. Has anyone re-examined those cells to determine in this type of manner whether λ pgal has been incorporated into cellular DNA?

Dr. D. Baltimore: That is a good question.

Dr. L. Philipson: I think it was reported in *Nature* that they had fibroblasts with an incorporated lambda DNA in which GPU transferase activity was expressed for quite some time. As far as I know, it has been absolutely impossible for anyone to confirm that finding. I don't think the original thesis still exists. I think the general feeling is that we do not know whether they got it in there or not.

Dr. D. Baltimore: But I think that what Dr. Nathans has just told us makes it not unlikely that lambda can be put in directly. Whether it would be expressed is another question.

Dr. D. Nathans: But, Dr. Baltimore, I do not know that it has any bearing on this question of whether lambda can itself integrate with SV40 function. It is really an open question right now as to whether early function of SV40 is needed. I know of no evidence that speaks to that question.

Dr. D. Baltimore: Neither do I.

Dr. P. H. Kourilsky: I would like to ask a somewhat irrelevant question which I understand our active chairman might postpone until the end of the session. A rumor circulated around Paris that some of the steps in making recombinant DNA molecules had been patented. Is this true? What is the possible impact on safety?

Dr. D. Baltimore: Is it a fact? Since I do not know whether it is a fact, it is hard to know its impact on safety.

Are there any further questions that relate to safety issues, especially as raised by Dr. Blattner's presentation? Dr. Philipson, you have two orientations of your Ad2-*Eco*RI-B fragment. In one orientation, that fragment is transcribed 100% in both directions. In the other orientation, the same sequence information is not totally transcribed. This suggests that there is some difference in the lambda promoter from which the transcription initiates as to whether it can make it all the way through that fragment. Do you have any comment on that?

Dr. L. Philipson: This is a difficult point to resolve. We are absolutely clear about the data. The only way I could explain it is if the termination governing the leftward promoter is different from the termination governing the rightward promoter. I do not know what the "lambdologists" think about this, but that is the only way we can explain the results we have with the two-directional fragments.

Dr. F. E. Young: Is it possible at this stage of development to predict the frequency of mutation of the newly inserted heterologous piece—possibly in regard to the comment by Dr. Nathans or others—in relation to the rate of mutation of the homologous piece in the organism from which it was isolated? A couple of contributors stated that the DNA piece seemed to remain intact during the cloning procedures. What are the limits of detection and what are the frequencies of mutation vis-à-vis the homologous situation?

Dr. S. P. Goff: We do not really know the answer to that yet. We have looked only at restriction fragment sizes of cuts internal to the insert and know that those do not seem to change.

We are in the process of mapping by S-1, as used by Tom Shenk, for base changes in that DNA and hope that will help us learn how frequently base changes arise in the inserted DNA.

Dr. L. Philipson: The type of lambda phage we have been using provides a selective advantage for keeping the inserted fragment. I reported that when we use a high ratio of inserted to vector fragments in the original ligation, we get trimers or dimers of the inserted fragment. Obviously, recognitive functions are required in order to segregate them down to monomers. But there is no indication, although we have not looked at them very carefully, whether the monomer fragments are gradually degraded to minimal size.

Mr. P. Youderian: On lambda SV40, have you looked for expression of lambda genes, say by fluorescent-labeled antibody against the J-protein, which is the tail antigen of lambda in your transformed cells?

Dr. D. Nathans: No, we have not. We are currently looking for transcripts, mapping this more thoroughly, and we hope to set up a better transformant to look for structural proteins.

Mr. P. Youderian: If lambda is expressed, you would have a way of screening cells that could be transformed only by lambda DNA. Is that correct?

Dr. D. Nathans: Yes.

Dr. E. N. Jackson: I have two questions about expression of inserted pieces under heterologous promoters. Do Dr. Murray or others know how many examples have so far been examined for expression of eukaryotic DNA linked to lambda?

Dr. K. Murray: I cannot give you a very satisfactory answer. I don't think any serious attempts are being made at such a search.

Dr. E. N. Jackson: The second question is addressed to Dr. Goff: Why, in view of the transcription of the *E. coli* segment in Dr. Hamer's insert, was his not transcribed?

Dr. S. P. Goff: Again, we certainly do not know the reason. I think Dr. Hamer covered the possibilities. It may simply be that the RNA is unstable and the tRNA region, having more double-stranded structure, is more stable. Further, perhaps something irregular about the dA-dT joints of the segment interfers with the stability.

We think that that second possibility is less likely, although not disproven, because we have made SV40 mutations using dA-dT insert regions and one of those is at the *Hpa* II site. This molecule, which is not really a hybrid but simply an insert of a dA-dT region, is perfectly viable. At least transcription through that occurs and the normal late functions are expressed.

So, at least in that case, the presence of the dA-dT joint does not confer stability on the RNA. But of course it may be that, when linked to bacterial DNA, it becomes unstable.

Dr. E. N. Jackson: I was confused as to whether you had seen transcription but RNA was unstable, or whether you simply had not detected any. You were assuming either there was none—

Dr. S. P. Goff: We do not know whether it is never made or whether, in fact, it is made and subsequently degraded. But the indications, I think, are that at least transcription probably occurs into that region from Dean's results and from the expected transcription of the normal 19S late mRNA in that region. Therefore, I think it is more likely that it is degraded, but we do not know that it is made at all.

Dr. D. Baltimore: Dr. Goff, a back-of-the envelope calculation in my head suggested from the attempt to drive DNA with the RNA from the cells that a transcript could not be stable for more than about 30 sec in order to yield your results.

Dr. S. P. Goff: If the level of transcription is comparable to the SV40—exactly.

Dr. D. Baltimore: So I would suggest that maybe you are not transcribing at all.

Dr. S. P. Goff: It may be. We do not know.

Dr. D. Nathans: Didn't your inserts have two dA-dT joints, compared to the deletions (which had one), so that there might be a loop out in your case?

Dr. S. P. Goff: That could easily be the situation.

Dr. G. C. Fareed: I would like to comment about the transcription studies we have performed, in a preliminary form, on an immunity region containing hybrid; these studies were also negative. These experiments involved filter hybridization analysis of labeled cellular RNA from cells infected with both of the lambda SV40 hybrids which I described, one of which contained the leftward operator. Our preliminary studies have been unable to demonstrate the synthesis of any specific RNA.

Mr. B. M. Kacsinski: Dr. Fareed, is there any evidence that the inserted lambda segment is itself repeated in SV40 reiteration mutants that you mentioned?

Dr. G. C. Fareed: In one of the two hybrids, the 520-base pair fragment was eventually reiterated in a tandem fashion with the vector segment in a triplicated structure. In the larger hybrid, the 2,400-base pair fragment was not reiterated.

Ms. F. R. Warshaw: I too am concerned about the comment that Dr. Kourilsky made earlier. We have also heard the rumor that someone has tried to file a patent for some of these techniques. I believe that the rumor originated somewhere on the West Coast and that probably it is someone in the group from Stanford Medical School.

I imagine someone here knows more about it than I do. Would some person give us some information about it?

Dr. D. Baltimore: Now that you mention it, I know something about it, and I think it is a dead issue; but if Stanley Cohen is here he might like to comment.

Dr. S. N. Cohen: Stanford University and the University of California have arranged to apply for a patent for possible commercial use of some of these techniques. I understand that similar steps have been taken in Edinburgh and Ken Murray may have some comments about this.

For the record, I would like to state that any proprietary rights that I may have had from research contributions to this area have been waived in favor of the University. If a patent is granted, and if there is commercial exploitation of these techniques, all income will go to Stanford and to the University of California; I will receive no royalties. Since I know there have been rumors about the patent application, I want to clarify this point.

Stanford and UC have had, for some years, an agreement with HEW and other agencies within the federal government, which allows the universities to receive income from patented material emanating from federally supported research. I believe a number of universities in the United States have similar arrangements. In any case, I do not believe the decision by Stanford and UC to apply for a patent represents an unusual arrangement.

Dr. D. Baltimore: Fine, that is certainly true. I would like Dr. Murray to comment on the situation in Edinburgh.

Dr. K. Murray: I can describe this very briefly. In common, I think, with most of you, in work that is carried out with grant support, the grant sponsors may wish to make use of any patent rights that may be available. In the United Kingdom, work supported by either the Medical Research Council or the Science Research Council—these two groups have been supporting the work done by Noreen and me—involves a governmental agency, the National Research Development Corporation, that handles patentable material for the investigators.

The NRDC examined the work that we described but decided that there was no case for filing a patent, due to prior disclosure of the information. Since we had talked about this work freely at scientific meetings and published it, under UK regulations it cannot be patented. We were mindful of the millions of dollars of royalties the British government did not realize following the development of penicillin.

Dr. D. Baltimore: That's right. I think under United States patent law one can file within 1 year after public disclosure. I presume that Stanford–UC has filed within that time.

Ms. F. R. Warshaw: I would like to repeat Dr. Kourilsky's question: Is it possible that some problems of safety could evolve because of this patenting procedure? I am not questioning the fact that universities often patent inventions developed by their departments. That is a conversation that one might have elsewhere, not here. But if this were patented, would any disasters, due to a lack of safety, arise?

Dr. D. Baltimore: To whom are you directing your question?

Ms. F. R. Warshaw: To anyone who would like to answer it, especially Dr. Murray and Dr. Cohen.

Dr. K. Murray: I can answer for myself. I cannot imagine such a consequence because anyone who would be exploiting anything patentable would, I am sure, be working under the conditions specified by the national government concerned.

In Britain, I think any commercial organization using this technology would be much more aware of the safety aspects than would most academic institutions. I hope that was clear from the chapters of Dr. Dart and Dr. Richards.

We are also protected by legislation called the Health and Safety at Work Act. Comparable to that in the United States is something called OSHA, which I am sure would be a fairly powerful means of ensuring that any commercial work carried out was rigorously monitored as to laboratory conditions.

Ms. F. R. Warshaw: So you think that patenting these inventions would result in increased rather than in less safety?

Dr. K. Murray: I just think it is irrelevant. I cannot see why it need affect the issue either way.

Dr. D. Baltimore: Lest there be any confusion that patenting involves any sort of secrecy, that is not true. Patents are, in fact, freely available as written documents.

Dr. S. N. Cohen: I would like to make two more points here. It is my understanding from the information I have received from Stanford that if the patent is granted to Stanford and UC, it will involve only commercial use of the procedures covered. The patent will not cover not-for-profit use of the procedures in any university or industrial laboratory.

Secondly, one of the points made to me by the Stanford Licensing Office handling the matter is that a patent can be used to ensure that commercial applications of DNA cloning will proceed in accordance with biohazard guidelines. Through their exercise of the patent, Stanford and UC would have the ability to stipulate levels of containment to be used in commercial applications by organizations not covered by the NIH guidelines.

So, to that extent, your question may be relevant. The patent may possibly lead to increased safety, but I can imagine no way in which a patent might result in decreased safety. If you know of any, I would be interested in hearing about this because, frankly, it was with some reluctance that I agreed to go along with the university in this whole process.

From the beginning, I have had some general conceptual misgivings about the patentability of advances in basic research and about the desirability of patents for research conducted in university laboratories. Only when I was convinced that this was a widespread and common practice by universities and that it would result in much-needed income to the universities did I agree to go ahead with this. Herb Boyer, in whose name the University of California has applied for a patent, has had the same kinds of concerns.

Dr. P. Kourilsky: Dr. Cohen, may I make a comment?

Dr. S. N. Cohen: Yes.

Dr. P. Kourilsky: Isn't it a possible concern that some private industry in Europe, for instance, would initiate recombinant DNA research and, not wishing to obtain licenses under the patent, use a system that is not as safe or well defined as that in the patent?

Dr. Cohen: I do not know how to respond to that. I have not thought about the point that you raise. Your point, as I understand it, is: someone might choose a less safe arrangement that might be patent-free rather than use a safer arrangement that happens to be patented. Is that the idea?

Dr. D. Baltimore: Terribly hypothetical. I think Dr. Philipson has a comment.

Dr. Philipson: I am a little hesitant now when you raise that point because, if I remember correctly, the NIH guidelines state that all vectors—and possibly also the clones—should be freely available. How is it possible to have a patented technique which involves this and still follows the NIH regulations?

Dr. Cohen: The vectors and clones constructed in my laboratory have been freely available, and many people in this audience have received them. We have been sending out plasmids and clones freely but have been requiring two things: one, that anything we send out be used in accordance with the Asilomar and subsequent guidelines; two, that they not be distributed secondarily to other laboratories so that we can maintain a record of distribution. We routinely indicate to recipients that we will be glad to send the same materials to other qualified laboratories that will agree to these stipulations.

Our materials have been made freely available to a number of industrial and university laboratories. In sending out samples, I have made no distinction between recipients. So I don't think that the questions that Dr. Philipson has raised really present a problem. As far as I know, a patent issued to the University would not prevent future distribution of vectors. It would just enable the University to share in the income derived from commercial applications of DNA cloning techniques.

Dr. F. R. Blattner: I would like to know if this could be used as a tool to force the people who use the vectors and the rest of the technology to publish the results. One of the things that Dr. Wald mentioned was that industrial operations often conduct their work in secret. While you were talking, it occurred to me that if you add the stipulation that the results be made freely available, it would help us all and possibly keep the thing a little more democratic.

Dr. S. N. Cohen: I wish you wouldn't use the term "you." I have no control over what is done by the University, and I indicated that I have no financial interest in the patent.

Dr. F. R. Blattner: Let's say I was not addressing you.

Dr. S. N. Cohen: I certainly can make suggestions to the University along these lines, and I think your suggestion is a good one. But I do want to make the point again that the University's rights in this area are derived from an agreement that Stanford and many other universities have had with HEW. This is not something that I am involved in. Obviously, Herb Boyer and I provided the technical information required by our respective universities in proceeding with this. But the patent application originated at the initiative of the University, and the University has acted independently in what it does along these lines.

Ms. F.R. Warshaw: I would like to read a petition our group has prepared. May I proceed?

Dr. D. Baltimore: Yes.

Ms. F. R. Warshaw: Thank you.

Dear Dr. Frederickson:

While attending the Miles Symposium on Genetic Engineering, we were disturbed by statements and actions which tended to discourage public participation.

To insure the continuation of democratic procedures, we ask the National Institutes of Health to explicitly request each institution conducting genetic engineering to call a series of public meetings for all employees to present the issues. Those meetings should be organized by the local safety committees which are mandated in the guidelines.

We further request that the guidelines specify that such local safety committees be representative of all segments of the work force employed in the laboratory buildings involved.

We will have copies of this petition at our table and urge you to consider signing it. If you are interested in talking about it, we will be discussing this petition among other things at the open discussion meeting to be held this evening at 7:30 in the west lounge of the Student Center.

Dr. D. Roy: After a series of really remarkable technical presentations, I was particularly struck with the absence of a broader philosophical discussion of how recombinant molecule engineering will possibly affect human development. There are fairly broad domains of discourse in contemporary society that are highly untechnical but that are quite concerned with work being done in genetics. I believe it will become increasingly important in coming years to establish bridges of effective dialog between the technical circles of activity and those broader circles of humanistic concern.

Moreover, I believe it will also be increasingly necessary for each of the individual sciences to develop its own center of matter reflection. I believe it is important to make these remarks because I think an opportunity has been missed at this international symposium. I believe it will become increasingly necessary to assuage and minimize the fears of many groups of people in our society who, at times, get fantastic projections as to the effects of what is being done in genetics. I think it would be highly important to set the story as broadly and deeply straight as can be done.

Mr. T. Moore: Drs. Goff, Hamer, and Fareed, you are constructing DNA molecules containing regions derived from both bacterial phages or bacterial animal viruses. Do you see any danger in case these molecules infect bacteria and recombine with the bacterial genome, especially keeping in mind that your animal virus vectors contain the region responsible for transformation? Do you see any possibility that these hybrid DNA molecules could infect that area, and if they do, do you see dangers in that?

Dr. D. Baltimore: Would any of you care to describe your precautions along those lines?

Dr. G. C. Fareed: I shall comment briefly on the type of hybrids I described. These were obtained from defective mutants of SV40 which lacked transformability. These segments, the particular hybrids, lacked replicative functions of the prokaryotic DNA. So it seems to me that if there is a potential for dissemination through microbial contamination, it would be minimized by working with hybrids of that type.

Dr. D. Baltimore: Does anybody else want to comment?

Dr. D. Nathans: I think clearly this is of concern and was certainly a concern to the DNA Recombinant Advisory Committee; although I was not a member of it, I knew of the deliberations of the subcommittee dealing with animal virus vectors. That kind of concern led to guidelines that were finally proposed.

Dr. S. P. Goff: This is indeed a possibility, and therefore one works with containment facilities that are thought adequate to lower, to an acceptable level, the probability of such an event; and that is why one works in a P3 facility. Under such an environment, it must escape from the lab into a bacterium, recombine with a lambda, and somehow integrate in the cell. We think that the products of all the probabilities of those events are extremely low.

Dr. D. Baltimore: May I ask a specific question along those lines? Is it possible to grow an *E. coli* culture in a solution of this DNA to maximize the opportunity for *E. coli* to pick up the DNA? Has anyone done anything like that? I know elaborate methodology is required to get DNA into *E. coli*. Can you demonstrate in some straightforward manner that *E. coli* placed in a solution containing 10^{10} DNA molecules per milliliter will not take up the DNA?

Dr. F. E. Young: I can speak to that relative to a couple of transformation systems but not to *E. coli*. Ottolenghi and Hotchkiss first showed that it is possible to do this with *Pneumococcus* and that the *Pneumococcus* excretes DNA during growth. *Bacillus subtilis* can take up DNA during its growth and, apparently, excrete it.

It has also been shown in *in vivo* experiments—again, I believe, in Hotchkiss' laboratory—that in the germ-free animal one could demonstrate exchange between the *Streptococcus* and *Pneumococcus*.

To my knowledge, this has not been done in *E. coli*. When it has been examined using transformable species, these *in vitro* exchanges do certainly occur without the addition of isolated DNA and then adding them to the cells.

Dr. D. Baltimore: That underlies the critical need for autoclaving everything that one uses, including DNA molecules.

Dr. F. E. Young: A very important point, Dr. Baltimore, that one cannot casually dispose of DNA—as has been done for years in many laboratory situations.

Dr. D. Baltimore: Dr. Ehrlich, you had a comment specifically on that?

Dr. S. D. Ehrlich: When *B. subtilis* is used in the experiments that Dr. Young described, the frequency of occurrence is very low; it is hard to detect.

Dr. D. Baltimore: It is hard, but possible, to detect.

Dr. S. D. Ehrlich: It is possible, but about 6 logs below that which can be detected optimally.

Dr. D. Baltimore: From the way his head is moving, Dr. Young does not think it is that low.

Ms. F. R. Warshaw: Are we talking about lambda DNA with sticky ends?

Dr. D. Baltimore: I think we are talking about random DNA.

Ms. F. R. Warshaw: If you did have lambda DNA, any wild-type lambda could act as a helper in a helper transfection. That would certainly increase the chance of getting the DNA into the cells.

Dr. S. P. Goff: The molecules that have been described do not have the lambda origin, nor do they have either left or right promoters.

Ms. F. R. Warshaw: Earlier in this volume Dr. Szybalski suggested that it would be much better to use disarmed *E. coli* as a host, rather than search for a new host. His rationale was that we know so much about this bacterium and it is better to work with something that we know about than to choose a new system. It seems to me that if perhaps five laboratories chose a new system and spent 5 years studying it, we might know just as much about the new system as we know know about *E. coli,* having the benefit of using *E. coli* techniques in the new system. Would anybody comment about that?

Dr. D. Baltimore: The most cogent comment I have heard on that question came from Roy Curtiss—who might want to repeat it here. He said that by the time we knew as much about the new strain as we do about *E. coli,* we would see problems that are just as big.

Mr. W. D. Roof: I would like to address the question that Dr. Young just raised about the possibility of having a foreign DNA taken up by a species that is transformable or perhaps could be made transformable. I think in most studies with *B. subtilis* and *Pneumococcus* spp., foreign DNAs that are taken up are rapidly degraded unless integrated; but the possibility exists that if a hybrid molecule is put into a cell, the molecule has complementarity or sequence homology for one region, and the molecule is not degraded. In fact, that has been adequately demonstrated by Gary Wilson and Frank Young with the *B. subtilis* system; others have demonstrated it with *Streptococcus* and *Pneumococcus.*

Dr. H. V. Aposhian: I would like to ask Dr. Young to elaborate on one of his points because it was unclear.

Many years ago, I think, Bayreuter and Romig showed that polyoma DNA was taken up by *Bacillus subtilis,* that plaques were formed, and that plaque-forming polyoma virus was made in *Bacillus subtilis.*

Has anyone ever been able to confirm these experiments, which I think are pertinent?

Dr. F. E. Young: Two experiments were performed, the one that you mentioned with polyoma, and another, I believe, with vaccinia by Trautner. In both of those instances the results could not be confirmed, and Trautner later said that in his own hands they could not be repeated. Under those circumstances, at that time, they were not confirmed. Whether it was there but not detected adequately, I think, has not been examined recently. Trautner officially withdrew his claim but Romig did not.

Dr. D. Baltimore: Pamela Abel was doing those experiments in Dr. Trautner's lab. She later tried to repeat those experiments in Dr. McAuslan's laboratory at Princeton but I believe was unable to. There is a long discussion, in fact, in a chapter by McAuslan in Hilton Levy's *Biochemistry of Viruses,* describing the difficulties they had in trying to repeat these results. Basically, McAuslan said they are not reproducible phenomena.

Dr. V. Sgaramella: We have heard about physical, biological, and chemical containment. Could something come from this meeting that would provide any kind of social containment? It seems to me the kind of suggestion contained in the petition that Scientists for People are planning to send to Dr. Frederickson could, in some form, provide a type of social containment for this area of research and would probably be extremely welcome.

Dr. D. Baltimore: Would you elaborate on what you mean by social containment?

Dr. V. Sgaramella: I think the adjective expressed what I mean. It seems to me that this is a possible form of social containment. What I suggest is whether it is possible to include social containment in the same manner that P1, P2, etc. are physical containments.

Dr. D. Baltimore: I think in the Asilomar discussion and guidelines there is extensive expression of the need for individual education, if you wish. The extent to which that is social containment may be a matter of definition, but I believe that a number of people have realized from the outset that the major barrier involved in maintaining these vehicles within the laboratory is the barrier of individual activity. Individuals have to conform to a certain degree of social containment, if you wish, containing their own impulses if any of these guidelines are to have any force. It is certainly appropriate that these issues be discussed as widely as possible on all campuses and at all sites where this kind of work is performed.

That highlights the importance to every bench scientist of the need for maintaining appropriate control over what is going on.

Dr. R. J. Roberts: I have one comment with regard to the transformation by DNA. Unfortunately, the data we have at the moment refer only to items which we can select. No numbers are available that might give an indication of how frequently DNA is integrated, without selection, into cellular genomes. I think the techniques are now at hand—namely, the Southern blotting technique—to examine that question in detail. It would be fun to do so, and some people should do that to obtain some data.

The second question regards a rather provocative comment Dr. Baltimore made that we should all autoclave our DNA. We all know of renaturation, but I would appreciate other people's views as to whether they think autoclaving DNA is the solution. In my lab, we prefer to use bleach.

Dr. D. Baltimore: Autoclaving was a euphemism for bleach.

Dr. R. J. Roberts: We routinely use both. Normally, we use Chlorox®, but as a test we autoclaved some supercoiled SV40 DNA and tested its infectivity. The infectivity was reduced by more than 6 logs but could have been more than that; we had no more DNA to test.

I know that Ron Davis has done similar tests with lambda DNA. So we think that autoclaving is at least fairly good.

Dr. S. Lederberg: I think we have been dancing around an item of the following type: we have seen curves for the inactivation of the lambda vector, specifically the tail-bonding protein, thereby preventing the sorption of phage by bacteria. Unfortunately, we do not know about the survival of the genetic elements of those vectors.

Dr. D. Baltimore: I think that is clearly true.

Dr. W. Szybalski: If DNA is autoclaved and the pH lowered slightly, then it is degraded very effectively.

Dr. D. Baltimore: Depurination.

Thank you all very much for your attention.

Recombinant Molecules: Impact on Science and Society, edited by R. F. Beers, Jr. and E. G. Bassett. Raven Press, New York © 1977.

32. Introduction to Section E: Cloning of Eukaryotic DNA

Charles A. Thomas, Jr.

Department of Biological Chemistry, Harvard Medical School, Boston, Massachusetts 02115

This session is devoted to the cloning of eukaryotic DNA. Cloning technology reaches its ultimate and most useful application in the analysis of eukaryotic DNA, which represents such a large and diverse collection of nucleotide sequences. With the availability of an expanding library of restriction endonucleases, together with new, rapid methods for the determination of DNA sequences, it is now reasonable to expect that we can know the nucleotide sequence of any segment we choose from any organism.

Knowledge of base sequence will not be enough. We will need to know how the sequence is interpreted by the cell. Of course, we already know how the coding sequence is translated to amino acid sequence, but most of the DNA in higher cells is not coding for proteins. How are these sequences interpreted and what role do they play in the determination of the differentiated state? It is likely that we will again rely on gene expression to disassemble the problem. One promising approach is to reintroduce a defined eukaryotic DNA segment into eukaryotic cells by means of a transducing virus such as SV40. During a cycle of viral growth, one might devise experiments to test for the expression of the defined segment. It will be possible to alter the segment in known ways (by deletions and insertions) and to determine what effects these changes may have on its expression.

Whether or not eukaryotic genes can function in bacteria will be of great practical interest. Given the vast differences that one might expect in the transcription, processing, and translation steps, a beginner might expect no expression at all! However, one should not be too pessimistic.

Recombinant Molecules: Impact on Science and Society, edited by R. F. Beers, Jr. and E. G. Bassett. Raven Press, New York © 1977.

33. The Construction and Use of Hybrid Plasmid Gene Banks in *Escherichia coli*

John Carbon, Louise Clarke, Christine Ilgen, and Barry Ratzkin

Department of Biological Sciences, University of California, Santa Barbara, California 93106

INTRODUCTION

The recent development of techniques for the cloning and amplification of DNA segments linked by biochemical methods into bacterial plasmid or phage vectors has now made it possible to isolate particular gene systems from any source for further study [see Nathans and Smith (19) for a review of restriction endonucleases and their use in restructuring and cloning of DNA]. Basically, DNA cloning is accomplished by the joining of a DNA segment (or a group of segments) *in vitro* to a DNA vector (plasmid or virus) capable of replication in a suitable cell, infecting the host organism with the hybrid DNA, and then isolating individual clones of cells (or virus) that are each the progeny of a single infected cell.

Although several different eukaryotic gene systems have been established on hybrid plasmids in *E. coli* cells, it is still uncertain whether any functionally active protein product can be made from eukaryotic genes in the prokaryotic environment (2,13,18). A definitive test for meaningful expression of the cloned DNA segment would be complementation of an auxotrophic mutation in the bacterial host cell, such that the foreign DNA is producing an enzymatically active protein capable of relieving the metabolic defect in the host. Recently, Struhl et al. (23) have reported that a segment of yeast DNA cloned onto a bacteriophage λ vector is capable of complementing the *his*B463 mutation in *E. coli.* Although the exact nature of this complementation has not yet been established, it seems possible that certain segments of eukaryotic DNA may yield functionally active protein products in the *E. coli* cell.

Several experimental requirements must be met in order to set up a really definitive test for the ability of any given eukaryotic gene system to be expressed and to complement an auxotrophic mutation in *E. coli.* Thus, the efficiency of the cloning procedure used must be high enough to ensure that sufficient transformant clones containing hybrid DNA plasmids are obtained to be representative of the entire genome of the organism under study. Secondly, it is preferable to use DNA segments produced by *random scission* of the parent DNA by hydrodynamic shear rather than by restriction endonuclease action, in order to ensure that the desired gene system remains intact on at least a portion of the cleaved DNA segments. In addition, a reasonably rapid and convenient *screening procedure* should be available to test recombinant DNA plasmids for the ability to complement bacterial mutations. The experimental system used must at least permit the isolation of a large number of *E. coli* gene systems on hybrid plasmids, in order to show in a convincing way that the cloning and complementation procedures are adequate for the task in hand.

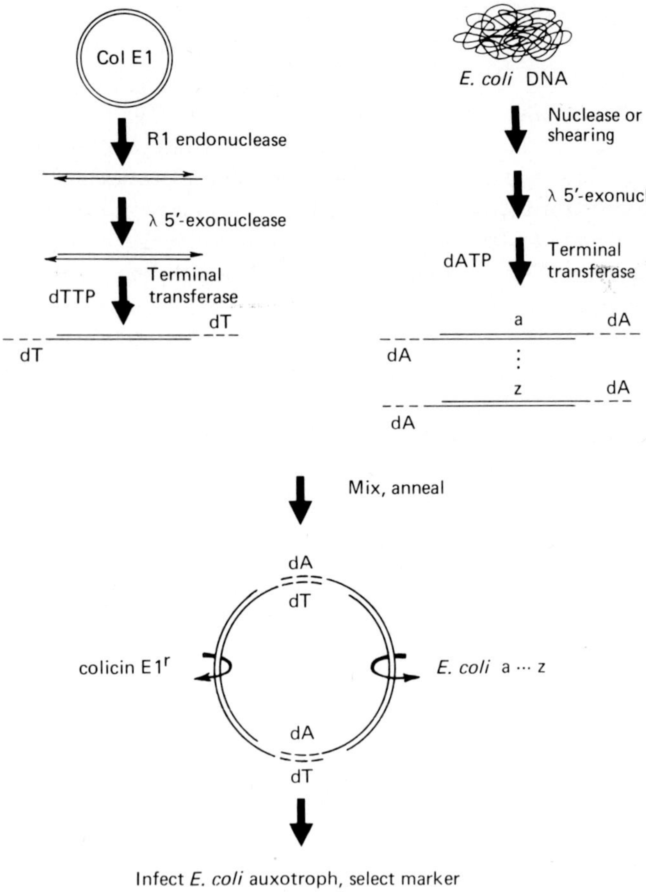

FIG. 1. Method for construction and selection of hybrid Co1E1-*E. coli* DNA plasmids. (From ref. 3.)

For the above reasons, we have used the poly(dA · dT) "connector" method for joining DNA (Fig. 1), a procedure that can give a high yield of recombinant DNA circles *in vitro,* and that can readily be used with sheared DNA samples (3,4,12,15,24). With this method, we have prepared hybrid circular DNA *in vitro* from poly(dT)-tailed plasmid ColE1 DNA (*Eco*RI-generated linear molecules) and randomly sheared segments of *E. coli,* yeast (*S. cerevisiae*), or *Drosophila* DNAs tailed with poly(dA). Transformation efficiencies using hybrid DNAs prepared in this manner are extremely high (about 10^3 unique transformants per microgram DNA), with essentially all of the transformants containing hybrid plasmids. Thus it has been possible to establish transformant colony banks containing a large number of clones that carry different hybrid plasmids representative of most of the genome of the parent organism.

E. coli strains harboring the fertility plasmid, F, along with the non-transmissible plasmid ColE1 transfer both plasmids with high efficiency to an F^- recipient (5,8,10). We find that hybrid ColE1 plasmids are also transferred efficiently from F^+ donors to auxotrophic F^- recipients, and this F-mediated transfer provides a convenient screening technique for the identification of specific hybrid plasmid clones in the transformant banks. This procedure has permitted the identification of over a hundred different hybrid plasmid clones carrying about 60 known *E. coli* gene systems on hybrid ColE1 plasmids.

Using preparations of mixed hybrid plasmid DNA purified from the total colony banks, we have transformed suitable *E. coli* auxotrophs and selected for and isolated a number of specific segments of *E. coli* DNA; more importantly, we have been able to isolate and characterize segments of yeast DNA that complement the *hisB* and *leu-6* mutations in *E. coli.* In principle, the methods described should permit the ready cloning of any segment of DNA for which a selection or mutation usable in complementation analysis is available.

MATERIALS AND METHODS

Bacterial Strains

The following strains, all derivatives of *E. coli* K-12, were used as recipients for transformations or in bacterial mating experiments: JA198 (*ΔtrpE5 recA thr-1 leu-6 lacY* str^r), JA199 (F^+ *ΔtrpE5 leu-6* hsr_K^- hsm_K^+), JA200 (F^+ *ΔtrpE5 recA thr-1 leu-6 lacY*), JA208 (*ΔaraC766 ΔlacZ-Y514 recA*), MV10 (*ΔtrpE5 thr-1 leu-6 lacY*)(11), MV12 (*ΔtrpE5 recA thr-1 leu-6 lacY*)(11), DM493 (W3110 *ΔtrpE5*), and CH754 (*argH metE xyl trpA36 recA56*). The F-plasmid was derived from *E. coli* strain Ymel.

Construction of Hybrid ColE1-DNA (*E. coli*) Annealed Circles

Linear poly(dT)-tailed plasmid Co1E1 DNA [$(L_{RI}$exo)-$(dT)_{150}$] was prepared as previously described (3). High molecular weight *E. coli* DNA was purified from strain CS520 (HfrC *trpA58 metB glyVsu58*) as described (3), resuspended at 100 μg/ml in 0.01 M Tris-HCl (pH 7.5), 0.01 M NaCl, 0.001 M Na_2EDTA (STE), and fragmented by hydrodynamic shearing. DNA was sheared in a stainless steel cup (capacity = 1 ml) at 0°C for 45 min using a setting of 4.5 (approximately 5,400 rpm) on a Tri-R Stir-R motor (model S63C) fitted with a Virtis shaft and micro homogenizer blades. The fragmented CS520 DNA (average MW = 8.4×10^6 daltons) was treated with λ 5′-exonuclease, and poly$(dA)_{150}$ extensions were added to the 3′-ends of the DNA with the calf thymus deoxynucleotidyl terminal transferase (3,12,15). An equimolar mixture of Co1E1 DNA [$(L_{RI}$exo)-$(dT)_{150}$] and CS520 DNA [$(L_{sh}$exo)-$(dA)_{150}$] was annealed, and the DNA was concentrated by precipitation in 67% ethanol, dissolved in STE, and examined under the electron microscope (3). The annealed mixture contained approximately 25% hybrid DNA circular molecules.

Establishment of an *E. coli* Hybrid Plasmid Colony Bank

Twelve micrograms of annealed hybrid DNA were used to transform strain JA200 (C600 *recA*/F^+) according to a modification of the method of Mandel and Higa (16) described by Wensink et al. (24), except that after exposure to DNA, cells were diluted 10-fold into L-broth and grown for 30 min. The cells were plated in the presence of colicin E1 as previously described (3,4).

A separate control culture of JA200 was simultancously treated in an identical manner to the transformed culture described above, except that cells were not exposed to DNA. From the number of colicin E1-resistant cells in this mock culture, it was estimated that about 67% of the colicin-resistant cells in the transformed culture contained Co1E1 plasmids, and the remaining clones were colicin tolerant but contained no plasmid (6). Twelve out of fifteen clones from the transformant culture were resistant to colicin E1 and sensitive to colicin E2 (11), and the remainder were resistant to both colicins. Use of the Triton lysis procedure (4) on 10 clones from a pilot experiment identical to the one described here and electrophoretic analysis of supercoiled plasmid DNAs in 1.2% agarose gels revealed that about 90% of the clones in the transformant culture (which were resistant to colicin E1 and sensitive to colicin E2) contained plasmids appreciably larger than Co1E1, whereas the remaining 10% contained no plasmid. From these data, we conclude that about 70% of the clones in the colony bank contain hybrid plasmids.

The colonies in the collection were arranged in a grid pattern of 48 per plate such that they could easily be transferred via a wooden block of 48 needles to standard 96-well MicroTest II dishes (Falcon Plastics), with each well containing 0.2 ml L-broth, colicin E1, and 8% dimethyl sulfoxide (20). The colonies in the collection could, therefore, be individually maintained, permanently stored at −80°C, and used repeatedly by inoculating fresh plates.

Isolation of hybrid plasmid DNA, and subsequent transformations were carried out as previously described (3,4). Standard techniques for bacterial selections, use of indicator plates, and liquid and replica matings were employed (17).

Establishment of the Yeast and *Drosophila* Hybrid Plasmid Banks

Yeast DNA was isolated from *S. cerevisiae,* strain X2180-1Aa SUC 2 *mal gal2* CUP1. Preparations of DNA from *D. melanogaster* were generously supplied by Dr. Norman Davidson (California Institute of Technology) and by Dr. David Finnegan and Dr. David Hogness (Stanford University). DNA was sheared to an average size of 8 to 9 × 10^6 daltons and tailed with poly$(dA)_{100}$ as described above. After annealing with Co1E1 DNA[L_{RI}exo-$(dT)_{100}$], transformations of strain JA199 and establishment of colony banks were as described above.

Electron Microscopy

DNA was spread by the aqueous method of Davis, Simon, and Davidson (7), and all contour lengths were measured relative to reference Co1E1 relaxed circles or *Eco*RI-cleaved Co1E1 DNA linears on the same grid.

RESULTS

Establishment of a Collection of *E. coli* Clones Containing Hybrid ColE1-DNA (*E. coli*) Plasmids

We have determined the transformant colony bank size needed to obtain a plasmid collection which represents 90 to 99% of the *E. coli* genome as follows: given a preparation of cell DNA that has been fragmented to a size such that each fragment represents a fraction (f) of the total genome, then the probability (p) that a given unique DNA sequence is present in a collection of n transformant colonies is given by the expression:

$$p = 1 - (1 - f)^n$$

or

$$n = \frac{\ln(1 - p)}{\ln(1 - f)}$$

Thus, if a preparation of *E. coli* DNA were randomly sheared to an average size of 8.5×10^6 daltons for the construction of annealed hybrid circular DNA, a colony bank of only 720 transformants would be adequate to give a probability of 90% that any *E. coli* gene would be on a hybrid plasmid in one of the clones (Table 1; it is assumed that the desired gene is small in comparison with the size of the cloned fragments). At a probability level of 99%, the colony bank size (n) is only about 1,400 colonies for *E. coli*. As the genetic complexity of the organism increases, n at high probability levels increases dramatically (Table 1); at $p = 99\%$, the bank size for yeast is 4,600 colonies, whereas for *Drosophila* it is 46,000 colonies.

The construction of hybrid ColE1-DNA (*E. coli*) annealed circles using poly(dA-dT) connectors was carried out as previously described (3), except that the *E. coli* DNA was fragmented by hydrodynamic shearing (instead of restriction endonuclease cleavage) to an average size of $8.4 \pm 3.0 \times 10^6$ daltons, determined by measuring 47 molecules in the electron microscope. The preparation of annealed hybrid ColE1-DNA (*E. coli*) contained approximately 25% circles, 12% branched circles, 53% linears, 2% branched linears, and 8% unscorable tangles (100 molecules scored by electron microscopy).

The annealed DNA preparations can be used to transform various *E. coli* auxotrophs, selecting directly for the desired plasmid by complementation. For example, in Table 2 are shown transformation data obtained in this type of experiment (3). ColE1 circular plasmid DNA alone, when used to transform strain SB2(C600) to colicin E1 immunity, routinely yielded approximately 6×10^4 transformants per microgram DNA on L-broth–colicin E1 plates. The transformants were resistant to colicin E1 but sensitive to colicin E2 (11). Annealed hybrid DNA, when used in similar trans-

TABLE 1. *Colony "bank" sizes* (n) *needed to contain a particular hybrid plasmid transformant at various probability levels*

DNA source	Average size of DNA fragment cloned (daltons)	"Bank" size (n), no. of colonies		
		$p = 0.90$	$p = 0.95$	$p = 0.99$
E. coli	8.5×10^6	720	940	1,440
Yeast	1×10^7	2,300	3,000	4,600
Drosophila	1×10^7	23,000	30,000	46,000

The above calculations are based on the formula, $p = 1 - (1 - f)^n$, and assume that each transformant colony in the "bank" arises from an independent transformation event and that each hybrid molecule transforms with the same efficiency. It is also assumed that the length (x) of the desired DNA segment is small in comparison with the length (L) of the DNA fragment actually cloned, in order to minimize the effect of random breaks occurring within the desired length. More accurately, a corrected f value (f^*) could be obtained from the expression, $f^* = \left(1 - \frac{x}{L}\right)f$, and substituted for f in the above probability equation. (From ref. 4.)

TABLE 2. *Transformation efficiency of annealed circular hybrid DNA*[a]

DNA	Recipient	Selection	Transformants/μg DNA
ColE1 (I)	SB2(C600)	Colicin E1^r	6×10^4
ColE1 (L_{RI} exo)-$(dT)_{150}$	SB2(C600)	Colicin E1$_r$	<1
ColE1 (L_{RI} exo)-$(dT)_{150}$ + *E. coli* (L_{RI} exo)-$(dA)_{150}$	SB2(C600)	Colicin E1^r	$\sim 10^3$
"	NL20-127 (K12 Δ*araC766* *recA*)	Ara$^+$	0.2 (2/2 colicin E1^r)
"	MV10 (C600 Δ*trpE5 rec*$^+$)	Trp$^+$	4 (12/45 colicin E1^5)

[a] In these experiments, the *E. coli* DNA was fragmented by digestion with endonuclease *Eco*RI, rather than by hydrodynamic shear. (From ref. 3.)

formations, gave approximately 10^3 transformants per microgram DNA. Again, these transformants were resistant to colicin E1 but sensitive to colicin E2. The ColE1[(L_{RI}exo)-$(dT)_{150}$] DNA gave fewer than one colicin E1-resistant transformant per microgram when used alone. Hybrid DNA was then used to transform a Δ*trpE5* strain (the *trpE* region is deleted) to Trp$^+$, or a Δ*araC* strain to Ara$^+$. In both cases, transformants were readily obtained which could be shown to contain hybrid plasmids carrying the *trp* operon (pLC5; 15×10^6 daltons) or the *ara-leu* region (pLC3; 18×10^6 daltons).

We used the same amount (12 μg) of annealed hybrid ColE1-DNA (*E. coli*) that was used to directly select for a specific hybrid plasmid-containing strain (3) to transform *E. coli* strain JA200 (C600 Δ*trpE5 recA*/F$^+$). In this experiment we selected instead for the vector determinant, colicin E1 resistance. Thus, in a single transformation and selection, a collection or "bank" of clones was obtained which carry different hybrid ColE1-DNA (*E. coli*) plasmids representative of a large portion of the bacterial genome. After exposure to DNA, the transformed cell culture was not incubated long enough to permit significant cell division before plating. Thus, each transformed clone was the consequence of a distinct and separate transformation event. In addition, the recipient for transformation was a strain harboring the sex factor, F, in order to permit the identification of a clone carrying a particular hybrid plasmid by F-mediated transfer of the hybrid through replica mating of the colony collection to a particular *E. coli* F$^-$ auxotroph (*see below*). A *recA* transformation recipient was chosen to avoid recombination of hybrid plasmids with host chromosomal DNA.

Approximately 2,100 colicin E1-resistant transformants were picked and transferred to plates in a grid array of 48 colonies per plate. The colonies were also individually maintained and stored as 8% DMSO cultures in MicroTest dishes at −80°C. Use of a Triton lysis procedure (4; H. Boyer, *personal communication*) on 10 clones from a pilot experiment and analysis of supercoiled plasmid DNAs in 1.2% agarose gels revealed that about 70% of the clones in the collection (1,400 transformant colonies) contained plasmids appreciably larger than plasmid ColE1. The remaining 30% of the colonies were tolerant to colicin E1 but probably contained no plasmid (6). Many of the latter colonies were also resistant to colicin E2. In more recent experiments, we have been able to reduce the background of colicin E1-tolerant cells by pregrowing the transformation recipient in 0.05% sodium deoxycholate (C. Ilgen and J. Carbon, *unpublished observations*).

Use of F-Mediated Transfer for the Identification of Specific Hybrid Plasmid-Bearing Clones in the Colony Bank

Once a transformant colony bank is established, it is essential to have a simple, rapid way to identify a desired hybrid plasmid-bearing clone within the collection. It has been shown that strains which harbor the fertility plasmid F and which also contain the nontransmissible plasmid Co1E1 will transfer both plasmids to an F^- recipient with high efficiency (5,8,10). Preliminary reconstruction experiments using strain MV12/pLC19 (Co1E1-*trp* plasmid, ref. 3) to which an F factor had been transferred indicated that this Co1E1-DNA (*E. coli*) hybrid transferred readily to an F^-/*trp* recipient (see Table 5). Furthermore, the transfer was of high enough efficiency to be easily detected by replica mating on plates, using appropriate selections and counterselections.

A number of replica mating experiments were therefore performed using the entire colony collection and various F^- auxotrophs. Some of these experiments were carried out in our laboratory and will be discussed in detail below. The remainder were done by other laboratories using the same colony collection. Table 3 lists the hybrid plasmids tentatively identified to date and some of the markers carried by these plasmids (Fig. 2). On the average, three hybrid plasmid containing clones were identified in the total collection for each marker sought in a replica mating experiment. The list includes about 40 known *E. coli* genes, some of which were assigned after clones harboring hybrid plasmids which carried neighboring genes were identified. For example, the four clones in the collection which were found by replica mating to contain Co1E1-*xyl* plasmids were tested for overproduction of glycine-tRNA synthetase, the product of the closely neighboring *glyS* gene (Fig. 2), and two of the four clones did produce elevated levels of this enzyme (G. Nagel, *personal communication*).

The transformant colony collection appears to contain hybrid plasmids representative of nearly the entire *E. coli* genome, in that the probability of finding any cloned gene system chosen at random appears to be high

TABLE 3. *Hybrid plasmids identified in the colony bank*

Approximate map location (min)	*E. coli* markers complemented	pLC code
1	*araC*	24–41
4	*dnaE*[a]	26–43
7	*proA*[b]	28–33 44–11
9	*lacZY*	20–30
12	*dnaZ*[a]	5–1 5–2 6–2 10–24 10–26 30–3 30–4
22–26	*flaKLM*[c]	24–46 35–44 36–11
27	*trpE*	4–6 5–23 29–41 32–12 32–27 41–15
35–38	*flaGH, cheB*[c]	21–2 24–15
	flaGH, cheB, mot[c]	1–28 1–29
	mot, cheA, flaI[c]	27–20 38–14 38–36
35–38	*flaD*[c]	7–18 13–12
	flaD, hag, flaN[c]	24–16 26–7
	flaN, flaBCOE, flaAPQR[c]	41–7
38	*his*[b]	14–29 26–21
43	*glpT*[d]	3–46 8–12 8–24 8–29 14–12 19–24 42–17
51	*recA, srl*[b]	17–43 18–42 21–33 22–40 24–32

(*Table 3 continues* →)

TABLE 3 (*Continued*)

Approximate map location (min)	*E. coli* markers complemented	pLC code
	recA[b]	17–38
		24–27
		30–20
70	*glyS, xyl*[e]	1–3
		44–22
	xyl	10–15
		32–9
75	*ilv*[b]	21–35
		22–3
		22–31
		26–3
		27–15
		30–15
		30–17
		44–7
	cya[f]	23–3
		29–5
		36–14
		41–4
		43–44
79	*argH*	20–10
		41–13
81	*dnaB*[a]	11–9
		44–14
89	*dnaC*[a]	4–39
		8–9
		25–8
		30–24
		31–39
90	*serB, trpR*[f]	32–33
	trpR, thr[f]	35–1
		35–21

[a] R. McMacken (*personal communication*).
[b] L. Margossiane and A. J. Clark (*personal communication*).
[c] M. Silverman and M. Simon (*personal communication*).
[d] J. Weiner (*personal communication*).
[e] G. Nagel (*personal communication*).
[f] J. Schrenk and D. Morse (*personal communication*).

Genes on hybrid plasmids identified as complementing specific *E. coli* markers in this table are not necessarily the only *E. coli* genes carried by the plasmids. In addition, plasmids listed here as complementing a specific marker do not always represent all the plasmids in the total collection which carry that particular *E. coli* gene. For example, the Co1E1-*lacZY* hybrid plasmid (pLC20–30) was identified by screening as the only plasmid in the collection capable of complementing a *lacY* mutation (see Table 4). pLC20–30 was later found to also complement *lacZ,* but other Co1E1-*lacZ* plasmids may occur in the colony bank. (From ref. 4.)

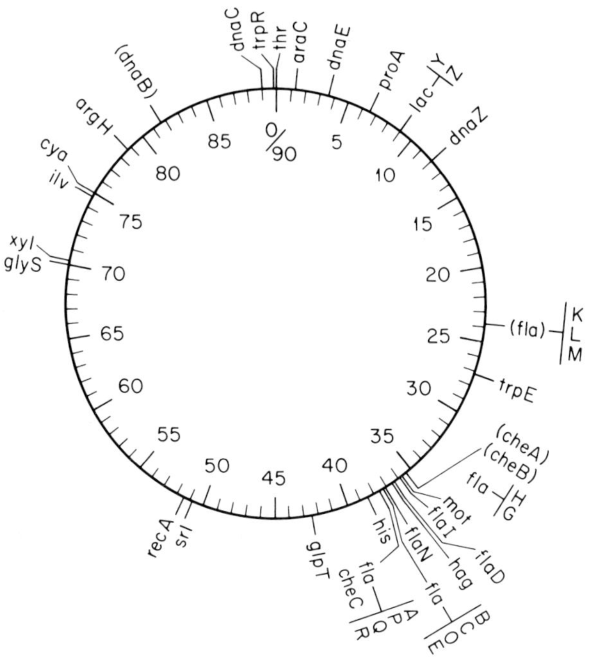

FIG. 2. Relative positions on the *E. coli* genetic map of gene systems complemented by hybrid Co1E1-DNA (*E. coli*) plasmids. For references, see Table 3. (From ref. 4.)

(about 80%). For example, in our laboratories the collection was screened for hybrid plasmids capable of complementing any of six mutations chosen at random (*araC, lacY, trpE, argH, xyl,* and *metE*). Of these, only the *metE* mutation could not be complemented by any hybrid plasmid from the bank (see Table 3). All of the known *E. coli* motility gene systems were found on hybrid plasmids in this same collection (M. Silverman and M. Simon, *personal communication*).

Characterization of Several ColE1-DNA (*E. coli*) Hybrid Plasmids from the Colony Bank

A semiconfluent to confluent patch on selective plates in the mating experiments described above was scored as representing a clone carrying the desired hybrid plasmid. Phenotypic reversion of a recipient marker could, however, have been a consequence of suppression if the recipient marker was a point mutation and the hybrid plasmid carried a suppressor transfer RNA gene, or a consequence of suppression possibly resulting from overproduction of an RNA or protein whose gene was carried by a hybrid plasmid. Complementation of a recipient marker in the F-mediated replica matings could also have been the result of Hfr formation (if the recipient was *recA*$^+$ or if the hybrid plasmid carried a *recA*$^+$ region) or F′ formation and subsequent mobilization of donor chromosomal genes. However, chromosomal mobilization generally occurs at a much lower frequency than F-mediated hybrid plasmid transfer (see Table 5) and could usually be distinguished from it.

To help establish that clones tentatively identified in the colony bank did carry the desired hybrid plasmids, we further characterized a number of these strains and their plasmids. Table 4 lists the hybrid plasmid strains identified by four replica mating experiments selecting for hybrid plasmids carrying the *ara, trp, arg,* and *xyl* regions. From one to six candidates were obtained from each screening of the bank. Since the parent bank strain was *lacY,* the entire colony collection was also screened for ColE1-*lacY* bearing clones by replica plating the bank onto indicator plates. One Lac$^+$ candidate was found by this method.

Covalently closed, supercoiled hybrid plasmid DNA was purified from at least one representative of each marker class in Table 4. These DNAs were then used to transform either the original recipients or a recipient with a mutation in the same gene as the original. In all cases tested, a high frequency of transformation to the expected phenotype was obtained (Table 4), and all transformants screened were both resistant to colicin E1 and resistant to fr male-specific phage. These results indicated that the genes responsible for reversion of recipient markers in the original mating experiments were carried by ColE1-DNA (*E. coli*) hybrid plasmids. It is improbable in the case of the *ara, trp,* and *lac* plasmids that reversion was due to suppressor tRNA genes carried by the hybrids, since the mutations complemented by these plasmids were deletions. Suppression by overproduction of a gene product, such as a tRNA, a protein with weak catalytic activity or a ribosomal component, is difficult to rule out in any case. However, two of the ColE1-*xyl* plasmids were shown to carry the closely neighboring *glyS* gene (G. Nagel, *personal communication*), and the ColE1-*lac* plasmid, originally detected in a *lacY* strain, transformed to Lac$^+$ a strain which carries a deletion in at least *lac Y* and *Z*. Thus, these three hybrid plasmids certainly carry the desired portion of *E. coli* DNA.

The contour lengths of a number of hybrid plasmids were measured by electron microscopy of relaxed circles using ColE1 DNA (4.2×10^6 daltons; ref. 1) as a standard on the same grid. Data for four of these measurements are found in Table 4. Other plasmids measured were pLC18-42 ($11.9 \pm 0.1 \times 10^6$ daltons), pLC21-33 ($10.3 \pm 0.1 \times 10^6$ daltons), pLC32-33 ($18.1 \pm 0.1 \times 10^6$ daltons), pLC36-14 ($14.9 \pm 0.2 \times 10^6$ daltons), and pLC50 (ColE1-*lac,* $13.6 \pm 0.1 \times 10^6$ daltons; isolated with the same annealed hybrid DNA but in a separate transformation). The average size of the *E. coli* inserts based on this relatively small sample is $9.9 \pm 3.4 \times 10^6$ daltons, as com-

TABLE 4. *Characterization of several hybrid plasmids*

Mating recipient	Selection	pLC code	Plasmid size ($\times 10^{-6}$ daltons)	Transformation recipient	Selection	Transformants/ μg DNA
JA208 (Δ*araC766*)	Ara^+	24–41	—[a]	JA208 (Δ*araC766*)	Ara^+	3×10^4
DM493 (Δ*trpE5*)	Trp^+	4–6	—			—
		5–23	—	MV12 (Δ*trpE5*)	Trp^+	—
		29–41	—			4×10^4
		32–12	—			4×10^4
		32–27	—			—
		41–15	—			—
CH754 (*argH*)	Arg^+	20–10	11.8 ± 0.2	CH754 (*argH*)	Arg^+	$>2 \times 10^5$
		41–13	10.0 ± 0.1			$>3 \times 10^5$
CH754 (*xyl*)	Xyl^+	1–3	—	CH754 (*xyl*)	Xyl^+	—
		10–15	17.0 ± 0.1			$>2 \times 10^5$
		32–9	—			$>10^4$
		44–22	—			—
MV (*lacY*)/ F^+/ColE1-DNA (*E. coli*) (screened on MacConkey-lactose plates)	Lac^+	20–30	19.4 ± 0.1	JA208 (Δ*lacZ-Y 514*)	Lac^+	$>3 \times 10^3$

[a] A dash indicates "not measured." (From ref. 4.)

pared with the average size of the sheared CS520 DNA ($8.4 \pm 3.0 \times 10^6$ daltons) used originally to construct the plasmids.

Efficiency of the F-Mediated Transfer of ColE1 Hybrid Plasmids

The F-mediated transfer of three ColE1-*trp* plasmids was studied (Table 5). Strain MV12/pLC19(F^-) has been described (3) and harbors a ColE1-*trp* plasmid which carries all or most of the *trp* operon genes. This strain was made F^+ by mating with strain Ymel and designated JA200/pLC19. Strains JA200/pLC29-41 and JA200/pLC32-12 are F^+ clones from the colony bank which were identified as carrying ColE1-*trpE* plasmids (see Tables 3 and 4). The F^- strains, MV12/pLC29-41 and MV12/pLC32-12, were constructed by transformation of strain MV12(F^-) with purified pLC29-41 and pLC32-12 DNAs isolated from their respective strains in the colony bank (see Table 4). These six strains were used as donors in mating experiments described in the legend to Table 5, with strain DM493 (Δ*trpE5*/F^-) serving as the recipient in all cases. All of the ColE1-*trp* plasmids were donated with high efficiency from strains that also contained F. No transfer was detected from strains that contained only the hybrid plasmids. Of the recipients in the matings receiving hybrid plasmids, 88 to 98% of them also received F and were able to further transfer ColE1-*trp* plasmids. A small fraction of the original recipients, however, picked up only the hybrid plasmids, since they could no longer transfer these plasmids and were resistant to phage fr. Several of the ColE1-*arg* and ColE1-*xyl* plasmids were also very efficiently transferred from F^+ strains (D. Richardson and J. Carbon, *unpublished observations*). These data indicate that

TABLE 5. *Efficiency of transfer of ColE1*-trp *plasmids from* F^+ *and* F^- *strains*

Donors	Recipient	% Donors transferring hybrid plasmid in 1 hr	% Trp^+ recipients that were F^+
MV12/pLC19 (F^-)	DM493 (Δ*trpE5*/F^-)	<0.004	–
MV12/pLC29-41 (F^-)	"	<0.002	–
MV12/pLC32-12 (F^-)	"	<0.002	–
JA200/pLC19 (F^+)	"	64	88
JA200/pLC29-41 (F^+)	"	46	98
JA200/pLC32-12 (F^+)	"	40	96

Liquid matings were performed as described by Miller (17) at a 10:1 ratio of recipients to donors. A 1-hr incubation with gentle shaking at 37°C was followed by Vortex interruption, dilution, and plating onto selective plates. The fraction of Trp^+ recipients that were also F^+ was determined by picking 50 recipients from each mating, replica mating these as donors, using JA198 (Δ*trpE5str*^{r}) as a recipient in each case, and selecting for Trp^+. Those donors which did not transfer the Trp^+ phenotype were confirmed as being F^- by their resistance to phage fr. (From ref. 4.)

ColE1-DNA (*E. coli*) plasmids behave in a manner similar to plasmid ColE1 in an F^+ background and that the hybrid plasmids and the F factor act as physically independent units.

Mixed Hybrid Plasmid DNA Preparations Representing a Large Portion of the *E. coli* Genome

An alternative way to maintain the genetic information of a complete genome as a collection of hybrid plasmids is in the form of a covalently closed, hybrid plasmid DNA preparation. Such a preparation was made by scraping a set of master replica plates containing all the clones in the colony bank, allowing the cells to grow nonselectively for approximately two generations, amplifying plasmid DNA in the presence of chloramphenicol, and purifying supercoiled total plasmid DNA. To identify individual plasmids, we then used this DNA to transform the five recipients listed in Table 6 to the Ara^+, Trp^+, Arg^+, Xyl^+, and Lac^+ phenotypes. The observed and calculated transformation frequencies are given in Table 5. At the DNA levels used, no Ara^+ or Lac^+ transformants were found, but the experiment yielded a disproportionately large number of Arg^+ transformants. Thus, this purified DNA preparation was not as representative of the bacterial genome as the colony bank and appeared to contain a relatively large amount of ColEl-*arg* DNA but less ColEl-*ara* and ColEl-*lac* DNAs.

Generally, we have observed that strains harboring various hybrid plasmids may grow at different growth rates and that certain hybrid plasmids segregate at high frequency and others not at all. These observations have indicated to us that the most suitable way to maintain a hybrid plasmid

TABLE 6. *Efficiency of transformation by mixed plasmid DNA prepared from the entire colony bank*

Transformation recipient	Selection	Transformants/μg DNA: Observed	Transformants/μg DNA: Expected
JA208(Δ*araC766*)	Ara^+	<0.5	20
MV12(Δ*trpE5*)	Trp^+	6.3×10^2	2×10^2
CH754(*argH*)	Arg^+	3.3×10^4	4×10^2
CH754(*xyl*)	Xyl^+	5.9×10^3	8×10^2
JA208(Δ*lacZ-Y514*)	Lac^+	<0.5	>2

Transformations and selections were carried out as described (3,4) using 2 μg DNA for each experiment. The expected number of transformants per microgram DNA was calculated from known efficiencies of transformation of purified individual plasmid DNAs using these same recipients (Table 4) and from the number of representatives of each plasmid in the colony bank, assuming all plasmids were approximately the same size. (From ref. 4.)

collection without loss of specific plasmids is as a set of individual clones, each harboring a unique plasmid.

Establishment of Hybrid Plasmid Colony Banks Derived from Yeast (*S. cerevisiae*) and *Drosophila* DNAs

The methods described above have also been applied to DNA prepared from either yeast (*S. cerevisiae*) or *Drosophila*. The DNA samples were sheared by high-speed stirring as described in Materials and Methods before reaction with the λ exonuclease and the terminal transferase. Using 25 μg of annealed hybrid DNA (Co1E1 DNA as vector) to transform *E. coli* strain JA199 (hsr_K^- hsm_K^+), we have obtained about 30,000 *unique* transformants to colicin E1 resistance in each case. Of these, at least 80% could be shown to contain hybrid plasmid DNA, although in both cases many of the hybrids contained a relatively small insert of about 1 kilobase (kb) or less in length. Only 20% of the clones contained relatively large hybrid plasmids with inserts ranging from 2.5 to 20 kb. The reason for the large number of relatively small inserts in these banks is still unclear.

These collections are of sufficient number to contain essentially all of the yeast genome, but because of the small size of most of the hybrid Co1E1-Dm DNA plasmids, it is likely that only 10 to 20% of the *Drosophila* genome is represented in this initial collection. The collections have been stored in three forms: (a) as individual colonies (4,300 for yeast; 10,000 for *Drosophila*) in 8% DMSO suspension at −80°C; (b) as a mixture of transformed cells; and (c) as a mixture of hybrid plasmid DNA extracted from all of the transformants.

Complementation of *E. coli* Mutations by Hybrid ColE1-Yeast DNA Plasmids

THE leu-6 *MUTATION*

The *E. coli* strain that was transformed with joined Co1E1-yeast DNA to form the hybrid plasmid colony bank carried two auxotrophic mutations, Δ*trpE5* and *leu-6* (strain JA199). As a first step toward the screening of the hybrid plasmids for ability to complement *E. coli* mutations, we plated a mixture of 15,000 colonies from the original transformation selection plates onto supplemented minimal media, selecting for growth in the absence of either tryptophan or leucine.

On minimal salts-glucose media supplemented with tryptophan but without leucine, healthy Leu^+ colonies were observed occurring at a frequency of one Leu^+ colony per 10^6 cells plated. Sixteen of the slower growing Leu^+ colonies were purified and subjected to further investigation. All 16 of the Leu^+ transformants were resistant to colicin E1 and sensitive to colicin E2, a property of cells carrying hybrid Co1E1 plasmids (11). In addition, all transferred the Leu^+ phenotype to suitable F^- *leu-6* recipients at high frequency (about 10^{-3}). Plasmid DNA was isolated and purified by CsC1-ethidium bromide banding from eight of the Leu^+ transformants. Seven of these plasmid DNAs were capable of transforming strain JA199 to Leu^+. One plasmid, pYe*leu*10, which produced the fastest growing transformants, was investigated further. Circular pYe*leu*10 DNA has a molecular weight of 13.4×10^6 daltons, as determined by direct length measurements on electron photomicrographs using Co1E1 DNA (4.2×10^6 daltons) as a standard.

The data presented in Table 7 indicate that pYe*leu*10 DNA is capable of transforming strain JA199 (*leu-6*) to both colicin E1 resistance and Leu^+ with high frequency ($>10^5$ transformants per microgram DNA). Thus, it is clear that the hybrid plasmid pYe*leu*10 carries genetic information capable of complementing the *leu-6* mutation in *E. coli.*

The segment of cloned DNA in pYe*leu*10 was shown to be yeast DNA by measuring the rates of reassociation of labeled single-stranded plasmid DNA fragments in the presence of various unlabeled single-stranded DNAs. In Fig. 3 are shown the rates of reassociation of single-stranded pYe*leu*10 DNA in the presence of a large excess of single-stranded salmon sperm DNA, *E. coli* DNA, and yeast DNA. Although salmon sperm and *E. coli* DNAs have little or no effect on the rate of reassociation of the pYe*leu*10 sequences, single strands of yeast DNA greatly increase the rate, as would be expected if a portion of pYe*leu*10 DNA is derived from yeast DNA. Reassociation kinetics data of this type indicate that about 70% of the pYe*leu*10 DNA is derived from yeast DNA. This is in excellent agreement with the length measurements of pYe*leu*10 DNA, from which a molecular weight of 9.2×10^6 daltons was derived for the cloned yeast DNA segment, or 69% of the total plasmid DNA (assuming a complete 4.2×10^6 daltons Co1E1 segment is present in pYe*leu*10). Assuming that a single copy of the DNA segment cloned in pYe*leu*10 is present in the yeast genome, the reassocia-

TABLE 7. *Transformation efficiencies of Co1E1-yeast DNA hybrid plasmids*

	Transformants/μg DNA		
Plasmid DNA	Colicin $E1^R$	Leu^+	His^+
pYe*leu*10	2.7×10^5	3.1×10^5	
pYe*his*1	1.5×10^4		2.2×10^3
pYe*his*2	2.6×10^4		2.5×10^3
pYe*his*3	2.4×10^4		2.5×10^3

Hybrid plasmid DNAs were isolated and purified as previously described (3,4). The transformation recipients were strain JA199 (for pYe*leu*10) and K12 *hisB463* (for the pYe*his* plasmid DNAs).

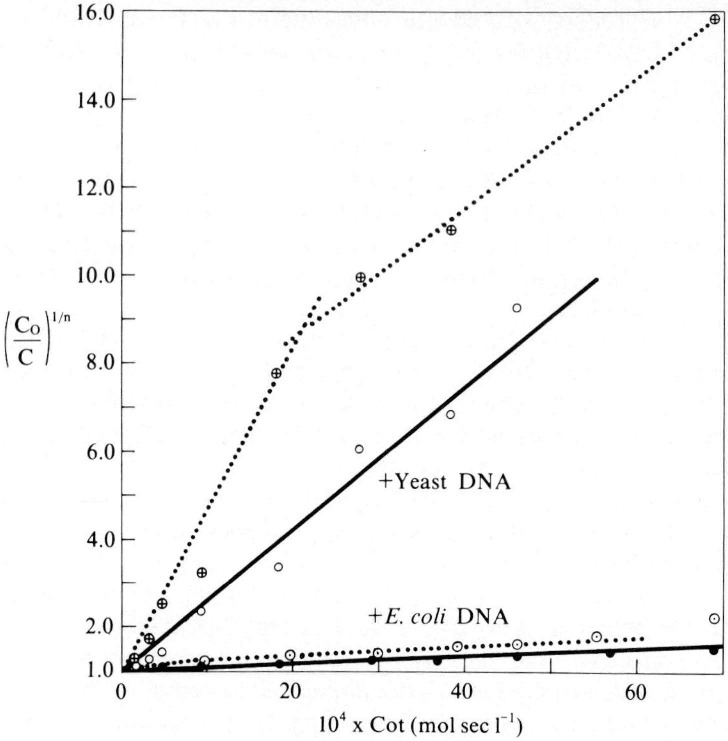

FIG. 3. Reassociation kinetics of labeled single-stranded pYe*leu*10 DNA fragments (0.085 μg/ml) in the presence of: salmon sperm DNA, 1,000 μg/ml (●——●); *E. coli* DNA, 196 μg/ml (⊙----⊙); yeast (*S. cerevisiae*) DNA, 176 μg/ml (○——○); and yeast DNA, 526 μg/ml (⊕----⊕). The pYe*leu*10 DNA was ^{3}H-labeled to a specific activity of 2.4×10^6 cpm/μg by nick translation in the presence of [^{3}H]dATP (8.33 c/mmole) and [^{3}H]dTTP (8.36 c/mmole) by the method of M. Dieckmann and P. Berg (*personal communication*), as described by Schachat and Hogness (21). The reassociation rate experiments were carried out as previously described (21,24).

tion rates give a haploid genome size of 4.7×10^9 daltons, in excellent agreement with recent estimates (14).

THE hisb463 *MUTATION*

Mixed hybrid plasmid DNA isolated from a mixture of all transformants in the yeast hybrid plasmid bank can be used to transform *E. coli* auxotrophs, selecting for the desired complementation. Since Struhl, Cameron, and Davis (23) have recently reported complementation of the *E. coli hisB463* mutation by a yeast DNA segment cloned on a bacteriophage λ vector, we have attempted the isolation of a hybrid Co1E1-yeast DNA plasmid capable of complementing this same mutation. Transformation of a K-12 *hisB463* strain with 2 μg of mixed hybrid plasmid DNAs derived from the total collection gave His$^+$ transformants at a frequency of 3 His$^+$ colonies per 10^5 transformants to colicin E1 resistance. No His$^+$ colonies were obtained in the absence of hybrid plasmid DNA, in line with the previous observation that the *hisB463* mutation is nonrevertable and is most probably a deletion that affects IGP dehydratase activity (23).

All of the His$^+$ transformants appeared to contain hybrid ColE1 plasmids and were colicin E1 resistant but sensitive to colicin E2. Plasmid DNA was isolated and purified from three of these clones (designated pY*ehis*1, pY*ehis*2, and pY*ehis*3). These three plasmids appeared to be identical in length when measured on electron photomicrographs using ColE1 DNA as a standard. The total molecular weight of the pY*ehis* plasmid is 10.7×10^6 daltons, indicating that a segment of DNA of molecular weight 6.5×10^6 daltons had been cloned.

Purified pY*ehis* plasmid DNAs were capable of transforming a *hisB463* *E. coli* strain to colicin E1 resistance and to His$^+$ with high frequency (see Table 7). Direct selection for His$^+$ consistently yielded 10-fold fewer transformants than when colicin E1 resistance was selected (Table 7). This observation is similar to that reported by Struhl et al. (23) for the λgt-Sc*his* phage containing a *hisB* complementing segment of yeast DNA. The purified phage yields His$^+$ transformants at only 1% of the expected value, possibly due to inefficient expression or functioning of the yeast gene product.

Reassociation rate experiments have shown that the segment of yeast DNA in pY*ehis*2 is, in part, identical to that cloned by Struhl et al. (23) in λgt-Sc*his* (see Fig. 4). Labeled single-stranded fragments of pY*ehis*2

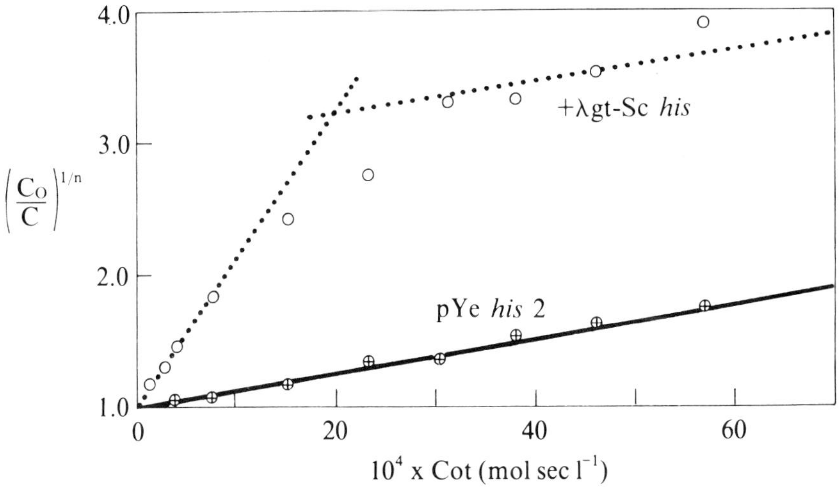

FIG. 4. Reassociation kinetics of labeled single-stranded pY*ehis*2 DNA fragments (0.07 μg/ml) in the presence of: salmon sperm DNA, 1,000 μg/ml (⊕——⊕); and λgt-Sc*his* DNA, 125 μg/ml (○-----○). The pY*ehis*2 DNA was nick-translated to a specific activity of 2.2×10^6 cpm/μg as described in the legend to Fig. 3.

DNA reassociate appreciably faster in the presence of unlabeled single-stranded λgt-Sc*his* DNA. The experiment indicates that about 40% of the pYe*his*2 DNA or 4.3×10^6 daltons of DNA is identical to the yeast DNA segment cloned in λgt-Sc*his*. The remaining 2.2×10^6 dalton segment of cloned DNA in pYe*his*2 is not present on λgt-Sc*his*, although the phage apparently contains other sequences not present on the plasmid. Since the yeast DNA segment in λgt-Sc*his* is an *Eco*RI-generated fragment whereas the segment in pYe*his*2 was produced by random scission, it is not surprising that there is only a partial overlap between the two segments. pYe*his*2 DNA contains a single *Eco*RI cleavage site, which is apparently identical to one of the two sites defining the yeast DNA segment in λgt-Sc*his*.

DISCUSSION

Hybrid Plasmid Colony Banks

Many of the advantages of having particular gene systems isolated and cloned on plasmids present in multiple copies per bacterial cell have been described (3,4,11). This chapter reports procedures by which most of the genetic information contained in a bacterial or simple eukaryote genome can be established as collections of hybrid Col E1 plasmids, each of which contains a unique segment of DNA of approximate molecular weight 8 to 10×10^6 daltons. We have maintained these plasmid collections in two ways: first, as colony banks of separate transformant clones, each harboring a unique hybrid plasmid, and, secondly, as covalently closed, supercoiled DNA preparations isolated from the total colony banks.

There are a number of ways to identify specific plasmids of interest within these collections. In this chapter we have used complementation of various *E. coli* mutations to isolate plasmids carrying specific genes by F-mediated transfer of hybrid plasmids from the F^+ clones in the colony bank to F^- auxotrophs on selective plates, or by transformation of auxotrophs and direct selection using the mixed plasmid DNA preparation. In several cases, neighboring genes have been found to occur on the same hybrid plasmid.

Maintaining a plasmid collection representative of most of the *E. coli* genome in the form of a colony bank permits the use of additional methods to identify specific plasmids. The presence of hybrid Col E1-DNA (*E. coli*) plasmids in multiple copies per cell frequently leads to overproduction of the products of genes carried by these plasmids (11). Overproduction of a protein—a repressor, for example—might result in pleiotrophy or auxotrophy in cells harboring specific plasmids because the gene system(s) under the control of this repressor might not be inducible. Thus, screening the clones in the colony bank for auxotrophy may result in further assignment of regulatory genes to plasmids or reveal hitherto unknown regulatory phenomena.

Another method for identification of specific genes on plasmids in colony banks is the hybridization technique described by Grunstein and Hogness (9). Many gene systems cannot be identified by F-mediated transfer and selection, either because a suitable auxotrophic recipient is unavailable (e.g., most tRNA and ribosomal RNA genes) or because the transformant would not be viable. For example, strains carrying a tRNA-derived suppressor mutation on an amplifying plasmid may not survive because of overproduction of a deleterious *su*$^{+}$ gene product. We are presently using hybridization techniques to detect plasmids in the collections that carry *E. coli* and eukaryotic tRNA and ribosomal RNA genes. Recently, the spacing of tRNA genes on cloned segments of *Drosophila* DNA has been described (P. Yen, C. Ilgen, J. Carbon, and N. Davidson, *unpublished observations*).

Several difficulties arise in attempting to maintain colony banks of the type described in this chapter. In our experience, hybrid plasmids derived from *E. coli* DNA seem to be readily maintained in *recA* strains if selective pressure, either colicin E1 resistance or complementation of a host marker, is kept on the strains. In the absence of selective pressure many of these plasmids segregate, however. In *recA*$^{+}$ strains some plasmids appear to be lost, presumably through recombination with the genome or through segregation, and others become smaller, possibly by looping out segments of the *E. coli* DNA. For these reasons, we have maintained the colony collections as sets of individual clones, have attempted to keep selective pressure on plasmid-bearing strains, and have stored the original transformant collections in the form of separate cultures in 8% DMSO at −80°C. We have gone back to these original cultures to inoculate fresh plates for the experiments described here and have avoided continued growth of clones and replication of plasmids.

Expression of Eukaryotic DNA in *E. coli*

Although several laboratories have established eukaryotic DNA segments on suitable cloning vectors in *E. coli,* little evidence has been presented regarding the meaningful expression of the cloned DNA to form enzymatically active protein products. Recently, Struhl et al. (23) have reported that *E. coli hisB* mutations are complemented by an *Eco*RI fragment of yeast DNA cloned on a phage λ vector; however, the mechanism of the complementation is still unclear.

When a large fraction of the yeast genome is represented in a collection of hybrid plasmids derived from randomly sheared DNA, it is surprisingly easy to find hybrid plasmids capable of complementing *E. coli* auxotrophic mutations. Thus, the yeast-ColE1 hybrid plasmids complementing the *hisB* and *leu-6* mutations were found without the need to screen the collection for complementation of hundreds of different *E. coli* mutations. In fact, the "success rate," in terms of finding *E. coli* mutations that are complemented by hybrid plasmids from the yeast DNA-ColE1 hybrid collection, was about 20% (two successful, *leu-6* and *hisB,* out of only nine different auxotrophic mutations attempted). It is possible that the "success rate" is even higher than 20% since apparent complementation of mutations other than *hisB* and *leu-6* has been observed in preliminary experiments but not verified as yet by isolation of pure plasmid DNAs for transformation tests.

The mechanism of the expression of yeast DNA segments in *E. coli* is still unclear. Thus, we do not know that the normal yeast transcription and translation initiation signals are being used in the bacterial system. One possibility might be that random transcription products are made from fortuitous promoter sequences on the yeast DNA, and that the *E. coli* ribosomes are capable of initiating at or near the normal ribosome-binding sequences on the resulting RNA. Transcription and read-through from vector promoter sequences do not seem likely since Struhl et al. (23) have shown that the yeast DNA segment in λgt-Sc*his* can be inserted in either orientation without affecting expression.

The exact nature of the complementation phenomena we are observing remains to be explained. In the case of the *hisB463* complementation, the mutation is probably a deletion, and thus some form of suppression seems unlikely (23). The *leu-6* mutation does revert with measurable frequency, however, so that in this case suppression effects have not been ruled out. In addition, we have not yet determined which of the four genes in the *leu* operon (*leuA, B, C,* or *D*) is defective in the *leu-6* strain. Additional studies with a set of well-defined and mapped *leu* mutants (22) should clarify this point.

Finally, the relative ease with which we have found yeast DNA segments capable of complementing *E. coli* mutations suggests that it should now be possible to isolate many different yeast gene systems on hybrid plasmids for studies on the genetic organization and expression of the yeast genome. It is possible that yeast DNA, being derived from a relatively primitive eukaryote, can be expressed in the prokaryotic host cell, but that DNA from higher eukaryotes would be mostly inert. Our current studies with *Drosophila melanogaster* DNA are designed to answer this important question.

Biohazard Considerations

F-mediated transfer of ColE1-like plasmids provides a valuable tool for the identification of specific hybrid plasmids constructed of ColE1 DNA and the DNAs of prokaryotic and simple eukaryotic organisms. However, the procedure[1] converts normally nontransmissible hybrid plasmids, such

[1] The current NIH guidelines on recombinant DNA research specifically prohibit the cloning of eukaryotic DNA on transmissible plasmid vectors. Thus, we are no longer performing cloning experiments in *E. coli* cells that contain the fertility plasmid F.

as those derived from Col E1, into a state which permits their ready transfer to other bacterial hosts. Because of biohazard considerations, we do not recommend the establishment in an F^+ background of strains or banks of clones containing hybrid plasmids constructed of Col E1 DNA and random fragments of DNA from higher eukaryotic organisms until the biological properties of such plasmids have been further characterized.

ACKNOWLEDGMENTS

The authors would like to thank Denise Richardson for excellent technical assistance. We also are indebted to R. Ratliff for a generous gift of deoxynucleotidyl terminal transferase, and to K. Struhl and R. Davis for a sample of λgt-Sc*his* DNA. This work was supported by NIH Grants CA-11034 and CA-15941. One of us (B.R.) received support by a grant from Abbott Laboratories. Portions of this work have been published elsewhere (3,4).

REFERENCES

1. Bazaral, M., and Helinski, D. R. (1968): Circular DNA forms of colicinogenic factors E1, E2 and E3 from Escherichia coli. *J. Mol. Biol.,* 36:185–194.
2. Chang, A. C. Y., Lansman, R. A., Clayton, D. A., and Cohen, S. N. (1975): Studies of mouse mitochondrial DNA in Escherichia coli: structure and function of the eukaryotic-prokaryotic chimeric plasmids. *Cell,* 6:231–244.
3. Clarke, L., and Carbon, J. (1975): Biochemical construction and selection of hybrid plasmids containing specific segments of the Escherichia coli genome. *Proc. Natl. Acad. Sci. U.S.A.,* 72:4361–4365.
4. Clarke, L., and Carbon, J. (1976): A colony bank containing synthetic Col E1 hybrid plasmids representative of the entire Escherichia coli genome. *Cell,* 9:91–99.
5. Clowes, R. C. (1964): Transfert génétique des facteurs colicinogénes. (Symp. on Bacteriocinogy, Paris, May, 1975.) *Ann. Inst. Pasteur,* 107 (Suppl. 5):74–92.
6. Davis, J. K., and Reeves, P. (1975): Genetics of resistance to colicins in Escherichia coli K-12: cross-resistance among colicins of group A. *J. Bacteriol.,* 123:102–117.
7. Davis, R., Simon, J., and Davidson, N. (1971): Electron microscope heteroduplex methods for mapping regions of base sequence homology in nucleic acids. *Methods Enzymol.,* 21:413–428.
8. Fredericq, P., and Betz-Bareau, M. (1953): Transfert génétique de la propriété colicinogéne chez Escherichia coli. *C. V. Seanc. Soc. Biol. Paris,* 147:2043.
9. Grunstein, M., and Hogness, D. S. (1975): Colony hybridization: a method for the isolation of cloned DNAs that contain a specific gene. *Proc. Natl. Acad. Sci. U.S.A.,* 72:3961–3965.
10. Hardy, K. B. (1975): Colicinogeny and related phenomena. *Bacteriol. Rev.,* 39:464–515.
11. Hershfield, V., Boyer, H. W., Yanofsky, C., Lovett, M. A., and Helinski, D. R. (1974): Plasmid Col E1 as a molecular vehicle for cloning and amplification of DNA. *Proc. Natl. Acad. Sci. U.S.A.,* 71:3455–3459.
12. Jackson, D. A., Symons, R. H., and Berg, P. (1972): Biochemical method for inserting new genetic information into DNA of simian virus 40: circular DNA molecules containing lambda phage genes and the galactose operon of Escherichia coli. *Proc. Natl. Acad. Sci. U.S.A.,* 72:2904–2909.
13. Kedes, L. H., Chang, A. C. Y., Houseman, D., and Cohen, S. N. (1975): Isolation of histone genes from unfractionated sea urchin DNA by subculture cloning in E. coli. *Nature,* 255:533–538.

14. Lauer, G. D., and Klotz, L. C. (1975): Determination of the molecular weight of Saccharomyces cerevisiae nuclear DNA. *J. Mol. Biol.,* 95:309–326.
15. Lobban, P. E., and Kaiser, A. D. (1973): Enzymatic end-to-end joining of DNA molecules. *J. Mol. Biol.,* 78:453–471.
16. Mandel, M., and Higa, A. (1970): Calcium-dependent bacteriophage DNA infection. *J. Mol. Biol.,* 53:159–162.
17. Miller, J. (1972): *Experiments in Molecular Genetics.* Cold Spring Harbor Laboratory, Cold Spring Harbor, New York.
18. Morrow, J. F., Cohen, S. N., Chang, A. C. Y., Boyer, H. W., Goodman, H. M., and Helling, R. B. (1974): Replication and transcription of eukaryotic DNA in Escherichia coli. *Proc. Natl. Acad. Sci. U.S.A.,* 71:1743–1747.
19. Nathans, D., and Smith, H. O. (1975): Restriction endonucleases in the analysis and restructuring of DNA molecules. *Annu. Rev. Biochem.,* 44:273–293.
20. Roth, J. (1970): Genetic techniques in studies of bacterial metabolism. *Methods Enzymol.,* 17:3–35.
21. Schachat, F. H., and Hogness, D. S. (1973): Repetitive sequences in isolated Thomas circles from Drosophila melanogaster. *Cold Spring Harbor Symp. Quant. Biol.,* 38:371–381.
22. Somers, J. M., Amzallag, A., and Middleton, R. B. (1973): Genetic fine structure of the leucine operon of Escherichia coli. *J. Bacteriol.,* 113:1268–1272.
23. Struhl, K., Cameron, J. R., and Davis, R. W. (1976): Functional genetic expression of eukaryotic DNA in Escherichia coli. *Proc. Natl. Acad. Sci. U.S.A.,* 73:1471–1475.
24. Wensink, P. C., Finnegan, D. J., Donelson, J. E., and Hogness, D. S. (1974): A system for mapping DNA sequences in the chromosomes of Drosophila melanogaster. *Cell,* 3:315–325.

Recombinant Molecules: Impact on Science and Society, edited by R. F. Beers, Jr. and E. G. Bassett. Raven Press, New York © 1977.

34. Organization of Members Within the Repeating Families of the Genes Coding for Ribosomal RNA in *Xenopus laevis* and *Drosophila melanogaster*

*Peter K. Wellauer and Igor B. Dawid

Department of Embryology, Carnegie Institution of Washington, Baltimore, Maryland 21210

INTRODUCTION

This chapter reviews studies on the structural organization of the ribosomal genes in two animals, the frog *Xenopus* and the fly *Drosophila.* The experiments reported here demonstrate the usefulness as well as some limitations of the bacterial cloning procedure for analysis of the structural arrangement of sequences in repetitive gene families.

The ribosomal DNA (rDNA) of both animal species studied consists of repeating units of tandem genes coding for 28S and 18S ribosomal RNA (rRNA) alternating with spacers of unknown function (1,4,13,19,33,39). In *Xenopus* it has been shown that spacer sequences evolve rapidly (5) suggesting that the function of spacers does not depend on a specific nucleotide sequence. However, all the spacers within a *Xenopus* species are closely related to each other since they have evolved together. We describe experiments that have been designed to obtain detailed information about the structural organization of spacer sequences and to distinguish among models that have been proposed for "horizontal or tandem" evolution of genes and spacers in *Xenopus* rDNA. The two general mechanisms that will be considered for this "horizontal" evolution are: (a) *sudden correction mechanisms,* such as the "master-slave" hypothesis (7), and (b) *gradual correction mechanisms,* such as multiple unequal crossing-over events between homologues and/or sister chromatids (3,27). The present chapter starts from the observation that spacers in *Xenopus* rDNA display a limited length heterogeneity (36,37), and reviews some recent work on the molecular basis of spacer length heterogeneity, its arrangement in chromosomal

* Present address: Swiss Cancer Research Institute, 21 rue du Bugnon, 1011 Lausanne, Switzerland.

and amplified rDNA, and the manner in which spacer patterns are inherited in a single generation of frogs (22,35,38).

In the second part of this paper we review recent experiments on the sequence organization in *Drosophila* rDNA (34). Two kinds of length heterogeneity exist in this rDNA. Heterogeneity is caused either by insertions of different lengths of DNA within the 28S gene or by variable-length nontranscribed spacer regions. The location and arrangement of these different repeats within the nucleolus organizer of X and Y chromosomes has been studied (31). We discuss the implications of these observations for possible mechanisms of tandem evolution of gene families.

ORGANIZATION OF SEQUENCES WITHIN *XENOPUS LAEVIS* RIBOSOMAL DNA

Molecular Basis for Length Heterogeneity in rDNA

In this section we summarize recent experiments on the sequence organization within the nontranscribed spacer of *Xenopus* rDNA. Approximately 500 repeating units are arranged in tandem in the nucleolus organizer region of one of the 18 *X. laevis* chromosomes. Each repeating unit consists of a nontranscribed spacer region plus a gene region. The gene region is transcribed into a 40S precursor rRNA, which is then processed through a series of cleavage steps into mature 28S and 18S rRNA (11,33).

Digestion of *X. laevis* rDNA with the restriction endonuclease *Eco*RI of *E. coli* yields a number of bands as visualized by agarose gel electrophoresis (Fig. 1). Based on secondary structure mapping of *Eco*RI fragments, it was concluded that the enzyme makes only two cuts in each rDNA repeating unit; one cut is within the 18S gene, the other within the 28S gene (36,37). One 3.0×10^6 fragment plus one of the larger fragments (3.8 to 7.5×10^6) add up to one rDNA repeat (Fig. 2). The 3.0×10^6 fragment is homogeneous in size and contains exclusively transcribed sequences, whereas the heterogeneous larger fragments contain some gene sequences at each end plus all of the nontranscribed spacer (Fig. 2). The length variation of these larger fragments is caused by heterogeneity within the nontranscribed spacer. This has been demonstrated by examining heteroduplexed molecules formed between different spacer-containing *Eco*RI fragments.

To study the sequence arrangement in the nontranscribed spacer, we recombined several spacer-containing *Eco*RI fragments to a bacterial plasmid (pSC101) and then cloned them in *E. coli* strain HB101 by the procedure of Morrow et al. (20). Four cloned spacer-containing fragments of various sizes were chosen for analysis. Two of these with molecular weights of 3.9×10^6 (CD30) and 4.2×10^6 (CD42) were originally generated by Morrow and co-workers (20) in *E. coli* C600, which is a *recA*$^+$ strain, and were

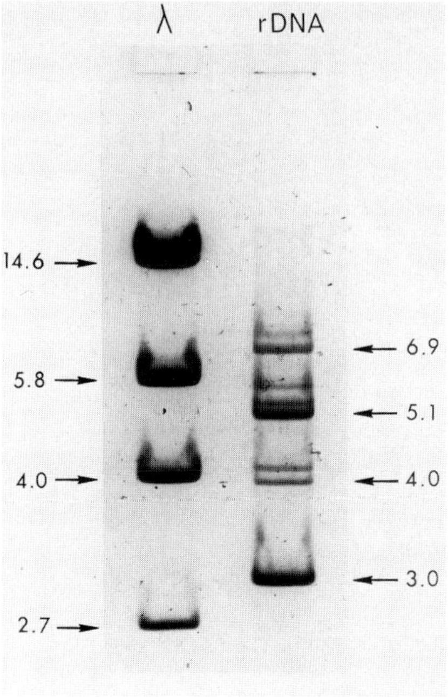

FIG. 1. Electrophoresis of *Eco*RI-digested amplified rDNA of *X. laevis*. The DNA was isolated from ovaries of young frogs; rDNA was purified on CsCl gradients, digested with *Eco*RI restriction endonuclease, run on a 1% agarose gel, and stained with ethidium bromide. In addition to the 3.0×10^6 fragments, at least seven bands of higher molecular weight fragments contain the nontranscribed spacer. Molecular weights $\times 10^{-6}$ are indicated by numbers and were determined by comparison with fragments generated by *Hin*dIII restriction of phage λ DNA.

transformed for the present study to the *recA*$^-$ strain HB101. The other two fragments with molecular weights of 5.4×10^6 (Xlr4) and 6.6×10^6 (Xlr5) were cloned directly in HB101 (35). The gel in Fig. 3 presents the results of the cloning experiment. Lanes a through d show the electrophoretic patterns of *Eco*RI digests of hybrid plasmids; intact hybrid plasmids are shown in lanes c through k. At the resolution of the gel, we conclude that no additional length heterogeneity is introduced during cloning. Control experiments showed that, even after prolonged replication of the Xlr plasmid series in *E. coli* C600 for up to 260 generations, no length variation of the rDNA fragments was observed. The molecular weights of the cloned *Eco*RI were determined from gels and by electron microscopy and are listed in Table 1.

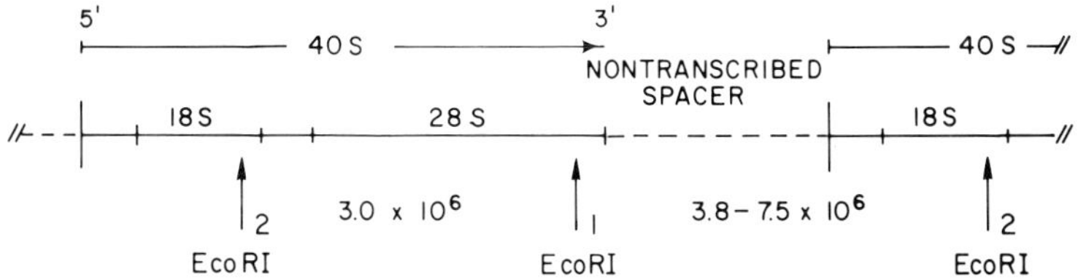

FIG. 2. Summary of the sequence organization in one repeating unit of *X. laevis* rDNA. The various regions described in the text are indicated, and the two *Eco*RI cleavage sites are labeled with arrows. The 5′ to 3′ polarity of the 40S precursor rRNA is based on recent experiments (11,23).

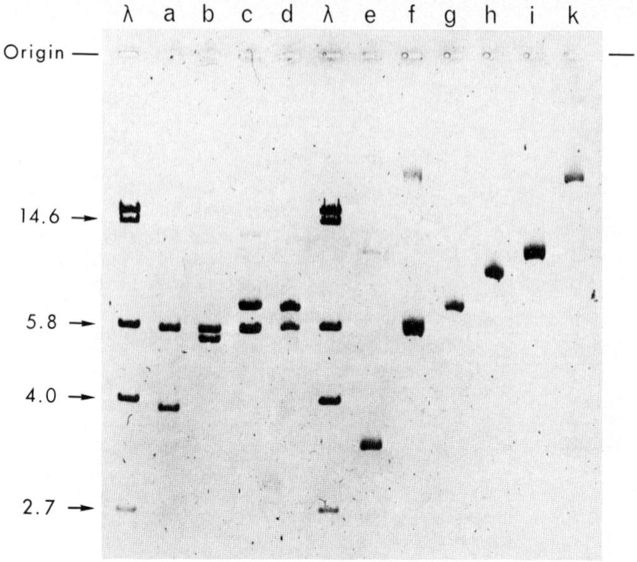

FIG. 3. Gel electrophoresis in 0.8% agarose gels of hybrid plasmids after digestion with *Eco*RI (a through d), and intact hybrid plasmids in closed-circular form (e through k). Fragments produced by *Hin*dIII restriction of phage λ DNA were used as molecular weight standards and their molecular weights $\times 10^{-6}$ are shown. Hybrid plasmids were obtained by ligation of *Eco*RI fragments of purified rDNA and pSC101 plasmid DNA and subsequent cloning in the presence of tetracycline in *recA*$^-$ strain HB101 of *E. coli* (20). In lanes a to d the band at 5.8×10^6 is pSC101 DNA. **a,** CD30; **b,** Xlr4; **c,** Xlr5; **d,** Xlr3 (obtained from plasmid Xlr3 which contains two tandemly linked 6.6×10^6 *Eco*RI fragments of rDNA). Circles: **e,** pSC101; **f,** CD18 (a plasmid containing the 3.0×10^6 rDNA fragment); **g,** CD30; **h,** pXlr4; **i,** pXlr5; **k,** pXlr3. In e and f, the slower moving bands are open-circular forms of the plasmid. The molecular weights of *Eco*RI rDNA fragments and intact hybrid plasmids are listed in Table 1.

These cloned homogeneous rDNA fragments were then analyzed by homoduplex analysis in the electron microscope (35). Whereas most of the reassociated molecules were perfectly matched DNA duplexes, some of the molecules contained two or more single-stranded deletion loops in the region of the nontranscribed spacer. Characteristic examples of such molecules are shown in Fig. 4a, b, e, and f. In all molecules with two loops, they are of equal size and the DNA duplex is shortened by the amount of DNA present in these loops. The frequency of imperfect homoduplexes (i.e., molecules with loops) was dependent on the stringency of the reassociation conditions used and the size of the fragment. In Xlr4, for example, 58% of the homoduplexes showed loops when reassociated at 60°C below the melting temperature (Tm), but only 18% were imperfect structures when reassociation occurred at 30°C below the Tm. The smaller fragments reformed perfect duplexes to a greater extent than did the larger fragments.

About 2% of all homoduplexes had only one loop. These molecules fell

TABLE 1. *Molecular weight oi rLNA fragments and plasmid DNAs*

	MW ± SD × 10^{-6}		
	*Eco*RI rDNA fragment		Intact plasmid DNA (circles)
DNA	Electron microscope[a]	Gel[b]	Gel[b]
CD18[c]	2.98 ± 0.05 (110)	3.0[d]	8.8
CD30	3.88 ± 0.11 (106)	3.8 ± 0.1	9.7
CD42	4.18 ± 0.15 (103)	4.2[d]	—
Xlr4	5.38 ± 0.11 (102)	5.4 ± 0.1	11.2
Xlr5	6.64 ± 0.17 (102)	6.6 ± 0.1	12.5

[a] MW and SD are given, with the number of molecules in parentheses. MWs were determined by using SV40 DNA (3.28 × 10^6) as length standard. In each experiment 60–90 molecules of the internal standard were measured.

[b] MWs were calculated by using *Hin*dIII fragments of λ DNA as a marker for linear DNAs (36). For less than three determinations no SD is given. For closed-circular DNAs pSC101 (5.8 × 10^6), *X. laevis* mitochondrial DNA (1.14 × 10^7) and pXlr3 were used as standards. The MW of pXlr3 was measured by electron microscope as 1.925 × 10^7.

[c] This fragment contains transcribed sequences only (see Fig. 2).

[d] From ref. 20.

into two length classes; in one of these the duplex was shorter than the original fragment by the length of the loop, whereas in the other one the duplex length was equal to the original length (Fig. 4c and d). These molecules, therefore, contain deletions or insertions. Because of their low frequency these molecules were not detected by the gel analysis of Fig. 3. Their rarity has also precluded further experiments on their origin for the present.

The occurrence of homoduplexes with two equally sized loops can be explained if regions of the nontranscribed spacer contain multiple repeats of a simple sequence. Figure 5b shows a simplified model to demonstrate how internally repetitious regions can reassociate out of register by looping-out of one subrepeat or multiples thereof. The model also predicts that loops will not occur necessarily at fixed positions but will migrate along the DNA axis throughout the repetitious regions. This phenomenon of loop migration is demonstrated more clearly in heteroduplexes between cloned spacer-containing fragments of different lengths. If two cloned rDNA fragments of different sizes had identical, unique sequences except for an extra and different DNA sequence which accounted for their difference, a heteroduplex between the two molecules should always result in a double-stranded region corresponding in length to the short strand, and a single-stranded deletion loop at a fixed position in the nontranscribed spacer region at the point where an additional unique sequence is present in the longer strand. However, heteroduplexes between cloned rDNA fragments contain multiple loops at variable positions. Representative electron micrographs of heteroduplexes between CD30 and Xlr4 (a and b) and CD30 Xlr5 (c and d) are shown in Fig. 6. The two ends of these molecules are always perfectly

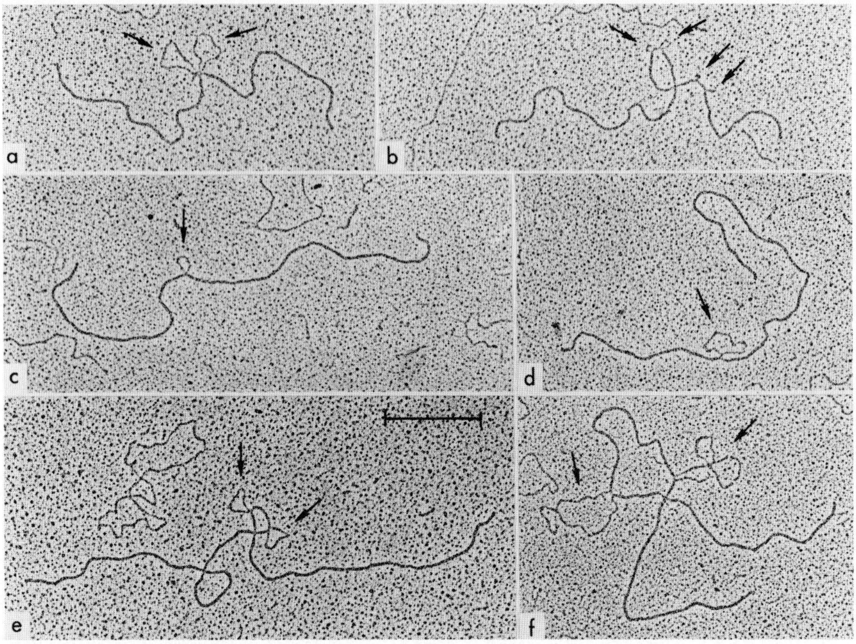

FIG. 4. Electron micrographs of homoduplex molecules of cloned rDNA fragments. To form homoduplexes, rDNA fragments were first denatured, then reannealed in formamide solutions at various temperatures, and were spread for electron microscopy by the formamide modification of the Kleinschmidt technique. To eliminate secondary structure in this G+C–rich rDNA, molecules were treated with glyoxal before spreading. **a** through **d:** CD30; **e, f:** Xlr4. **a, e, f:** Homoduplexes with two large loops of equal size. **b:** Molecule with 4 loops, 2 pairs of equal-sized loops. These molecules are formed by looping-out of repetitious regions (see text and Fig. 5). **c:** "Deletion" molecule; **d:** "insertion" molecule. The bar is 0.3 μm.

matched as one would expect, since they contain unique gene sequences. The long duplex region which is shown to the left in these heteroduplexes contains most of the 18S gene plus additional transcribed sequences (transcribed spacer). The shorter duplex region on the opposite end of the heteroduplex has 28S gene sequences.

The model in Fig. 5a demonstrates how variable loop patterns in heteroduplex molecules can be explained in theory. The model presents only a simplified situation for loop migration in a heteroduplex with a single loop. In practice, most molecules contain multiple loops of variable lengths and at variable positions. An additional complication is that both DNA strands may loop out as suggested by the homoduplex analysis described above.

A detailed statistical analysis of a large number of heteroduplex molecules formed between all the cloned rDNA fragments listed in Table 1 demonstrates that loops are confined to two regions of the nontranscribed spacer

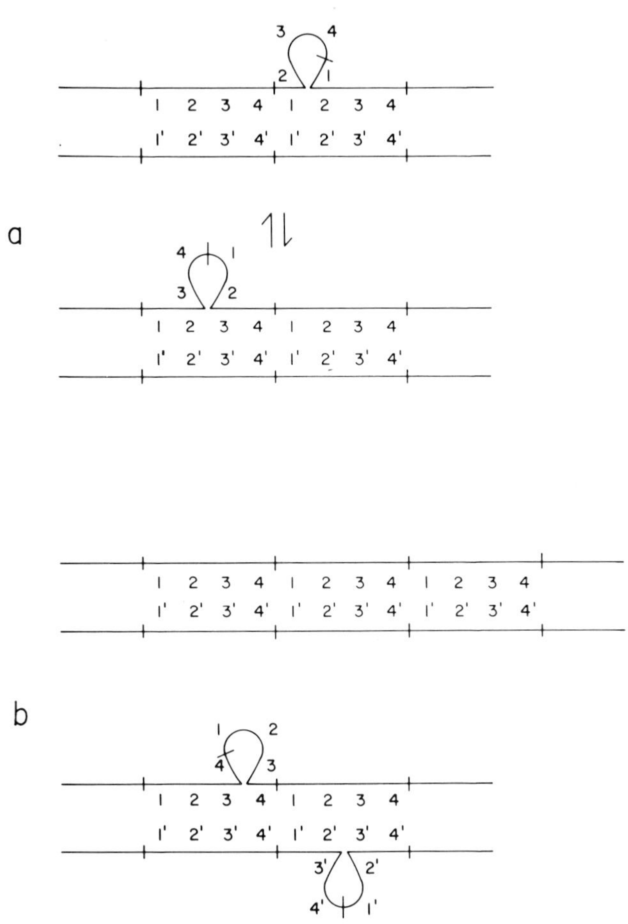

FIG. 5. Models of duplex molecules containing repetitious regions. The digits represent single bases or groups of bases that are not internally repetitious. **a:** Heteroduplex between one strand with 2 and one strand with 3 repetitious elements; the two forms have equal stability. **b:** Homoduplex molecules between strands with 3 repetitious elements. The "perfect" duplex is more stable than the molecule with loops. In order to allow in register pairing in the duplex region, it is necessary for two loops of equal size containing integral multiples of the subrepeat to be formed.

(35). The boundaries of these two repetitious regions (B and D in Fig. 7) are defined by the limits of loop migration within them. These repetitious regions are separated by a DNA segment without loops (region C in Fig. 7), and an additional presumably nonrepetitious region (A in Fig. 7) is located adjacent to the 3′ end of the transcription unit, i.e., the 28S gene sequence. Thus, length heterogeneity in the nontranscribed spacers is confined to two internally repetitious regions, B and D. Long nontranscribed spacers have more copies of subrepeats than do short spacers. In all four cloned rDNA

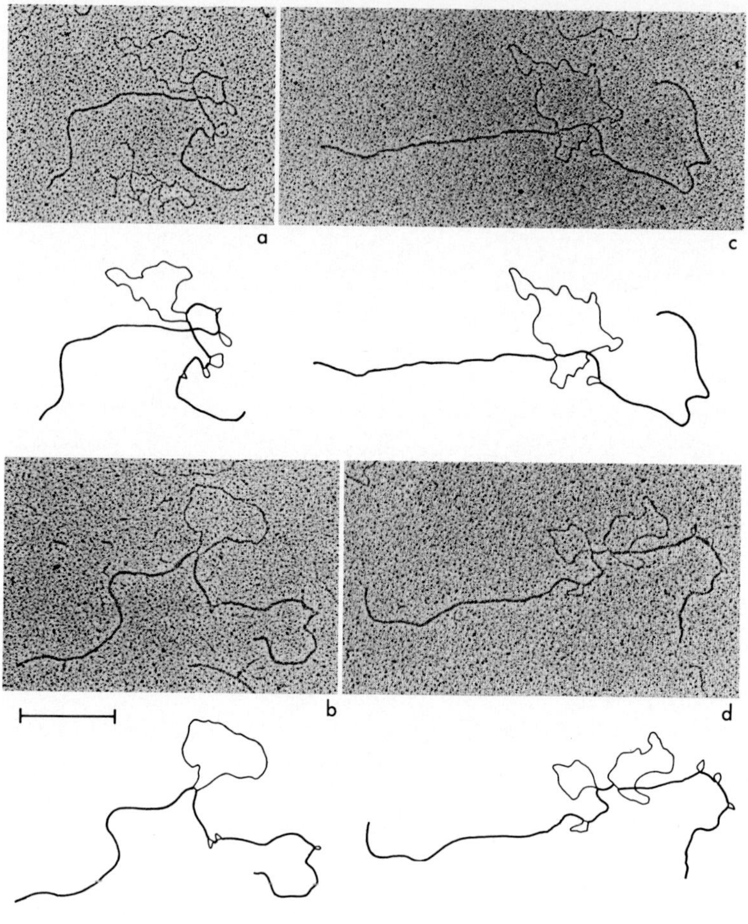

Fig. 6. Electron micrographs and tracings of heteroduplex molecules between CD30 and Xlr4 **(a, b)** and CD30 and Xlr5 **(c, d).** These molecules were prepared for electron microscopy as indicated in the legend to Fig. 4. The bar is 0.3 μm.

fragments, most of the spacer length variation is accounted for by varying amounts of subrepeats within repetitious region D (Fig. 7).

The presence of internal repeats in the nontranscribed spacer is supported by optical melting studies on cloned rDNA fragments (35). The thermal denaturation profiles of cloned rDNA fragments of different length differ in one region which melts with a very sharp transition. The fraction of hyperchromicity present in this transition increases with increasing length of the cloned fragments. Such hypersharp melting curves are characteristic for simple sequence DNAs such as satellite DNAs.

At present we do not know the length of one subrepeat in regions B and D. The smallest deletion loops that were detected by electron microscopy are about 50 base pairs long. This would indicate that the basic subrepeat is 50

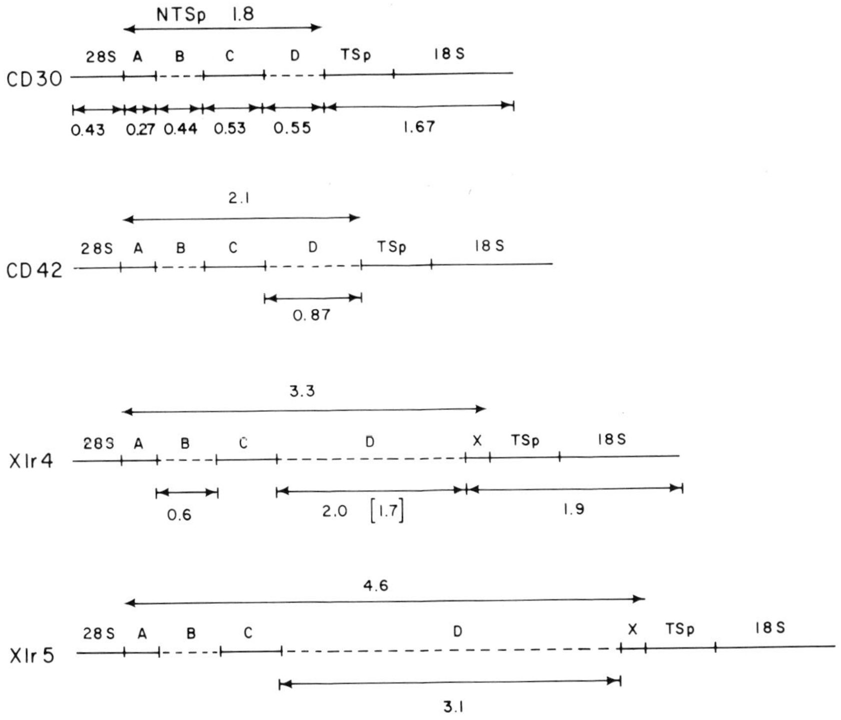

FIG. 7. Summary of sequence organization in 4 cloned rDNA fragments. The repetitious regions B and D are represented by dashed lines. All sizes are given as double-stranded molecular weights $\times 10^{-6}$ and are based on measurements of a large number of molecules. All molecular weights are shown in CD30. Molecular weights for regions in larger fragments are shown only if they differ from their counterpart in CD30. The sizes and boundaries of various regions were determined by measuring loop positions in these molecules. Note that size differences between cloned fragments are due to greater or fewer numbers of repetitious elements in region D.

base pairs or shorter. Recent experiments by Carroll and Brown (8) demonstrate that the genes coding for 5S RNA contain heterogeneous spacer lengths as well. The length variation in this family of repetitive genes is due to fewer or greater numbers of a basic subrepeat of 15 base pairs (6). Other repetitive genes that have been isolated, such as the genes coding for histone proteins (2,17,32) and those coding for transfer RNAs (10), are also associated with spacer sequences. To date it is not yet known whether these spacers contain repetitious elements.

ARRANGEMENT OF LENGTH HETEROGENEITY IN AMPLIFIED AND CHROMOSOMAL rDNA

Does length heterogeneity occur in individual frogs and within a single nucleolus organizer, and, if so, are repeating units of different lengths arranged in blocks of identical repeats or are they scrambled? The answer to this question will distinguish between "sudden" and "gradual" mechanisms for their evolution. Furthermore, the arrangement and variation of different size classes of spacers will be compared in chromosomal and amplified rDNA (38).

Comparative electrophoretic studies on *Eco*RI digests of chromosomal and amplified rDNA preparations pooled from a large number of frogs demonstrated that spacer length heterogeneity is present in both types of rDNA (36). A similar analysis of individual heterozygous frogs which carry the anucleolate mutation showed that length heterogeneity exists in a single nucleolus organizer (38). Different nucleolus organizers contain different size classes of spacer lengths ranging in molecular weight from about 1.8 to 5.5×10^6. From one to seven classes of spacer lengths have been observed in a single nucleolus organizer.

A comparison of the distribution of spacer lengths in amplified and chromosomal rDNA from individual frogs gave the following results (38). Amplified and chromosomal rDNA of the same frog contain the same classes of spacer lengths. This indicates that no additional heterogeneity is introduced during the amplification process. However, the relative abundance of certain spacer classes can be drastically different in the two rDNAs. Frequently, a class of spacer lengths which is abundant in chromosomal rDNA may be rare in amplified rDNA from the same animal or vice versa. This indicates that strong selection for certain repeat lengths can occur during amplification. Although the mechanism(s) that govern selective amplification are not understood, it is clear that selection must be determined genetically. Sibling frogs which have a single nucleolus organizer that was inherited from the same parental nucleolus organizer amplify their rDNA in the same way.

These observations raise the question whether the intramolecular arrangement of length heterogeneity is the same in amplified and chromosomal rDNA. The study of sequence organization in rDNA at higher levels points out the limitations of the cloning procedure as the only means of analysis. Because of the large size of the rDNA repeating units, it would be difficult and laborious to obtain a sufficient number of hybrid plasmids containing two or more tandemly linked repeats. However, combination of molecular cloning with isolation of high molecular weight rDNA by classic gradient centrifugation methods represents a powerful tool for such an analysis. Long single strands from two preparations each of amplified and chromosomal rDNA were hybridized with an excess of the cloned 3.9×10^6 spacer-containing rDNA fragment (CD30) and analyzed in the electron microscope (38). A large number of molecules contained two or more tandemly arranged heteroduplex regions which were separated by a segment of single-stranded DNA. This single-stranded stretch of DNA has a double-strand molecular weight of 3.0×10^6 and contains exclusively transcribed sequences (see Fig. 2 and Table 1). One heteroduplex region plus one single-stranded region add up to one rDNA repeating unit.

In order to determine whether adjacent repeats within amplified and chromosomal rDNA have identical or nonidentical spacer lengths, we carried out a statistical evaluation of nearest neighbor pairs. Since heteroduplexes between cloned rDNA fragments revealed that loop patterns vary considerably due to internal repetitiousness of the nontranscribed spacer, specific loop patterns would not be expected in adjacent repeats whether or not they were identical in size. Therefore, the analysis was based on length measurements of heteroduplex regions in neighboring repeats (38). The results of these experiments are listed in Table 2. The data show that, within the limits of this technique, 87 to 98% of nearest neighbor repeats are identical in size within one particular strand of amplified rDNA. On the other hand, only 32 to 50% of adjacent repeating units in one particular molecule of chromosomal rDNA are identical in size. From quantitative gel analysis of the two chromosomal rDNA preparations, we would have expected 30% identical nearest neighbor repeats for both samples if repeats of different size classes were arranged at random along the DNA molecule. Thus, in one of the chromosomal rDNA preparations analyzed, the arrangement of length heterogeneity is compatible with random assortment, whereas in the other one some clustering of identical spacer lengths may occur. Interspersion of different length classes is extensive in both samples.

A major conclusion from this experiment is that the arrangement of repeats with different lengths is fundamentally different along molecules of

TABLE 2. *Frequencies of adjacent repeating units of equal length in preparations of amplified and chromosomal rDNA of* X. laevis

Type	Source	No. of nearest neighbor repeat pairs measured	% identical neighbors
A amplified	Many animals	54	98
B amplified	Siblings; sample of two nucleolar organizers[a]	62	87
B chromosomal	Same as above	60	50
C chromosomal	Siblings; one nucleolar organizer[b]	82	32

[a] The blood (chromosomal rDNA) and ovaries (amplified rDNA) were pooled from several sibling animals that were progeny of a mating between two parents which were heterozygous for the anucleolate mutation. The rDNA of the sibling group is derived therefore from two nucleolar organizers only. The rDNA from individual frogs of this group showed one of two distinct patterns of *Eco*RI digestion products. The pattern of the pooled rDNA is a mixture of these two distinct patterns.

[b] Pooled rDNA from two sibling animals which were heterozygous for the anucleolate mutation. The rDNAs from these two frogs had identical fragment patterns after *Eco*RI digestion, and thus represent the same nucleolar organizer.

amplified and chromosomal rDNA. The fact that most, if not all, repeating units within one molecule of amplified rDNA are homogeneous in size supports the rolling circle mechanism for amplification (14,15,25). The data also suggest that predominantly single repeats are copied or excised from chromosomal rDNA during the initial amplification event. In contrast to amplified rDNA, chromosomal rDNA shows extensive scrambling of repeats with different lengths. Although some clustering may occur, and a higher order of regularity involving more than two repeats could not have been detected in this analysis, the data exclude *sudden correction mechanisms* in their extreme form as the means by which tandemly repeated sequences evolve together, since sudden correction predicts tandem homogeneity. The results favor *gradual correction mechanisms* which allow scrambling of heterogeneity between adjacent repeats (3,27). Carroll and Brown (9) have shown that adjacent repeating units in 5S DNA are frequently of different lengths. Observations on length heterogeneity and its arrangement in *Drosophila* rDNA are summarized below.

Inheritance of rDNA in *Xenopus*

Individual frogs differ in the length classes of their rDNA repeats. Therefore, the *Eco*RI pattern of a frog's rDNA can be used as a marker and may be followed in crosses. Such experiments have been carried out in order to test whether each cluster of rDNA repeats is transmitted as a unit and whether information on recombination between nucleolus organizers could be obtained. The results show that, in general, spacer patterns are inherited unchanged and each nucleolus organizer behaves as a mendelian unit (22). In two individual cases a change in the pattern of *Eco*RI fragments has been observed, but these changes could not be interpreted unambiguously in terms of recombination or any other model.

ORGANIZATION OF SEQUENCES WITHIN *Drosophila melanogaster* RIBOSOMAL DNA

Structural Differences in the Ribosomal Genes from X and Y Chromosomes

Generalizations about the structure and evolution of eukaryotic ribosomal genes can be based only on comparative studies with rDNAs from several organisms. The fly *Drosophila melanogaster* is a unique system in which to study eukaryotic rDNA organization since this organism has the advantage of many well-characterized mutants which affect the ribosomal locus and shows the interesting phenomena of rDNA compensation and magnification (30). There are 200 to 250 copies of tandem genes coding for rRNA in *Drosophila* (24,29). These genes are located within two different nucleolus organizers, one in the X chromosome and one in the Y chromosome.

Tartof and Dawid (31) reported that the ribosomal genes from the two ribosomal loci differ in structure. *Drosophila* rDNA was purified from embryos by either of two methods. The first method involves two cycles of centrifugation in cesium sulfate in the presence of mercury ions plus one centrifugation in cesium chloride containing actinomycin D. The second purification procedure is based on the observation by R. L. White and D. S. Hogness (*personal communication*) that, in high concentrations of formamide and at the proper temperature, RNA can displace one of the strands in duplex DNA and form a hybrid with its complementary sequence on the other DNA strand (R-loop). Since the three-stranded RNA/DNA hybrid is more dense than native DNA, it can be separated from duplex DNA by centrifugation in cesium sulfate. *Drosophila* rDNA prepared by this method was essentially pure after two rounds of centrifugation in cesium sulfate (34).

Purified rDNA analyzed by gel electrophoresis after digestion with the *Eco*RI restriction endonuclease yields a rather complex pattern of bands (31,34). This fact is illustrated by gel a in Fig. 8 of an *Eco*RI digest of rDNA from wild-type male and female flies (strain Ore-R). There are two

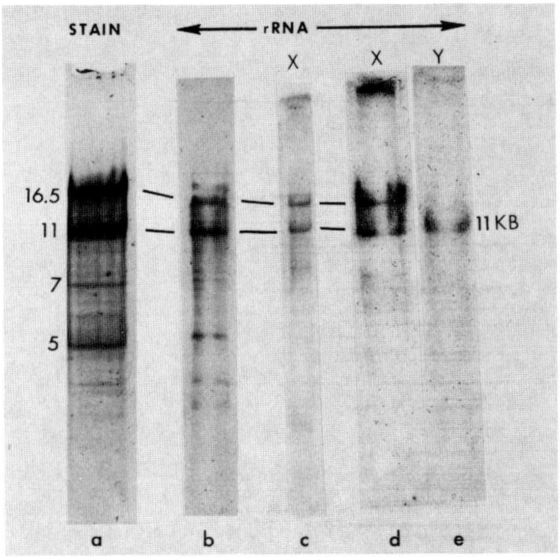

FIG. 8. Gel electrophoresis of *Drosophila* rDNA digested with *Eco*RI restriction endonuclease. Lane **a,** agarose gel of purified rDNA from wild-type flies stained with ethidium bromide; lane **b,** autoradiogram of purified wild-type rDNA *Eco*RI digest hybridized with labeled RNA. Lanes **c** and **d,** autoradiograms of rDNA from flies which carry a lethal *bobbed* mutation on their Y chromosome so that at least 95% of the rDNA is derived from the X chromosome; lane **e,** autoradiogram of rDNA from flies of the genotype sc^4-sc^8/Y in which the nucleolus organizer is deleted from the X chromosome. The lengths of rDNA fragments are given in kb and were determined by comparison to *Hin*dIII fragments of phage λ DNA.

prominent size classes of fragments with lengths of 16.5 and 11 kilobases or kilobase pairs (kb), two relatively abundant bands at 7 and 5 kb, plus a number of minor bands. If rDNA fragments of such a gel are transferred to nitrocellulose filters and hybridized with labeled rRNA (28), most of the bands hybridize as shown in the radioautogram of Fig. 8b. From mapping studies in the electron microscope (see below), we know that the 16.5- and 11-kb fragments are full-length rDNA repeating units. The fragments smaller than 11 kb are at least in part due to additional *Eco*RI cuts in some of the rDNA repeats (31,34). These fragments are ignored in the following discussion.

Purified rDNA from *Drosophila* females (X/X) yielded a pattern of *Eco*RI fragments indistinguishable from that of rDNA of wild-type flies (Fig. 8c and d), i.e., both the 16.5- and 11-kb repeating units are present in the X chromosome. In contrast, the *Eco*RI digest of rDNA from *Drosophila* males carrying only the Y chromosome nucleolus organizer (Fig. 8e) showed a characteristic difference in that 11-kb repeating units are present as the only major band, and 16.5-kb units are not detectable (31).

Electron Microscopy Reveals a Novel Form of Length Heterogeneity in rDNA Repeating Units of *Drosophila*

Drosophila rDNA yields two major classes of fragments after digestion with *Eco*RI restriction endonuclease. These fragments have been studied both after direct isolation and after cloning in *E. coli;* some of the minor *Eco*RI fragments were also analyzed (31,34; and Glover, White, and Hogness; Pellegrini, Manning, and Davidson, *personal communication*). Our subsequent account summarizes the results of Wellauer and Dawid (34).

Three classes of *Eco*RI fragments were chosen for analysis in the electron microscope, one containing 16.5-kb fragments, one with 11-kb fragments, and one containing all the fragments in between the 11- and 16.5-kb bands. The DNA was denatured and hybridized in solution with 28S and/or 18S rRNA to form RNA/DNA hybrids. Analysis of hybrid molecules between rRNA and single-strand 11-kb rDNA fragments showed that these molecules are full-length repeating units. The structural arrangement of sequences within the 11-kb repeats (B repeats) is shown in Fig. 9. The 18S and 28S rRNA gene regions are separated by a short spacer segment, which is presumed to be transcribed (Sp1), and by a longer spacer region, which must contain mostly nontranscribed sequences (Sp2). The DNA region hybridized with 28S rRNA has a short single-stranded gap of DNA, about 140 bases long, near the middle of the 28S gene sequence. It is known that 28S rRNA from many insects, including *Drosophila,* contains a "hidden break" since, upon denaturation, 28S rRNA migrates as a single component at about 18S in aqueous gels (16,26). Since this break is observed as a gap in RNA/DNA hybrids, it must be generated by removal of a short sequence

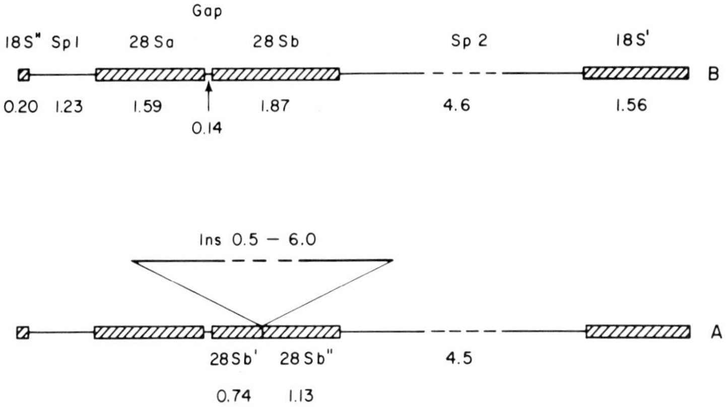

FIG. 9. Models of *Eco*RI fragments of *Drosophila* rDNA. These fragments are full-length repeating units. The *Eco*RI cleavage site is in the 18S RNA gene region. Sp1 is a spacer region which is most likely transcribed (transcribed spacer), whereas Sp2 contains mostly nontranscribed sequences (nontranscribed spacer). The gap in the 28S RNA hybrid region is the position where a "hidden break" occurs in 28S rRNA. Sp2 is shown as a broken line since length heterogeneity occurs in this region. Model B shows the 11-kb repeating unit; the numbers are the sizes of the different regions in kb. Model A shows the class of repeats which contain an insertion within the 28S gene region; only those regions are labeled which have not the same length as in model B. Insertions occur in size classes of 0.5 to 6.0 kb and are multiples of 0.5 kb (see text).

from 28S rRNA rather than by a single "nicking" event. Gel electrophoresis on denaturing formamide gels as well as electron microscopy reveal that the two 28S rRNA fragments are actually slightly different in size. The measurements on RNA/DNA hybrids indicate that the larger of the two RNA fragments hybridizes to a region adjacent to the nontranscribed spacer region in rDNA (Fig. 9B). The *Eco*RI restriction endonuclease cleaves *Drosophila* rDNA at a site within the 18S gene. This conclusion is based on the observations that in linear RNA/DNA hybrids a short single-stranded RNA tail extends beyond the 18S gene region (18S″ in Fig. 9) and that in a substantial number of hybrids this RNA tail hybridizes to its complementary DNA sequence at the other end of the DNA fragment to form circular molecules.

The RNA/DNA hybrids from the 16.5-kb and the intermediate size fraction all contain insertions of various sizes within the 28S gene region. The insertions, which range in size from about 0.5 to 6.0 kb and occur in distinct size classes (which are multiples of a basic unit of about 0.5 kb), account for the length difference between the 11-kb repeat and the larger repeating units in this rDNA (34). The structure of these molecules is summarized in the map of Fig. 9A. All insertions are located at the same site within the 28S rRNA gene region. This position is not the same as that at which the break occurs in 28S rRNA.

The observation that a large fraction of rDNA repeating units in *Drosophila* contains insertions within the 28S rRNA gene raises questions as to whether these repeats are transcriptionally active at any stage of development in this fly, and, if so, whether they yield functional transcription products. At present we cannot answer these questions.

Length Heterogeneity in the Nontranscribed Spacer of rDNA Repeats

The existence of insertions in the 28S gene of *Drosophila* rDNA constitutes a novel form of length heterogeneity which has not been observed in rDNA from other organisms. Length heterogeneity is also present in the nontranscribed spacer region of *Drosophila* rDNA and is similar to that originally detected in *Xenopus* rDNA. This is evident from examination of heteroduplex molecules of *Eco*RI rDNA fragments in the electron microscope (34). Heteroduplexes that were formed by self-reannealing of denatured 11-kb, 16.5-kb, or intermediate-size fragments often displayed one or several single-stranded deletion loops in the nontranscribed spacer region. Length heterogeneity is unambiguously demonstrated only in molecules with a single deletion loop since it is not known whether part of the spacer in *Drosophila* rDNA is internally repetitious as it is in *Xenopus* rDNA. Reassociation kinetics of a cloned rDNA fragment suggested that part of the DNA is internally repetitious (12). Most of the length differences are confined to a relatively narrow range of 0.5 to 1.0 kb, although a few spacers may differ in length by as much as 3.5 kb. Thus, length heterogeneity in *Drosophila* rDNA repeats is predominantly accounted for by insertions of various sizes in the 28S gene and to a lesser degree by different spacer lengths.

Intramolecular Arrangement of Length Heterogeneity in rDNA Repeating Units

Wellauer and Dawid (31) tested whether rDNA repeats without insertions and with insertions of different size occur in blocks or whether they are interspersed along the chromosome. High molecular weight rDNA was isolated from wild-type flies, denatured and hybridized to saturation with 28S and 18S rRNA, and RNA/DNA hybrids were analyzed in the electron microscope.

In 63 molecules containing 149 rDNA repeats, approximately one-third of all repeats (35.6%) were "standard" 11-kb units, another third (34.9%) were inserted repeats from the 16.5-kb class, and the rest (29.5%) consisted of intermediate-size repeats. Thus, approximately two-thirds of all repeating units in these flies contain insertions in the 28S gene. From gel analysis we know that the 11-kb repeating units in rDNA from wild-type flies originate from both the X and Y chromosome, whereas the 16.5-kb units are present in the X chromosome only (see page 391). However, in this particular stock of flies, the Y chromosome contains a partial deletion in the ribosomal locus so that only about 13% of all repeats are contributed by the Y chromosome to the total rDNA.

Nearest neighbor frequencies for different pairs of repeat lengths were determined from this analysis. From the relative abundance of various repeat lengths in total rDNA (see above), one can calculate the probabilities with which two repeats of identical lengths would be nearest neighbors and how often they would be adjacent to repeats of different lengths if they were arranged at random in this DNA. The results demonstrated that repeats of different lengths are interspersed, and that the nearest neighbor frequencies are close to those expected from a random assortment of the different length classes (34). Repeats of 11 kb may occur next to 11-kb, 16.5-kb, or intermediate-length repeats, and the same holds true for all classes of repeats. Again, a higher order of regularity in rDNA would not have been detected in this analysis, and one cannot determine whether the distribution of repeat lengths is fully random or not. However, it is clear that extensive scrambling does occur in *Drosophila* rDNA.

Cloning of *Drosophila* rDNA

The earliest report of a recombinant plasmid including *Drosophila* rDNA is by Glover and his colleagues (12). We have undertaken additional cloning experiments with the primary goal of a detailed comparison of rDNA from the X and Y chromosomes (I. B. Dawid, K. D. Tartof, and P. K. Wellauer, *unpublished observations*). For this purpose we started with partially purified *Eco*RI fragments of rDNA and linked them to the plasmid vector pMB9. From X chromosome rDNA we selected two clones of 16 to 17 kb and two clones of about 11 kb. Two 11-kb clones were obtained from the Y chromosome rDNA. In addition, a clone of 4 and 7.4 kb was obtained from the X, and a 4 and 6 kb clone from the Y rDNA. These latter results confirm earlier observations (31) that additional *Eco*RI sites are present in some repeats in both the X and Y nucleolus organizers.

In addition, we have selected a series of clones from wild-type rDNA. Different sizes were obtained reflecting the size distribution in the total population of rDNA *Eco*RI fragments. Using these cloned fragments, we have mapped the sites in rDNA for the restriction endonuclease *Sma*I. This enzyme has two sites within the insertion, allowing the isolation of pure insertion sequences. We are in the process of using the cloned material in a detailed comparison of X and Y rDNA sequences.

CONCLUSIONS

These observations complicate our understanding of the molecular correction mechanisms which operate during tandem evolution of repetitive genes for the following reason. In *Drosophila*, genetic exchange between the ribosomal loci of X and Y chromosomes is rare (21). The absence of 16.5-kb repeating units from the Y chromosome is compatible with the view that X-Y exchanges are too infrequent to affect the evolution of rDNA. If extensive recombination occurred between the X and Y chromosome genetic ribosomal loci, one would expect the 16.5-kb repeat to spread to the Y chromosome, unless a selective mechanism exists that prevents these repeats from being maintained on the Y chromosome. If there is indeed little recombination between X and Y nucleolus organizers, how are the structural similarities between them conserved during evolution? These similarities are extensive since not only the genes coding for rRNA are presumed to be identical (18), but also the nontranscribed spacer sequences of the 11-kb repeats appear to be closely related in rDNA from X and Y nucleolus organizers. Therefore, we must either postulate mechanisms that reconcile the observations on *Drosophila* rDNA with crossing-over between the X and Y nucleolus organizers or consider alternate models, such as the effect of selective pressure, to explain horizontal evolution of rDNA repeats.

ACKNOWLEDGMENTS

We thank Drs. R. H. Reeder, D. D. Brown, and K. D. Tartof for their contributions to the original work summarized in this chapter and for their critical reading of this manuscript. We also thank Drs. D. M. Glover, R. L. White, and D. S. Hogness for communication of results prior to publication. Peter K. Wellauer is a recipient of a research fellowship from the Cystic Fibrosis Foundation.

REFERENCES

1. Birnstiel, M., Speirs, J., Purdom, T., Jones, K., and Loening, U. E. (1968): Properties and composition of the isolated ribosomal DNA satellite of Xenopus laevis. *Nature,* 219:454–464.
2. Birnstiel, M., Telford, J., Weinberg, E., and Stafford, D. (1974): Isolation and some properties of the genes coding for histone proteins. *Proc. Natl. Acad. Sci. U.S.A.,* 71:2900–2904.
3. Brown, D. D., and Sugimoto, K. (1973): The structure and evolution of ribosomal and 5S DNAs in Xenopus laevis and Xenopus mulleri. *Cold Spring Harbor Symp. Quant. Biol.,* 38:501–505.
4. Brown, D. D., and Weber, C. S. (1968): Gene linkage by RNA/DNA hybridization. *J. Mol. Biol.,* 34:681–697.
5. Brown, D. D., Wensink, P. C., and Jordan, E. (1972): A comparison of the ribosomal DNAs of Xenopus laevis and Xenopus mulleri: the evolution of tandem genes. *J. Mol. Biol.,* 63:57–73.
6. Brownlee, G. G., Cartwright, E. M., and Brown, D. D. (1974): Sequence studies of the 5S DNA of Xenopus laevis. *J. Mol. Biol.,* 89:703–718.
7. Callan, H. G. (1967): The organization of genetic units in chromosomes. *J. Cell Sci.,* 2:1–7.
8. Carroll, D., and Brown, D. D. (1976*a*): Repeating units of Xenopus laevis oocyte-type 5S DNA are heterogenous in length. *Cell,* 7:467–475.
9. Carroll, D., and Brown, D. D. (1976*b*): Adjacent repeating units of Xenopus laevis 5S DNA can be heterogenous in length. *Cell,* 7:477–486.

10. Clarkson, S. G., Birnstiel, M. L., and Purdom, T. F. (1973): Clustering of transfer RNA genes of Xenopus laevis. *J. Mol. Biol.,* 79:411–429.
11. Dawid, I. B., and Wellauer, P. K. (1976): A reinvestigation of 5′-to-3′ polarity in 40S ribosomal RNA precursor of Xenopus laevis. *Cell,* 8:443–448.
12. Glover, D. M., White, R. L., Finnegan, D. J., and Hogness, D. S. (1975): Characterization of six cloned DNAs from Drosophila melanogaster including one that contains the genes for rRNA. *Cell,* 5:149–157.
13. Hamkalo, B. A., Miller, O. L., Jr., and Bakken, A. H. (1973): Ultrastructure of active eukaryotic genomes. *Cold Spring Harbor Symp. Quant. Biol.,* 38:915–919.
14. Hourcade, D., Dressler, D., and Wolfson, J. (1973*a*): The nucleolus and the rolling circle. *Cold Spring Harbor Symp. Quant. Biol.,* 38:537–550.
15. Hourcade, D., Dressler, D., and Wolfson, J. (1973*b*): The amplification of ribosomal RNA genes involves a rolling circle intermediate. *Proc. Natl. Acad. Sci. U.S.A.,* 70:2926–2930.
16. Jordan, B. R., Jourdan, R., and Jacq, B. (1976): Late steps in the maturation of Drosophila 26S ribosomal RNA: generation of 5.8S and 2S RNAs by cleavages occurring in the cytoplasm. *J. Mol. Biol.,* 101:85–105.
17. Kedes, L. H., Cohn, R. H., Lowry, J. C., Chang, A. C. Y., and Cohen, S. N. (1975): The organization of sea urchin histone genes. *Cell,* 6:359–369.
18. Maden, B. E. H., and Tartof, K. (1974): Nature of the ribosomal RNA transcribed from the X and Y chromosomes of Drosophila melanogaster. *J. Mol. Biol.,* 90:51–64.
19. Miller, O. L., Jr., and Beatty, B. R. (1969): Extrachromosomal nucleolar genes in amphibian oocytes. *Genetics Suppl.,* 61:133–143.
20. Morrow, J. F., Cohen, S. N., Chang, A. C. Y., Boyer, H. W., Goodman, H. M., and Helling, R. B. (1974): Replication and transcription of eukaryotic DNA in Escherichia coli. *Proc. Natl. Acad. Sci. U.S.A.,* 71:1743–1747.
21. Palumbo, G., Caizzi, R., and Ritossa, F. (1973): Relative orientation with respect to the centromere of ribosomal RNA genes of the X and Y chromosomes of Drosophila melanogaster. *Proc. Natl. Acad. Sci. U.S.A.,* 70:1883–1885.
22. Reeder, R. H., Brown, D. D., Wellauer, P. K., and Dawid, I. B. (1976): Patterns of ribosomal DNA spacer lengths are inherited. *J. Mol. Biol.,* 105:507–516.
23. Reeder, R. H., Higashinakagawa, T., and Miller, O. L., Jr. (1976): The 5′-to-3′ polarity of the Xenopus ribosomal RNA precursor molecule. *Cell,* 8:449–454.
24. Ritossa, F. M., Atwood, K. C., Lindsley, D. L., and Spiegelman, S. (1966): On the chromosomal distribution of DNA complementary to ribosomal and soluble RNA. *Natl. Cancer Inst. Monogr.,* 23:449–472.
25. Rochaix, J.-D., Bird, A., and Bakken, S. (1974): Ribosomal RNA gene amplification by rolling circles. *J. Mol. Biol.,* 87:473–487.
26. Shine, J., and Dalgarno, L. (1973): Occurrence of heat-dissociable ribosomal RNA in insects: the presence of three polynucleotide chains in 26S RNA from cultured Aedes aegypti cells. *J. Mol. Biol.,* 75:57–72.
27. Smith, G. P. (1973): Unequal crossover and the evolution of multigene families. *Cold Spring Harbor Symp. Quant. Biol.,* 38:507–513.
28. Southern, E. M. (1975): Detection of specific sequences among DNA fragments separated by gel electrophoresis. *J. Mol. Biol.,* 98:503–517.
29. Tartof, K. D. (1971): Increasing the multiplicity of ribosomal RNA genes in Drosophila melanogaster. *Science,* 171:294–297.
30. Tartof, K. D. (1973): Unequal mitotic sister chromatid exchange and disproportionate replication as mechanisms regulating ribosomal RNA gene redundancy. *Cold Spring Harbor Symp. Quant. Biol.,* 38:491–500.
31. Tartof, K. D., and Dawid, I. B. (1976): Similarities and differences in the structure of X and Y chromosome rRNA genes of Drosophila. *Nature,* 263:27–30.
32. Weinberg, E. S., Overton, G. C., Shutt, R. H., and Reeder, R. H. (1975): Histone gene arrangements in the sea urchin, Strongylocentrotus purpuratus. *Proc. Natl. Acad. Sci. U.S.A.,* 72:4815–4819.
33. Wellauer, P. K., and Dawid, I. B. (1974): Secondary structure maps of ribosomal RNA and DNA. *J. Mol. Biol.,* 89:379–395.
34. Wellauer, P. K., and Dawid, I. B. (1977): Structural organization of ribosomal DNA in Drosophila melanogaster. *Cell,* 10:193–212.
35. Wellauer, P. K., Dawid, I. B., Brown, D. D., and Reeder, R. H. (1976): The molecular

basis for length heterogeneity in ribosomal DNA from Xenopus laevis. *J. Mol. Biol.* 105:461–486.
36. Wellauer, P. K., and Reeder, R. H. (1975): A comparison of the structural organization of amplified ribosomal DNA from Xenopus mulleri and Xenopus laevis. *J. Mol. Biol.*, 94:151–161.
37. Wellauer, P. K., Reeder, R. H., Carroll, D., Brown, D. D., Deutch, A., Higashinakagawa, T., and Dawid, I. B. (1974): Amplified ribosomal DNA from Xenopus laevis has heterogenous spacer lengths. *Proc. Natl. Acad. Sci. U.S.A.*, 71:2823–2827.
38. Wellauer, P. K., Reeder, R. H., Dawid, I. B., and Brown, D. D. (1976): The arrangement of length heterogeneity in repeating units of amplified and chromosomal ribosomal DNA from Xenopus laevis. *J. Mol. Biol.*, 105:487–505.
39. Wensink, P. C., and Brown, D. D. (1971): Denaturation map of the ribosomal DNA of Xenopus laevis. *J. Mol. Biol.*, 60:235–247.

Recombinant Molecules: Impact on Science and Society, edited by R. F. Beers, Jr. and E. G. Bassett. Raven Press, New York © 1977.

35. The Application of Recombinant DNA Cloning for the Analysis of Sea Urchin (*Strongylocentrotus purpuratus*) Histone Genes

Laurence H. Kedes

Department of Medicine, Division of Hematology, Stanford University School of Medicine and Veterans Administration Hospital, Palo Alto, California 94304

INTRODUCTION

A major key to our understanding of genetic regulation in animal cells must come from knowledge gained through molecular analyses of specific genes. Heretofore, the only expressed animal cell genes which have been amenable to such detailed analyses have been repetitive DNA sequences coding for structural RNA constituents of the cell. For example, the experiments of Birnstiel (1,2,24) and Brown (5,6) and their co-workers led to the purification and extensive analysis of the ribosomal cistrons and advanced considerably our understanding of the organization of these repetitive, clustered genes. Genes coding for proteins, on the other hand, have been less amenable to purification and direct examination. The major reason for this is that most animal cell genes coding for structural proteins are single copies and represent such a low fraction of the genome that their isolation becomes akin to looking for the proverbial needle in the haystack.

The advent of molecular cloning techniques (8,9,20) has presented those interested in the regulation of eukaryotic protein synthesizing genes with a technological breakthrough allowing such analyses. Other contributors to this volume have presented in great detail the methodology used to clone fragments of foreign DNA, including eukaryotic DNA, into bacterial episomes or phages. Thus, we now can isolate and analyze specific segments of eukaryotic coding sequences.

My colleagues and I have been interested in the organization of the eukaryotic histone genes and have taken advantage of the molecular cloning technology to study the organization of these interesting DNA sequences. Sea urchin histone genes are known to be repetitive and clustered (15). This is probably a feature of histone genes of most other eukaryotes as well (14). Since there are at least five coding sequences for the five known histone proteins, one of our first goals was to determine whether the genes for the five proteins were interspersed in one repeat unit of the DNA or arranged as blocks of tandem sequences coding for a single histone adjacent to a second block of repetitive sequences coding for a second histone, etc. Our interest in determining the organization of the sea urchin histone genes was also aroused because of their unique and peculiar characteristics relative to the cell cycle and to the production of histone proteins: histones and histone messenger RNAs are synthesized only during "s" phase of the cell cycle (3,4). In addition, four of the histone proteins accumulated in precise, stoichiometrically equivalent amounts in animal cell chromatin (18,21). The physical organization of sea urchin histone genes must be compatible with these biological observations, and a careful analysis of histone DNA may even shed light on the mechanisms whereby histone genes are so precisely regulated both in cell cycle and during development.

RESULTS AND DISCUSSION

Cloning of Histone Genes

We devised a strategy (16) to isolate sea urchin histone genes by cloning them into the *E. coli* plasmid pSC101 (9). We were able to detect sea urchin histone DNA sequences by molecular hybridization with histone messenger RNA. We reasoned that if we inserted restriction endonuclease fragments of sea urchin DNA into the *Eco*RI site of pSC101, then the fraction of chimeric plasmids containing sea urchin histone genes should approximate the percentage of the sea urchin genome containing histone DNA sequences. There are several hundred copies of histone genes per haploid genome in the sea urchin, although the number varies from species to species. In the sea urchin *Strongylocentrotus purpuratus,* approximately 0.25% of the genome represents histone genes (11). Accordingly, we estimated that 1 in 400 chimeric plasmids will contain histone DNA. In our initial experiments, about 10% of the transforming DNA molecules were chimeras; the remainder were recircularized pSC101. Thus, in our initial transformation only one in several thousand (i.e., 1 in 4,000) bacteria was expected to contain histone DNA.

In order to select bacteria containing sea urchin histone genes, we developed a method based on the subculture cloning technique originally described by Cavalli-Sforza and Lederberg (7). In this technique, small numbers of colonies of bacteria were pooled and a large number of such pools analyzed separately for the presence of the appropriate DNA sequence by molecular hybridization with radioactive histone messenger RNA. Consider, for example, an initial culture of *E. coli* containing 1 bacterium with histone genes for every 4,000 without. If we set up 10 cultures, each containing 400 bacteria as measured by dilution, only 1 bacterium out of the entire 4,000 inoculated would contain histone DNA. (This example is obviously simplified and does not take into account the statistical probabilities that of 4,000 bacteria several, or none, might contain histone genes.) When each of these 10 cultures was allowed to grow and aliquots were sampled, only 1 of the 10 cultures would be expected to harbor bacteria containing histone DNA. And in that culture, 1 in 400 bacteria would contain histone genes, a 10-fold enrichment. If we then set up 10 new cultures inoculated with 40 bacteria from the enriched culture and repeated the growth, sampling, and assays, 1 of the 10 cultures would be expected to harbor histone gene containing *E. coli*. This would represent an additional 10-fold enrichment. If from this positive culture 40 or more individual colonies are grown and tested, a histone gene containing *E. coli* plasmid should be found. In this way we have isolated several independent clones of *E. coli* containing *Eco*RI-restricted *S. purpuratus* histone genes.

Characterization of Histone Genes

Only two kinds of *Eco*RI eukaryotic histone fragments have been isolated to date from the sea urchin *S. purpuratus* (16). One is 2.0 kilobases long and the other is 4.6 kilobases long (Fig. 1). These two fragments have been shown to be adjacent segments of a 6.54-kilobase, histone gene repeat unit of *S. purpuratus* (17). We have determined the precise head-to-tail arrangement of these two fragments in native histone genes by restriction endonuclease mapping of total sea urchin DNA (17). Cleaved DNA was separated on agarose gels and transferred to nitrocellulose filter paper strips by blotting (10). ^{32}P-labeled histone genes were used as hybridization probes with the filter-bound total sea urchin DNA. Only DNA corresponding to predicted size classes had hybridized. These results not only oriented the two cloned fragments relative to each other but also demonstrated that (a) the histone gene repeat units have no detectable length heterogeneity and (b) the spacer sequences present in the DNA are not detectably shared with any other sea urchin genes. We have found that between them the two eukaroytic fragments contain sequences which hybridize to each of the five histone messenger RNAs (17). We have concluded therefore, and with certainty, that the histone genes are arranged in an interspersed manner in the repeat unit.

The Order of the Five Coding Sequences

A large number of restriction endonuclease–cleaved fragments have been mapped on the two eukaryotic segments. Three such fragments have been cloned into other plasmid vectors capable of carrying *Hin*dIII and *Bam*I restriction pieces (10). This subset of cloned fragments, as well as other restriction fragments isolated from agarose gels by electrophoresis, have been isolated, mapped, and used to determine the order of the various

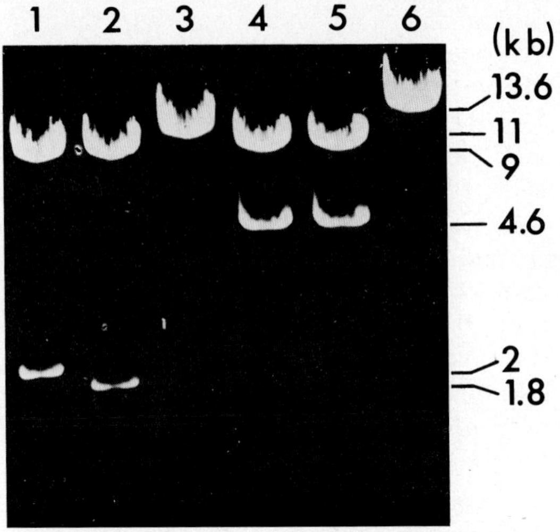

FIG. 1. Agarose gel electrophoresis of restriction endonuclease–cleaved chimeric plasmids containing the two *Eco*RI segments of *S. purpuratus* histone genes. Covalently closed, circular DNA from *E. coli* containing the plasmids pSp17 and pSp2 was isolated by ethidium bromide-CsCl equilibrium centrifugation as described previously (16). The DNAs were digested with either or both *Eco*RI and *Hin*dIII and the digests electrophoresed on 0.8% agarose slab gels, stained with ethidium bromide and photographed with long-wave UV transillumination as described elsewhere (17). **1:** pSp17 cleaved with *Eco*RI. **2:** pSp17 cleaved with *Hin*dIII plus *Eco* R1; 0.2-kilobase (kb) and 0.03-kb fragments are also liberated (17) but are not visible on this gel. **3:** pSp17 cleaved with *Hin*dIII; a 0.2-kb fragment is also liberated (17) but is not visible on this gel. **4:** pSp2 cleaved with *Eco*RI. **5:** pSp2 cleaved with *Eco*RI plus *Hin*dIII. **6:** pSp2 cleaved with *Hin*dIII alone; fragment lengths are given in kilobase pairs (kb). (From ref. 17.)

histone messenger RNA coding regions on the two cloned fragments. We have taken two approaches: we have used purified histone messenger RNAs to hybridize to one or another of the isolated fragments. Alternatively, we have used the isolated DNA fragments bound to nitrocellulose filters to pull out specific messenger RNA sequences by hybridization with total sea urchin RNA populations. Such hybridization is highly specific and selective. As seen in Fig. 2, the hybridized RNA can be eluted from the filters intact, analyzed on acrylamide gels alongside the starting material, and a specific hybridizing mRNA identified accordingly. In this way we have determined that the relative order of the histone-coding regions is histone H1, histone H4, histone H2B, histone H3, and histone H2A. The H1 and H4 coding specificities have been determined by *in vitro* translation of specific mRNAs (19). The H2B coding specificity was determined by the synthesis *in vitro* of H2B-like protein after using a specific DNA fragment as template in a linked transcription-translation system (R. Mulligan, B. Roberts, and J. Lazar, *unpublished observations*). The H2A sequence specificity has been determined by DNA sequence analysis of a coding region with nucleo-

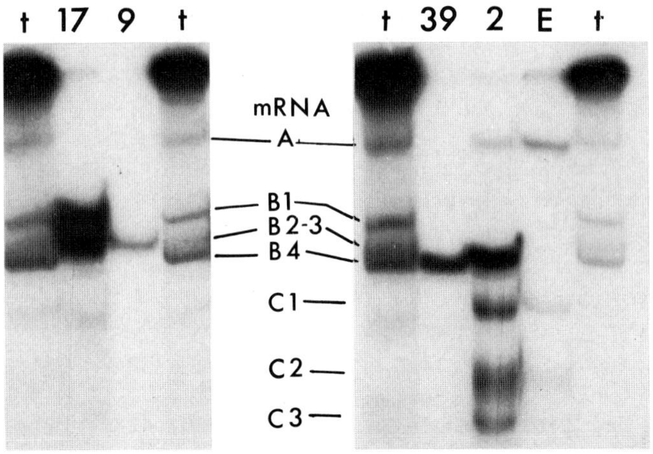

FIG. 2. Polyacrylamide gel electrophoresis of sea urchin (*S. purpuratus*) polysomal mRNAs which hybridize to various fragments of histone genes. Five DNA segments of the histone genes were isolated by cloning in *E. coli* plasmids or by separation of restriction endonuclease–cleaved DNAs on agarose gels. The DNAs were denatured, bound to nitrocellulose filters, and hybridized with total ^{3}H-uridine-labeled polysomal RNA from cleavage stage sea urchin embryos. The filters were washed, and the hybridized RNAs were eluted and electrophoresed on 6% polyacrylamide-SDS gels along with the total polysomal RNA as marker. The gel was dried and autofluorographed. Experimental details are given in Cohn et al. (10). The RNA bands are identified as described previously (17). Left to right, **t:** total polysomal RNA; **17:** RNA hybridizing to the plasmid with the 2.0-kb eukaryotic fragment (Fig. 1) contains only two major bands, B1 and B2–3. (The autofluorograph is overexposed; shorter exposures show the two bands clearly and especially that B4 is not present.) **9:** RNA hybridizing to a plasmid containing only a 0.2-kb fragment of pSp17 eukaryotic DNA; only RNA co-migrating with B2–3 hybridizes. **39:** RNA hybridizing to a plasmid containing a 1-kb fragment of pSp2 eukaryotic DNA; only RNA co-migrating with B4 hybridizes. **2:** RNA hybridizing to the plasmid containing the 4.6-kb eukaryotic fragment (Fig. 1) contains three mRNAs: A (H1) B4 (H2B) and the three isocoding C RNAs (H4). **E:** A 3.58-kb eukaryotic fragment of pSp2 DNA, isolated from agarose gel electrophoresis of *Bam*I digested pSp2, hybridizes two mRNAs: A and the three isocoding C RNAs. The mapping positions of these plasmids are depicted in Fig. 4. Fragment E is all of the pSp2 eukaryotic fragment excluding pRC39 DNA. (From ref. 10.)

tides matching the expected codons of 23 amino acids of the H2A histone (22). The fifth coding sequence was assigned to H3 by inference.

Strandedness and Polarity of the Histone Genes

We have also been able to determine that all five sequences are on the same coding strand and their 5′–3′ orientation. In order to obtain these results, we made the plasmids containing the two kinds of eukaryotic fragments into linear molecules by restriction endonuclease cleavage near one of the junctions between the eukaryotic and prokaryotic segments. We used the *E. coli* enzyme exonuclease III, which digests from 3′ to 5′ and attacks only double-stranded DNA, to determine the orientation of the coding regions. Before exonuclease digestion, the linear duplex DNA molecules were incapable of hybridizing with histone messenger RNA as might be expected. After digestion, both sets of molecules were able to hybridize extensively with sea urchin histone messenger RNA, suggesting that the noncoding strand had been digested and that histone genes, at the 5′ end of the linear molecule, had been made accessible for hybridization. This was confirmed by an important control experiment in which an identical digestion was carried out on a linear plasmid which had the histone DNA at one end but inserted in the reverse orientation. In this case, digestion with exonuclease III would destroy the opposite DNA strand. Such digestion was found to completely eliminate the ability of the DNA to hybridize with histone messenger RNA. Our conclusion is that the histone genes are transcribed in the order listed in the preceding paragraph. We do not yet know whether there is one or several RNA polymerase initiation sites for these genes.

Spacer Sequences Are Interspersed with Coding Sequences

In collaboration with David Holmes, Madeline Wu, and Norman Davidson of the California Institute of Technology, we have approached the problem of histone gene organization in an entirely different way and have been able to extend these observations quite graphically. Histone messenger RNA has been hybridized to the eukaryotic plasmids or purified native histone genes and examined by electron microscopy (13,26). The DNA-RNA hybrids were treated with the gene 32 protein from T4 phage, which definitively distinguishes single-stranded and double-stranded regions (25). The cloned genes were seen to contain five hybridized regions of different lengths intermingled with five noncoding regions (spacer DNA) of varying lengths. Several adjacent repeat sets have been visualized by the gene 32 technique on native DNA. Figure 3, an electron micrograph prepared by David Holmes, shows a DNA strand containing slightly more than two repeat units. The location of the five coding regions on the same strand and the identification of several of them based on the lengths of the histone messenger RNAs (11,12) are in complete agreement with the data obtained from the biochemical hybridization studies (10,17). When the messenger RNA was hybridized to the plasmids under conditions in which the plasmid DNA was not completely denatured but RNA:DNA hybridization was favored, displacement of the noncoding DNA strand created an electron microscopic "R loop" (23). The data obtained by such R loop analysis demonstrated five coding regions, nearly equal in length to those observed by the gene 32 technique, interspersed with noncoding regions (13).

A Map of the Histone Gene Repeat Unit

The combined data of the electron microscope analyses and the RNA: DNA biochemical hybridizations have allowed us to construct a detailed

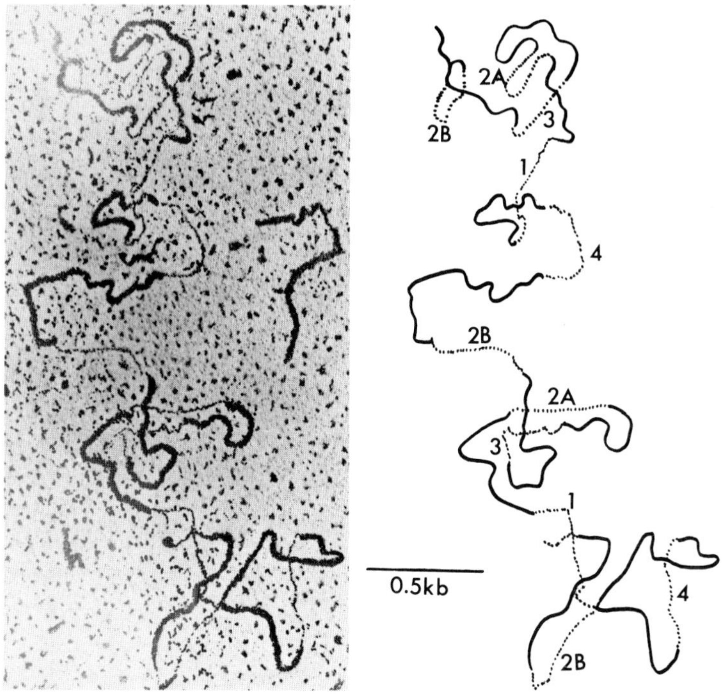

FIG. 3. Electron micrograph of native sea urchin histone DNA hybridized with histone mRNAs. The photograph (kindly provided by Dr. David Holmes) shows a 14- to 15-kb length of single-stranded histone DNA [isolated by actinomycin D-CsCl centrifugation (17)], hybridized with 9S-sized, cleavage stage sea urchin polysomal RNAs (19), stained by the gene 32-ethidium bromide procedure (25), and spread for electron microscope analysis. The staining method shows single-stranded DNA as thick segments and the hybridized regions as thin segments. The location of the genes is shown in the tracing. The identification of genes is based on data derived from similar studies on cloned histone gene fragments (13,26). The line on the tracing represents approximately 500 bases.

map of histone gene organization in *S. purpuratus* (Fig. 4). The general features of this map are that the five histone genes are present once per repeat and are interspersed with noncoding spacer DNA. All five coding regions are on the same DNA strand and their 5′-3′ orientation has been determined and is indicated on the map.

One obvious way to look for regions of interest in eukaryotic DNA segments would be to examine the primary DNA sequence. We have begun such sequencing experiments and one result is that the DNA region coding for messenger RNA B2–3 has been found to contain nucleotide sequences which would code for histone H2A (22). Such sequencing data, coupled with messenger RNA sequencing data for *S. purpuratus* H4 template (12), should lay to rest any doubt that the cloned DNA fragments are indeed the genes for histones.

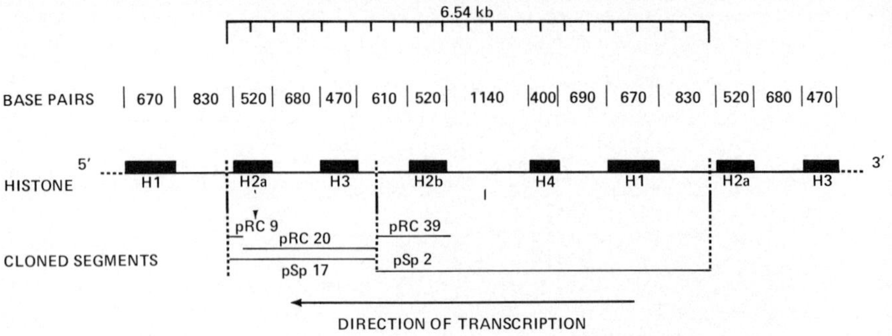

FIG. 4. A map of a histone gene repeat unit of the sea urchin *S. purpuratus*. The data used to arrive at this map are presented in detail elsewhere (10,13,14,17,26). Fragments of the repeat unit cloned in plasmids are also shown. All lengths are in kilobase (kb) pairs. (Adapted from ref. 10.)

It would be easy to speculate, based on the structure of histone genes elucidated here, how that structure might accommodate the stringent regulation of histone protein synthesis exhibited by eukaryotic cells. However, no data directly bearing on those points are yet available.

ACKNOWLEDGMENTS

Many of the experiments described in this paper were done in collaboration with Ron Cohn and Jean Lowry, Annie Chang and Stanley Cohen, and David Holmes, Madeline Wu, and Norman Davidson. This work was supported in part by grants from the NIH and the Veterans Administration. The author is an Investigator of the Howard Hughes Medical Institute.

REFERENCES

1. Birnstiel, M. L., Speirs, J., Purdom, I., Jones, K., and Loening, U. E. (1968): Properties and composition of the isolated ribosomal DNA satellite of Xenopus laevis. *Nature*, 219:454–463.
2. Birnstiel, M. L., Wallace, H., Sirlin, J. L., and Fischberg, M. (1966): Localization of the ribosomal DNA complements in the nucleolar organizer region of Xenopus laevis. *Natl. Cancer Inst. Monogr.*, 23:431–447.
3. Borun, T. W., Scharff, M. D., and Robbins, E. (1967): Rapidly labeled, polyribosome-associated RNA having the properties of histone messenger. *Proc. Natl. Acad. Sci. U.S.A.*, 58:1977–1983.
4. Breindl, M., and Gallwitz, D. (1974): On the translational control of histone synthesis. *Eur. J. Biochem.*, 45:91–97.
5. Brown, D. D., and Weber, C. S. (1968*a*): Gene linkage by RNA-DNA hybridization. I. Unique DNA sequences homologous to 4s RNA, 5s RNA and ribosomal RNA. *J. Mol. Biol.*, 34:661–680.
6. Brown, D. D., and Weber, C. S. (1968*b*): II. Arrangement of the redundant gene sequences for 28s and 18s ribosomal RNA. *J. Mol. Biol.*, 34:681–697.
7. Cavalli-Sforza, L. L., and Lederberg, J. (1956): Isolation of pre-adaptive mutants in bacteria by sib selection. *Genetics*, 41:367–381.

8. Cohen, S. N., and Chang, A. C. Y. (1974): A method for selective cloning of eukaryotic DNA fragments in Escherichia coli by repeated transformation. *Mol. Gen. Genet.,* 134:133–141.
9. Cohen, S. N., Chang, A. C. Y., Boyer, H. W., and Helling, R. (1973): Construction of biologically functional bacterial plasmids in vitro. *Proc. Natl. Acad. Sci. U.S.A.,* 70:3240–3244.
10. Cohn, R. H., Lowry, J. C., and Kedes, L. H. (1976): Histone genes of the sea urchin (S. purpuratus) cloned in E. coli: Order, polarity and strandedness of the five histone coding and spacer regions. *Cell,* 9:147–161.
11. Grunstein, M., and Schedl, P. (1976): Isolation and sequence analysis of sea urchin H4 ($f2a_1$) messenger RNA. *J. Mol. Biol.,* 104:323–350.
12. Grunstein, M., Schedl, P., and Kedes, L. H. (1976): Sequence analysis and evolution of sea urchin histone H4 ($f2a_1$) messenger RNAs. *J. Mol. Biol.,* 104:351–370.
13. Holmes, D. S., Cohn, R. H., Kedes, L. H., and Davidson, N. (1977): Positions of sea urchin (Strongylocentrotus purpuratus) histone genes relative to restriction endonuclease sites on the chimeric plasmids pSp2 and pSp17. *Biochemistry,* 16:1504–1512.
14. Kedes, L. H. (1976): Histone messengers and histone genes. *Cell,* 8:321–331.
15. Kedes, L. H., and Birnstiel, M. L. (1971): Reiteration and clustering of DNA sequences complementary to histone messenger RNA. *Nature* [*New Biol.*], 230:165–169.
16. Kedes, L. H., Chang, A. C. Y., Housman, D., and Cohen, S. N. (1975): Isolation of histone genes from unfractionated sea urchin DNA by subculture cloning in E. coli. *Nature,* 255:533–538.
17. Kedes, L. H., Cohn, R. H., Lowry, J. C., Chang, A. C. Y., and Cohen, S. N. (1975): The organization of sea urchin histone genes. *Cell,* 6:359–369.
18. Kornberg, R. D. (1974): Chromatin structure: A repeating unit of histones and DNA. *Science,* 184:868–871.
19. Levy, S., Wood, P., Grunstein, M., and Kedes, L. (1975): Individual histone messenger RNAs: Identification by template activity. *Cell,* 4:239–248.
20. Morrow, J. F., Cohen, S. N., Chang, A. C. Y., Boyer, H. W., Goodman, H. M., and Helling, R. B. (1974): Replication and transcription of eukaryotic DNA in Escherichia coli. *Proc. Natl. Acad. Sci. U.S.A.,* 71:1743–1747.
21. Noll, M. (1974): Subunit structure of chromatin. *Nature,* 251:249–251.
22. Sures, I., Maxam, A., Cohn, R. H., and Kedes, L. H. (1976): Identification and location of the histone H2A and H3 gene by sequence analysis of sea urchin (S. purpuratus) DNA cloned in E. coli. *Cell,* 9:495–502.
23. Thomas, M., White, R., and Davis, R. N. (1976): Hybridization of RNA to double stranded DNA: Formation of R-loops. *Proc. Natl. Acad. Sci. U.S.A.,* 73:2294–2298.
24. Wallace, H., and Birnstiel, M. L. (1966): Ribosomal cistrons and the nucleolar organizer. *Biochem. Biophys. Acta,* 114:296–310.
25. Wu, M., and Davidson, N. (1975): Use of gene 32 protein staining of single strand polynucleotides for gene mapping by electron microscopy: Application to the $80d_3$ ilvsu+7 system. *Proc. Natl. Acad. Sci. U.S.A.,* 72:4506–4510.
26. Wu, M., Holmes, S., Davidson, N., Cohn, R., and Kedes, L. H. (1976): The relative positions of sea urchin histone genes on the chimeric plasmids pSp2 and pSp17 as studied by electron microscopy. *Cell,* 9:163–169.

Recombinant Molecules: Impact on Science and Society, edited by R. F. Beers, Jr. and E. G. Bassett. Raven Press, New York © 1977.

36. Studies on the Silk Fibroin Gene

John F. Morrow, John M. Wozney, and *Argiris Efstratiadis

*Department of Biological Chemistry, Harvard Medical School, Boston, Massachusetts 02115; and *Biological Laboratories, Harvard University, Cambridge, Massachusetts 02138*

INTRODUCTION

Differential gene expression in the various cell types of higher organisms is a fundamental aspect of their development. Biochemical studies of eukaryotic genes and of their transcription should be helpful in elucidating the mechanisms of differential gene expression. However, most structural genes occur only once in the haploid genome (4). Consequently, a particular gene is a very small portion of the DNA of a eukaryote. This portion varies from about one part in thirty thousand to one in ten million, depending on the size of the particular gene and on the size of the genome in question. In any case, purification of a unique gene from eukaryotic DNA is a formidable task. Even if carried out it could yield only minute quantities of DNA, too small for biochemical analysis.

Fortunately, DNA cloning enables us to replicate particular segments of DNA in bacteria (2,3,9–11,15,17,19,20,22,26; *this volume*). The rapid multiplication of the laboratory strains of *Escherichia coli* used for this purpose permits convenient production of defined DNA segments in very large quantities. Several reports in this volume have documented the importance of recombinant DNA methods in the study of repeated gene families of animal genomes.

We are interested in studying the silk fibroin gene and the regulation of its expression. Silk is produced by a variety of arthropods including most Lepidopteran larvae. Silk consists of protein, mainly of a single fibrous protein called fibroin. About 90% of the amino acid sequence of fibroin from the commercial silkworm, *Bombyx mori* (L.), is repetitions of the sequence Gly-Ala-Gly-Ala-Gly-Ser (13). Consequently, the fibroin messenger RNA (mRNA) consists mainly of repeats of an 18-nucleotide sequence (23). Some of the nucleotides of the repeated sequence may vary because of the degeneracy of the genetic code (7).

The fibroin gene is present in a single copy per haploid genome (7,24), and its activity is regulated in two ways. First, fibroin is synthesized only in the posterior silk gland, not in other tissues. Secondly, fibroin mRNA disappears from the posterior silk gland during the molting periods, presumably in response to the secretion of ecdysone, a steroid hormone (25). The fibroin gene is an appealing model system for the study of differential gene expression because it is very active in a tissue which is available in gram quantities and consists of a single cell type.

This chapter describes the construction and amplification of hybrid plasmids carrying synthetic fibroin gene sequence. This specific DNA sequence, now available in large amounts, can be used for direct DNA sequencing or as a probe for (a) isolation of the chromosomal gene together with its flanking sequences, (b) quantification of fibroin mRNA sequences at different developmental stages, and (c) study of fibroin mRNA precursors (12). We followed the approach taken for the construction of plasmids containing the DNA sequence of the rabbit β-globin gene (5,6,15). It involves reverse transcription of a messenger RNA under conditions favoring the synthesis of long DNA products. The complementary DNA strand is then synthesized. Homopolymeric tails are added by terminal transferase. Annealing this DNA with a suitable plasmid DNA, to which complementary tails have been added, forms circular hydrogen-bonded molecules. Introduction of these into *E. coli* permits formation of viable, closed-circular recombinant plasmids (3,26).

MATERIALS AND METHODS

Materials and Strains

Bombyx mori Japanese inbred strain p22 was obtained from Dr. Y. Tazima and fed mulberry leaves. Fibroin messenger RNA was purified on the basis of its large size by the method of Greene et al. (8), which involves gel filtration, heating to disaggregate RNA, and sucrose gradient centrifugation. Fibroin mRNA prepared in this way is about 90% pure, by weight. *E. coli* strain HB101, which is *hsr⁻recA⁻*, was obtained from H. W. Boyer; pMB9 was from K. Backman.

Synthesis and Purification of DNAs

Synthetic fibroin DNA was prepared as described for globin DNA (15). *Bombyx mori* DNA was prepared from silk glands by a modification of the method of Suzuki et al. (24). Covalently closed, circular plasmid DNA was purified as described by Carroll and Brown (2).

Miscellaneous Methods

Fibroin mRNA was labeled with ^{125}I by the method of Prensky et al. (18). DNA-RNA hybridization was performed as described by Suzuki et al. (24).

Transformation of *E. coli* by annealed mixtures of plasmid DNA and synthetic DNA was done according to Wensink et al. (26). Electrophoresis of DNA restriction fragments was done by a modification of the method of Aaij and Borst (1); DNA was stained afterward by incubation of the gel in 1 μg/ml ethidium bromide in electrophoresis buffer. Agarose slab gels 1.5 mm in thickness were used.

The experiments were performed under the following containment conditions: physical, P2; biological, EK1.

RESULTS

Using fibroin mRNA as a template, we synthesized double-stranded DNA by the sequential actions of avian myeloblastosis virus reverse transcriptase and *E. coli* DNA polymerase I (5,6,15; Fig. 1). After incubation with single-strand-specific nuclease S1, to cleave the hairpin loop in the product, we examined the duplex DNA by electrophoresis in a 3% polyacrylamide slab gel. *Hin*dII + III restriction endonuclease fragments of $\phi\chi$174 RF DNA were used as molecular weight standards (14). An autoradiogram revealed that duplex molecules as large as 2,000 base pairs (bp) had been synthesized. The greatest amount of material was in the vicinity of 1,000 bp in length, although some smaller than 200 bp was found. Discrete bands of synthetic DNA were seen at about 860, 760, 700, 610, and 490 bp, superimposed on a background.

Two size classes of DNA, 1100 to 900 and 700 to 500 bp, were eluted from the gel. An average of 200 deoxythymidylate residues was added to each 3′-terminus, using terminal transferase in the presence of $CoCl_2$ (21). Each pool was annealed with *Eco*RI-digested pMB9 DNA to which about 240 deoxyadenylate residues had been added. *E. coli* strain HB101 (deficient

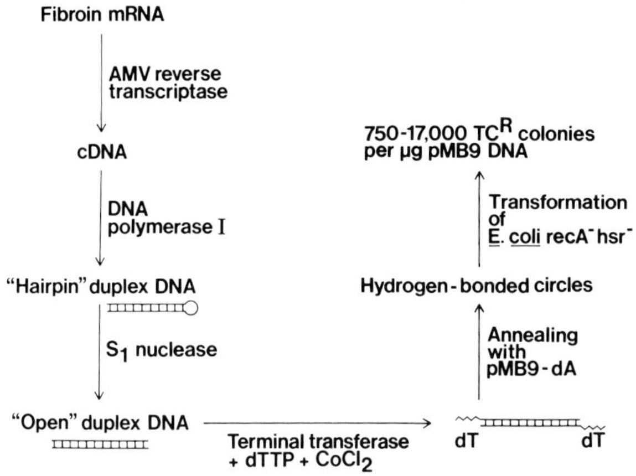

FIG. 1. Outline of the method of synthesis of recombinant plasmids. For details, see Materials and Methods.

in DNA restriction and recombination) was transformed with each annealed preparation. The number of independently derived tetracycline-resistant colonies obtained was 5 to 20 times that resulting from an equimolar quantity of poly(dA)-terminated pMB9 DNA alone.

The colonies resulting from transformation by annealed DNA preparations were screened for DNA complementary to silk fibroin messenger RNA. Partially purified plasmid DNA preparations were denatured, bound to nitrocellulose filters, and hybridized with fibroin mRNA. The results using synthetic DNA about 1,000 bp in length are shown (Table 1). Of 20 colonies examined, 12 (60%) showed evidence of fibroin DNA sequences. The frequency of positive colonies from synthetic DNA about 600 bp long was similar (six of nine).

TABLE 1. *Screening transformed bacteria for fibroin DNA sequences*[a]

DNA applied to filter		Counts/min hybridized sample
Colony pBF	21	2,519
	22	−10
	23	− 7
	24	5,353
	25	18
	26	1,424
	27	−15
	28	− 9
	29	1,948
	30	− 2
	31	1,949
	32	6,213
	33	2,496
	34	3,590
	35	− 4
	36	3,566
	37	28
	38	2,979
	39	3,603
	40	1,382
pMB9		5
Bombyx DNA:	19 μg	542
	38 μg	1,019
	58 μg	1,483

[a] Three percent of the volume of a partially purified plasmid DNA preparation from a 5-ml culture was bound to each filter. Filters were incubated with 0.002 μg/ml fibroin mRNA labeled with ^{125}I (specific activity 5.9×10^6 cpm/μg), in a volume of 0.3 ml/filter. Unfractionated *Bombyx mori* DNA of the indicated amounts was bound to three filters. The radioactivity of blank filters (195 cpm) has been subtracted. These plasmids are called pBF to indicate *Bombyx* fibroin.

Closed-circular plasmid DNA was purified from each of the positive cultures by ethidium bromide–cesium chloride density gradient centrifugation. The average yield was about 1 mg of plasmid DNA per liter of culture. These DNAs were examined by electrophoresis in 1% agarose gels, with closed-circular pMB9 (3.3×10^6 daltons; 5,000 bp) and pML2 (9) (8.7×10^6 daltons) as molecular weight standards (16).

One class of plasmids, from the synthetic DNA preparation of about 1,000 bp, electrophoresed at the rate expected for covalently closed, circular DNAs about 6,400 bp in length. These included pBF21,24,31,36,38,39, and 40. Digestion by *Hin*dIII endonuclease and electrophoresis in 1.4% agarose gels showed they were cleaved to unit-length linear molecules of about 6,400 bp [λ phage *Hin*dIII fragments were used as molecular weight standards (2)]. This was expected since their parent plasmid, pMB9, has a single *Hin*dIII site 310 bp from its *Eco*RI site (15).

Another class of plasmids capable of hybridizing with fibroin mRNA electrophoresed in 1% agarose gels at the rate expected for supercoiled DNA molecules of about 13,000 bp (16). This class includes pBF26,29,32,33, and 34. Digestion of each by *Hin*dIII followed by electrophoresis in a 1.4% agarose slab gel showed a single DNA band with the mobility characteristic of linear molecules of about 6,400 bp. These results suggest that these plasmids are head-to-tail dimers, each containing two pMB9 molecules and two inserted segments.

The structures of several of these plasmids were examined in more detail. *Hin*cII endonuclease alone digested pMB9 DNA to two fragments of lengths 3,030 and 1,970 bp. Slab gel electrophoresis revealed that *Hin*dIII cleaved the larger of these two fragments to 2,480- and 550-bp segments (Table 2). When recombinant plasmids were digested with *Hin*cII and *Hin*dIII, the smaller two fragments of pMB9 were preserved in all cases, whereas the largest fragment was replaced by an even larger one (Table 2). Since the synthetic DNA was inserted at the *Eco*RI site of pMB9, it is possible to orient the *Hin*dIII and *Hin*cII sites on a map of the recombinant plasmids (Fig. 2). The difference in length between a recombinant plasmid and pMB9 is the amount of inserted DNA (assuming that no deletion of pMB9 DNA occurred in this region). This difference is about 1,400 bp, on the average. This corresponds well to the average length of the synthetic duplex DNA, 1,000 bp, plus the expected length of the dA:dT duplexes on each side of it (Fig. 2).

The recombinant plasmids contain the repeating nucleotide sequence which makes up most of the length of the fibroin gene. This was shown by hybridization of fibroin mRNA with an excess of pBF5 DNA bound to nitrocellulose filters. pBF5 is a recombinant plasmid obtained from the synthetic duplex DNA 500 to 700 bp long. We find that pBF5 contains about 700 bp more than pMB9, by the technique described above. For this experiment 1.2 μg of pBF5 DNA digested by *Hin*dIII was bound to each

TABLE 2. Hinc*II* + Hind*II* *fragments of plasmid DNAs*[a]

Plasmid	Fragment lengths (base pairs)	Difference from pMB9
pMB9	2,480, 1,970, 550	0
pBF21	3,830, 1,970, 550	+1,350
pBF32	3,940, 1,970, 550	+1,460
pBF34	3,940, 1,970, 550	+1,460
pBF39	3,810, 1,970, 550	+1,330

[a] Plasmid DNAs digested simultaneously by *Hinc*II and *Hind*III restriction endonucleases were electrophoresed in both 1.4% and 2% agarose gels. Their molecular weights were estimated by comparison of their electrophoretic mobilities to those of fragments of λ phage DNA digested by *Hind*III (1). Molecular weights were taken from the linear ranges of plots of mobility versus the logarithm of molecular weight.

filter. Fibroin ^{125}I-mRNA (0.002 μg in 1 ml) was incubated sequentially with two DNA filters, 12 hr with each. When filters were washed repeatedly but not incubated with ribonuclease, 84% of the RNA bound to the first DNA filter and 3% to the second, so that 87% of the RNA bound in two hybridization incubations. Filters with no DNA bound 0.3% of the RNA, and no additional binding to pMB9 DNA was observed. When incubation with 20 μg/ml ribonuclease A (23°, 30 min, in 0.3 M NaCl + 0.01 M Tris, pH 8) was included in the washing procedure, the pBF5 filters still bound 67% of the total RNA radioactivity. This demonstrates that a DNA sequence less than 700 nucleotides long, in pBF5, can anneal to most of the sequences in the 16,000 nucleotides of fibroin mRNA. This experiment shows that pBF5 contains the major repeating sequence of fibroin mRNA, and not merely a unique, nonrepeated sequence represented once in the mRNA. Other recombinant plasmids also hybridized rapidly with fibroin mRNA present

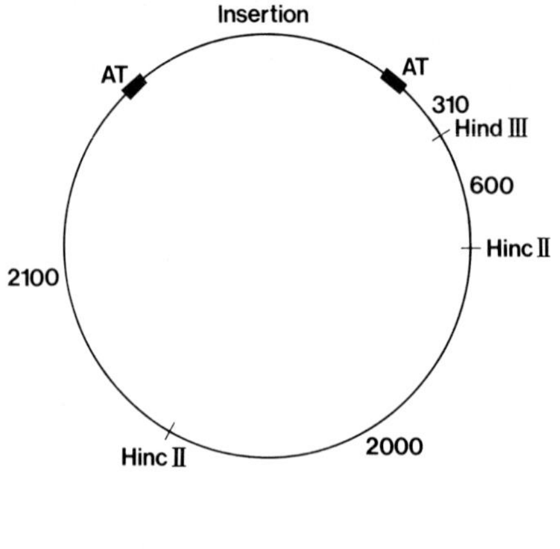

FIG. 2. Restriction endonuclease cleavage sites in fibroin gene recombinant plasmids. The distances between cleavage sites, in base pairs, are shown. AT indicates the polydeoxyadenylate:polydeoxythymidylate tails (15,26). These segments and the synthetic fibroin DNA sequence constitute the insertion.

at a low concentration (Table 1). This suggests that they also contain the major repeating sequence of fibroin mRNA. Finally, the high percentage of the mRNA hybridized indicates that these plasmids do not result from reverse transcripts of minor contaminants in the mRNA preparation.

DISCUSSION

We have carried out reverse transcription of fibroin mRNA, synthesis of the complementary DNA strand, and insertion of the synthetic DNA into a suitable bacterial plasmid. We have shown that this procedure yields viable recombinant plasmids containing the predominant repeating sequence of the silk fibroin gene. This is evidence for the generality of the reverse transcription method of cloning gene sequences. Related methods have been used previously to clone globin gene sequences (10,15,19,20).

We were interested in synthesizing full-length reverse transcripts of fibroin mRNA, but this was not possible by available techniques, although molecules three times as long as the globin gene were produced (2,000 versus 650 bp).

These recombinant plasmids contain a tandemly repeated DNA sequence. (The fibroin gene contains many tandem *internal* repeats encoding -Gly-Ala-Gly-Ala-Gly-Ser-, although the genome contains only one copy of the entire gene.) In spite of this repetition, the inserted sequences are rather stable during growth in *E. coli*. From the duplex DNA preparation about 1,000 bp in length with 200 deoxythymidylate residues added to each end, plasmid insertions of about 1,400 bp resulted (Table 2). This indicates that little or no deletion or unequal recombination occurred during growth of these recombination-deficient strains. The DNA preparations of Table 2 were from bacteria grown from single cells into colonies twice after transformation. At least 80 cell doublings occurred between introduction of the DNA into bacteria and purification of recombinant plasmid DNA. We are presently studying the stability of these plasmids during longer periods of growth in *E. coli*.

ACKNOWLEDGMENTS

We thank Anne C. Richards for outstanding assistance. J. F. M. thanks Donald D. Brown, for introducing him to research on silk biosynthesis, and Karen U. Sprague and Joan A. Steitz, for generous help. This work was supported in part by a grant from the American Cancer Society, Massachusetts Division, Inc., and by National Institutes of Health Grant GM22383. J. M. W. acknowledges the support of a U.S. Public Health Service Predoctoral Traineeship. A. E. was supported by an Intermediate Fellowship of the Harvard Society of Fellows.

REFERENCES

1. Aaij, C., and Borst, P. (1972): The gel electrophoresis of DNA. *Biochim. Biophys. Acta,* 269:192–200.
2. Carroll, D., and Brown, D. D. (1976): Adjacent repeating units of Xenopus laevis 5s DNA can be heterogeneous in length. *Cell,* 7:477–486.
3. Cohen, S. N., Chang, A. C. Y., Boyer, H. W., and Helling, R. B. (1973): Construction of biologically functional bacterial plasmids in vitro. *Proc. Natl. Acad. Sci. U.S.A.,* 70:3240–3244.
4. Davidson, E. H., and Britten, R. J. (1973): Organization, transcription, and regulation in the animal genome. *Q. Rev. Biol.,* 48:565–613.
5. Efstratiadis, A., Kafatos, F. C., Maxam, A. M., and Maniatis, T. (1976): Enzymatic in vitro synthesis of globin genes. *Cell,* 7:279–288.
6. Efstratiadis, A., Maniatis, T., Kafatos, F. C., Jeffrey, A., and Vournakis, J. N. (1975): Full length and discrete partial reverse transcripts of globin and chorion mRNAs. *Cell,* 4:367–378.
7. Gage, L. P., and Manning, R. F. (1976): Determination of the multiplicity of the silk fibroin gene and detection of fibroin gene-related DNA in the genome of Bombyx mori. *J. Mol. Biol.,* 101:327–348.
8. Greene, R. A., Morgan, M., Shatkin, A. J., and Gage, L. P. (1975): Translation of silk fibroin messenger RNA in an Ehrlich ascites cell-free extract. *J. Biol. Chem.,* 250:5114–5121.
9. Hershfield, V., Boyer, H. W., Yanofsky, C., Lovett, M. A., and Helinski, D. R. (1974): Plasmid ColE1 as a molecular vehicle for cloning and amplification of DNA. *Proc. Natl. Acad. Sci. U.S.A.,* 71:3455–3459.
10. Higuchi, R., Paddock, G. V., Wall, R., and Salser, W. (1976): A general method for cloning eukaryotic structural gene sequences. *Proc. Natl. Acad. Sci. U.S.A.,* 73:3146–3150.
11. Kedes, L. H., Chang, A. C. Y., Houseman, D., and Cohen, S. N. (1975): Isolation of histone genes from unfractionated sea urchin DNA by subculture cloning in E. coli. *Nature,* 255:533–538.
12. Lizardi, P. M. (1976): The size of pulse-labeled fibroin messenger RNA. *Cell,* 7:239–246.
13. Lucas, F., and Rudall, K. M. (1968): Extracellular fibrous proteins: the silks. In: *Comprehensive Biochemistry, Vol. 26B: Extracellular and Supporting Structures,* edited by M. Florkin and E. H. Stotz. Elsevier, New York.
14. Maniatis, T., Jeffrey, A., and van de Sande, H. (1975): Chain length determination of small double- and single-stranded DNA molecules by polyacrylamide gel electrophoresis. *Biochemistry,* 14:3787–3793.
15. Maniatis, T., Kee, S. G., Efstratiadis, A., and Kafatos, F. C. (1976): Amplification and characterization of a β-globin gene synthesized in vitro. *Cell,* 8:163–182.
16. Martin, M. A., Howley, P. M., Byrne, J. C., and Garon, C. F. (1976): Characterization of supercoiled oligomeric SV40 DNA molecules in productively infected cells. *Virology,* 71:28–40.
17. Morrow, J. F., Cohen, S. N., Chang, A. C. Y., Boyer, H. W., Goodman, H. M., and Helling, R. B. (1974): Replication and transcription of eukaryotic DNA in Escherichia coli. *Proc. Natl. Acad. Sci. U.S.A.,* 71:1743–1747.
18. Prensky, W., Steffensen, D. M., and Hughes, W. L. (1973): The use of iodinated RNA for gene localization. *Proc. Natl. Acad. Sci. U.S.A.,* 70:1860–1864.
19. Rabbitts, T. H. (1976): Bacterial cloning of plasmids carrying copies of rabbit globin messenger RNA. *Nature,* 260:221–225.
20. Rougeon, F., Kourilsky, P., and Mach, B. (1975): Insertion of a rabbit β-globin gene sequence into an E. coli plasmid. *Nucleic Acids Res.,* 2:2365–2378.
21. Roychoudhury, R., Jay, E., and Wu, R. (1976): Terminal labeling and addition of homopolymer tracts to duplex DNA fragments by terminal deoxynucleotidyl transferase. *Nucleic Acids Res.,* 3:101–116.
22. Schaffner, W., Gross, K., Telford, J., and Birnstiel, M. (1975): Molecular analysis of the histone gene cluster of Psammechinus miliaris. II. The arrangement of the histone-coding and spacer sequences. *Cell,* 8:471–478.
23. Suzuki, Y., and Brown, D. D. (1972): Isolation and identification of the messenger RNA for silk fibroin from Bombyx mori. *J. Mol. Biol.,* 63:409–429.

24. Suzuki, Y., Gage, L. P., and Brown, D. D. (1972): The genes for silk fibroin in Bombyx mori. *J. Mol. Biol.,* 70:637–649.
25. Suzuki, Y., and Suzuki, E. (1974): Quantitative measurements of fibroin messenger RNA synthesis in the posterior silk gland of normal and mutant Bombyx mori. *J. Mol. Biol.,* 88:393–407.
26. Wensink, P. C., Finnegan, D. J., Donelson, J. E., and Hogness, D. S. (1974): A system for mapping DNA sequences in the chromosomes of Drosophila melanogaster. *Cell,* 3:315–325.

Recombinant Molecules: Impact on Science and Society, edited by R. F. Beers, Jr. and E. G. Bassett. Raven Press, New York © 1977.

37. Discussion

Moderator: Charles A. Thomas, Jr.

Dr. A. B. Haberman: Dr. Carbon, do you have any way to design experiments to determine whether your expressed yeast DNAs are being read from promoters on a plasmid or from their own promoters?

Dr. J. A. Carbon: We are just starting to get into that now. The idea, of course, would be to isolate the protein, the active enzyme, from the hybrid and compare it with wild-type. We are in the process of doing those experiments but have nothing that I can tell you about.

Mr. D. P. Taylor: Have you observed any instabilities or rearrangements of plasmids containing *Drosophila* yeast or DNA?

Dr. J. A. Carbon: They appear to be very stable. They segregate. Grown on nonselective media through many generations—that is, from a small inoculum to the stationary phase—the cells will segregate the entire plasmid at a frequency of 2 to 5%. I have noted no other instability of the DNA, although we have some careful studies underway because obviously that is an important point.

Mr. D. P. Taylor: How about rearrangements?

Dr. J. A. Carbon: We have not looked at that aspect. I will mention, however, that the restriction map Davis did of the lambda *his* DNA and the restriction map from our DNA work contained pieces that were identical with respect to their mobility. It appears that there were no rearrangements in the two labs using two different methods.

Dr. C. A. Thomas, Jr.: Dr. Wellauer, you did not comment on the repeat length on the nontranscribed spacer portion that gives rise to these loops. Do the loops have modular lengths?

Dr. P. Wellauer: Basically, since the subrepeat is only 15 base pairs long, electron microscopy is a hopeless way of telling whether or not there are multiples of one repeat.

Dr. C. A. Thomas, Jr.: How many repeats do you think there are in the C region?

Dr. P. Wellauer: The one that expands in size?

Dr. C. A. Thomas, Jr.: Yes.

Dr. Wellauer: It depends on which fragment you look at. If it is 15 repeating units, you just divide the molecular weight by 15.

Dr. C. A. Thomas, Jr.: Jerry Feitelson has asked me if he could make an announcement.

Mr. J. S. Feitelson: I would like to make a request to investigators who are busily generating restriction maps of phages, plasmids, and other naturally occurring DNAs. It might be interesting to centralize these data into a computer bank so that we can distribute it to interested investigators. I am a graduate student working for Professor Joshua Lederberg. He is currently principal investigator for the SUMEX-AIM program, a national resource funded by the NIH to serve as a computer basis for artificial intelligence in medicine.

I just spoke to Dr. Lederberg and he thought it might be a good idea to centralize the data if everyone could send him restriction maps either published or unpublished at the current time to the Department of Genetics, Stanford University Medical Center, Stanford, California 94305. Thank you.

Dr. C. A. Thomas, Jr.: I consider Dr. Lederberg's enterprise to be of great importance because some of you understand that the total number of DNA sequences that are possible to construct vastly exceeds, by many orders of magnitude, the number of nucleotide sequences that exist. Therefore, it does not stretch one's imagination that various living organisms plagiarize bits and pieces of one another's DNA in a way to make their own composite genomes. This enterprise is directed toward a beginning of an understanding of whether this may be true or not.

Mr. D. H. Hamer: Dr. Kedes, I would like to know if you have data or even any speculation on the locations of promoters on the histone cluster and ribosome binding positions. Also, what system do you use for *in vitro* translation studies? Have you ever tried translating these messages in a mammalian system?

Dr. L. H. Kedes: In answer to the first question, we have some speculation and soft data regarding the presence of high molecular weight precursor RNA which is a transcript of several of the coding and spacer regions. We do not know where the beginning and end of it are, and as far as ribosome binding sites go, no information is available.

As you may know, in my lab a couple of years ago Mike Grunstein did a sequence analysis of the H-4 messenger RNA; we have some ribosome binding studies on the messenger RNA, but we have not sequenced that part of the DNA.

The second part of the question: what was the translation system? The initial translation of the isolated messenger RNA was by Levy in my lab a few years ago using the ascites system. The link system I mentioned earlier that Mulligan and Roberts have used in these studies requires wheat germ as the protein-synthesizing system. We do not use the ascites system here.

Dr. L. H. Kedes: Dr. Morrow, you said approximately 75% of your colonies contained DNA which hybridized to fibroin messenger RNA. You did not say what the other 25% contained. Did they contain fragments of DNA other than the vector itself?

Dr. J. C. Morrow: Yes. In doing this we had not vigorously eliminated nicked circular plasmid molecules from the preparation, so we anticipated there would be some PMB-9 without insertions. In fact, there were. But we have since found other plasmids with insertions, also approximately 900 base pairs long, which apparently do not hybridize fibroin messenger RNA under our conditions. The point is that the message hybridized rapidly because it has repetition. We do our hybridizations under conditions where only a repeated sequence would hybridize.

If the plasmid contained only a unique sequence from the 3′ end of the message, we would not have picked it up, at that point, by hybridization. We would have to incubate at higher RNA concentrations or for longer periods of time. We do not believe, certainly, that the fibroin message used for this experiment is 100% pure.

For those who are interested in biohazard considerations, I might comment that a provision in the current guidelines allows lowering the containment by one step for purified sequences. We did not use that provision in this work because we did not believe that the purity of messenger RNA justified it.

Mr. D. A. Konkel: Since you do have the message available, why are you doing shotgun-type experiments rather than trying to pick out the particular sequence?

Dr. J. C. Morrow: I assume you mean by picking out a particular sequence –

Mr. D. A. Konkel: By c-DNA column or something like this.

Dr. J. C. Morrow: We are trying that first. The purification approach was around before we started thinking about putting the plasmid DNA on a column, for instance, and picking it out. That is an older approach, but the approach you suggest would probably be quicker, actually.

Mr. D. A. Konkel: Wouldn't it probably be safer?

Dr. J. C. Morrow: According to the considerations that people usually apply to these experiments, namely, producing as few uncharacterized clones as possible, it would be safer.

Mr. A. H. Deutch: As you mentioned, Dr. Morrow, there have been a number of reports, a few of complete c-DNA transcripts made. What do you feel the secret is since many of the messengers are rather large. What is the secret to making full-length c-DNA?

Dr. J. C. Morrow: I think the reports you are talking about are exclusively with globin messenger RNA, although you may well know things that I do not.

Mr. A. H. Deutch: I think O'Malley has done some work on ovalbumin.

Dr. J. C. Morrow: The reports on the globin mRNA depict it as 650 bases long, if I recall, whereas our messages are 16,000. I stated that the significant portion of the duplex DNA was 2,000 bases long, so it would be three times as long as a full-length globin reverse transcriptase. I believe it is just a linkage matter.

Dr. C. A. Thomas, Jr.: I think it is important that Dr. Morrow should be a contributor to this volume because, as you will recall, he, together with others, was responsible for the first experiments that inserted a eukaryotic segment of DNA into *E. coli.*

I wanted to ask Philippe Kourilsky to give us a report from Paris.

Dr. P. H. Kourilsky: I would like to describe briefly some progress we have made with respect to this problem. This work was done in collaboration with Rougeon and Mach in Geneva.

In the first experiments we inserted synthetic c-DNA. We used a complicated approach in which both strands were synthesized with the help of primers that were added to the terminal transferase. The first experiment yielded five globin clones, that is, clones carrying part of the globin sequence from rabbit. The final analysis disclosed clones that have relatively short sequences.

Rougeon modified the method and finally used one which is similar to that devised by Maniatis and his group.

Finally, I would like to say that the technique to make complete reverse transcripts has now yielded a number of plasmid species, essentially complete chains of rabbit alpha and beta globin. Rougeon has done mouse globin chain and recently has inserted the total reverse transcript made from 14S immunoglobulin light-chain RNA.

One special feature of technique was employed by Rougeon and Mach and me. The special feature is that the insertion into the plasmid is not as simple as that which was presented. The *Eco*R1 sites are reconstituted before insertion. This is done as follows. The recipient plasmid is opened to the restriction enzyme, in this case *Eco*R1, and the short sticky ends are filled with reverse transcriptase or DNA primer before elongation in such a way as to reconstitute *Eco*R1 sites on both sides of integrated sequence. This does not work as well as one would hope. Among the plasmids that were created, as Dr. Morrow's chapter indicated, the process was quite efficient. Thousands of plasmids can be obtained easily with a few nanograms of material.

When these plasmids are examined, only a few of them have the *Eco*R1 sites on both sides. We do not know why this is so. It is possible that our *E. coli* enzyme is contaminated with an exonuclease which created a mutation, if you like, in the site. Anyway, some of the plasmids do carry two *Eco*R1 sites, both sides having integrated sequence.

One of the globin plasmids which is rabbit beta-chain transcript is of this sort. Plasmid PR-19 is cut with *Eco*R1, and it yields the parent plasmid plus two small plasmids of 200 and 400 base pairs, respectively.

The nice thing about this is that we can transfer these fragments into lambda phage. We transferred into lambda, which has only one *Eco*R1 site in the beta chain. We can play with fragments in the region which is the end of the enzyme control, that is, in the region transformed from the very efficient P_L promoter.

This was done and illustrates what I think is the general approach: to use two different biological systems to study a single gene. This will become obvious when you know that transcripts of globin sequence in *E. coli,* a very simple affair, are not a hybridization of the RNA made *in vivo* from the lambda lysogen with plasmid DNA. We have no cross-hybridization. The plasmid DNA is a pure probe which we can produce in large amounts.

The expected result was obtained. In dealing with a lysogen there is no transcription of the globin plasmid if lysogen is repressed. If we raise the temperature in the C-1-857 lysogen, we get some transcription of globin plasmid. We should expect this from genetic arguments because the genes that are downstream are expressed, since the phage can, in fact, lysogenize, integrate, and excise.

We wanted to know whether we could make an antigen. Indeed, this experiment is difficult to do with the globin system because globin is poorly antigenic. As a matter of fact, this experiment has not yet been done. We have not tried to do it even with globin. We are waiting for immunoglobulin or some other genes which are cooking right now to do this experiment.

I would like to say, however, that in order to do this experiment, we devised methods that included antibodies in the plate. With a special method of sandwiching, we can observe phage which produces an antigen. A second antibody coupled to peroxidase is used to reveal peroxidase antibodies. The final result is a nice red halo. But we have not performed the final experiment.

I would like to stress that this method is very interesting, and if antigen is not produced with these means we shall certainly look for mutants that produce the antigen.

Dr. C. A. Thomas, Jr.: Thank you, Dr. Kourilsky, for your comments.

Recombinant Molecules: Impact on Science and Society, edited by R. F. Beers, Jr. and E. G. Bassett. Raven Press, New York © 1977.

38. Introduction to Section F: Societal Impact—Issues and Policies

Kenneth Murray

Department of Molecular Biology, University of Edinburgh, Edinburgh EH9 3JR, Scotland, U.K.

Anyone who believes that fundamental scientific research is in the broadest sense beneficial will, I am sure, agree that the new opportunities offered to biological research by the advances that have been discussed in the previous chapters of this volume deserve the intensive attention they are now receiving. There can be no doubt that the enhanced facility for transfer and manipulation of genetic information between species and the exploitation of the new opportunities for enhanced gene expression will aid our understanding of the way in which biological cells work. In turn, this may well eventually provide many opportunities for exploitation of the newly constructed organisms for beneficial use. On the other hand, it is argued, the new techniques provide similar opportunities for those who would wish to misuse them, but this situation is hardly novel; it has, after all, been with mankind since the invention of steel and probably for a very long period before that. Some of the benefits that may accrue from the type of research discussed in this volume are rather obvious and center around the production of useful proteins or polypeptides by fermentation or cell culture processes. These have been well rehearsed and I do not intend to dwell on them here, particularly since we may encounter some aspects of this topic later. It was the intention of the contributors to this volume to discuss instead some of the broader aspects raised by this type of research and some of the attitudes toward it, bringing especially points of view from areas outside the United States.

The great surge of interest and enthusiasm in the scientific community for research in this area stems from a feature that is often believed to be uniquely novel, i.e., the ability to cross the natural barriers to breeding between different biological species and to effect transfer of genetic information between organisms that do not mate naturally. With such opportunities there is ample scope for imagination of some possible consequences of laboratory experiments, both intentional and otherwise. Although I believe a consensus of informed opinion would judge the potential benefits of research in this area to outweigh the speculative hazards, to quote the Ashby Report, I think it would be a brave man indeed, or perhaps a fool, who would insist that production of a new form of organism that is harmful is totally impossible. The causes of apprehension cover quite a spectrum. Most of us, I am sure, would ignore the more bizarre areas of science fiction concerned with the creation of monsters. We may also in general believe that serious problems would arise only if one were to produce an organism that were both nasty and possessed of a selective advantage, but at the other end of the scale are subtler changes the effect of which are difficult, if not impossible, to assess. The problem that confronts us is that of attempting to make judgments in a total absence of information. The reasonable policy, therefore, is to proceed with caution, and this has been the attitude of most of the scientific community over this new phase of biological research.

Apprehension arose initially because the organism that has been the workhorse of molecular biology, and therefore the vehicle for the initial experiments, is the bacterium *Escherichia coli,* which inhabits the human intestine. The introduction of new and, for the large part, unknown genetic information into an organism so intimately concerned with the human population obviously brings the problem close to home, but what are the opportunities for expression of the new genes to the detriment of the animal in which they are functioning? What are the possibilities of disseminating new organisms among the human population, and more broadly, into the environment? What are the opportunities for further change of these new organisms? For that matter, are these experiments really as novel as we think? Since *E. coli* and other bacteria are to be found in large numbers in animals and since bacteria can quite readily take up DNA although not always with enormous efficiency, are these exchanges of genetic information across the species barrier as novel as we might like to think? When a rat dies in a sewer or a sheep dies on the mountainside, there is no lack of opportunity for bacteria living on the decaying animal cells to acquire new sequences of their host DNA. We would not normally observe such an assimilation, however, but this does not mean that it does not take place. Many would argue that such changes have and do occur all the time, albeit with low frequency, and perhaps sequences for which the bacterium has little use are rapidly lost. On these aspects we have little information, but we do have information about the ecology and behavior of *E. coli,* its promiscuity in the human intestine as well as in the laboratory, its interactions with antibiotics, and the effect of these on its life cycle. We also have information about parasites on the bacteria themselves, and about means of containing and controlling these organisms by physical and biological methods.

In the first chapter in this section, Mark Richmond discusses the natural history of *E. coli* and the transfer of its genetic material *in vivo,* and hence why there may be a need to curb the free dissemination of new organisms throughout the environment. To this end is discussed the formulation of guidelines for the safe handling of these organisms in the United States by the National Institutes of Health through their various advisers; then Dr. Tooze summarizes the attitudes and policies toward this research as he sees them emerging in European countries. The second part of the section concerns some of the broader social implications for this type of research, including its relevance to human genetics and genetic diseases, and also discussed are related problems that may exist in other areas of biological research which are now widely accepted, established subjects. Two industrial colleagues then outline the way in which hazards can be assessed for particular operations and how steps can be, or have been, taken to contain dangerous materials. I think it is instructive to see what we can learn from this type of philosophy in the context of biological research generally.

The second section concludes with a viewpoint that I have entitled "Counterpoint" expressed by a local group of people who feel that in the face of such ignorance of the possible consequences of some of the new research, a far more stringent attitude toward its pursuit should prevail.

Recombinant Molecules: Impact on Science and Society, edited by R. F. Beers, Jr. and E. G. Bassett. Raven Press, New York © 1977.

39. *Escherichia coli* K-12 and Its Use for Genetic Engineering Purposes

Mark H. Richmond

Department of Bacteriology, University of Bristol, Bristol BS8 1TD, England

INTRODUCTION

Although there have been suggestions that *Bacillus subtilis* may emerge as the bacterial species of choice for genetic engineering experiments (F. E. Young, *this volume*), most of the bacterial work has been carried out in *Escherichia coli* K-12 up to this point. Furthermore, since so much is known about the genetics and biochemistry of this strain, it seems highly probable that it will continue to be used in this way, as long, that is, as its use does not pose unacceptable risks.

One of the reasons for proposing that *B. subtilis* should be used in place of *E. coli* in genetic engineering experiments was the worry that *E. coli* might be better able to survive in the gut flora of the scientists carrying out the experimental work than could the *Bacillus* species. The gut of man is certainly rich in "*E. coli*" strains (5,18), and certainly one might expect *E. coli* K-12 to be more at home under the highly anaerobic conditions prevailing in the human gut than a highly aerobic *Bacillus* sp.

In practice, the hazards that could arise from the escape of a laboratory strain of *E. coli* K-12 are of three broad types: first, the strain itself might become established as part of the gut flora of the experimenter and from there it might infect others. Secondly, the *E. coli* already in the gut of the experimenter might carry the necessary information to repair the genetic lesions deliberately inserted into the plasmid used for engineering work, thus facilitating the escape of that recombinant element to the outside world. For example, the plasmid concerned might become self-transmissible. The third worry is that the *E. coli* of the gut flora might carry phages—and in particular lambdoid phages (16)—that could repair the genetic defects deliberately inserted into the defective lambda phages used for engineering purposes.

In this chapter we first try to assess the factors influencing the probability with which *E. coli* K-12 strains may become part of the flora of the experimenter and thence infect others. Then we consider the chance that the plasmids used for engineering experiments might recombine either with a resident plasmid in the *E. coli* flora of the gut, or with a lambdoid phage, or even with lambda itself. In fact too little information is yet available to reach more than tentative conclusions on any of these topics, but, in reviewing what is known, it may be possible to see what needs to be investigated further.

OBSERVATIONS

E. coli K-12 and the Human Gut Flora

The majority of *E. coli* strains isolated from human fecal samples are "smooth" (18), that is, they carry surface lipopolysaccharides (*O*-antigens) complete with the specific chains of monosaccharide units characteristic of those molecules (19). In contrast, *E. coli* K-12 is "rough"; indeed, it is sometimes even called "deep rough." In this strain only the "core" of the lipopolysaccharide is expressed and no *O*-antigen side chains are present (19). To date, five core types of *E. coli* have been recognized and these are designated R1 through R5 (24). *E. coli* K-12 has an R5 core.

The absence of lipopolysaccharide side chains on "rough" *E. coli* means that they do not react with *O*-antisera prepared against surface antigens from *E. coli* strains [more than 150 distinct *O*-antigens are known (8,18)] and, indeed, "rough" strains autoagglutinate in the standard *O*-serotyping procedures (28). Thus, an approximate guide to the incidence of "rough" *E. coli* in the normal human fecal flora can be gained by *O*-antigen typing of individual isolates. In practice, autoagglutinable strains do not usually account for more than 1 to 2% of the human fecal flora at any one time (12); although "rough" strains are present in the human gut, they are much less common than "smooth" *E. coli*. Among the "rough" *E. coli*, it is not known what proportion carry each of the five R core types. In practice, this may be an important point to establish since both the *O*-antigen side chains of the lipopolysaccharide and the core type do have some effect on the ability of an *E. coli* strain to participate in fruitful matings with other *E. coli* (*see next section*).

Although it is commonly accepted that the relative incidence of a given *O*-type in the human gut flora measures the ability of that strain to survive in man, the evidence for this is far from perfect. Furthermore, most of the information available refers to "smooth" *E. coli*, and it would be unwise to assume that *E. coli* K-12 necessarily survives poorly in the human gut because it is relatively uncommon in fecal samples.

Perhaps the best experimental evidence for the survivability of *E. coli* K-12 comes from the experiments by Smith (27) and by Anderson (1) in which appropriately marked *E. coli* K-12 strains were fed to human volunteers and the excretion of the marked strains followed. In both series of experiments, ingested *E. coli* K-12 did not persist in gut of volunteers for more than 4 days, even when 10^{10} bacteria were fed.

An even more indirect measure of the ability of *E. coli* K-12 to establish itself in the gut flora of experimenters using plasmid-carrying strains comes from some routine monitoring of experimental workers in our Department of Bacteriology (14). In this case, the experimenters were working with *E. coli* K-12 carrying R-plasmids of various types. Those concerned have so far been followed for 517 man-days, and although R-factors have been encountered on at least 51 occasions during this period, they have never, so far, appeared in autoagglutinable strains of *E. coli.* It is important to stress that these experiments are complex in that they reflect both on the technique of the experimenters *and* on the colonization properties of *E. coli* K-12. Nevertheless, they are particularly relevant in the genetic engineering context since they suggest that normal low-containment techniques recommended (22) for use in bacteriological laboratories (e.g., free work in an open laboratory but no mouth pipetting, no eating or smoking in the laboratory, wearing laboratory overalls, autoclaving all cultures after use) do not lead to massive cross-infection of experimenters with *E. coli* K-12. Furthermore, none of the experimenters was found to harbor any of the plasmids with which they were working in *E. coli* other than K-12 (14), and apparently the chance of *E. coli* K-12 passing R-plasmids on to other gut *E. coli* is also not large.

On the basis of these—admittedly limited—experiments, therefore, it seems that *E. coli* strains of the type such as K-12 are not very potent colonizers of the human gut, nor do they seem to pass on any self-transmissible R-plasmids they carry at high frequency.

"Rough" and "Smooth" *E. coli* and Their Ability to Mate

"Rough" *E. coli* (such as K-12) are generally much more effective partners in a conjugation experiment than are "smooth" strains, at least when studied under laboratory conditions (26). Moreover, there are some claims that "smooth" *E. coli in vivo* do not participate to a significant extent in mating experiments (17).

Experiments in this laboratory and elsewhere have shown that transfer frequencies between "rough" and "smooth" *E. coli,* and vice versa, and experiments in which different self-transmissible plasmids are used in such crosses, give a wide range of different transfer frequencies depending on the particular pair of strains concerned and the nature of the plasmid involved. Table 1 summarizes a typical set of results obtained during our studies. All the R-plasmids tested transferred to rough R1 core *E. coli* at about the same frequency as found with *E. coli* K-12 (R5 core), but rough R2 core strains (otherwise isogenic with rough R1 core strains) showed an appreciably

TABLE 1. *Transfer of R-plasmids from* E. coli *K-12 to various* E. coli *recipients*

Phenotype			Plasmid transfer frequencies				
Donor	Recipient		RP1	R100–1	R64–11	R46	R386
E. coli K-12 (R5) rough	*E. coli* (R1)	Rough	1×10^{-5}	8×10^{-3}	3×10^{-4}	4×10^{-5}	2×10
" "	*E. coli* (R1:08)	Smooth	6×10^{-6}	0	3×10^{-4}	0	0
" "	*E. coli* (R2)	Rough	0	2×10^{-3}	6×10^{-4}	0	0
" "	*E. coli* (R2:08)	Smooth	0	0	2×10^{-4}	0	0
" "	*E. coli* (R1:07)	Smooth	0	0	0	0	0
" "	*E. coli* (R5)	Rough	6×10^{-5}	8×10^{-3}	1×10^{-3}	2×10^{-5}	1×10

All matings were carried out for 1 hr and the exconjugants then plated on appropriate selective agar. Frequencies are recorded per donor bacterium. The bracketed strains are isogenic save for their variation in lipopolysaccharide constitution (C. L. Hartley and M. H. Richmond, *unpublished observations;* see also ref. 29).

lower recipient ability with some of the R-plasmids. Similar conclusions were reached, with a smaller range of strains, by Wiedemann and Schmidt (29). Since the introduction of side chains onto the lipopolysaccharide to give a smooth phenotype almost always leads to a reduced recipient ability (in certain cases the decrease is dramatic), there is some support for the view that smooth strains are less efficient recipients in R-plasmid transfer, but it would be dangerous to conclude that similar events never occurred *in vivo*.

R-Plasmid Transfer in the Human Gut

The evidence that plasmid transfer does, in fact, occur in the human gut comes from two types of studies: (a) experimental observations where marked strains have been fed to volunteers, and (b) epidemiological data from which it can be inferred with considerable certainty that R-plasmid transfer must have occurred.

Feeding experiments have been carried out with *E. coli* K-12 with the object of studying the survival of this strain in the human gut and also its ability to mate with the resident *E. coli* flora of the volunteers. The first experiments of this type were performed by Smith on himself (25). First he established a nalidixic acid–resistant, hemolytic, "smooth" *E. coli* as a predominant member of his own gut flora, a situation which he was able to maintain for over 2 years. Once this strain was established, he ingested appropriately marked "smooth" *E. coli* strains in doses of 10^6 to 10^{10} organisms, some carrying R-plasmids and others not. The first conclusion to be reached from these experiments is that it is not easy to establish *E. coli* strains as major components of the human gut flora. In fact, the longest persistence of any strain was 18 days, and for this to be achieved daily doses of 10^{10} organisms for a week were needed. Against this background, his ability to establish the nalidixic acid–resistant, hemolytic strains is clearly anomalous and stresses the caution with which statements on bacterial colonization of the human gut must be treated. If small numbers of organisms were ingested, no R-plasmid transfer could be demonstrated *in vivo* even when the plasmids involved had been shown to be self-transmissible. When 10^{10} organisms were fed on a daily regime for 7 days, however, some evidence for transfer was obtained. However, the exconjugants generated in this way never survived in the gut flora for more than 1 or 2 days (25). It follows from these experiments, therefore, that R-plasmid transfer between smooth strains of *E. coli* can occur when massive doses of donor strains are fed, but (on the basis of these limited studies) the progeny do not survive well. From the point of view of laboratory accidents, it is worth stressing that large infective doses were needed in these experiments—larger on the whole than are likely to be ingested by accident.

In this laboratory we have fed an appropriately marked, "rough," R-plasmid–carrying *E. coli* (not *E. coli* K-12) to a volunteer and succeeded in establishing the ingested strain for a period of about 5 days (2,3,15). We observed, however, that R-plasmid transfer to the resident flora was undetectable unless an antibiotic that could select for one of the resistance markers on the fed plasmid was administered on the day after feeding. These experiments confirm those of Smith (25) in that the feeding of a rough *E. coli* carrying an R-plasmid did lead to transfer of it to the resident flora (the recipient was a "smooth" serotype O2 strain in our case). However, in our experiment the survival of the exconjugant line was so poor that it could be detected only when the person was subsequently treated with antibiotics.

Although these experiments are reassuring in the sense that resistant exconjugants do not arise with great abundance, they do have their less satisfactory aspect. When antibiotic selection was applied, the fed R-plasmid could be isolated in two forms: first, unchanged in the O2 *E. coli* line, and, secondly, as part of a recombinant plasmid in an *E. coli* O9 strain (2,3). So we must conclude that both R-plasmid transfer *and* plasmid recombination can occur *in vivo* with a rough strain as donor, but the process does not commonly seem to lead to a predominantly resistant flora, at least in the absence of antibiotic selection. This last point is important. The administration of therapeutic courses of antibiotics could greatly potentiate the hazards of using *E. coli* for engineering experiments. Perhaps experimenters taking antibiotics should be excluded from genetic engineering laboratories while their course of treatment continues.

Recent feeding experiments have employed large quantities of *E. coli* K-12 as the ingested strain (1,27). In Anderson's experiments (1), the organisms were fed in milk, a procedure known to potentiate colonization by *E. coli* strains (2). In neither series of trials did the fed *E. coli* K-12 persist for more than 4 days. The bacterial counts published by Smith (27) suggest that in his experiments the K-12 bacteria (not protected by milk in this case) hardly multiplied before being voided. In Anderson's experiments, on the other hand, some limited multiplication did occur, and on one occasion a transfer event (that of a tetracycline-resistant derivative of F) was also detected (1).

EPIDEMIOLOGICAL EVIDENCE

The feeding experiments just described are often criticized for the relatively massive dosing used to demonstrate transfer and the fact that colonization is helped by milk and other adjuvants. In this they hardly mimic typical laboratory accidents. To avoid these criticisms we have monitored one individual who has been carrying an O18 *E. coli* strain as a major component of his fecal flora for over 8 months (13,20,21). The strain concerned was resistant to tetracycline, streptomycin, and sulfonamide and carried an R-plasmid of FII compatibility group specifying resistance to these antibiotics. During the observation period the individual received no therapeutic antibiotics. We wished to determine whether the FII R-plasmid in the O18 *E. coli* transferred to any other *E. coli* strain in the gut flora. Table 2 shows part of this study. It relates to the period when the O18-resistant strain had already been prevalent in the gut flora for about 6 months. On day 202 after the commencement of this study, an O88 *E. coli* with the same resistance pattern as the O18 strain appeared in the gut flora, reappeared briefly a few days later, and then disappeared. Comparison of the R-plasmids in the O18 and O88 strains showed that they were extremely similar. Perhaps the most clear evidence for this comes from the *Eco*RI restriction endonuclease digestion patterns of the R-plasmids from the two strains (Fig. 1). The patterns are indistinguishable and this is strong evidence that R-plasmid transfer occurred from the O18 strain to establish the resistant O88 line in the gut of the person being studied. Similarly, *Hin*dIII hydrolysis patterns of the plasmids from the two sources are indistinguishable.

This experiment is valuable in a number of respects. First, it shows beyond reasonable doubt that R-plasmid transfer can occur under completely normal conditions in the human gut; and secondly (and perhaps more importantly), it stresses that such transfers do not seem to occur abundantly. The potential plasmid donor was present at levels of above 10^5/g feces in the individual under study for about 6 months before a transfer event was actually recorded—despite careful searches of samples taken on the average of every 3 days. Moreover, the "new" resistant strain did not survive well and was eliminated rapidly.

Plasmid Transfer from Nonreplicating Bacteria

A number of proposals have been made to reduce the potential hazard of genetic engineering experiments involving *E. coli* by constructing mutated strains in which bacterial replication is impossible unless exacting environmental conditions (which are unlikely to be encountered outside the laboratory) are met (R. Curtiss, III, *this volume*).

Under these circumstances it becomes highly important to clarify the conditions under which *E. coli* will act as a donor, since transfer potential is not necessarily coupled tightly to cell growth. This question has been examined at length by Curtiss and his colleagues (6). It emerges that bacterial replication is unnecessary for plasmid transfer to occur—presumably because the types of DNA synthesis needed for transfer are distinct from that involved in normal reproductive replication. On the other hand, *E. coli* K-12 does need an energy source to be a potential donor, and protein synthesis is needed for transfer replication to be initiated. In practice, it is unlikely that *E. coli* multiplies rapidly in the human intestine—presumably because the available nutrients are likely to be largely exhausted. Nevertheless, transfer

TABLE 2. *Sequence of* E. coli *strains in the fecal flora of one person over 257 days*

Strain	Days after start:	1	2	3	4	5	6	7	8	9	11	12	13	14	17	18	20	21	22	27	29	31	32	33	35	37
011	TcSu[a]	4	2			4	4	4	2	3																
018	TcSmSu	6	8	10	9	5	5	6	6	3	2	[b]	[b]	3	7	7	2	10	6	8	4	6	5	10	4	1
086	ApSmSuCm											3	8	7												
0108											3															
NT												2	2													

Strain	Days after start:	39	41	42	43	45	46	64	65	67	68	69	72	74	76	78	80	81	82	84	87	88	89	90	94	95
09								5	1																	
011																		1	1							
015					3														1				1			
017																		4								
018	TcSmSu	10	6	6	6	5	6	[b]	9	10	3	4	1	5	6	10	10	1	3	[b]	9	10	9	10	4	10
018	SmSu													1							1					
021								2	1																	
024						1																				
050																		2								
064																									3	
NT								2																		

Strain	Days after start:	96	97	100	103	105	112	115	117	118	122	123	125	129	130	132	137	138
011																		1
015		1	1															
018	TcSmSu	9	9		2	3	10	8	[b]	[b]	5	10	6	[b]	1	6	10	[b]
018	SmSu														1			
020									1	4								
024									4	4								
027									2									
050									3	1				9	6			

Strain	Days after start:	140	143	147	150	151	153	168	171	175	183	185	188	189	193	197	200	202
015			2										3	6	10	10		
018	TcSmSu		5	[b]	[b]	[b]	[b]	[b]	[b]	2	2	1	4	4	[b]	[b]	10	2
018	SmSu			2		10	6	10	5	8	4							
088	TcSmSu																	8

Strain	Days after start:	208	209	210	216	223	228	230	235	237	240	242	244	246	249	256	257
09							1										
015						8											
018	TcSmSu	[b]	2	[b]	[b]	[b]											
018	SmSu		2				3	2	10	1	5	1		10	10	5	5
021		1					2						1				
024													3				
040					3												
075					1												
086			5														
088	TcSmSu											6					
0101						2											
0122					4												
NT					2						1	1	4				

[a] TC, tetracycline; Sm, streptomycin; Su, sulfonamide; Ap, ampicillin; Cm, chloramphenicol; NT, not typable with available sera. In the instance where value is <10, the remaining isolates proved not to be *E. coli.*
[b] Culture on selective agar showed presence of strain as a substantial minor component of flora.

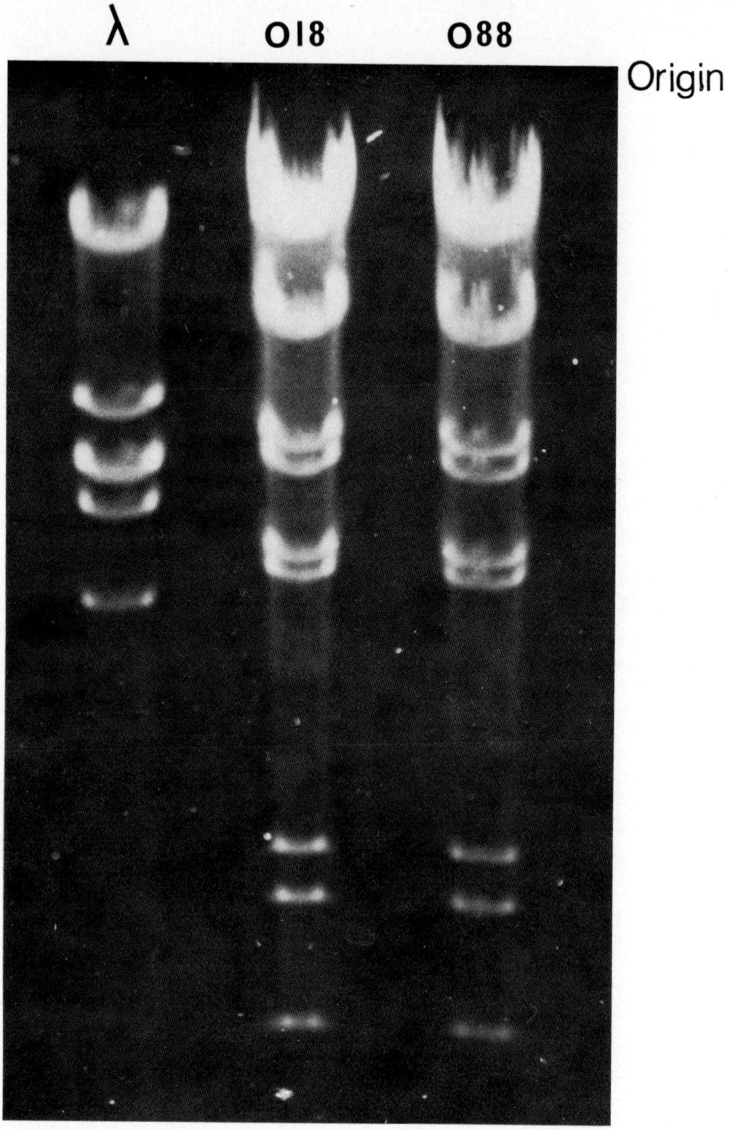

FIG. 1. *Eco*RI restriction endonuclease digest patterns from phage lambda DNA and from plasmid DNA from the O18 and O88 strains of *E. coli* isolated from the fecal flora of a single individual.

does occur. Under these circumstances it would be unwise to place too much reliance on "safe" mutants of *E. coli* K-12 if the mutations involved merely result in an inability of the strain to multiply except under bizarre growth conditions.

The Incidence of Plasmids in Naturally Occurring *E. coli*

One of the factors that bears on the possible hazard of using *E. coli* K-12 in genetic engineering experiments is the possibility that plasmids present in the coliform flora of the experimenter may provide the necessary genetic information to render the constructed hybrid DNA infectious. We have no real idea of how likely such an event may be, largely because we have little idea of the nature, incidence, and diversity of plasmid DNA in naturally occurring *E. coli* strains. Probably it is very common. When bacterial plasmids were detected by their centrifugal properties in the Spinco Model E (9), the presence of satellite DNA in a bacterial strain seemed uncommon, but with the advent of the dye/buoyant density method of detecting covalently closed circular (CCC) DNA (23), plasmids now seem ubiquitous. In our laboratory at one period we had to look at about 20 natural isolates of *E. coli* before we found one that appeared to have no CCC DNA on dye/buoyant density centrifugation, and this seems to have been a common experience in other laboratories. This suggests that the majority of *E. coli* carry one or more plasmids, although the detailed properties of these elements are usually unknown.

Another approach to this problem is to determine how many naturally occurring human *E. coli* strains carry R-plasmids. Studies in our Department of Bacteriology show that R-plasmid–carrying strains constituted about 2% of the 4,000 or so independent human *E. coli* isolates we have examined routinely by serotyping (12,20). However, it must be stressed that the use of antibiotics may effectively select the gut flora of an individual to the point where virtually all the coliforms are carrying R-plasmids.

The Incidence of Co1E1 in *E. coli* of Natural Origin

Derivatives of the colicinogenic factor Co1E1 have been used in a number of genetic engineering experiments (D. Helinski, *this volume*), and it is particularly important to know the incidence of this plasmid in naturally occurring *E. coli* since such plasmids could be particularly effective at repairing any genetic deficiencies deliberately inserted into these Co1E1 plasmids used for engineering purposes. Again our information is scanty. A survey of the literature suggests that the proportion of naturally occurring *E. coli* strains that produce some type of colicine is wide (10). In many publications, however, it is difficult to determine which particular colicine is implicated. Meynell and his colleagues (11) have shown that colicines can be divided broadly into two categories, i.e., small proteins (e.g., Co1E1) and larger molecules such as colicine V; at present, it is far from certain that all the "small" colicines are evolutionarily related to one another, although Meynell presents some evidence to suggest that they may be (11). As with the lambdoid phages (*see below*), it is probably important not to rely heavily on the published literature but to institute a study of the incidence of Co1E1 itself in naturally occurring *E. coli* strains.

The Incidence of Phage Lambda and Lambdoid Phages in *E. coli*

The existence of Co1E1 plasmids in the normal *E. coli* flora of an experimenter may give the opportunity for recombination to repair any genetic lesions deliberately introduced into the mutated derivatives of these plasmids used for genetic engineering work. In a like manner, similar repair of defective lambda phages could occur if the gut flora of the experimenter was rich in wild-type lambda or carried phages with lambda sequence homology.

Very little is known about the incidence of lambda and of lambdoid phages in naturally occurring *E. coli*. Practically the only published information on this topic is the 1957 paper by Jacob and Wollman (16), stating that about 30 phages in this category were encountered among 500 independently isolated *E. coli* strains of natural origin provided by Leon Le Minor. If these data are widely representative, it seems that lambda and related phages are far from rare, and consequently the chance of repair of the lambda phages used for engineering purposes must be considered as a serious possibility.

In practice, it is perhaps a little unwise to rely too much on the frequencies found in these studies at the Institut Pasteur. First, the work was carried out nearly 20 years ago, and the situation may have changed by now. Secondly, it is not certain to what extent the *E. coli* strains used in that study were really epidemiologically distinct. In short, another survey on the incidence of strains liable to recombine with the defective lambdas used for genetic engineering purposes is overdue. Perhaps the most reassuring aspect of this problem is that any recombinative event leading to a lambda phage having a wild-type replicative function might be likely to excise the exogenous DNA inserted into the lambda by the engineering procedures. It would be unwise, however, to assume that such a replacement must inevitably occur during such a recombination step.

DISCUSSION

Perhaps the one firm conclusion that we can reach in this difficult area is that experimenters being treated with antibiotics that would be expected to select for the resistance markers carried on experimental plasmids or strains should be barred from a genetic engineering laboratory until their course of treatment is complete. Antibiotic therapy often increases enormously the abundance of resistant *E. coli* strains in the human gut, and such therapy must therefore increase substantially the risk of survival or transfer.

Apart from this we do not have enough information to reach any really sound conclusions. Certainly it is possible to calculate probability values for the chance of various untoward events taking place, but any such calculations must be very insecure—if only because one can never be sure that there is no unsuspected selective advantage that will greatly enhance the survival of the strain under consideration. Evidence is slowly accumulating which suggests that *E. coli* K-12 does not colonize the human gut well and that plasmids carried by isolates of this particular *E. coli* strain are not very potent donors of plasmids by conjugation. But the large number of different *E. coli* strains that occur in man and the considerable variation among individual experimenters make it hazardous to make categorical statements in this context (7). The best plan at the moment seems to be for the scientific community to proceed with its experiments in this field, but always taking care that the level of containment used for the experiment is commensurate with the risk that might be anticipated if the strains used for the studies actually escaped and became established in the gut flora of the experimental workers or their contacts.

The Ashby Committee (4), which in 1974 reported on the problems of genetic engineering to the British Minister concerned with the problem, recommended that experimenters in this field should be subjected to bacteriological monitoring on a routine basis. We have been carrying out such studies on a limited scale in this Department, but one has the impression that such studies are not widespread. This seems a pity. It is true that the information gained from such studies is unlikely to be of much help to the individual in whom any accident occurs, but the steady accumulation of information about the properties of *E. coli* K-12, or of plasmids being used for experimental purposes, in the gut flora of the experimenters themselves is of the greatest value if we are to assess the behavior of these strains of *E. coli.*

The other important need at the moment is for basic research into the factors that influence the survival of *E. coli* strains in the alimentary tract of humans. Quite apart from considerations of genetic engineering, this information is important for our understanding of the etiology of human disease. Little information exists about the influences that determine whether a given strain persists or is eliminated, and in the engineering context, studies of this kind may help us to determine which "safe vectors" are really safe.

SUMMARY

The preliminary information about the behavior of *E. coli* K-12 in the human gut does not suggest that its use for genetic engineering is excessively hazardous—provided adequate containment procedures are enforced. This is reassuring since the enormous accumulation of genetic and biochemical information about this strain is of the greatest use in designing genetic engineering experiments. This relatively sanguine conclusion does not mean, however, that we should not exercise the greatest possible vigilance, and that information about the behavior of this strain in those working in the field should not be assiduously collected as experimentation continues.

ACKNOWLEDGMENTS

I would like to thank Dr. Christine Hartley for many helpful discussions in connection with the work described in this chapter. I would also like to thank Drs. Schmidt and Jann of the Max Planck Immunobiology Laboratory, Freiburg, West Germany for the kind gift of some strains. The original work described here was supported by a grant from the Medical Research Council for Molecular and Epidemiological Studies in R-factors and Other Plasmids.

REFERENCES

1. Anderson, E. S. (1975): Viability of, and transfer of a plasmid from Escherichia coli K-12 in the human intestine. *Nature,* 255:502–504.
2. Anderson, J. D., Gillespie, W. A., and Richmond, M. H. (1974): Chemotherapy and antibiotic resistance transfer between Enterobacteriaceae in the human gastrointestinal tract. *J. Med. Microbiol.,* 6:461–473.
3. Anderson, J. D., Ingram, L. C., Richmond, M. H., and Wiedemann, B. (1974): Studies of the nature of the plasmids arising in the human gastro-intestinal tract. *J. Med. Microbiol.,* 6:475–486.
4. Ashby, Lord (1974): *Report of a Working Party on the Experimental Manipulation of the Genetic Composition of Micro-organisms.* Her Majesty's Stationery Office, London.
5. Cooke, E. M. (1974): *Escherichia coli and Man.* Churchill Livingstone, Edinburgh.
6. Curtiss, R., III, Charamalla, L. J., Stallions, D. R., and Mays, J. A. (1968): Parental factors during conjugation in Escherichia coli. *Bacteriol. Rev.,* 32:320–348.
7. Dupont, H. L., Formal, S. B., Hornick, R. B., Snyder, M. J., Libonati, J. P., Sheahan, D. G., Labrec, E. H., and Kalas, J. P. (1971): Pathogenesis of Escherichia coli diarrhea. *N. Engl. J. Med.,* 285:3–11.
8. Ewing, W. H., and Davis, B. R. (1961): *The O-antigen Groups of Escherichia coli from Various Sources.* Center for Disease Control, Atlanta.
9. Falkow, S., Wohlheiter, J. A., Citarella, R. V., and Baron, L. (1964): Transfer of episomic elements to Proteus. *J. Bacteriol.,* 87:209–219.
10. Fredericq, P. (1948): L'antibiose chez les Enterobacteriaceae. *Rev. Belge Pathol. Med. Exp.* [*Suppl.*], 4:5–107.
11. Hardy, K. G., Meynell, G. G., Dowman, J. E., and Spratt, B. G. (1974): Two major groups of colicin factors: their evolutionary significance. *Mol. Gen. Genet.,* 125:217–230.
12. Hartley, C. L., Howe, K., Linton, A. H., Linton, K. B., and Richmond, M. H. (1975): Distribution of R-plasmids among the O-antigen types of Escherichia coli isolated from human and animal sources. *Antimicrob. Agents Chemother.,* 8:122–131.
13. Hartley, C. L., and Richmond, M. H. (1975): Antibiotic resistance and the survival of Escherichia coli in the alimentary tract. *Br. Med. J.,* 4:71–74.
14. Hartley, C. L., Petrocheilou, V., and Richmond, M. H. (1975): Antibiotic resistance in laboratory workers. *Nature,* 255:502–504.
15. Ingram, L. C., Anderson, J. D., Arrand, J., and Richmond, M. H. (1974): A probable example of R-factor recombination in the human gastro-intestinal tract. *J. Med. Microbiol.,* 7:251–257.
16. Jacob, F., and Wollman, E. L. (1957): Sur les processus de conjugaison et de recombinaison chez Escherichia coli. I. Induction par conjugaison ou induction zygotique. *Ann. Inst. Pasteur* (*Paris*), 91:486–510.
17. Jarolmen, H. (1972): Experimental transfer of antibiotic resistance in swine. *Ann. N.Y. Acad. Sci.,* 182:72–79.
18. Kaufmann, F. (1966): *The Bacteriology of the Enterobacteriaceae.* Munksgaard, Copenhagen.
19. Lüderitz, O., Jann, K., and Wheat, R. (1968): Somatic and capsular antigens in Gram-negative bacteria. In: *Comprehensive Biochemistry, Vol. 26a,* edited by M. Florkin and E. H. Stotz. Elsevier, Amsterdam.

20. Petrocheilou, V., and Richmond, M. H. (1976): Distribution of R-plasmids among the O-antigen types of Escherichia coli isolated from various clinical sources. *Antimicrob. Agents Chemother.,* 9:1–5.
21. Petrocheilou, V., Grinsted, J., and Richmond, M. H. (1976): R-plasmid transfer in vivo in the absence of antibiotic selection pressure. *Antimicrob. Agents Chemother.,* 10:753–761.
22. Public Health Laboratory Service Monograph, No. 6 (1974): *The Prevention of Laboratory Acquired Infection.* Her Majesty's Stationery Office, London.
23. Radloff, R., Bauer, W., and Vonograd, J. (1967): A dye/buoyant density method for the detection and isolation of closed circular duplex DNA: the closed circular DNA of HeLa cells. *Proc. Natl. Acad. Sci. U.S.A.,* 75:1514–1521.
24. Schmidt, G., Jann, B., and Jann, K. (1974): Genetic and immunochemical studies on Escherichia coli 014:K7:H^-. *Eur. J. Biochem.,* 42:303–309.
25. Smith, H. W. (1969): Transfer of antibiotic resistance from farm animal and human strains of Escherichia coli to resident Escherichia coli in the alimentary tract of man. *Lancet,* 1:1174–1176.
26. Smith, H. W. (1973): Chloramphenicol resistance of Escherichia coli. *J. Med. Microbiol.,* 6:347–350.
27. Smith, H. W. (1975): The fate of orally administered E. coli K12 in the alimentary tract of man. *Nature,* 255:500–502.
28. Wiedemann, B., and Knothe, H. (1969): Unterschungen über die stabilität der Koliflora des gesunden Menschen. I. Uber das Vorkommen permanenter und passanter Typen. *Arch. Hyg. Bakteriol.,* 4:342–348.
29. Wiedemann, B., and Schmidt, G. (1972): Structure and recipient ability in E. coli mutants. *Ann. N.Y. Acad. Sci.,* 182:123–125.

Recombinant Molecules: Impact on Science and Society, edited by R. F. Beers, Jr. and E. G. Bassett. Raven Press, New York © 1977.

40. The Role of the National Institutes of Health in Rulemaking

Leon Jacobs

Office of Collaborative Research, National Institutes of Health, Bethesda, Maryland 20014

People at NIH are fond of saying that NIH is a research institution, not a regulatory organization. This is indeed true now, and research is what NIH does and supports best. Therefore, NIH approaches rulemaking the way porcupines must make love—gingerly. However, NIH does strongly influence the scientific and medical care community, almost entirely by informal means.

If one visualizes NIH's waves of influence as concentric circles, there is only a tiny inner circle of true regulation—and that indirectly, on behalf of the Public Health Service—to license the importation of primates from India, and to approve the conditions for transportation of certain hazardous biological materials. Later I will describe the manner in which regulations are made, and the legalities of regulations compared with other types of NIH's published policies and procedures. Such documents represent a second somewhat larger circle of NIH influence, where the policies and procedures are made conditions of grants and contracts. They include the procedures by which we provide for the use of certain national or regional research facilities by medical investigators, the requirements for the care of laboratory animals, the guidances for review of research involving human subjects, peer review of grant applications, and many more. Guidelines for research on recombinant DNA are the latest example. In some cases, our informal mechanisms have been replaced by the formal promulgation of regulations, and I will talk about those briefly later.

An outer circle of NIH influence affects everyone, particularly health providers. This influence results from the NIH conduct and support of clinical trials and the validation of new forms of therapy as well as long-accepted modes of therapy. Other influence stems from the development of vaccines and advice on the use of such preventive measures, subjects about which NIH works closely with the Bureau of Biologics and the Communicable Disease Center.

All of the regulations and guidelines are important. In their absence,

- India would cut off supplies of monkeys.
- Airline pilots would refuse to transport, on passenger aircraft, materials essential to the conduct of research and to the diagnosis of disease.
- Either we would put research patients unnecessarily at risk, or the public would rise up and stop clinical research in its tracks.
- We would continually be involved in litigation against the use of animal models of human disease.
- The decisions on support of scientifically meritorious work would be compromised or locked into a bureaucratic system.

All of these measures, regulations, guidelines, and operating procedures have become increasingly the objects of public scrutiny and involvement, the result of heightened sensitivity to their possible effect. We do not make rules in isolation. As I describe the development of the NIH guidelines for research on recombinant DNA molecules, you will see how open that process has been. We have no objection to such openness because it is perfectly proper for the public to be involved. However, public involvement does complicate the process, and unless all parties exercise a high degree of responsibility the process can be thrown into confusion.

The Secretary of the Department of Health, Education, and Welfare has general authority to promulgate rules for the conduct of Departmental business. When he uses this authority, the rules are published in the Federal Register, as a notice of proposed rulemaking. A period of 60 days or more is allowed for comment, the comments are studied, and revised rules are then published, with an explanation of the reasons for acceptance or rejection of comments. When the final rules are published, they have the force of law.

The Secretary has delegated special authority to the Food and Drug Administration for certain rules and regulations regarding drugs and biologics because of the large regulatory responsibilities of the FDA. NIH has no such delegations; as I have pointed out, we have no significant responsibility for regulating. However, we do follow a large set of rules, which we make known to the communities we deal with, concerning the receipt and review of grant applications and all of the other subjects mentioned previously. These are internal documents, and when they are approved by the Public Health Service and published in the Grants Administration Manual, the academic community is notified and we conduct our business with them in accordance with these rules. Although these guidelines do not have the same force of law as do published regulations, a recent court decision gives them a degree of authority. Whether other courts would rule similarly, in relation to other issues in our operations, is always a question.

In some cases NIH guidelines do become regulations. For example, NIH originally developed, in 1956, policies and procedures for its intramural programs regarding the protection of human subjects of research. In the 1960's, similar policies and procedures were developed for the conduct of clinical research supported by grants and, later, by contracts. These were subsequently adopted by the Public Health Service for all of its agencies. The publication of these policies and procedures announced the requirements that PHS agencies would impose on institutions before support would be provided for clinical research projects. Ultimately they were published as formal regulations.

Another rulemaking about to be promulgated formally concerns peer review of grant applications and contract projects for biomedical and behavioral research to be administered by NIH and its sister agency, the Alcohol, Drug Abuse, and Mental Health Administration (ADAMHA). We have been using peer review for years for both grant applications and contract proposals. However, in the early years of this decade, peer review came under questioning from certain sectors of the Executive Branch. (Actually, this questioning was originally generated by disgruntled scientists who were unsuccessful in obtaining grants because of the financial crunch in the late 1960's and early 1970's. When the funding situation improved, the complaints of the scientists were largely stilled.) The Congress, to protect our system, wrote a requirement for formal peer review regulations into the Cancer Act Amendments of 1974. NIH took the lead in writing the proposed rules, and ADAMHA participated. The rules were passed up to the PHS, the Assistant Secretary for Health, through the domains of a number of other Assistant Secretaries, and to the Under Secretary, and appeared as a notice of proposed rulemaking in the Federal Register. Publication of proposed rules took about 18 months from the time the law was passed. After all comments by the public and appropriate staff in the agencies are considered, the rules will be promulgated in final form, again with an explanation given for all comments accepted and rejected.

NIH has been involved in regulations in the past, specifically, from 1902 when it became responsible under the Public Health Service Act for the purity, safety, and efficacy of biologics, until 1972 when the Division of Biologics Standards (DBS) and its functions were transferred to the agency of the Public Health Service whose mission is mainly regulatory—the Food and Drug Administration. The DBS became the Bureau of Biologics (BoB), FDA. Every time a new vaccine or other biological product is developed, it is necessary for the BoB to publish a rule describing the standards of production and the tests for purity, safety, and efficacy that would be required before a manufacturer could be licensed to distribute the product in interstate commerce, and before any individual lot of the product could be released by a manufacturer for such distribution. Again the process requires the publication of proposed rules in the Federal Register, analysis of comments, and publication of a final set of rules. The BoB remains on the NIH campus and interacts with NIH scientists engaged in work on infectious diseases and the development of vaccines and biological diagnostic products. It has to do research and be close to research because its staff must become skilled and knowledgeable about new biologics that are under development.

The first step in rulemaking by the Bureau of Biologics has generally been to hold a workshop or a series of workshops on the state of the art in a particular problem area. When a consensus is reached that a vaccine or a biological diagnostic reagent or test system has been developed sufficiently so that the criteria for efficacy, purity, and safety can be described and tests in human beings can be conducted, the process of rulemaking becomes formal.

Beyond the workshops, the Bureau of Biologics has standing committees to provide the BoB Director and staff with advice on the development of standards for the product under consideration. The tests and the rulemaking then proceed.

We have been following a similar course in regard to research on recombinant DNA molecules; and, to some extent, we are informally following the rulemaking process. This has involved at least a year and a half of debate at open committee meetings and public comment, as I describe our involvement in this topic.

When the NAS committee chaired by Dr. Paul Berg issued its statement on DNA recombinant research in July, 1974, NIH responded rapidly to the requests the committee addressed to it. Arrangements for the Asilomar conference were made, with funds from the National Cancer Institute and a contribution from the National Science Foundation. The NIH Recombinant DNA Molecules Program Advisory Committee was established by the Secretary, HEW, on our initiative. The relative speed with which this committee received approval indicates the high priority placed on it by both NIH and HEW. The functions of the committee, as stated in the charter, are to provide advice concerning a program for:

1. Evaluation of potential biological and ecological hazards of DNA recombinants of various types;
2. Development of procedures that will minimize the spread of such molecules within the human and other populations;
3. Development of guidelines to be followed by investigators working with potentially hazardous recombinants.

NIH has had, for some years, some relatively simple and easily applied procedures for "flagging" grant applications on projects involving naturally occurring hazardous microorganisms and for assuring the granting agency that the investigator has adequate facilities and expertise to protect him and his staff from infection. As early as 1972, we recognized the need for some control of work on laboratory-produced hybrid viruses that could spread in the general population if they escaped from the laboratory. Dr. Robert Berliner, then Deputy Director for Science of NIH, had forbidden work on competent adenovirus 2-SV40 hybrids, except in high-containment facilities; and I was engaged in writing NIH rules for managing grants and contractual projects involving such agents. Thus, we were prepared to use the advice of the new committee. We will address here only that function of the committee which relates somewhat to rulemaking.

I am sure you are all familiar with what transpired at the Asilomar conference on DNA recombinants that was held in February, 1975. I will not extend my comments on that meeting, except to say that it was one of the most interesting and exciting that I have ever attended. It reflected the great moral responsibility of the participants and especially the organizers, who had, the previous July, urged all investigators to join them in the commitment to forego experiments with recombinants until the meeting could be held and the hazards discussed. It came up with a product—a set of general recommendations that identified certain types of experiments as too hazardous to be performed—and recommended that other experiments be done only under certain levels of physical containment. It emphasized a new idea—biological containment, the development of hosts and vectors that would be unable to survive outside the laboratory environment. In some respects, the conference was analogous to the type of workshops I have mentioned as the initial step in rulemaking by the Bureau of Biologics. When our advisory committee met on February 28, 1975, immediately following the conference, it was decided to accept the recommendations of the Asilomar conference on a provisional basis, and to concentrate on a set of guidelines that would be more specific than those of Asilomar and would be used to inform our study section members, our grant applicants, and our scientists administering grants in this field of:

- What experiments should not be supported under any conditions at the present time.
- What levels of physical containment should be used for the various kinds of experiments.
- What levels of biological containment should be required for these experiments.
- The roles and responsibilities of the investigator, the institution, and the granting agency in the review and monitoring of research involving recombinant DNA molecules.

The committee has labored long and hard on these guidelines. There have been three drafts. The first, prepared by a subcommittee chaired by Dr. David Hogness, was reviewed by the full committee at a meeting in Woods Hole in July, 1975. The draft set up the physical and biological containment levels for research on recombinant DNA derived from various types of organisms. The resulting document came to be known as the Woods Hole draft. It was distributed to the members, and a number of other scientists became informed of its contents. The result was a flood of letters into Dr.

DeWitt Stetten's office, most of them charging that the guidelines were too lenient and permissive. Dr. Stetten, as chairman of the committee, reproduced these letters and distributed them to all the committee members. This precipitated a second wave of letters arguing that the guidelines were far too restrictive. There was nothing to do but to devote the next meeting of the committee to further review of the guidelines. Dr. Stetten asked another committee member, Dr. Elizabeth Kutter, to chair a new subcommittee to produce a third draft, taking into account all of the communications and other information received in extensive telephone discussions with committee members and many other interested scientists. There seemed to be little hope of all these people coming to an agreement because of the wide range of opinion and the completely opposite views of extremists on both sides of the argument.

The Kutter draft became available before a fourth meeting of the committee in December, 1975, at La Jolla, California. We went through every major item of difference in the drafts, leaving only trivial editorial matters to be resolved by the chairman, and worked hard for 2 days and nights. Eventually, the committee members, finally really having worked as a committee, prepared a document that represented more of a consensus than we could have hoped for originally.

Probably the reasons for agreement were many. One, undoubtedly, was the gradual fusion of the many individualists on the committee, during the previous meetings and innumerable private discussions, into a working group with a united purpose. The purpose was to construct a set of rules that would be adequate to protect human beings and the rest of the ecosystem from theoretical risks, while assuring that opportunities for research in this most important area were left open for the eager investigators who wanted to explore it. Another reason was apparent at La Jolla: our committee met after a workshop on the design and testing of safer prokaryotic vehicles and bacterial hosts for research on recombinant DNA molecules, which was held there earlier that month. (Indeed, our reason for holding our committee meeting at that site was for the convenience to committee members who also attended the workshop.) It became apparent at the workshop that safer bacteriophage vectors and plasmid vectors were being developed, and that a safer bacterial host had been constructed because of a really all-out effort on the part of one of our committee members, Dr. Roy Curtiss, and his associates. These developments indicated that the requirements for EK2 and EK3 host-vector systems, which the committee was considering recommending for various types of experiments, were indeed realistic, and at least a few could be expected to be available in the near future. A third reason was, very likely, the realization that although it is possible to clone segments of eukaryotic DNA in prokaryotic hosts, transcription of eukaryotic genes might prove difficult in these hosts. Thus, the scenarios of horrendous dangers resulting from the introduction into the biosphere of microorganisms with new capabilities, which have been dreamed up by many people, seemed less likely to be staged.

I am reminded of similar scenarios composed about 15 years ago when President Kennedy announced the program to send men to the moon. At about that time, someone had sterilized the outside of a fragment of a meteorite, transferred it to a germ-free chamber, crushed it, and inoculated the powdered product into a series of different kinds of bacteriological culture media. As I remember the story, in one type of medium some small filamentous material appeared. This was never proved to be a microorganism. However, it may have contributed to expressions of concern that exploration of space might result in the introduction into this planet of organisms that could destroy our agriculture or otherwise damage our ecosystem. Discussions at the National Academy of Sciences resulted in more scenarios. The exploration of the lunar surface was followed, when the astronauts returned to earth, by attempts at containment of the men and the specimens they brought back and a period of quarantine. I believe it was prudent to do these things. However, to my knowledge, no organism was found in or on the astronauts or the specimens they collected that was not a known earthly entity. We will do well to be prudent, but we should not be paralyzed by fear of the dark. We should be willing to enter it with caution and searchlights. The guidelines, I believe, so equip us.

After the La Jolla meeting, the proposed guidelines were prepared in a final draft and presented to Dr. Donald S. Fredrickson, Director of NIH, for his study. He convened in February, 1976, a special meeting of the Advisory Committee to the Director, a broadly expert group itself, and supplemented by leaders in law, ethics, and consumer affairs, to review the guidelines and to determine whether, in their judgment, they balanced scientific responsibility to the public with scientific freedom to pursue new knowledge. Other members of the scientific community were informed of the meeting and told that they would have the opportunity to present their views. Similarly, some public interest representatives were informed directly of the opportunity to participate, and a notice was published in the Federal Register announcing the meeting and its open nature. These steps were taken in the hope that the scientific community can maintain the public's confidence that the goals of this important research accord respect to the equally important ethical and social values of our society and to the laws designed on that base.

The meeting brought up a number of points that required further study. Dr. Fredrickson reviewed all of these and then asked the DNA Recombinant Advisory Committee to consider, at their meeting of April 1 and 2, a number of selected issues raised by the commentators. Dr. Fredrickson has taken those issues, the response of the Recombinant Advisory Committee, and numerous written comments into account in arriving at his own decision on the guidelines. He has prepared an analysis of the issues and the basis for his decision, which will be published along with the final guidelines.

As I mentioned earlier, NIH can publish the guidelines on its own authority. This will be merely a statement of rules to be followed, but it will not have the force of law. However, we can operate within these rules if we obtain voluntary compliance by the institutions that receive our grants and if we are not challenged in the courts.

A later step will probably be publication of an NIH manual issuance, which will have to be approved by the Public Health Service, because it bears on grants administration policy. The manual issuance will serve to instruct institutions as to procedures, as well as the underlying policy, by which grant projects involving recombinant DNA will be submitted, reviewed, and monitored. The guidelines need not be incorporated into the manual, but merely must be cited as a reference. Here again, they will not have the force of law.

We hope to gain considerable experience with the guidelines and the way they are used in the next few years. If we are successful with them, we may not have to proceed with the formal rulemaking.

Beyond these activities, NIH has attempted to apprise all federal government agencies and departments of the research on recombinant DNA, the expected benefits and possible risks of such research, and the guidelines being developed. Dr. Fredrickson held a meeting of representatives of industrial concerns that may be or may become involved in such research; these firms are involved in pharmaceutical research, industrial fermentation, and similar endeavors. Here again, we hope to foster communications and obtain voluntary compliance with the guidelines.

I believe that the process has been, from the start, responsive to the concerns of interested people and responsible in the way the problems have been addressed. Now I would like to turn to what may go awry in the process.

In recent years, we have had a number of problems at NIH involving dissident scientists who have taken their cases to the newspapers and whose lawyers have lobbied for Congressional hearings about issues which are really scientific matters and do not benefit by public debate. I am talking about issues other than recombinant DNA. Frequently, the arguments made public are incomplete; the information displayed is lacking in data contrary to the points the dissidents wish to make. Judgments arrived at by peer review groups are ignored. Quite apart from the nuisance of handling such cases, the litigation involved, the inquiries from reporters for more information, and so on, these events hurt the scientific reputation of the NIH, of some of its staff, and of some of the extramural community of scientists that NIH serves.

I am well aware of Dr. Alvin Weinberg's papers on science and trans-science. In the particular cases I have in mind, the issues are resolvable as scientific issues. The tactics of the dissidents are unfair, in my mind, and represent the machinations of unreasonable men. Because of the political furor they have sometimes been able to raise, individuals and organizations have suffered seriously, and the disputes recrudesce.

I have some grave doubts about the way we do some of our business now. Nobody can question the sincerity of many of the public interest groups seeking to preserve our environment, etc., and of their right of participation when the issues are transscientific. However, I would hope that they all behave in as responsible a manner as possible. In the case of research on recombinant DNA molecules, there is ample evidence that the expertise and the public responsibility reside in the scientists who raised the issues in the first place, and who have labored hard to provide us with some guidelines on how studies involving recombinants of various types should be done. We have also asked for public participation in the work. We cannot hope that public representatives will be able to understand all of the science and make their own assessments of risk, but they can make sure that the scientists are making rules in a fair way.

To people who say we are too restrictive, we can say it is best to proceed with caution. To those who say we are not restrictive enough, we can say that there is only a theoretical risk of damaging the environment. There is a risk, admittedly also theoretical, that failure to continue research with recombinant DNA may delay many important advances, such as one that may avert disaster in regard to our ability to produce enough food for the world's burgeoning population. There is a *real* risk in that clamor against research can lead to serious consequences in the arenas of politics and law. Statements have been made by politicians that if scientists disagree, then the politicians will have to intervene. There are good reasons for politicians and the public in general to become involved in some disputes, as Dr. Weinberg has so well explained in his discussion of transscience.

So far as recombinant DNA research is concerned, we may be in an area of transscience at this time. However, I am fearful that if dissident scientists, in public forum, engage in irresponsible attacks on the rules we have so painstakingly tried to write, they will stimulate political intervention that will block free inquiry and put such constraints and restraints on research that American scientists will have to emigrate to other countries to pursue their studies. The Democratic Party of Washtenaw County, Michigan, has already adopted (on May 20) a resolution that research on recombinant DNA be restricted by law to a limited number of facilities "equipped to prevent the spread of recombinant DNA." I might note here that it is expected that Great Britain will probably lay down some rules in consonance with the Ashby Report, which states that, "provided precautions are taken, the potential hazards need not cause public concern." We have also had communication from the European Molecular Biologists' Organization that it plans to adopt our guidelines for its own purposes, providing the rules are not made more restrictive than in the La Jolla edition. We will hear more directly on that point from Dr. Tooze.

I guess I am revealing the discomfiture that Dr. Weinberg mentions, which scientists feel when they must engage in public debate of issues. I am in complete agreement with him that the public has a right to participate in the debate when questions arise that cannot be settled by scientific means. As he points out, we have a responsibility to define, in such debate, how much we know and what we cannot know, where science ends and trans-science begins. I am not worried about that debate; I am worried about the tactics that may confound honest debate, and I am stating my hope that they will not be used regarding recombinant DNA research.

Felix Frankfurter once wrote to a friend that what constitutes friendship is not agreement on opinions, but harmony of aims. I hope we have that harmony of aims that is essential to the support of the scientific endeavor. I hope, too, that we can manage with our rulemaking process, making adjustments as needs become apparent.

Recombinant Molecules: Impact on Science and Society, edited by R. F. Beers, Jr. and E. G. Bassett. Raven Press, New York © 1977.

41. Emerging Attitudes and Policies in Europe*

John Tooze

European Molecular Biology Organization, Heidelberg, Federal Republic of Germany

INTRODUCTION

The techniques for recombining DNA molecules in cell-free systems are essentially an American innovation as is the concomitant public debate about the scientific benefits and possible biological risks that may be associated with such experiments. In Europe, so far, much of the science and most of the debate about the impact of this research on science and society have been derivative; Europe has played variations on the American theme. Here I shall try to review the various European responses to the discussion that was initiated in the United States and suggest how the situation may develop in the near future.

I have used the plural, European responses, advisedly. Although Europe may be a geographic unit, it includes a set of sovereign nation states with different languages, histories, and cultures. Science, and especially molecular biology, is of course international; as a group, molecular biologists must be among the world's most widely traveled people, but in Europe the political and administrative decisions which control the development of science remain jealously guarded national prerogatives. Each country has its own set of research councils, academies, ministries of health, science, technology, education, and so on. There is no organization in Europe in any way comparable to the National Institutes of Health, which, with great resources of money and people, can make and execute policies affecting a population in excess of 200 million. There are, of course, European international scientific organizations, such as the European Centre for Nuclear Research (CERN), the European Southern Observatory (ESO), the European Space Agency (ESA), and the European Molecular Biology Laboratory (EMBL), which were established essentially as economy measures, but the policies of these organizations are decided by councils of delegates representing the sovereign member nations. National delegates can in-

* The opinions expressed here are those of the author. They do not necessarily represent the opinions of the European Molecular Biology Organization.

fluence one another, but in the last analysis they have no power over each other.

The response in Europe to the advent of recombinant DNA research has followed the characteristic European pattern. Most countries have or are now establishing national administrative structures to deal with the issues that have been raised and given a thorough public airing in the United States. The duplication of effort has been great: more time has probably been spent holding virtually identical discussions in the several countries of Europe than has been spent actually doing experiments. That is the inevitable price of national sovereignty, and I believe these debates might have been much longer if the United States had not had such hegemony in this field. Once a consensus had been reached in the United States, the molecular biologists and science administrators in European countries felt that, for the short term at least, they had little option except to follow, more or less reluctantly, the American lead. Only in the United Kingdom has there been an extended independent inquiry at the governmental level.

EUROPEAN RESPONSES TO THE RECOMBINANT DNA DEBATE

The United Kingdom

THE ASHBY REPORT

Considering the strength of molecular biology in the United Kingdom, it is perhaps not surprising that the first official European reaction to the publication of the statement of the National Academy of Sciences' Committee on Recombinant DNA in July, 1974 came from the British government. Within weeks of this publication a working party, under the chairmanship of Lord Ashby, was established by the Advisory Board of the Research Councils. This working party, after hearing many witnesses, submitted its report to the Secretary of State for Education and Science on December 13, 1974. The Ashby Report was widely welcomed on both sides of the Atlantic, and it was clearly of great help to the Organizing Committee of the Asilomar Conference. It focused the debate by drawing attention to the great benefits likely to accrue from recombinant DNA research, by acknowledging that there were possible biological risks involved, and by emphasizing that our ignorance of the properties of recombinant DNA was virtually total so that estimates of risk were necessarily mostly subjective. The Ashby Working Party advised, therefore, that special precautions should be taken initially to allow the research to proceed, and it made a series of points that have gained universal acceptance, for example:

1. That the precautions taken should match our best estimate of the possible risks.

2. That in addition to physical containment there is the additional possibility of biological containment.
3. That working with large volumes of cultures is more risky than working with small volumes.
4. That experimenters should be familiar with the techniques for handling pathogens commonly used by medical microbiologists.
5. That the individual investigator is primarily responsible for judging the possible risks and maintaining the appropriate standard of laboratory hygiene.

By chance, while the Ashby working party was considering the genetic manipulation of microorganisms, a separate working party chaired by Sir George Godber, which had been established in November, 1973, was preparing a report for the British Government on the laboratory use of dangerous pathogens, following an accidental release from a London laboratory of smallpox virus which had fatal consequences. The Godber working party presented its report in May, 1975, five months after the Ashby Report.

THE WILLIAMS WORKING PARTY

The British Government felt that the subjects of these two reports were sufficiently closely related to justify establishing a third working party to consider how to implement the recommendations of both reports. However, it was not until August 6, 1975, eight months after the Ashby Report had been submitted, that the Department of Education and Science (DES) announced that the third working party was to be chaired by Professor R. E. O. Williams. The brief included: "to draft a central code of practice and to make recommendations for the establishment of a central advisory service for laboratories using the techniques available for such genetic manipulation (recombinant DNA techniques), and for the provision of training facilities" and also to consider controls for work with dangerous pathogens. The report of the Williams working party is in press, and its publication is awaited with great interest in Europe, not least because in the interim the remainder of the countries of Western Europe have agreed in principle to follow the draft NIH guidelines which emerged from the December, 1975 meeting of the NIH Recombinant DNA Program Advisory Committee.

After such a fast response by the Ashby working party, many British molecular biologists, who are eager to begin recombinant DNA experiments, have been disappointed and frustrated by their government's tardiness in acting on its recommendations. To be sure, the DES press release of August, 1975, which announced the formation of the Williams working party, stated that "The government for its part agrees that the potential benefits identified in the Ashby report are such that relevant work should continue in appropriate places provided that adequately stringent precautions for containment are taken." But in practice, the research councils in the United Kingdom appear to have discouraged new experiments involving recombinant DNA molecules pending the publication of the report of the Williams working party.

If, as is expected, the Williams working party reports to the British government at the end of May or the beginning of June, 18 months will have elapsed and presumably further time will pass before whatever the Williams working party recommends can be implemented. The more sanguine British molecular biologists have sought consolation for the delays in the thought that the simultaneous consideration of research with recombinant DNA and research with dangerous pathogens might have had the great advantage of putting the conjectural biological risks of the former into a realistic perspective.

Continental Europe

In the other countries of Western Europe there was no official reaction to the publication of the letter of the National Academy of Sciences' Committee in July, 1974, but in the aftermath of the Asilomar Conference in the early summer of 1975, several countries established national committees.

FRANCE

In June, 1975 the Délégation Générale à la Recherche Scientifique et Technique (DGRST), an organization responsible for coordinating the work of the French Research Councils, established two committees concerned with aspects of *in vitro* genetic manipulation. The first committee, Commission Reflexion dite d'Ethique ou d'Appel, chaired by Professor Jean Bernard, has the responsibility of considering philosophical, ethical, moral, and legal questions posed by recombinant DNA research. This is a committee of mandarins, and at least to date its meetings have been infrequent.

The French Control Commission. The second French committee, the Commission de Contrôle (Control Commission), chaired by Professor J. F. Miquel of the DGRST, has the day-to-day responsibility of reviewing research proposals involving recombinant DNA and classifying them according to possible risk. It must then recommend the appropriate containment conditions and other safety precautions. Initially, the commission followed the guidelines of the Asilomar Report, but it has subsequently used the December, 1975 draft NIH guidelines. This commission does not judge scientific priorities; that remains the prerogative of the various French research councils and universities. Meeting virtually every month, the commission has already considered some 10 major research proposals, and its experiences and initiatives are, I think, instructive. It is worthwhile mentioning, for example, that when the 16 members in addition to the chairman were appointed, efforts were made to include medical men, trade union representatives, and other educated laymen, as well as molecular biologists working with recombinant DNA. The laudable aim was to involve laymen in the decisions on safety precautions for specific experiments. However, and perhaps predictably and understandably, the attendance of these lay members at commission meetings had decreased with time. Apparently, this is also the experience in the Netherlands. In practice it may be difficult to avoid the situation in which molecular biologists act as judge and jury of their own cases.

The French Control Commission, in addition to its routine work of assessing the safety aspects of particular experiments, has taken two related policy initiatives that will undoubtedly influence other European countries. First, it has drafted a convention or declaration of agreement under the auspices of the DGRST (*see Appendix I*), which in essence will require all heads of French governmental, academic, military, and industrial laboratories to submit in advance projects involving recombinant DNA to the Control Commission. Although the convention will not have the force of law, its existence—and it will be widely publicized—will have moral force and perhaps even legal force if an unreported experiment leads to legal proceedings.

Secondly, the French Control Commission is to establish a national registry of recombinant DNA research based on answers to a questionnaire that has been drafted. The Commission has from the outset sought, and apparently gained, the cooperation of both the pharmaceutical industry's association and the military. The French pharmaceutical industry, for example, has agreed to join the registry when it begins recombinant DNA projects. If such close cooperation is maintained, in spite of industrial and military secrecy, the French Control Commission will have established a valuable precedent for the remainder of Europe and perhaps for the United States as well.

In short, in France a central administrative body for overseeing the safety aspects of recombinant DNA research is established and working. Developments in several other continental Western European countries are similar to those in France.

THE NETHERLANDS

In the Netherlands, the Royal Academy of Science established an *ad hoc* committee after the Asilomar conference, which in August, 1975 recommended to the Minister of Science that a permanent committee responsible to the ministry should be established. As a result, Professor D. Bootsma was appointed chairman of a Commission in Charge of the Control over Genetic Engineering. The membership includes representatives not only of molecular biology and other biological sciences but also of sociology and epidemiology. The Dutch commission's brief includes the responsibilities for establishing a national registry, advising on safety precautions for particular experiments (to date some 10 have been considered), advising on possible international agreements or treaties, and advising the government on any national legislation that may be required.

It is noteworthy that the Dutch commission is contemplating recommending legislation to ensure that all laboratories in the Netherlands, irrespective of the source of their financial support, register their recombinant DNA research projects.

WEST GERMANY

Shortly after the Asilomar conference, the Deutsches Forschungs Gemeinschaft (the German Research Association) established a Senatskomission für Sicherheitsfragen bei der Neukombination von Genen (Senate Commission for safety questions posed by new gene combinations), which has responsibilities for setting up guidelines for recombinant DNA research, advising on the construction and financing of containment laboratories, advising on possible legal actions, and advising on international cooperation and collaboration. The commission's membership includes molecular biologists, other biological scientists, and representatives of ministries and the Volkswagen Foundation; it works in close cooperation with German industry and the government. At least two pharmaceutical companies in Germany are contemplating projects involving recombinant DNA, and at least one of them is seeking government financial support. Although the establishment of a national registry is not explicitly included in the terms of reference, it is a development that can be anticipated.

SCANDINAVIA

In Sweden, an *ad hoc* committee of the Royal Swedish Academy of Sciences was established initially, but this has now been superseded by a committee, under the auspices of the Swedish research councils and chaired by Professor P. Reichard. It has representatives of the science, medical, and agricultural research councils, defense research, and the ministry of health and lay members, including politicians; it met for the first time in May, 1976.

The Norwegian Research Council for Science and the Humanities is currently establishing a national committee, while in Denmark no initiative has been taken as yet; we may expect the research councils of the latter country to recommend the formation of a committee.

Other West European Countries

In Switzerland, a Commission for Experimental Genetics was established by the Swiss Academy of Medical Sciences at the end of 1975. It is chaired by Professor W. Arber and has representatives of the biological and medical sciences of the universities, industrial representatives, and representatives of the Federal Department of Health and of Science and Research. A committee of the Italian Society for Biophysics and Molecular Biology, chaired by Professor E. Calef, has been recently established to report to the Italian government. An *ad hoc* Committee of the Medical Research Council of Ireland has been established. As far as I am aware, national committees have not yet been established in Austria, Belgium, Finland, Spain, or Greece.

Eastern Europe

Developments in Eastern Europe are more obscure, but East European molecular biologists attended the Asilomar Conference and presumably the general recommendations made in the Asilomar Report have been accepted in East Europe. Unfortunately, contacts with Eastern Europe are not frequent and perhaps the most satisfactory way of influencing developments there as well as in Africa and Asia is through global scientific associations such as the international scientific unions and the International Council of Scientific Unions (ICSU) and, at the governmental level, through organizations such as the World Health Organization (WHO). The ICSU established in the autumn of 1975 an *ad hoc* Recombinant DNA Committee which will meet for the first time in July in Heidelberg; it has both Russian and Japanese members. The Advisory Committee on Medical Research of the WHO has a subcommittee to consider safety in the handling of microorganisms and cells employed in research. Through either or both of these organizations, it is possible to relay information about developments and decisions in the United States and Western Europe.

INITIATIVES AT THE EUROPEAN LEVEL

Many European countries have established national committees to oversee the safety aspects of recombinant DNA research. Most are under the auspices of national research councils, although in the Netherlands the committee is responsible directly to a ministry, whereas in Switzerland it is under the auspices of an academy. It is clearly important that all these committees should, despite national differences, adopt the same or similar guidelines, and it is equally important that the common guidelines are similarly interpreted and applied in the various countries. Such harmonization will not be easy to achieve. Fortunately, however, there exist in Europe four international organizations which can play useful roles.

THE EUROPEAN ORGANIZATIONS

The organizations are

1. The European Molecular Biology Organisation (EMBO), a private organization of some 300 European molecular biologists, which has been in existence since 1963.
2. The European Molecular Biology Conference (EMBC), an intergovernmental organization supported by 14 West European governments and Israel, which has existed since 1970 and provides EMBO with its budget.
3. The European Molecular Biology Laboratory, supported by 10 of the 15 governments of EMBC, which began its legal existence in 1974.
4. The European Science Foundation (ESF), established in 1974 and represented by the research councils and academies of 16 European countries including Yugoslavia.

These organizations can be used as channels of communication at various levels: EMBO at the level of individual molecular biologists, EMBC at the level of governments, and ESF at the level of research councils (the extent to which the policies of governments and research councils differ varies from country to country). The EMBL, of course, provides a central research facility.

The EMBO Standing Advisory Committee

All of these regional international bodies have taken initiatives. With the financial and moral support of the EMBC and EMBO, there was established in 1975 an *ad hoc* Advisory Committee on Recombinant DNA, which in 1976 became a standing committee. The EMBO Committee is an expert committee with a purely advisory role; it has no legislative or licensing roles but exists to advise, on request, governments, research councils, national committees, institutes, and individuals on technical questions arising from recombinant DNA research. Its first formal meeting was held in February, 1976, the chief item on the agenda being to consider the draft NIH guidelines that emerged at La Jolla in December, 1975. Its first report is reproduced in Appendix 2. This report went immediately to the governments of the EMBC, the EMBO members, the ESF, the NIH Advisory Committee, and the press.

The European Science Foundation

The European Science Foundation, which provides a forum in which European research councils can discuss their policies, plans, and priorities, established in April, 1975 and *ad hoc* Committee on Genetic Manipulation with the tasks of surveying European initiatives relating to recombinant DNA research, and considering the scientific, social, legal, and philosophical implications of this research so as to facilitate the development of a common European attitude to the various issues. The EMBO and ESF committees quickly established close cooperation, the former assuming the role of technical adviser to the latter. As a result of this cooperation, the ESF committee was in February able to submit to the Executive Council of the ESF, and through it to the national research councils, a series of interim recommendations. The most important recommendation was that European bodies responsible for drawing up the appropriate safety guidelines for this research should keep in close contact with each other and with the equivalent NIH committee. Other recommendations included promotion of recombinant DNA research through fellowships and training courses, the establishment of national registries for all research in this area, including industrial research, the provision of training for safe handling of pathogens, the provision of one or a few European strain and type collections, and so on. The interim report was on the whole favorably received by the Executive Council of the ESF and through that channel brought to the attention of the national research councils.

U.S. Guidelines in Europe

At first sight, there may seem to be little substance to the first recommendation of the ESF committee, but one must bear in mind that the ESF was aiming its report at a set of sovereign national bodies who respond much more favorably to diplomatic language and allusion than to forthright language and explicit statements. With the exception of the United Kingdom, whose position is to be revealed in the report of the Williams working party, the European research councils (and the national committees that they have established) have proved willing to adopt, at least for the time being, the draft version of the NIH guidelines produced in December, 1975. Indeed, both the French and the Dutch commissions have been using the draft NIH guidelines when making recommendations on particular proposals.

Any attempt to estimate the risks associated with recombinant DNA research is today based on prejudice and conjecture rather than knowledge. Guesswork and intuition rather than objectivity have to be the order of the day and, therefore, it seemed to many Europeans pointless to duplicate the task of assessing risks when it was being pursued so energetically in the United States. Moreover, the consensus represented by the draft guidelines seemed to be a sufficiently stringent, workable compromise that had the virtue of internal consistency; as the EMBO committee quickly discovered, this consistency is lost if attempts are made to change particular risk assessments that seem less reasonable than others. These sorts of considerations made the European acceptance of the draft guidelines for the present a matter of little trauma.

It should scarce need saying, however, that although the European countries have been prepared to adopt in principle and, for the immediate future, the draft NIH guidelines, they cannot adopt all of the detailed implementation procedures designed for the United States. Each country will develop methods of implementation best suited to the structure and organization of its national scientific research. In the larger countries, the end result may well resemble that arrived at in the United States with a central national advisory committee and perhaps even regional or institute committees. But such a structure is not necessarily feasible or desirable in a country with a population of only a few million and perhaps only three or four molecular biology laboratories. Indeed, the smaller European countries may well prefer to use the advice of the European organizations in many instances.

The European Molecular Biology Laboratory

While these discussions and policy decisions were being made, the Council of the EMBL decided that a high-risk (P4) containment laboratory specifically designed for recombinant DNA research should be constructed as an additional facility at the EMBL, Heidelberg. In Europe a small number of high-risk containment laboratories are available, and although several countries are building P3 facilities, there are no plans to build new national P4 laboratories. The EMBL, therefore, should become a central laboratory for high-risk experiments in Europe, and it will complement not only national P3 laboratories but also the few P4 facilities available such as the Microbial Research Establishment at Porton Down, England. As plans now stand, the EMBL containment laboratory is expected to be completed toward the end of 1977.

FUTURE DEVELOPMENTS IN EUROPE

At the time of this writing, neither the final version of the NIH guidelines nor the report of the Williams working party was available; both of those documents will have appeared and been widely discussed well before publication of this volume. I hesitate, therefore, to predict how the situation and policies will develop in Europe beyond making a few general comments.

Rumors of what the report of the Williams working party will say have naturally been rife. It has been suggested that a central advisory board and possibly regional advisory boards are being contemplated. The central board would presumably advise on safety measures but not on scientific priorities about which decisions would continue to be made by the research councils. One may expect the establishment of a comprehensive national registry, and perhaps in the United Kingdom greater reliance will be placed on physical containment than on biological containment; better the devil you know than the one you don't know. It also seems likely that in the United Kingdom, as befits its legal traditions, rigidly codified guidelines will not be laid down but instead each experiment will be considered individually and safety precautions recommended for it. This will result in the accumulation of a body of case law and precedents; at the outset, since there are few precedents to fall back on, a conservative stance will no doubt be adopted.

This case law approach is, I believe, fundamentally different from the approach followed in the United States, where an attempt has been made to envisage every conceivable class of experiment and to produce an all-embracing, detailed code of practice. Obviously, the national committees in Europe as well as the EMBO standing committee will wish to compare the report of the Williams working group and the definitive version of the NIH guidelines as soon as they appear. The British thinking will, I expect, have considerable influence on the policies of the other European countries. In practice, the two approaches to the question of safety precautions may lead to rather similar results, but should the case law approach prove to offer greater flexibility without sacrificing safety, it will surely gain widespread support.

I believe that it will be found to be both desirable and inevitable to establish yet another committee in Europe, a committee of the chairmen or representatives of the national committees. This seems to me to be the only feasible way to ensure that common guidelines are interpreted and applied in a fairly uniform way in so many different countries so that a flag of convenience state does not emerge. Furthermore, free communication among the European national committees and the NIH committee is highly desirable, but the channels for it have still to be established.

One can anticipate strong pressures, even legislation in some countries, to make registration of recombinant DNA experiments with the national committees an automatic first step. The European organizations may well also press for the exchange of national registries between the national safety committees, notwithstanding the counterpressures of academic confidentiality and industrial secrecy. The convention and questionnaire drawn up by the French Control Commission may well be used as models.

Numerous national P3 laboratories will certainly be constructed; in France, for example, four are to be built, and at least one (and possibly two or three) strain and type collections are likely to be established. The EMBL at Heidelberg is one obvious place for such a collection.

We can anticipate priority being given to experiments, many of them very simple (*see Appendix II*), which should yield information that will make it possible to assess more objectively the risks associated with recombinant DNA molecules. Some have already been done, others are still on the drawing board.

In the United States, I understand, the task of certifying biologically disarmed hosts and vectors will be the sole responsibility of the NIH Advisory Committee. In Europe there is no single equivalent of the NIH Committee. It seems likely that in the immediate future some EK2 strains will be obtained from the United States, but European molecular biologists cannot be expected to rest content, dependent on American investigators for disarmed hosts and vectors. The European national committees, with or without the advice of the EMBO Standing Committee, will no doubt decide to permit particular experiments with particular host and vector strains produced by the investigator, and in this way they will assume a certifying responsibility.

Finally, in Europe, as in the United States, there will surely be continuous pressure from those doing recombinant DNA experiments to revise the guidelines as soon as further information is accumulated. Although most European countries have so far agreed to follow closely the lead given by the NIH Advisory Committee and the American debate, Europe will certainly look for a closer dialogue with the United States when the question of revision of the guidelines arises, if the unison that has been achieved is to be maintained. For it may prove that unison is easier to achieve in the face of universal ignorance than in the face of partial knowledge.

Hopefully, however, as further information is obtained the debate will become less subjective, many of the present restrictions may in unison be relaxed, some of the committees may be disbanded, and some of the benefits this research offers may at last be realized.

Note added in proof—The report of the Williams Working Party (see p. 458 and 459 of this chapter) has been published: *Report of the Working Party on the Practice of Genetic Manipulation.* (Command 6600). Her Majesty's Stationery Office, London, August, 1976.

APPENDIX I. MINISTRY OF INDUSTRY AND RESEARCH, DÉLÉGATION GÉNÉRALE À LA RECHERCHE ET TECHNIQUE, COMMISSION OF CONTROL OF GENETIC RECOMBINATION *IN VITRO*, 35 RUE SAINT DOMINIQUE, PARIS 7

Agreement Concerning Genetic Recombination *in Vitro*

ARTICLE 1 *Object:*

The parties undertake to submit to a preliminary control, in accordance with the conditions hereafter defined, the experimental protocol of every project involving the creation of new genetic material through *in vitro* recombination, the manipulation of this, and the transfer of new genetic elements to hosts which are not natural to them. This control applies to projects liable to be carried out in their laboratories, or outside their laboratories, or to projects in which they have scientific and financial interests.

ARTICLE 2 A national technical Commission of Control has been created, the members of which are appointed by the Délégué Général à la Recherche Scientifique et Technique.

ARTICLE 3 This Commission includes scientists, currently representing the following disciplines:

1. Microbiology: medical bacteriology, virology
2. Microbial and molecular genetics
3. Enzymology
4. Epidemiology

as well as observers representing the laboratory staff.

ARTICLE 4 The parties undertake to submit to the Commission of Control, provided for in Article 2, before the commencement of any experiments, the research projects using techniques of *in vitro* genetic recombination. The projects should be accompanied by answers to a questionnaire, designed to this end by the Commission of Control.

Any publication in connection with research work, which has been examined by the Commission of Control and has received a favourable reply, should mention the following:

"The risks associated with the work described in this publication have formed the subject of a preliminary examination by the French national and technical Commission of Control."

ARTICLE 5 The Commission will express its decision on submitted projects as soon as possible. It will indicate, if necessary, the scientific and technical conditions which it considers should be met. To this end, it will take into account international and national recommendations known at the time of decision. It has the sole authority to assess the extent to which the development of techniques and materials will lead to the revision of these recommendations.

ARTICLE 6 The parties undertake to give to the Commission the means permitting the Commission to fulfill its duties.

ARTICLE 7 The parties undertake not to commence and not to finance projects until after their submission to the Commission of Control and the setting up of a committee responsible for supervising the experiments locally.

ARTICLE 8 The projects are to be carried out under the sole responsibility of the parties who initiate them.

ARTICLE 9 Until the expiry of the present Agreement, neither of the parties may withdraw from the obligations resulting from it. Requests for admission to the agreement may be received during the period of its validity.

ARTICLE 10 On the initiative of the Délégué Général à la Recherche Scientifique et Technique and with the agreement of the parties, additional clauses may modify, insofar as necessary, the conditions of the present agreement.

ARTICLE 11 The present agreement will be for a term of five years, commencing from the date of signature. The agreement is renewable by tacit agreement.

ARTICLE 12 In case of difficulty in the interpretation or execution of the present agreement, the parties undertake to strive to resolve their differences amicably.

Paris

APPENDIX II. REPORT ON THE FIRST MEETING OF THE EMBO STANDING ADVISORY COMMITTEE ON RECOMBINANT DNA HELD AT LONDON ON 14/15 FEBRUARY 1976

Introduction

1. This Committee was established by the Council of the EMBO with the endorsement of the EMBC, to advise, upon request, governments, organisations and individuals on the scientific and technical aspects of recombinant DNA research including the conditions under which this research should be done. At the request of the EMBO Council we have met to decide whether it is appropriate or necessary to elaborate guidelines, other than those currently available, for such research in Europe.

Considerations

2. We have considered, therefore:

   - the Ashby Report,
   - the provisional report of the Asilomar Conference,
   - the proposed guidelines for research involving recombinant DNA molecules of the National Institutes of Health Recombinant DNA Molecule Program Advisory Committee, drawn up at La Jolla on 4–5 December 1975 and currently being considered by the Director of NIH for adoption as official policy,
   - various other documents of that committee, together with other relevant memoranda.

   We fully support a number of general principles enunciated in these several reports, namely, that:

   a. recombinant DNA research promises great scientific and social benefits and, therefore, should proceed under appropriate safeguards;
   b. at present one cannot exclude the possibility that this work could entail undesirable biological side effects;
   c. therefore, physical and biological containment procedures should be instituted for certain types of experiments;
   d. the containment should be designed to match the best estimate of the possible risks involved.

3. This Committee feels that there are two categories of risk: one is the intentional creation of new genotypes containing combinations of genes known to specify harmful products or endow dangerous potentialities: for example, the introduction of genes for *Cl. botulinum* toxin into *E. coli;* the introduction of plasmids which increase the enteropathogenicity of *E. coli;* or the introduction of penicillin resistance into β-haemolytic streptococci or pneumococci. We believe that experiments in this category should in general be proscribed at the present time, as has been suggested in all previous draft guidelines drawn up by other committees.
4. The second category of risk arises from the possibility of adventitiously producing genotypes of unknown and undesirable properties. This risk depends upon chains of events, such as:

   a. unwitting introduction of a foreign genetic material with pathogenic potential into a micro-organism,
   b. the escape of this new micro-organism from the contained laboratory,

c. the establishment of the new micro-organism in a natural environment,
d. expression of this genetic information in the host micro-organism to the detriment of the ecosystem,
e. or, the transfer of the harmful genetic material in the new micro-organism to the genome of some other organism to the detriment of the environment.

Sequences of events, such as the above, are in our opinion very unlikely to be realised.

5. During the past year, recombinant DNA research has proceeded under draft guidelines set out in the Ashby Report and the Asilomar Conference Report. For the most part the evaluation of risks cannot, at present, be based on adequate data, as the work of several other committees has already shown. The EMBO Committee does not feel, therefore, that it should engage in drawing up a new estimate of risks, since it is unlikely to be any more accurate than its predecessors.
6. The most detailed code of practice yet to be drafted is that which is currently being considered for adoption by NIH as official policy. This could, we believe, provide a basis for an international code of practice but details of implementation and the precise classification for particular experiments are something on which opinion may vary. The current NIH proposed guidelines are still under debate and until a final version is approved we believe more detailed comment would be premature.

Recommendations

7. We feel that it is important to initiate an experimental analysis of the various parameters on which an objective assessment of the various putative risks might be based. To this end this Committee will strive to have experiments, such as those listed in the Addendum, performed as soon as possible. The results of such experiments should permit a more objective evaluation of possible problems.
8. Meanwhile experimental research should proceed with due caution and the national committees in Europe responsible for controlling this research should give serious consideration to the assessments of possible risks in various types of experiments and appropriate levels of containment that are proposed in the latest NIH proposed guidelines, which we believe represent the upper limit of stringency necessary.
9. Furthermore, we believe it is essential that those European national committees that are responsible for the control of recombinant DNA research should come into close contact, for instance through this EMBO Committee, through the EMBC and through the ESF to ensure that in Europe the conditions under which recombinant DNA research is allowed to proceed are as uniform as possible. We therefore recommend that procedures be established for the exchange of information and discussion of policy between European national committees.
10. This Committee is prepared to advise, on request, governmental and private institutions and individuals on scientific and technical aspects, including (a) the assessment of possible biohazards of particular experiments (b) the assessment of appropriate levels of physical and biological containment, so far as questions of principle are concerned. The EMBO Committee wishes to stress, however, that its role is only advisory, it has no juridical powers.
11. Drawing upon available expertise, this Committee intends as a matter of urgency to establish and implement procedures for evaluating biologically disarmed micro-organisms in relation to prevailing criteria of biological containment.

12. This Committee also intends to establish a voluntary registry of recombinant DNA research in Europe and to achieve this it will seek, within the limits set by confidentiality, the cooperation of individual scientists as well as research councils and other involved agencies.
13. We recommend that a small number of centres housing collections of strains of micro-organisms and animal cell lines used in recombinant DNA research be established in Europe.
14. Finally the Committee will continue to promote the organisation of courses on the safe handling and containment of pathogenic micro-organisms and on recombinant DNA technology.

Members of the Committee, present:	Professor Ch. Weissmann (Chairman) Professor E. S. Anderson Dr. K. Murray Professor L. Philipson Dr. J. Tooze Professor H. Zachau
present on 14th February only:	Professor W. F. Bodmer
absent:	Dr. S. Brenner Professor F. Gros
present as ICSU observer:	Dr. W. J. Whelan

Addendum

Examples of experiments which the EMBO Committee will attempt to have performed as soon as possible to provide data for one objective assessment of the biohazards of recombinant DNA research.

1. Measurements of the frequency of mobilisation of non-self transmissible plasmids in the human gut and of the frequency of lysogenisation of resident enterobacteria with phage lambda from *E. coli* K12 in the human gut.
2. Measurement of the efficiency of plating of lambda phage on wild type *E. coli* isolated trom the human gut.
3. Kinetics of the survival of DNA in the gut and on incubation with gut fluids.
4. Measurement of the survival, fate and the biological activity of micro-organism DNA introduced by various routes (eye, mouth, nose, etc.).
5. Measurement of the pathogenicity of polyoma virus DNA attached to a plasmid vector and within *E. coli* when introduced into mice.

Recombinant Molecules: Impact on Science and Society, edited by R. F. Beers, Jr. and E. G. Bassett. Raven Press, New York © 1977.

42. Beware the Lurking Virogene

Natalie M. Teich and Robin A. Weiss

Imperial Cancer Research Fund Laboratories, Lincoln's Inn Fields, London WC2A 3PX, England

INTRODUCTION

Most probably recombination between eukaryote and prokaryote DNA occurs naturally within our bodies. We eat several milligrams of plant and animal DNA daily. Although this DNA is digested efficiently by pancreatic DNase, gut bacteria must frequently be exposed to large molecular weight DNA, particularly during acute alimentary tract infections (e.g., *Salmonella* in the duodenum). Foreign DNA can penetrate bacteria, especially when the cell wall is damaged, and diverse restriction enzymes have surely evolved to afford some protection from the onslaught of foreign DNA. Bacteria also are exposed to our own DNA, for we shed grams of dead intestinal cells into the gut daily. There are further opportunities for bacteria to ingest human DNA in septicemic wounds and in decomposing corpses. Davis (5) has pointed out that the human gut is a huge chemostat, and it is almost inconceivable that human-bacterial hybrid DNA molecules have not occurred and been subject to natural selection during man's history.

Viral infection of animals offers obvious opportunities for recombination with host DNA, particularly with integrating species such as the papovaviruses, adenoviruses, and retroviruses (RNA tumor viruses), for integration is itself a recombinant event.

This volume is primarily concerned with the scientific and social impact of the manufacture in laboratories of recombinant DNA molecules by means of specific excision and ligation. The use of restriction endonucleases offers new and elegantly simple methods of preparing recombinant DNA molecules, and the novelty and artificial nature of their manufacture has led first the practitioners of the craft, and subsequently the community at large, to question whether such recombinant DNA molecules, when sequestered in their vectors, might not lead to the reproduction and evolution of new genetic elements or organisms that could be hazardous to health or environment.

Two major potential hazards have been postulated. The first is that artificially recombined DNA may yield replicating molecules that could not have arisen naturally, and whose behavior is unpredictable; the second is that nonselective recombination, the so-called shotgun experiment, could result in the cloning of genes that might be dangerous when transmitted by their vectors into a new host.

In this chapter we shall argue that the new recombinant DNA technology is not the only method by which unrelated genes may recombine. Such recombination also occurs naturally as well as by other commonly used laboratory procedures, so the fear that we have engineered an entirely novel situation may be unfounded. However, the results of these older natural and laboratory methods of recombination suggest that shotgun recombination does indeed lead to the production of infectious organisms, often with unsuspected properties. We feel, therefore, that the debate on the potential hazards of genetic recombination should be concerned less with the novelty of the technique used in obtaining recombinants (with its emotional overtones of "meddling with nature") than with the properties of the recombinant molecules obtained.

Many experiments have demonstrated that infectious RNA tumor viruses can be activated from a quiescent state when exposed to change in the host cell environment. Because many species of vertebrates contain latent RNA tumor viruses as proviral DNA copies (proviruses or virogenes), one must consider that these viruses are truly genetic elements of the host cell. Therefore, it is important that anyone considering DNA cloning experiments, particularly into or out of vertebrate cells, should envisage the possibility of transducing and/or activating latent viral gene expression or fully competent, infectious virus. Furthermore, we wish to illustrate that recombinant techniques practiced at the supramolecular level, such as somatic cell hybridization and the formation of chimeric animals, can activate or greatly amplify the production of latent tumor viruses, which are not necessarily themselves recombinant forms.

We hope that, in the sense of "forewarned is forearmed," cognizance of endogenous latent viruses will lead to an increased respect for the possibility that manipulation of eukaryotic cells may select virogene elements.

RETROVIRUSES—A BRIEF DESCRIPTION

The Retroviridae family is composed of three subfamilies: (a) Oncovirinae, the RNA tumor virus groups morphologically distinguished as types A, B, and C; (b) Spumavirinae, the foamy viruses; and (c) Lentivirinae, the visna and related viruses. This discussion is concerned predominantly with the type C retroviruses of the first subfamily which are associated with leukemias and sarcomas.

One reason for concern about retroviruses is that there is evidence for wide distribution of these viruses among vertebrates (Table 1). Only those families from which infectious replicating virus have been obtained are listed in the table. In some, the evidence for integrated proviral nucleic acid

TABLE 1. *Phylogenetic distribution of retroviruses among vertebrates*

Vertebrate	Known endogenous virogenes
Bony fishes	?
Reptiles	?
Birds	+
Mammals	
Mice	+
Rats	+
Hamsters	+
Guinea pigs	+
Cats	+
Pigs	+
Cattle	?
Monkeys and apes	?
Baboons	+

sequences is unknown. However, several vertebrate species, such as mice, rats, chickens, cats, pigs, and baboons, appear to carry complete copies of viral genomes integrated within the cellular DNA. In these species, however, viremia is uncommon and the viral genomes usually remain "silent" or only partially expressed. This has been analyzed in most detail in mice and chickens, where genetic factors, possibly located in the proviruses themselves, control viral gene expression at transcriptional and post-transcriptional levels. Some genes, such as the *gs* locus in chickens (23,36) and the *Mlv* locus in mice (32), determine an incomplete expression of virus, so that viral RNA and some of the viral antigens are synthesized but whole virus is not released from cells. Intrinsic factors such as the hormonal and physiological milieu may also affect levels of endogenous virus expression. Thus, in cells *in vivo* and *in vitro,* endogenous virus expression represents an interaction involving the genetic constitution of the cell and the physiological conditions of the internal and external environment. However, once the virus has been activated in some of the cells, other host genes affect the susceptibility of cells to secondary infection; for instance, the *Fv-1* locus in mice affects proviral DNA integration (14) and the receptor genes of chickens affect virus penetration (39). These genes influence the ultimate titer of virus production both *in vivo* and *in vitro.*

The finding that the DNA provirus of retroviruses has, on many occasions, become a heritable trait of the host perhaps represents the ultimate symbiosis of balanced host-virus adaptations. However, new endogenous virogenes have been created under laboratory conditions. Infection of mouse blastocysts with a murine leukemia virus, followed by implantation of the blastocysts into pseudo-pregnant mothers, results in the development of a mouse that contains new virogene sequences. In a proportion of the mice thus derived, there is a higher-than-normal incidence of leukemia. Interestingly, offspring of such mice also contain the new provirus, demonstrating that these virus sequences were integrated into the germ line cells and thus had become a host mendelian trait (13).

The latent proviruses inherited by the host may become active, especially when the host cells are manipulated in experimental procedures. Therefore, they constitute a recognizable potential hazard to indiscriminate experimental recombination of vertebrate DNA. Although the majority of the endogenous virogenes may not have any oncogenic potential, some are known to cause leukemia (8,24) or mammary carcinoma (2) or to be associated with autoimmune disease (17). For this reason, we shall use endogenous viral genomes as examples of the activation of unwanted infectious agents produced unwittingly during biomedical research.

XENOTROPIC VIRUSES

The endogenous virus genomes, when expressed as infectious replicating particles, show distinct host range patterns. The "ecotropic" viruses replicate readily in the cells of the species in which the viruses are transmitted as host genetic elements and replicate less well or not at all in cells of other species. On the other hand, some endogenous viruses after activation cannot infect their own host species but can infect cells of several other species (Fig. 1). These are called "xenotropic" viruses because they infect foreign host cells only (16). Xenotropism has important and perhaps far-reaching implications.

It is unlikely that xenotropic viruses are frequently transmitted from one species to another under natural conditions, although there are cases, such as

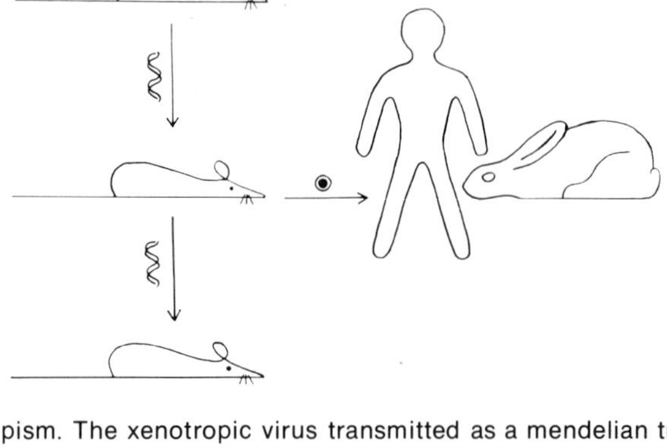

FIG. 1. Xenotropism. The xenotropic virus transmitted as a mendelian trait in mice can, upon activation, infect cells of other species. Xenotropic viruses have been induced from several vertebrate species.

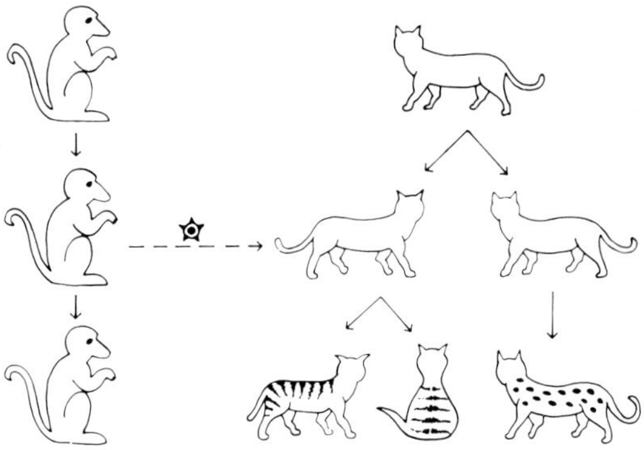

FIG. 2. Interspecific transmission of endogenous virogenes. Sequences related to the baboon endogenous virus became integrated into the germ line cells of some *Felis* species (*F. catus*, the domestic cat and *F. sylvestris*, the European wildcat), whereas other *Felis* species (*F. bengalensis*, the leopard cat) and more distantly related Felidae do not contain these sequences (3). Nucleic acid homology studies between endogenous viruses suggest that interspecific transmission of virogenes may have occurred in other instances.

the cat (3), in which provirus sequences have clearly been acquired from other unrelated host species (Fig. 2); it has been estimated that this interspecific transfer of virus occurred about 10 million years ago. The mechanism of this transmission is unknown, but the important feature is that a virus exogenous to a species may become integrated into the germ line of that species and be maintained as a genetic element during the prolonged period of evolution and speciation. The question that concerns us here is that cross-specific transfer may readily occur in laboratory experiments even when a natural zoonosis is unlikely to occur.

ACTIVATION AND AMPLIFICATION OF ENDOGENOUS VIRUSES

As mentioned above, the expression of endogenous viral genes is regulated by many intrinsic host cell factors. However, host control mechanisms may be modified or countermanded by a variety of extrinsic agents, resulting in the induction of endogenous virus synthesis (Table 2). Many of these agents

TABLE 2. *Agents that activate endogenous retroviruses in murine cells*

Treatment	Reference
Physical mutagens: X-rays, UV irradiation	25
Chemical mutagens: halogenated pyrimidines (BrdU, IdU), dimethylbenzanthracene, methylcholanthrene	21
Inhibitors of protein synthesis: cycloheximide, puromycin	1
B-lymphocyte mitogen: lipopolysaccharide	22
Immunological stimulation: graft-versus-host reactions	12,28

are quite efficacious in inducing quiescent viral genomes. The research worker treating cells with such agents may not realize the problems engendered by the viruses now introduced as infective agents in the system.

Table 3 summarizes some procedures routinely used in cell biology and oncology laboratories which activate or amplify the titer of xenotropic viruses. Such common procedures as the treatment of cells with cycloheximide (1), mixed culture of lymphocytes (12,28), or infection with ecotropic retrovirus (6) selectively activate xenotropic virus in mouse cells. The activated virus will not reinfect mouse cells but will grow avidly in human cells. Thus, it is possible that mouse cultures could be a source of infection for human cells and perhaps for humans, giving some cause for concern where, for example, mixed lymphocyte cultures are conducted in the same building that houses immunodeficient patients.

The greatest likelihood of xenotropic viruses gaining an infectious foothold that allows them to propagate to high titers is where cells of heterologous species come into close proximity. Thus, mixed cultures of two species, as with mouse and human cells, and somatic cell hybridization will permit amplification of xenotropic virus. If bromodeoxyuridine (BrdU) is used in the selection procedure for cell hybrids, the virus may be activated readily and efficiently; moreover, treatment of cells with fusion agents (such as inactivated Sendai virus) abrogates host range barriers of the cell surface to retroviruses (37,39). Hybrid cells are not inevitably susceptible to xenotropic virus as the resistance of the parent species may be dominant in the hybrid (7); however, one must consider the situations in which resistance may be a recessive trait, or in which interaction between genes in hybrid cells may itself lead to virus activation (Fig. 3).

Another opportunity for amplification of xenotropic virus is exemplified in the transplantation of human cells into nude or immunodeficient mice. Many mouse strains have a low spontaneous incidence of xenotropic virus activation, and further replication of the virus will be possible only in susceptible xenografts. During subsequent manipulations of the xenograft, the experimenter may be ignorant of the production of virus by his material (Fig. 4). Tests on a number of human tumors propagated in nude mice indicate that contamination by xenotropic virus is not an infrequent occurrence.

The conspectus is that a variety of chemicals and techniques commonly

TABLE 3. *Procedures that may activate or amplify production of xenotropic viruses*

1. Treatment of cells with BrdU or IdU, inhibitors of protein synthesis, glucocorticosteroids, membrane fusion agents, chemical carcinogens.
2. Mixed lymphocyte culture.
3. Somatic cell hybridization or co-cultivation of heterogeneic cells.
4. Transplantation of tissues into immunosuppressed animals.
5. Transfection of cells with cellular DNA from species containing endogenous proviruses.
6. Exogenous virus infection.

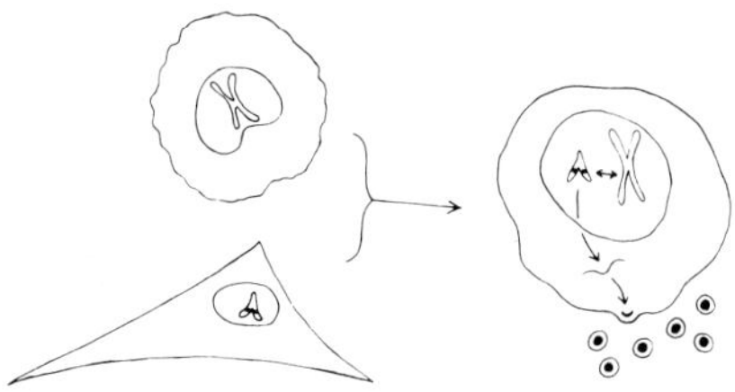

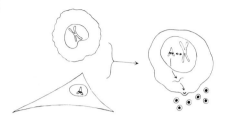

FIG. 3. Somatic cell hybridization. The production and selection of somatic cell hybrids may lead to the activation, amplification, and chronic replication of endogenous viruses. In particular, hybrids created from cells of heterologous species may select for xenotropic virus production.

used in research laboratories are capable of inducing latent proviruses from a wide range of vertebrate cells. These phenomena may not be particularly specific; slight perturbations of cellular DNA or metabolism may perhaps suffice to activate endogenous virogenic elements.

HYBRID VIRUSES

The opportunities for phenotypic and genotypic mixing with and among retroviruses are abundant in nature and in conditions of experimental design. As mentioned above, somatic cell fusion may create an environment suitable for the growth and amplification of xenotropic viruses. The same technique also can enhance the frequency of phenotypically mixed virus particles, as is illustrated in Fig. 5. If cells of homologous or heterologous species, each infected with a different retrovirus, are fused, then the progeny virus particles may exhibit phenotypic properties of both parental viruses. Most notable is that a high proportion of the particles may show a broadened host range due to mosaicism of the envelope glycoproteins. This phenomenon may also be observed as the result of dual infection of a single cell with two retroviruses (35,36). Indeed, phenotypic mixing of envelope

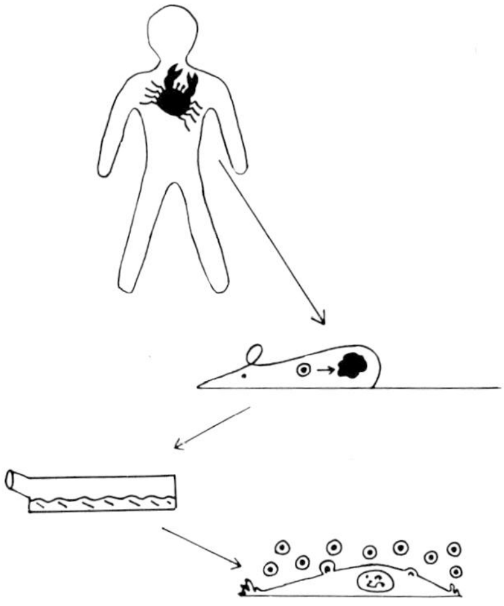

FIG. 4. Xenografts. Transplantation of human tumor cells into nude or immunodeficient mice frequently leads to expression of xenotropic murine viruses. Thereafter, infection of the human cells by the xenotropic virus leads to the integration of proviral sequences in the human cellular DNA and to chronic virus production.

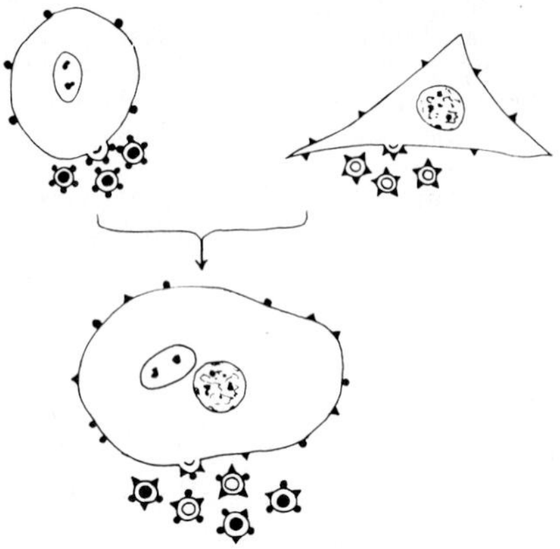

FIG. 5. Phenotypic mixing between retroviruses. When cells producing different retroviruses are fused together, the resulting hybrid cell may give rise to progeny virus particles which display phenotypic characteristics of both parental viruses. This illustration shows the production of particles that have mosaics of envelope glycoproteins.

properties has been demonstrated following infection with exogenous retrovirus of cells producing their endogenous virus (4) or of "virus-free" cells expressing endogenous viral glycoprotein only (9,37). In fact, this last example of phenotypic mixing between endogenous and exogenous viral components was one of the early indications of the existence of latent virogenes. As a consequence of phenotypic mixing, progeny virus particles may have altered antigenic specificities which change their neutralization, interference, and host range properties.

Complementation of defective virus particles by components of endogenous or exogenous viruses is also widespread. For example, the majority of mouse sarcoma viruses are defective for replication; a helper leukemia virus, providing envelope glycoprotein and perhaps internal structural and enzymatic proteins as well, is required for successful infection (34). Thus, the phenotype of sarcoma viruses may be modified by substitution of new helper viruses; this can be accomplished by superinfection with a different helper virus or by activation of endogenous helper viruses.

The ease with which various helper viruses can rescue defective viruses has led to the creation of still more unnatural pseudotypes. Phenotypically mixed particles of quite unrelated retroviruses, even between avian and mammalian type C particles, are readily obtained by double infection, cocultivation, or cell fusion (18,42). Although this may be an aesthetically interesting laboratory event, the potential risks must be evaluated. Recently, the production of a murine sarcoma virus pseudotype with a baboon endogenous virus helper was shown to cause tumors in dogs and monkeys (S. S. Kalter, *personal communication*). The baboon virus alone does not have demonstrable oncogenic potential, whereas the murine sarcoma virus (with a murine helper virus) is oncogenic in rodents only; the phenotypic mixing and complementation of the two viruses leads to an expanded host range of animals susceptible to the tumorigenic capacity.

Phenotypic mixing of viral glycoproteins has also been extended to enveloped viruses unrelated to retroviruses. The initial finding that vesicular stomatitis virus (VSV) propagated in cells producing avian or murine leukemia viruses forms pseudotypes bearing the envelope specificities of the retrovirus (43) has stimulated other laboratories to pursue this technique. Similarly, VSV pseudotypes are formed in normal, virus-free chick cells by incorporation of glycoproteins specified by the endogenous provirus (20,40). VSV may not be exceptional in its capacity to incorporate endogenous retrovirus proteins. This phenomenon might well affect the purity of virus vaccines prepared in eggs or in cultured cells which, although free of infectious virus, may be expressing endogenous viral glycoprotein as part of the cell membrane. Further studies have shown that reciprocal phenotypic mixing can also occur in which the VSV glycoprotein is assembled into the retrovirus particles (19,40); this could markedly broaden the host range of an endogenous virus and could be viewed as a potential biohazard.

Genetic recombination too has been demonstrated to occur between related exogenous and endogenous retroviruses (10,31,41). In the case of the avian retroviruses, recombination appears to proceed initially by the formation of viral heterozygotes on passage of exogenous virus through cells transcribing endogenous viral RNA; up to 10% of the progeny may be recombinants or heterozygotes (41). However, recombinant forms have not been detected in cells in which the level of endogenous viral transcription is low (41). Recombination between unrelated endogenous viral genomes can also occur; propagation in pheasant cells of the endogenous virus induced from chicken cells results in interspecies recombination with endogenous pheasant virogenes (33).

Recombination or complementation can occur at yet another level: between retroviruses and cellular DNA sequences. The best example of this phenomenon has been the derivation of many murine sarcoma viruses. The majority of murine sarcoma viruses have been obtained not from tumors in mice but from murine leukemia viruses passaged through rats *in vivo*. Nucleic acid hybridization experiments have demonstrated that the sequences specific for the sarcomagenic potential are homologous to sequences detected in the chromosomal DNA of normal rats (27). The results can be interpreted in two ways: these sarcoma-specific sequences may be part of an unidentified provirus or they may represent normal cellular sequences which are nononcogenic when in a repressed state but oncogenic when they "escape" from host control. Furthermore, it has been found that some sequences in normal avian cell DNA are closely related to the sarcoma-specific sequence (the *src* gene) of avian sarcoma viruses (30); this supports the concept that viral sarcomagenic potential may evolve from an interaction between retroviral and host nucleic acids.

Expression of endogenous viral function could lead to unusual recombinants with cell DNA. Two laboratories have recently claimed that, in persistent infections with measles and respiratory syncytial virus, proviral

DNA copies of these lytic viruses appear to integrate in the cellular DNA (29,44). Although other investigators have not been able to confirm these findings, the possibility exists that this phenomenon might occur occasionally in cells in which the reverse transcriptase of the endogenous retrovirus is particularly active. In a similar fashion, cellular RNA might also be transcribed into a DNA form.

In summary, these examples of phenotypic mixing, genetic complementation, genetic recombination, and possible unscheduled reverse transcription illustrate the wide variety of molecular interactions between retroviruses and cells or other viruses. These phenomena constitute recognizable problems for those who specifically choose to study retroviruses; they also pose difficulties for those working with vertebrate cells. The consequences of retroviral gene expression could complicate interpretation of studies of cell surface antigens, identification or origin of new viruses, establishment of persistent infections, exogenous virus infections, cell morphology or growth characteristics, etc. More important, however, is the need for all researchers to be aware of retroviruses as cellular genes, genes that may be latent, partially expressed, or activated to produce infective agents.

TRANSFECTION OF VIROGENIC OR ONCOGENIC DNA

Experiments involving transfection of cellular DNA of vertebrates have come into vogue in the past few years. The technique has been used to demonstrate the existence of and to recover the sarcoma virus information present in transformed nonproducer cells (11,15); this can be considered the rescue of viral oncogenes. Furthermore, transfection of DNA from one species to cell cultures of another (26) provides another opportunity for the activation and replication of xenotropic proviruses (Fig. 6). Once again, host constraints over the expression of endogenous virogenes have been ablated by transfer to a more permissive environment. The release of endogenous viruses from a quiescent state to a productive infection in new cell species also leads to the establishment of proviral copies in the new host. In a sense, DNA transfection represents the "shotgun" recombination experiment for which no restrictions or precautions have yet been promulgated. It is conceivable that the genetic sequences in normal cells that

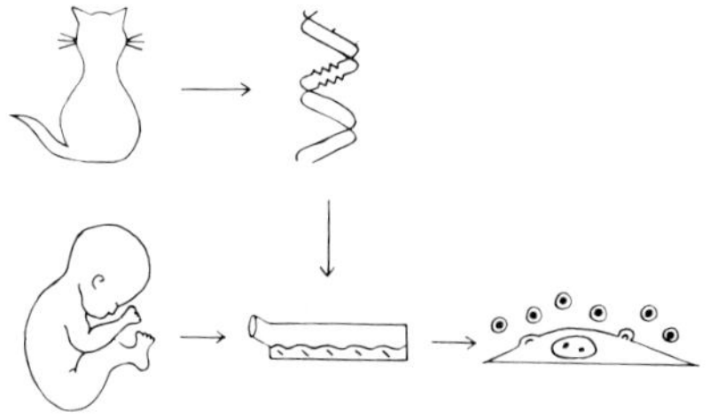

FIG. 6. Transfection of cellular DNA. The DNA from cells containing xenotropic proviruses can, upon transfection to cells permissive for the virus, give rise to infectious virions. The use of total chromosomal DNA, fragmented by physical shearing or restriction enzymes, provides a system analogous to "shotgun" recombination.

are related to transformation genes in sarcoma viruses (30) may be similarly transmitted.

CONCLUSIONS

Retrovirus genomes exist as endogenous genetic elements in the cells of many vertebrate species used in biomedical research. Many cell lines that spontaneously release virus and other cells are induced to do so by procedures commonly used in research laboratories. Because DNA cloning experiments could act as inducers or transducers of such endogenous virogenes, researchers should be aware of the necessity for monitoring, and perhaps containing, the potential risk of releasing virogenes or oncogenes into a new biosphere. This caveat is also prescribed for laboratories routinely using vertebrate cells. Endogenous viral expression may affect both the results and the safety of conduct of experiments that may have been set up for purposes other than to study latent tumor viruses. Some retroviruses endogenous to animals grow avidly in human cells; they are not known to be hazardous to man, but the exigency for further studies is indicated.

We hope that this discussion, intended for those not primarily interested in endogenous retroviruses, has illustrated some of the implications of the presence of tumor virus genes and their products in "uninfected" cell cultures.

REFERENCES

1. Aaronson, S. A., and Dunn, C. Y. (1974): High-frequency C-type virus induction by inhibitors of protein synthesis. *Science,* 183:422–424.
2. Bentvelzen, P., Timmermans, A., Daams, J. H., and van der Gugten, A. (1968): Genetic transmission of mammary tumor inciting viruses in mice: possible implications for murine leukemia. *Bibl. Haematol.,* 31:101–103.
3. Benveniste, R. E., and Todaro, G. J. (1974): Evolution of type C viral genes: Inheritance of exogenously acquired viral genes. *Nature,* 252:456–459.
4. Besmer, P., and Baltimore, D. (1977): Mechanism of restriction of ecotropic and xenotropic murine leukemia viruses and formation of pseudotypes between the two viruses. *J. Virol.,* 21:965–973.
5. Davis, B. D. (1976): Evolution, epidemiology, and recombinant DNA. *Science,* 193:442.
6. Fischinger, P. J., and Nomura, S. (1975): Efficient release of murine xenotropic oncornavirus after murine leukemia virus infection of mouse cells. *Virology,* 65:304–307.
7. Gazdar, A. F., Russell, E. K., and Minna, J. D. (1974): Replication of mouse-tropic and xenotropic strains of murine leukemia virus in human × mouse hybrid cells. *Proc. Natl. Acad. Sci. U.S.A.,* 71:2642–2645.
8. Greenberger, J. S., Stephenson, J. R., Moloney, W. C., and Aaronson, S. A. (1975): Different hematological diseases induced by type C viruses chemically activated from embryo cells of different mouse strains. *Cancer Res.,* 35:245–252.
9. Hanafusa, H., Miyamoto, T., and Hanafusa, T. (1970): A cell-associated factor essential for formation of an infectious form of Rous sarcoma virus. *Proc. Natl. Acad. Sci. U.S.A.,* 66:314–321.
10. Hayward, W. S., and Hanafusa, H. (1975): Recombination between endogenous and exogenous RNA tumor virus genes as analyzed by nucleic acid hybridization. *J. Virol.,* 15:1367–1377.

11. Hill, M., and Hillova, J. (1971): Production virale dans les fibroblastes de poule traités par l'acide désoxyribonucléique de cellules XC de rat transformées par le virus de Rous. *C. R. Acad. Sci.* [*D*] (*Paris*), 272:3094–3097.
12. Hirsch, M. S., Phillips, S. M., Solnik, C., Black, P. H., Schwartz, R. S., and Carpenter, C. B. (1972): Activation of leukemia viruses by graft-versus-host and mixed lymphocyte reactions in vitro. *Proc. Natl. Acad. Sci. U.S.A.*, 69:1069–1072.
13. Jaenisch, R. (1976): Germ line integration and Mendelian transmission of the exogenous Moloney leukemia virus. *Proc. Natl. Acad. Sci. U.S.A.*, 73:1260–1264.
14. Jolicoeur, P., and Baltimore, D. (1976): Effect of *Fv-1* gene product on proviral DNA formation and integration in cells infected with murine leukemia viruses. *Proc. Natl. Acad. Sci. U.S.A.*, 73:2236–2240.
15. Karpas, A., and Milstein, C. (1973): Recovery of the genome of murine sarcoma virus (MSV) after infection of cells with nuclear DNA from MSV transformed non-virus producing cells. *Eur. J. Cancer,* 9:295–299.
16. Levy, J. A. (1973): Xenotropic viruses: murine leukemia viruses associated with NIH-Swiss, NZB and other mouse strains. *Science,* 182:1151–1153.
17. Levy, J. A. (1975): Xenotropic C-type viruses and autoimmune disease. *J. Rheumatol.,* 2:135–148.
18. Levy, J. A. (1977): Murine xenotropic viruses. II. Phenotypic mixing with mouse and rat ecotropic C-type viruses. *Virology* (*in press*).
19. Livingston, D. M., Howard, T., and Spence, C. (1976): Identification of infectious virions which are vesicular stomatitis virus pseudotypes of murine type C virus. *Virology,* 70:432–439.
20. Love, D. N., and Weiss, R. A. (1974): Pseudotypes of vesicular stomatitis virus determined by exogenous and endogenous avian RNA tumor viruses. *Virology,* 57:271–278.
21. Lowy, D. R., Rowe, W. P., Teich, N., and Hartley, J. W. (1971): Murine leukemia virus: high-frequency activation in vitro by 5-iododeoxyuridine and 5-bromodeoxyuridine. *Science,* 174:155–156.
22. Moroni, C., and Schumann, G. (1975): Lipopolysaccharide induced C-type virus in short term cultures of BALB/c spleen cells. *Nature,* 254:60–61.
23. Payne, L. N., and Chubb, R. C. (1968): Studies on the nature and genetic control of an antigen in normal chick embryos which reacts in the COFAL test. *J. Gen. Virol.,* 3:379–391.
24. Rowe, W. P. (1973): Genetic factors in the natural history of murine leukemia virus infection. *Cancer Res.,* 33:3061–3068.
25. Rowe, W. P., Hartley, J. W., Lander, M. R., Pugh, W. E., and Teich, N. (1971): Noninfectious AKR mouse embryo cell lines in which each cell has the capacity to be activated to produce infectious leukemia virus. *Virology,* 46:866–876.
26. Scolnick, E. M., and Bumgarner, S. J. (1975): Isolation of infectious xenotropic mouse type C virus by transfection of a heterologous cell with DNA from a transformed mouse cell. *J. Virol.,* 15:1293–1296.
27. Scolnick, E. M., Rands, E., Williams, D., and Parks, W. P. (1973): Studies on the nucleic acid sequences of Kirsten sarcoma virus: a model for formation of a mammalian RNA-containing sarcoma virus. *J. Virol.,* 12:458–463.
28. Sherr, C. J., Lieber, M. M., and Todaro, G. J. (1974): Mixed splenocyte cultures and graft versus host reactions selectively induce an "S-tropic" murine type C virus. *Cell,* 1:55–58.
29. Simpson, R. W., and Iinuma, M. (1975): Recovery of infectious proviral DNA from mammalian cells infected with respiratory syncytial virus. *Proc. Natl. Acad. Sci. U.S.A.,* 72:3230–3236.
30. Stehelin, D., Varmus, H., Bishop, J. M., and Vogt, P. K. (1976): DNA related to the transforming gene(s) of avian sarcoma viruses is present in normal avian DNA. *Nature,* 260:170–173.
31. Stephenson, J. R., Anderson, G. R., Tronick, S. R., and Aaronson, S. A. (1974): Evidence for genetic recombination between endogenous and exogenous mouse RNA type C viruses. *Cell,* 2:87–94.
32. Taylor, B. A., Meier, M., and Huebner, R. J. (1973): Genetic control of the group-specific antigen of murine leukaemia virus. *Nature* [*New Biol.*], 241:184–186.
33. Temin, H. M., and Kassner, V. K. (1976): Avian leukosis viruses of different subgroups

and types isolated after passage of Rous sarcoma virus-associated virus-O in cells from different ring-necked pheasant embryos. *J. Virol.,* 19:302–312.
34. Tooze, J. (Ed.) (1973): *The Molecular Biology of Tumour Viruses.* Cold Spring Harbor Laboratory, Cold Spring Harbor, New York.
35. Vogt, P. K. (1967): Phenotypic mixing in the avian tumor virus group. *Virology,* 32:708–717.
36. Vogt, P. K. (1971): Genetically stable reassortment of markers during mixed infection with avian tumor viruses. *Virology,* 46:947–952.
37. Weiss, R. A. (1969): The host range of Bryan strain Rous sarcoma virus synthesized in the absence of helper virus. *J. Gen. Virol.,* 5:511–528.
38. Weiss, R. A. (1975): Genetic transmission of RNA tumor viruses. *Perspect. Virol.,* 9:165–205.
39. Weiss, R. A. (1976): Receptors for RNA tumor viruses. In: *Cell Membrane Receptors for Viruses, Antigens and Antibodies, Polypeptide Hormones and Small Molecules,* edited by R. F. Beers, Jr., and E. G. Bassett, pp. 237–251. Raven Press, New York.
40. Weiss, R. A., Boettiger, D., and Love, D. N. (1975): Phenotypic mixing between vesicular stomatitis virus and avian RNA tumor viruses. *Cold Spring Harbor Symp. Quant. Biol.,* 39:913–918.
41. Weiss, R. A., Mason, W. S., and Vogt, P. K. (1973): Genetic recombinants and heterozygotes derived from endogenous and exogenous avian RNA tumor viruses. *Virology,* 52:535–552.
42. Weiss, R. A., and Wong, A. C. H. (1977): Phenotypic mixing between avian and mammalian RNA tumor viruses: I. Envelope pseudotypes of Rous sarcoma virus. *Virology* 76:826–834.
43. Zavada, J. (1972): Pseudotypes of vesicular stomatitis virus with the coat of murine leukemia and of avian myeloblastosis viruses. *J. Gen. Virol.,* 15:183–191.
44. Zhdanov, V. M. (1975): Integration of viral genomes. *Nature,* 256:471–473.

Recombinant Molecules: Impact on Science and Society, edited by R. F. Beers, Jr. and E. G. Bassett. Raven Press, New York © 1977.

43. The Least Hazardous Course: Recombinant DNA Technology as an Option for Human Genetic, Viral, and Cancer Therapy

Seymour Lederberg

Division of Biology and Medicine, Brown University, Providence, Rhode Island 02912

INTRODUCTION

Techniques of molecular genetics developed during the past decade have enormously simplified the study of DNA molecules. Direct analyses of fragments produced by site-specific cleavage of purified DNA from animal viruses, bacteriophage, plasmids, and prokaryotic and eukaryotic chromosomes (2–5,13,15,17,21,26,29) have set the stage for a novel assault employing fragments of DNA recombined by ligase *in vitro* and selected and amplified by cloning *in vivo.* The striking features of this cut-and-splice technology include the ability to rearrange the topology of controlling and controlled regions of DNA and to recombine DNA regions between prokaryotic and eukaryotic domains (14) so that eukaryote genes replicate in prokaryotic cells (11,12,16,31,32,34,36) and prokaryote genes propagate in eukaryote cells (10,19).

This work heralds a new wave of basic research into the structure, replication, and expression of genomes, especially those of higher cells. It also forecasts attempts to use reconstructed genotypes to improve on genetic limitations in plant and animal species which nourish us and to correct genetic dysfunctions which burden our health.

The potential of recombinant DNA for advances in basic genetic studies and for applied research on reconstructed genotypes is the subject of an ongoing debate, one component of which concerns the risk that hybrid viruses, episomes, or bacteria with unexpected pathological properties may arise, escape, and proliferate, and thereby cause more harm than any intended benefit.

Actually, we can only guess at the nature of the first-order hazards and benefits that may await us. For example, in the early stages of automotive technology, 8,000 cars were registered in the United States at the turn of this century. The worst dangers of the motor car might reasonably have been expected to arise from horses stampeding in panic. A thoughtful response might have been the development of stronger bridle reins and carriage brakes to contain the runaway mishap. Instead, today frightened horses are blameless for the 40,000 to 50,000 automobile deaths a year we suffer in motor vehicle accidents, and for the air pollution of our high traffic density centers. Similarly, at that embryonic stage, predictions on the baneful or beneficial implications of the automobile for private transportation hardly could have appreciated its impact on commercial transport, agriculture, heavy construction, metallurgical industries, and warfare. I suggest our attempts to identify the worst and best possible outcomes of DNA technology will appear just as inadequate and be the solution of the wrong problems unless these outcomes are continuously examined and revised.

Considering this shortcoming, a commentary on the implications of recombinant DNA technology for human genetic and paragenetic problems might be excused for skepticism of the extreme positions on risks and benefits. With this apology, I intend to pursue the following possibilities:

1. That containment of recombinant molecules and their vectors will be substantial but imperfect.
2. That some of the imaginable hazards arising from accidental non-containment are plausible.
3. That a selected portion of genetic dysfunctions may be correctable by gene products or by genetic supplementation, but that many genetic disorders will be better treated by other means or remain intractable.
4. That our present focus on correcting simple genetic deficiencies may be dwarfed by spectacular opportunities for tumor virus experimentation and cancer immunotherapy of benefit to a much larger population.
5. That viral and cancer research programs which otherwise now expose us to real biological hazards may be pursued with DNA recombinants with reduced risk to the investigator and to the public—any risk-benefit equation will have to consider these antirisk consequences.

IMPERFECT CONTAINMENT AND ESCAPE

The hazards guessed at arise from unknown interactions of novel DNA molecules with unintended hosts. The proposed guidelines (18) reduce the range of possible hazards by preventing the funding at the present time of experiments involving genomes of unusual pathogenicity, virulence, toxicity, drug resistance, and ecological impact. Safeguards for permissible experiments are based first on appropriate microbiological techniques used in dealing with pure cultures and pathogenic microbes. They then rely on physical barriers which, in their highest level, limit laboratory access, use enclosed systems where aerosols are generated, have a negative air pressure environment, employ safety cabinets for handling biohazardous agents, and inactivate or remove potentially hazardous material from exhaust air, liquid, and solid wastes before their release. Biological barriers are meant to complete the containment by specifically limiting the infectivity, dissemination, and survival of the vector for recombinant DNA.

Animal cells in culture and defective viral genomes are representatives of eukaryotic host-vector systems whose biological containment relies on the fragility of animal cells and the inefficiency of reproduction of the viral vector without a complementing helper virus. The host-vector system chosen for many experiments and for much of the discussion of risk is *Escherichia coli* K-12 with nonconjugative plasmid vectors or with defective bacteriophage λ vectors. Further containment is proposed by use of mutants of *E. coli* which require specialized nutrients and which are inactivated at normal body temperatures, so that they can survive and multiply only in a stringent laboratory environment.

How effective are the biological barriers? Estimates of the frequency of survival of ingested *E. coli* and transfer of plasmids have been investigated by Anderson (1) and other workers (27,28; S. Falkow, *personal communication*). Within 4 to 6 days after human consumption of up to 3×10^{10} *E. coli* cells, fecal samples were free of viable *E. coli* organisms of the ingested type (1,27). Transfer of a tetracycline plasmid from ingested *E. coli* cells in the absence of antibiotics to resident *E. coli* strains *in vivo* or to co-ingested *E. coli* was not observed by Smith (27). However, in studies by Anderson (1) some replication of ingested *E. coli* was indicated in two of nine subjects. Transfer *in vivo* of a hybrid tetracycline-resistance plasmid was transiently detected at the low level of about 10^{-5} transfers per surviving donor cell 24 hr after ingestion of 10^{10} cells. The conjugate derivatives disappeared in the absence of antibiotic selection. Similar results in man and in calves are reported by Falkow (*personal communication*), who also observed the transfer of R plasmids from indigenous flora to ingested *E. coli* K-12. In none of these experiments could the ingested strains establish themselves as long-term stable components of the enteric flora of the volunteer subject.

Some qualifications, however, disturb our optimism. First, the titers of viable ingested *E. coli* K-12 excreted 1 day after the ingestion of 10^7 to 10^9 cells varies from 10^3 to 10^8 cells/g feces (1,27). This amounts to dissemination of 1 to 100% of ingested cells and cells acquiring plasmids *directly out of the laboratory and into public sewage courses.* Obviously, the disappearance of viable K-12 cells is not the end of the matter. As an indication of the ease of further escape, after heavy rainfalls, fecal contamination of upper Narragansett Bay by surface drainage and overloading of sewage treatment capacity pollutes shellfish beds sufficiently to quarantine their use for several days. Falkow (*personal communication*) notes that the fecal *E. coli* concentration in the Washington, D.C.–Potomac River area rose dramatically (10^5-fold) after a heavy rainfall, and he believes this is descriptive of many urban areas.

A second qualification is the selective advantage that antibiotics may confer on cells with drug resistance plasmids. The widespread use of antibiotics in livestock and poultry feed suggests that if bacteria-plasmid systems escape through human excretions, there may be ample selective pressures for their continued propagation.

A further qualification arises from the recent report by M. G. Smith that *E. coli* K-12 can multiply and transfer an R factor in the intestinal tract of sheep that have been briefly deprived of food (28). Clearly, we can expect health and nutritional conditions to influence drastically the viability and sexuality of ingested *E. coli* cells.

What confidence then can we muster over containment? Considering that Fort Detrick with the highest level of physical containment and specialized personnel has had serious laboratory infections (A. G. Wedum, *personal communication*), that most laboratories will have less-stringent physical barriers and will include novices in their personnel, that there will be 100 to 1,000 recombinant DNA laboratories, each routinely preparing perhaps 10^{11} to 10^{13} recombinant cells per culture, and that there are plausible biological routes of leakage to the environment, then it is not radical to conclude that our decisions on cloning technology are strongly helped by containment, but that we cannot rely on its absolute infallibility.

PLAUSIBLE HAZARDS OF RECOMBINANT DNA

Few people would argue that it is safe to reconstruct *E. coli* with genes for botulin or diphtheria toxin. Our real uncertainties arise instead from genetic reconstructions of unknown or benign genetic elements whose novel circumstances produce an unintended harmful effect. In this event we cannot refer to protection by gastric digestive processes, since the worst outcome might be the colonization of the lower intestine by a microbe producing a large amount of a specialized active protein such as a growth hormone, or carrying DNA regions capable of triggering a neoplastic process. The misplaced hormone factory assumes the recombinant eukaryote-bacterial RNA-protein translation barrier has been solved; the wayward virogen may require only transcription for its malignancy.

Most proteins are not absorbed as such in our alimentary canal and in any case are degraded by trypsin and chymotrypsin by individuals with normal pancreatic function. However, membrane surface-active polypeptides or hormones may be another matter, and our evolutionary survival as omnivores has never been tested by the persistence of an extra passenger pituitary gland in our lower bowel.

One source of confidence in the ineptness of foreign DNA stems from our past experimental handling of eukaryotic DNA and from our continuous exposure to DNA in foodstuffs. Again, these latter routes face barriers of gastric denaturation which would be avoided by an intestinal colonization.

What we need is information on the possible resistance of small circular DNA molecules and on the long-range fate of phage and phage *genomes* after their apparent inactivation following ingestion. Curiously, we apparently lack data on the recoverability of genetically active fragments of the DNA of these vectors from our excretions. We need information on the ability of soil, aquatic, marine, and symbiotic microorganisms to degrade the genetic information in new recombinant DNA molecules which may escape their laboratory containment.

The human, animal, and microbial models which allow for testing of these possibilities are obvious, and we should be able to reinforce or assuage our concerns over the significance of these hazards with experimental data. However, the problem of the time factor for expression after long latent periods suggests that it would be useful also to carry out epidemiological studies on the incidence of antiviral antibodies and neoplastic or regulatory disorders in researchers who have worked on phage, viruses, and animal cells for several years while we await the legacy of our previous exposures.

WHAT ABOUT THE BENEFITS?

Genetic Disorders

Probably the greatest value of recombinant technology will be in expanding a future knowledge base on genes and viruses, which in turn will create its own progeny of rewards. Beyond this, human genetic benefits have largely been discussed in terms of a bank of benign eukaryotic vectors reconstructed to contain specific genes whose dysfunction otherwise leads to serious clinical consequences, and of efficient bacterial factories housing eukaryote genes whose protein products permit direct therapeutic treatment of a genetic disorder such as factor VIII for hemophilia and human insulin for diabetes (30).

The gene therapy bank is patterned after transducing bacteriophages which correct host genome shortcomings, pseudovirions, and the Shope papilloma virus, which still holds prospects of use for human argininemia. The achievement of these benefits requires that we know whether the genetic defect is in a structural or regulatory gene; that the vector replicate, transcribe, and translate the recombinant genetic region but remain benign; that our regulatory controls remain in control of the expression of the recombinant genetic region; that post-translation modification of the gene product is carried out correctly; that this region be free of harmful sequences; and that the recombinant vector reach only the desired target tissue. Any one of these requirements poses a major research problem. Even then, the irreversibility of a therapeutic genetic infection calls for extensive prior experimentation on animals and human cells *in vitro*. As a useful clinical tool, gene therapy will more likely be a later than earlier benefit in most cases.

The prokaryotic-housed eukaryote gene transplant offers fewer human experimentation problems but faces major problems on the universality of controls of gene expression. Eukaryotic DNA replicates and is transcribed in bacterial cells (11,12,16,31,32,34,36), but the completeness of transcription and the fidelity of translation are uncertain. Correction of a *his*$^-$ bacterial requirement by cloned *his*$^+$ yeast-phage recombinant DNA has been reported (31). However, the meaning of this phenotypic suppression is still under investigation, and at least some other markers have not been successfully translated. The requirements for successful *in vivo* transcription and translation may be low molecular weight factors, prokaryotic recognition signals not present on eukaryotic DNA, or post-translation processing functions not normally present in bacteria. Possibly, sequences coding for bacterial ribosome binding sites may have to be incorporated into the neighborhood of the DNA recombinant ligation. A faithful transcription of eukaryote DNA by *E. coli in vitro* has been reported for a cell-free system which used SV40 DNA and *E. coli* RNA polymerase for transcription coupled with wheat germ extracts (21,22) for translation. Bacterial translation and processing components equivalent to those in wheat germ are both exciting research areas in themselves and possible barriers to a fuller exploitation of recombinant DNA potentials.

Even given a replacement product, only a fraction of the simple monogenic disorders which involve tissues accessible at postnatal stages are apt to yield to these approaches. The effective delivery of these genetic products to the patient is yet another challenge to biomedical engineering. Human plasma α_1-antitrypsin deficiency has not yet been successfully corrected although the plasma protein is readily available. On the other hand, Fabry's disease and Hunter syndrome, catabolic lysosomal storage disorders, may yield to enzyme replacement therapy (6). For other important disorders such as cystic fibrosis, the basic genetic function at fault remains elusive. Therefore, we can only guess that genetic technology is medically relevant to 0.01 to 1% of the newborn human population.

Gene correction by these methods is at the somatic level and so will affect gene frequency primarily by removing selection pressures against the procreative potential of affected individuals. Dysfunctional alleles will increase in number, but concern in this area is not more warranted than for the technological correction of inheritable forms of poor vision by eyeglasses, defects that would predispose one to early death in a hunting society.

Recombinant DNA techniques also provide an unusual opportunity to increase our understanding of the regulatory controls of eukaryote genes. The availability of easily purified globin mRNA has already generated individual clones for DNA containing regions coding for rabbit α- and β-globin chains using a plasmid vector in *E. coli* (7,20,25). At least some of the sequences controlling expression of the globin genes should be near the structural gene region and therefore cloned and amplified together. Examination of the requirements for *in vitro* and *in vivo* transcription and translation of the different globin genes may provide rapid progress toward treatment of some of the human thalassemias and may allow us to correct for sickle cell anemia by selective modulation of the relative levels of synthesis of β-, γ, and δ-globin chains.

Viral and Cancer Control

If eukaryote genes can be translated correctly in bacteria, then far wider distribution of benefits affecting all of us can be visualized for immunotherapy programs directed against viral disorders and cancers. Surprisingly, discussion of these possibilities has been overshadowed by prospects for genetic disorders. I suggest the use of specially restricted centers to deliberately recombine and clone, in *E. coli,* genetic regions coding for the coat proteins of selected pathogenic viruses which had been previously cultivated only in animal cells. Present viral vaccine methodology calls for large-scale cultivation of intact viruses which have to be purified from potential dangerous viral contaminants in the host cells. In such cases, pathogenicity is minimized by using attenuated virus or by physical or chemical inactivation. Recombination of attenuated viruses with related strains in nature, the appearance of virulent variants, incomplete inactivation, and passenger virus contamination offer significant risks to the public health. Compare the present technology with the possible biosynthesis of viral antigen in *E. coli* carrying recombinant DNA: in this event, *high concentrations of intact animal virions would be avoided,* and the nonviral DNA of the *E. coli* recombinant that is foreign to us would be relatively well characterized. Large-volume preparations would be simpler to carry out and would entail much less costly growth media and processing activities. If this approach can be coupled to RNA viruses by using reverse transcriptase and DNA polymerase to make a complementary DNA genome, then, for example, immune control of the influenza family could be attempted. Once the protocol for an influenza prototype is developed, and lead time needed for stockpiling antigens of variants should be shortened and potential epidemics thwarted earlier.

The genetic analysis of RNA viruses will be greatly facilitated by cloning DNA reverse transcripts (24) so that the resolving power of restriction nucleases and DNA sequencing techniques can be applied to the cloned and amplified recombinant. The characterization of animal tumor viruses capable of transforming human cells and the C-type viruses released from human myeloid leukemia cells (9,33) probably holds less risk when done on cloned fragments (as compared to the entire virus) since a fragment will have fewer regions of homology to human DNA and be defective in some functions. If cloned DNA regions from tumor viruses or from transformed cells allow a fuller analysis of the synthesis, properties, and significance of tumor-specific transplantation antigens, then it may be possible to formulate cancer immunotherapy programs based on recombinant DNA for such antigens and other virus-specific proteins accessible to immune defense systems.

ANTIRISK IN THE RISK-BENEFIT ASSESSMENT

Risks of recombinant DNA technology have been raised as though the questions under consideration were "yes" and "no" to such experimentation. I suggest a more accurate assessment of our options would be obtained by asking whether this technology provides a safer way of analyzing and exploiting properties of genes and viruses than the means currently in wide use. Curiously, nonrecombinant technology has received little debate in this country.

It is my impression that the majority of university laboratories involved in microbial genetics, nucleic acid biochemistry, or animal cell culture are open bench–type arrangements. If any environmental regulation exists, it is to prevent contaminants from entering the laboratory area and to rapidly dilute volatile chemicals into the outside air. Contamination of pure cultures and disposal of organic solvents are the major procedural concerns.

Many cultured cell lines in common use liberate C-type and other viruses that can propagate in cells of foreign species. With noteworthy insight, Robin Weiss and Natalie Teich have pointed out some of the potential hazards of activation of these xenotropic viruses in interspecific tissue grafts, in hybrid cell lines obtained by fusion of somatic cells of different species, in mixed viral infections, and in exposure of extracted DNA to heterospecific cells (35). Such techniques are quite common in contemporary experimental biology and are at the core of work on animal cell genetics and animal virology. Thousands of laboratories are so involved. On a different scale, large volumes of human lymphocytes are cultured to make antigens for antilymphocyte antiserum production, and diverse animal cell lines are used for cultivating viruses for vaccine production. In one instance, poliovirus preparations which had been widely distributed and used for vaccines were later discovered to contain SV40 virus, which had been present in the cell lines used for polio virus propagation (8).

The risk of some of these procedures with intact viruses and virogens should be compared with that of recombinant DNA technology. One can well counter that all viral and animal cell culture procedures should be brought under an Advisory Committee's supervision. However, even with such regulation, the basic question for a specific research program or application is whether it is safer to study an intact, potentially pathogenic viral genome, or a fragment of such a genome associated with a disarmed vector, or to use other techniques, or to refrain totally from a study. One responsibility is to consider these options in making our choice. I suggest that for many situations, the preferred course as a matter of safety is a recombinant DNA approach.

REFERENCES

1. Anderson, E. S. (1975): Viability of, and transfer of a plasmid from, E. coli K12 in the human intestine. *Nature,* 255:502–504.
2. Botchan, M., McKenna, G., and Sharp, P. A. (1973): Cleavage of mouse DNA by a restriction enzyme as a clue to the arrangement of genes. *Cold Spring Harbor Symp. Quant. Biol.,* 38:383–395.
3. Brown, W. M., and Vinograd, J. (1974): Restriction endonuclease cleavage maps of animal mitochondrial DNAs. *Proc. Natl. Acad. Sci. U.S.A.,* 71:4617–4621.
4. Cohen, S. N., Chang, A. C. Y., Boyer, H. W., and Helling R. B. (1973): Construction of biologically functional bacterial plasmids in vitro. *Proc. Natl. Acad. Sci. U.S.A.,* 70:3240–3244.
5. Danna, K., and Nathans, D. (1971): Specific cleavage of simian virus 40 DNA by restriction endonuclease of Hemophilus influenzae. *Proc. Natl. Acad. Sci. U.S.A.,* 68:2913–2917.
6. Dean, M. F., Muir, H., Benson, P. F., Button, L. R., Boylston, A., and Mowbray, J. (1976): Enzyme replacement therapy by fibroblast transplantation in a case of Hunter syndrome. *Nature,* 261:323–325.
7. Efstratiadis, A., Kafatos, F. C., Maxam, A. M., and Maniatis, T. (1976): Enzymatic in vitro synthesis of globin genes. *Cell,* 7:279–288.
8. Fraumenti, J. F., Jr., Ederer, F., and Miller, R. W. (1963): An evaluation of the carcinogenicity of simian virus 40 in man. *J.A.M.A.,* 185:713–718.
9. Gallagher, R. E., and Gallo, R. C. (1975): Type C RNA tumor virus isolated from cultured human acute myelogenous leukemia cells. *Science,* 187:350–353.
10. Ganem, D., Nussbaum, A. L., Davoli, D., and Fareed, G. C. (1976): Propagation of a segment of bacteriophage λ-DNA in monkey cells after covalent linkage to a defective simian virus 40 genome. *Cell,* 7:349–359.
11. Glover, D. M., White, R. L., Finnegan, D. J., and Hogness, D. S. (1975): Characterization of six cloned DNAs from Drosophila melanogaster, including one that contains the genes for rRNA. *Cell,* 5:149–157.
12. Hamer, D. H., and Thomas, C. A., Jr. (1976): Molecular cloning of DNA fragments produced by restriction endonucleases SalI and BamI. *Proc. Natl. Acad. Sci. U.S.A.,* 73:1537–1541.
13. Hedgpeth, J., Goodman, H. M., and Boyer, H. W. (1972): DNA nucleotide sequence restricted by the RI endonuclease. *Proc. Natl. Acad. Sci. U.S.A.,* 69:3448–3452.
14. Jackson, D. A., Symons, R. H., and Berg, P. (1972): Biochemical method for inserting new genetic information into DNA of simian virus 40: circular SV40 DNA molecules containing lambda phage genes and the galactose operon of Escherichia coli. *Proc. Natl. Acad. Sci. U.S.A.,* 69:2904–2909.
15. Kelly, T. J., Jr., and Smith, H. O. (1970): A restriction enzyme from Hemophilus influenzae. II. Base sequence of the recognition site. *J. Mol. Biol.,* 51:393–409.
16. Morrow, J. F., Cohen, S. N., Chang, A. C. Y., Boyer, H. W., Goodman, H. M., and Helling R. B. (1974): Replication and transcription of eukaryotic DNA in Escherichia coli. *Proc. Natl. Acad. Sci. U.S.A.,* 51:1743–1747.
17. Mowbray, S. L., and Landy, A. (1974): Generation of specific repeated fragments of eukaryote DNA. *Proc. Natl. Acad. Sci. U.S.A.,* 71:1920–1924.
18. National Institutes of Health (1976): Recombinant DNA research: Guidelines. *Fed. Register,* 41(131):27902–27943.
19. Nussbaum, A. L., Davoli, D., Ganem, D., and Fareed, G. C. (1976): Construction and propagation of a defective simian virus 40 genome bearing an operator from bacteriophage λ. *Proc. Natl. Acad. Sci. U.S.A.,* 43:1068–1072.
20. Rabbitts, T. H. (1976): Bacterial cloning of plasmids carrying copies of rabbit globin messenger RNA. *Nature,* 260:221–225.
21. Robberson, D. L., Clayton, D. A., and Morrow, J. F. (1974): Cleavage of replicating forms

of mitochondrial DNA by EcoRI endonuclease. *Proc. Natl. Acad. Sci. U.S.A.*, 71:4447–4451.
22. Roberts, B. E., and Paterson, B. M. (1973): Efficient translation of tobacco mosaic virus RNA and rabbit globin 9S RNA in a cell-free system from commercial wheat germ. *Proc. Natl. Acad. Sci. U.S.A.*, 70:2330–2334.
23. Roberts, B. E., Gorecki, M., Mulligan, R. C., Danna, K. J., Rozenblatt, S., and Rich, A. (1975): Simian virus 40 DNA directs synthesis of authentic viral polypeptides in a linked transcription-translation cell-free system. *Proc. Natl. Acad. Sci. U.S.A.*, 72:1922–1926.
24. Rothenberg, E., and Baltimore, D. (1976): Synthesis of long, representative DNA copies of the murine RNA tumor virus genome. *J. Virol.*, 17:168–174.
25. Rougeon, F., Kourilsky, P., and Mach, B. (1975): Insertion of a rabbit β-globin gene sequence into an E. coli plasmid. *Nucleic Acids Res.*, 2:2365–2378.
26. Smith, H. O., and Wilcox, K. W. (1970): A restriction enzyme from Hemophilus influenzae. *J. Mol. Biol.*, 51:379–391.
27. Smith, H. W. (1975): Survival of orally administered E. coli K12 in alimentary tract of man. *Nature*, 255:500–502.
28. Smith, M. G. (1976): R factor transfer in vivo in sheep with E. coli K12. *Nature*, 261:348.
29. Southern, E. M., and Roizes, G. (1973): The action of a restriction endonuclease on higher organism DNA. *Cold Spring Harbor Symp. Quant. Biol.*, 38:429–433.
30. Stetten, D., Jr. (1975): Freedom of enquiry. *Genetics*, 81:415–425.
31. Struhl, K., Cameron, J. R., and Davis, R. W. (1976): Functional genetic expression of eukaryotic DNA in Escherichia coli. *Proc. Natl. Acad. Sci. U.S.A.*, 73:1471–1475.
32. Tanaka, T., Weisblum, B., Schnos, M., and Inman, R. B. (1975): Construction and characterization of a chimeric plasmid composed of DNA from Escherichia coli and Drosophila melanogaster. *Biochemistry*, 14:2064–2072.
33. Teich, N. M., Weiss, R. A., Salahuddin, S. Z., Gallagher, R. E., Gillespie, D. H., and Gallo, R. C. (1975): Infective transmission and characterisation of a C-type virus released by cultured human myeloid leukaemia cells. *Nature*, 256:551–555.
34. Thomas, M., Cameron, J. R., and Davis, R. W. (1974): Viable molecular hybrids of bacteriophage lambda and eukaryotic DNA. *Proc. Natl. Acad. Sci. U.S.A.*, 256:551–555.
35. Weiss, R. (1975): Virological hazards in routine procedures. *Nature*, 255:445–447.
36. Wensink, P. C., Finnegan, D. J., Donelson, J. E., and Hogness, D. S. (1974): A system for mapping DNA sequences in the chromosomes of Drosophila melanogaster. *Cell*, 3:315–325.

Recombinant Molecules: Impact on Science and Society, edited by R. F. Beers, Jr. and E. G. Bassett. Raven Press, New York © 1977.

44. Industrial Risk Analysis

Edward C. Dart

Bioscience Group, Corporate Laboratory, Imperial Chemical Industries Ltd., Runcorn, Cheshire, England

In Imperial Chemical Industries, we found a group of Hazard and Reliability Engineers extremely helpful in designing and making even safer our genetic engineering program. I thought it would be worthwhile discussing the ways in which they have helped us. In mid-1974, when we first saw the tremendous potential of recombinant DNA research, we had the problem of deciding whether the risks to which our employees would be exposed fell within sufficient limits for the work to continue. Here the experience of the ICI team of Hazard and Reliability Engineers proved invaluable. They have spent years doing this sort of risk analysis for chemical plant construction in the design of safety measures. At first sight this might seem remote from molecular genetics, but their systematic approach can be applied to the sort of biological procedures with which we are all familiar.

We start with the premise that since we do not know the extent of the hazard, we must be sure that we make every effort to contain it. Step one, then, is to treat all gene transfer processes as potentially hazardous. The skills of the Hazard and Reliability Engineer come in at step two, in which he makes flow sheets of every operation associated with laboratory functions and working protocols. Every operation is then analyzed as quantitatively as possible for the chances of escape of potentially hazardous material (PHM). This helps identify those areas where PHM release is most likely to occur. In these areas, a series of fail-safe procedures can then be devised to minimize the overall chance of such a release. Finally, we arrive at a quantitative estimate of the chances of PHM release.

To this number we graft on one which relates to the chances of PHM causing a serious infection in one of our laboratory workers. In this way we can compare the risks he faces from a possible PHM release with other day-to-day hazards to which he is normally exposed.

Calculations such as ours can be criticized as being oversubjective, but by erring on the side of pessimism at every stage we believe the exercise becomes worthwhile.

Before summarizing our results I would like to discuss the two major potential causes of PHM release identified in our study. Unfortunately, I do not think the current safety guidelines devote sufficient attention to these.

The major cause of potential hazard we saw as human error, and we believe that fail-safe procedures can be designed to minimize such things as taking the wrong culture from the refrigerator or failing to adequately seal a transferred culture flask or safety cabinet.

The second most likely hazard we saw as ingestion of biological aerosols by inhalation. Escapes of this sort are not going to be contained by special airlocks and shower facilities. It is extremely important, therefore, that the design of working procedures in laboratories minimize the chances of aerosol formation. It is also important that some form of face mask or respiratory apparatus be worn to limit the likelihood of ingestion.

The results of our risk analysis are shown in Table 1. Here we compare the chance of one of our laboratory workers suffering a serious infection as a result of PHM release with some death statistics for U.K. males.

Based on these figures, we have decided we can continue safely and sensibly to realize the benefits of recombinant DNA research, but we still are subject to whatever guidelines the U.K. Government may impose. We are happy to help them shape their guidelines but will follow them nonetheless.

TABLE 1. *Comparative risks (per year)*

Chance of serious infection due to PHM release in:	
A laboratory worker	<1 in 10^6
A colleague outside the lab	$\ll 1$ in 10^6
U.K. death statistics (male 15–34):[a]	
Infectious disease	1 in 10^5
Motor accident	1 in 3×10^4
All causes	1 in 10^3

[a] From Social Trends No. 5, p. 136. HMSO/Central Statistical Office, 1974.

Recombinant Molecules: Impact on Science and Society, edited by R. F. Beers, Jr. and E. G. Bassett. Raven Press, New York

45. A Real Situation

Brian M. Richards

Department of Molecular Biological Research, Searle Research Laboratories, High Wycombe, Buckinghamshire, England

With regard to the problem of biohazards, I agree with Dr. Murray's statement that the things we have done in the past are going to be helpful in planning facilities for genetic manipulation. He also mentioned that we at Searle were faced a few years ago with the problem of how best to contain the purification procedures for the hepatitis surface antigen (HB_sag). The hepatitis agent is found in significant quantities in the blood of human carriers. Using large volumes of human blood, we employ the usual laboratory purification procedures of column chromatography and ultracentrifugation in preparation of concentrated hepatitis agent.

At the beginning we made an important decision: to strive for total containment of the agent and of the equipment, thereby leaving the workers hazard-free and hopefully safe. In conjunction with our suite of laboratories maintained under negative pressure, we originally designed shower facilities, air locks, and ancillary safeguards. But we never, in fact, have found it necessary to use these fully since we have so successfully contained the equipment. Taking into account the kind of hazard analysis (but not to the same degree) which Dr. Dart mentioned, we felt that, in dealing with hepatitis, the requirements for what are now called P4 containment conditions were not appropriate but could be easily reinstated if we wished.

We designed a series of containment cabinets designated A, B, and C. The A-type cabinet is quite similar to a glove box, but with the important distinction that it is capable of being completely sealed to within 0.5- to 1.0-inch water gauge negative pressure. In other words, it can maintain a negative pressure against an external negative pressure. Within this kind of cabinet, we can perform all operations normally done on a worktop area, for example, membrane filtration or, with suitable modification, even column chromatography employing 4- to 5-foot columns together with their fraction-collecting equipment. Figures 1 and 2 show examples of these type-A cabinets.

In the B-type containment facilities, more dynamic processes are going on: for example, we have totally contained an MSE 6L centrifuge and an

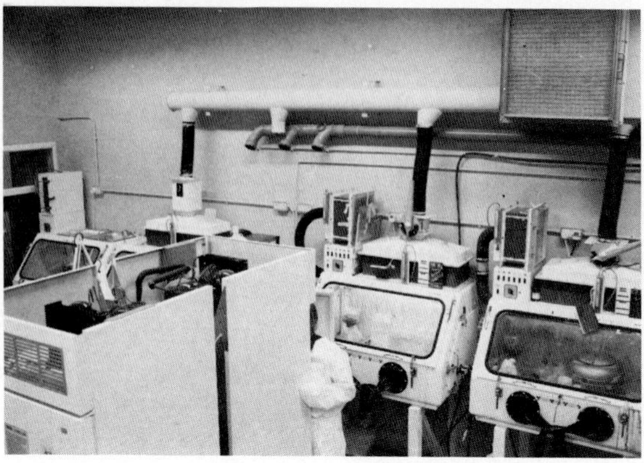

FIG. 1. Type-A containment cabinets.

MSE Superspeed 65 centrifuge operating with a batch-type zonal head. Figure 3 shows the interior of a cabinet containing a zonal ultracentrifuge with fraction-collecting equipment.

Finally, in the C-type cabinet—a major engineering exercise—we have contained an Electronucleonics RK continuous flow zonal centrifuge, the control panel for which is shown in Fig. 4. This machine incorporates an air-driven turbine motor, thus potentially capable of generating a massive

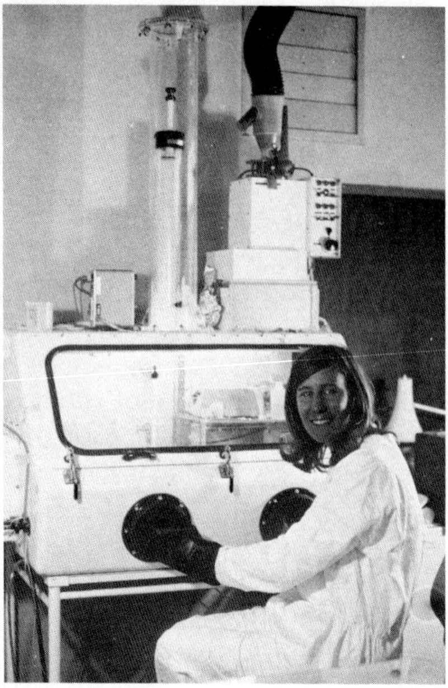

FIG. 2. Type-A containment cabinet.

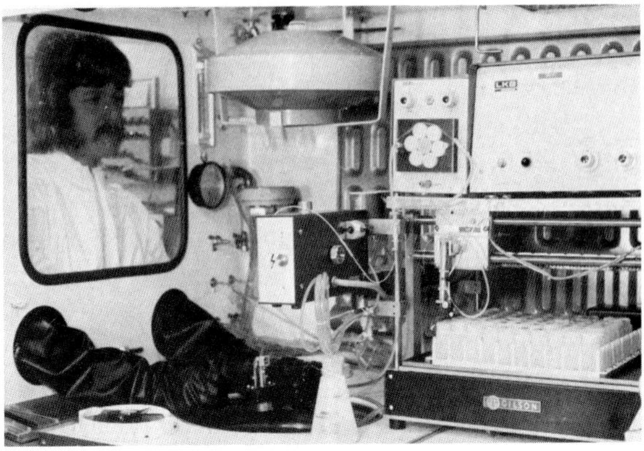

FIG. 3. Interior of a B-type cabinet containing zonal centrifuge and fraction-collecting equipment.

aerosol. This problem is well within the design capability of the containment. All characteristics of the type-A cabinet are incorporated in B- and C-type cabinets. All have absolute filtration on incoming and outgoing air, and most are now connected to a centralized exhaust system. Also, they can be sterilized on either a manual or semiautomatic or automatic cycle with ethylene oxide or some similar decontaminant. We have selected ethylene oxide because we found it to be the most suited for use in an online auto-

FIG. 4. Control panel of a continuous flow zonal centrifuge.

mated feed system. For a biological program which is now getting into production, we will have a totally programmable, fail-safe capability of cabinet-by-cabinet operation and decontamination. Each of the containment cabinets also has backing up ports so that we can introduce material and equipment under sterile conditions or remove contaminated material and dispose of it elsewhere. When we are working with centrifuges and associated fraction collectors in type-B cabinets, all the operating controls are externalized; we have developed semi-"idiot proof" systems and protocols for such operations.

Infrequently mentioned in discussions on safety are schedules for dealing with nonsystematic or catastrophic needs for decontamination. We have developed programs and protocols for each of these requirements on a routine basis. When the equipment needs servicing, we have systems and protocols which permit decontamination of not only all the exposed surfaces but also all the fluid systems, so that the repairman is protected. In the event of any catastrophic failure of the machine, such as a head explosion, we have designed procedures to cope with the resulting hazard without sacrificing the machine or endangering people.

We emphasize the need to take account of emergency procedures. We have ethylene oxide detectors and alarm systems and full air suits so that people who must enter an atmosphere of ethylene oxide are completely protected. Of predictable mishaps, a major one is the escape of ethylene oxide.

We think we have considered most of the other desirable aspects for a biohazard facility of this kind. All the people who work with this facility, and in addition many others, are regularly monitored; that, of course, is an easy thing to do with hepatitis. We also have an active and vocal Safety Committee consisting of representatives of the scientists, the technicians, and the support staff; these people meet regularly to review all our protocols, and it is within their power to ask for any outside expertise they may need. The system sometimes annoys me but it works well. I would like to mention a publication (1) on the containment system; unfortunately, it deals only with the containment of the batch-type zonal head of an MSE 65 centrifuge.

I think it fair to conclude that, in this instance, industry has tried to maximize its attention to safety. If we go back to one of the emerging attitudes in Europe which John Tooze mentioned (*this volume*), that "the clever people don't need containment, but people who are not so clever need them . . .", then we can, in this instance, aspire to the maximal idiocy.

REFERENCE

1. Webb, N. L., Richards, B. M., and Gooders, A. P. (1975): The total containment of a batch-type zonal centrifuge. *Biotechnol. Bioeng.*, 17:1313–1322.

Recombinant Molecules: Impact on Science and Society, edited by R. F. Beers, Jr. and E. G. Bassett. Raven Press, New York © 1977.

46. Gene Implantation: Proceed with Caution
Reservations Concerning Research in Recombinant DNA

Frances R. Warshaw

The Group on Genetics and Social Policy, Boston Area Science for the People, Cambridge, Massachusetts 02139

INTRODUCTION

This chapter represents the position of the Group on Genetics and Social Policy of the Boston Area Science for the People. For over 3 years our group has been active in raising questions about the social implications of research in genetics, in genetic screening, and in gene implantation (6,7,11,18).

When I was invited to prepare a paper on our reservations about research in gene implantation, I had no idea how difficult and absorbing a task it would be. One of the reasons that it has been difficult is that there has been little real dialogue between those who wish to do the work and those who wish the work to be restricted or even halted. An enormous antagonism has developed which forces the substitution of altercation for conversation. Arguments fly by, but few are heard (17). The subject has become an emotional one, and many people are ambivalent about it. Very often people say things that may be interpreted as conflicting, and it is possible to interpret statements to mean just the opposite of what the speaker intended.

I want to emphasize here that people on both sides of the issue are at fault, neither listening sufficiently carefully to the other. We hope that discussions such as this one will encourage more rational communication between the antagonists. We feel that the risks inherent in continuing research in recombinant DNA molecules warrant great caution, that we must pause and carefully think out the possible consequences, and what we should do about them.

What are the responsibilities of scientists in the selection of their research projects? There is a common belief that scientists are not responsible for the repercussions of their work. This follows from the belief that scientists should follow their research wherever it leads them, and that they should be free to do this. The rationale for this belief is that pure research could lead to great and useful discoveries, that outside controls would inhibit progress and turn science into a political tool, and that, since scientists arc made pure by their commitment to the pursuit of knowledge, such control should be superfluous.

Although this sounds a bit simplistic, we know that many scientists and much of the rest of society would like to believe it. (In fact, I believed it until quite recently.) Nevertheless, the direction of our research is already guided by outside pressures: the limitations of the budget, the priorities of the funding agencies, and the constraints imposed on them by societal interests such as public health, and by social and ethical considerations such as the restrictions on human experimentation (5,24,36).

How can we claim that scientists can choose what to study when it is clear that the granting agencies, variously persuaded by Congress, do a large part of the choosing? Also, professional pressures encourage us to investigate currently fashionable subjects; not to do so would jeopardize our chances for success and more grants.

Most scientists seem to believe that science somehow hovers above the social fabric; but if science is at all pure, it is only in the search for new knowledge. As long as society uses the products of science, science cannot be neutral because social actions are always moral issues (24 p. 355). There is no value-free science. Although some view science as entirely objective, many associate only a positive value to science and see its applications through rose-tinted glasses. They do not recognize that, without careful planning and forethought, science can lead to disaster as easily as to advantage.[1]

POLITICS AND PUBLIC DEBATE

Let me ask bluntly: Should we pursue the study of recombinant DNA? If we should, under what circumstances? And who is to decide?

Many biologists are not worried about interspecies gene transplantation. We are indebted to Professors Berg and Baltimore and to the rest of the Asilomar Conference organizers for having first pointed out the need for caution in this field of research. Indeed, so much valuable time has been spent in discussing this issue that, had I not been concerned to begin with, I should certainly be so now. I have wondered at what seems to be the lighthearted attitude of so many respected biologists about the prospect of transplanting genes across species barriers (3,10 p. 153).

The results of our research have direct bearing on the public welfare. It has been pointed out that there are both potential hazards and potential

---

[1] A survey taken in 1974 reports that 75% of the American public believes that science has changed things for the better (a point of view we do not dispute). Furthermore, 29% feel that science and technology have caused a few of our current problems. In addition, 23% believe that science will solve most of our problems, whereas 53% believe that only some of our problems will be solved by science. It is important to note that responses reflecting positive attitudes toward science correlate strongly with education and income (27).

benefits. The benefits are often painted in such an exaggerated fashion that many of the informed public and, indeed, many biologists have come to believe that this technology will save the world from disease and starvation. (Of course, equally exaggerated scenarios of disaster can also be imagined.)

Scientists are not the appropriate judges of the legitimacy of this research; the public is at risk. We as scientists may be best suited to calculate risks and to predict benefits, but does this mean that we are best able to decide whether or not to run the risks? Here the debate becomes one of public policy. A small number of scientists in influential positions should not have the power to shape the future. The public is at risk, the public should benefit, and the public should decide whether the benefits justify the risks.

Anyone can understand the issue. With few exceptions, the public is not in a position to make decisions about scientific matters due to the lack of social mechanisms to allow for public participation and to the mystification of the scientific endeavor which pervades our society. This, in turn, is due to an elitist attitude promulgated by academicians that only a few people are capable of higher learning. I do not want to debate this issue at length; however, my own experience is that anyone can understand the issues we are discussing here and can ask probing questions that are difficult to answer. Because this research has such impact on society, it is our responsibility to educate the public and to make the decision a shared one.

The debate has not been sufficiently public. It is often said that an attempt has been made to open the debate to the public. The examples generally given are the Asilomar Conference, which was reported in many newspapers, the NIH hearings in February 1976, which were called "public hearings," and the NIH Guidelines, which are supposed to be open to public scrutiny and criticism. We question the public nature of these forums. The fact that the Asilomar Conference was convened by inviting a group of prestigious individuals without extending an open invitation to interested persons from the general public is indication enough of the *semi*-public nature of this forum (N. Wade, *personal communication*).

The experts should not be only molecular biologists. There are obvious problems involved in making the NIH both the funding agency for the research as well as the body responsible for its regulation. In addition, why was the NIH Guidelines Advisory Committee so heavily staffed by molecular biologists (10 p. 169; 12 p. 69; 32,43)? It is quite true that important technical questions needed to be addressed in order to propose the Guidelines. Some of these involved the details of how the experiments would be done and, in this regard, the molecular biologists planning to do the experiments were indeed the appropriate group of experts to be consulted. But the much thornier, and more important, scientific questions are those of danger to public health and to the biosphere. Evolutionary biologists, ecologists, epidemiologists, infectious disease experts, and public health officials should have been on the Advisory Committee in force in order to deal adequately with these aspects of the application and repercussions of the research. Remember, too, that not all of the questions raised by gene implantation research are scientific in nature; there are moral and social dilemmas to be confronted as well. Philosophers and historians of science should have been present to assure careful consideration of the difficult global problems raised by the research (43).

The Guidelines have not been available to the public. When the Guidelines (11) were released, they were sent to major investigators, editors of scientific journals, and the like. Especially in the light of the public interest which had been displayed by news coverage and local city actions, why was no effort made to make the Guidelines accessible to the general public? They should have been distributed, complete with explanations and an expanded glossary, to every public library. At the very least, the Guidelines should have been made immediately available to scientific libraries so that interested and informed individuals not directly involved in the research would have easy access to them.

The environmental impact statement should have come earlier. The handling of the environmental impact statement for the Guidelines provides further demonstration of how public access has been limited. Publishing a Draft Environmental Impact Statement before issuing the Guidelines would have been in keeping with the spirit, although beyond the letter, of the National Environmental Protection Act. However, it has finally emerged $2\frac{1}{2}$ months after publication of the Guidelines. As required by law, time was allotted to receive public response, but a member of our group found that the month provided was inadequate even to obtain a copy of the Draft on request.

The use of language is important. Recently, I asked a friend who is trained as a philosopher of science if he would read an article in the *New York Times* about recombinant DNA. He doubted that he would, but said he would certainly read about transplanting genes which, after all, is what this is all about (W. Berkson, *personal communication*).

When Dr. Frederickson of the NIH speaks of the potential dangers and the promise of great benefits (10 p. 149), the bias of his viewpoint is clear. It is not clear to us that the benefits are any more likely to be realized than the dangers. What is clear is that, since there is potential danger of a disastrous and irreversible nature, the decision to continue this work is not the scientists' alone.

The consequences of this work are not predictable. An enlightened view held by many people has been well articulated by S. E. Luria (24 p. 355): "I personally believe that not all research is legitimate: its legitimacy has to be judged in terms of its clearly predictable consequences."

We would go a step further. One cannot predict with confidence the consequences of research in gene implantation. Under such circumstances, what is needed is a careful analysis of the possibilities. Even when one cannot easily assign probabilities to the projected consequences, if significant hazards can be projected it becomes absolutely necessary to open up the decision-making process to the public. We must not allow the natural bias of those involved in the research to predetermine the decision to proceed or not. Since it is impossible to be free of bias, the only way to allow the public to make balanced decisions is to make our biases known and explicit.

We cannot be free of bias. We do not mean to imply that only when the scientific questions cannot be answered does the public need be involved. It is often true that, even when the scientific implications are clear, moral and political implications must be considered. Decisions concerning social consequences of research must be made in the public sector.

JUSTIFICATIONS AND RISKS

In attempting to justify a particular avenue of research, it is usually sufficient to show that the experiments will be scientifically productive. When the research appears hazardous, it becomes necessary to balance the possible benefits against the possible dangers (10 p. 329). Those at risk must be involved in making the choice.

Two sorts of risks are entailed in carrying out the research: those run by the researchers themselves (26) and those which affect a broader population. When the risks incurred will affect only laboratory personnel, it is sufficient to have safety procedures that satisfy the workers involved.[2] However, the major risks of research in gene implantation are run not simply by the participating investigators but by the public as a whole; therefore, the decision to proceed must be a public one. Benefits which will accrue to the general public are needed to justify the risks. Moreover, this research, like animal virology, will be very expensive and should be shown to be a worthwhile investment (10 p. 90).

PREDICTIONS OF BENEFITS

Proponents of this research claim that it will yield far-reaching benefits to humankind. I have listed below some of the generally projected benefits of research in gene implantation (3,10 pp. 162–164). In a subsequent section I shall point out some of our objections.

1. Intellectual advances. We are told that invaluable knowledge will result from this research that cannot be gained any other way, and that

[2] Unfortunately, in many cases, the workers are neither informed of the dangers nor allowed to modify safety procedures. This is why we urge the organization of safety committees including students, technicians, custodians, dishwashers, and secretaries. At present, the principal investigator, who is responsible for maintaining safety, often does not enforce or even teach safety regulations (1).

this technology is the key to understanding the functions and control of DNA.

2. Progress in agriculture. The world's food supply is limited and many people are starving. By the use of genetic engineering we might create variants of grains which could fix atmospheric nitrogen, and thus be independent of fertilizer. The creation of a single grain which would be a source of complete protein might also be possible.

3. Treatment of genetic diseases. These may be cured (in somatic cells) or prevented (in germ cells) via genetic manipulations.

4. Progress in cancer research. The technology might allow us to unveil the mystery of cancer and, hence, to cure it.

5. Progress in drug production. Genetic engineering may provide us with an easy, inexpensive way to manufacture insulin, antibiotics, and other biologically active substances.

CRITIQUE OF THE PREDICTIONS

I would like to offer an alternate way of looking at these benefits. I shall discuss the items in *reverse* order and give a few examples to illustrate our points.

1. General medical advances. Self-reproducing, biochemically active substances are dangerous. The massive manufacture of biochemically active substances by the use of their DNA in self-reproducing form poses an incalculable and irreversible danger. This is especially true if the host of the DNA is *E. coli,* always present in the human gut and capable of being a human pathogen. Moreover, any organism which can exchange genetic material with *E. coli* is potentially as dangerous a host.

Drugs are not expensive to make. In any case, although such products as antibiotics are expensive to the consumer—that is, the patient—they are not really very expensive to make. The expense is due to excessive and unnecessary advertising and packaging costs, and to profit made by the manufacturers, distributors, and pharmacies.

Antibiotics have remained sufficiently cheap and abundant to lead to indiscriminate use, which, in itself, constitutes a hazard. You may have heard of the epidemic in the Yukon Territory in which many children have died or been brain damaged by bacterial meningitis (S. Falkow, *personal communication*). Bacterial meningitis is caused by *Haemophilus influenzae,* many strains of which are now resistant to ampicillin. The infection is now being treated with chloramphenicol. The side effects of chloramphenicol are of two kinds: direct effects to the patient in the form of anemia and depression of the bone marrow, and epidemiological effects. Indiscriminate use of chloramphenicol in Vietnam and in Mexico has resulted in chloramphenicol-resistant *Salmonella typhi,* which cause typhoid fever (10 pp. 23–24,30). A chloramphenicol-resistant strain of *Haemophilus influenzae* has been isolated in Paris (4,39; P. Anderson, *personal communication*). It is likely that there will soon be strains of *Haemophilus influenzae* which are resistant to both ampicillin and to chloramphenicol, which will make treatment of spinal meningitis very difficult. This is a flagrant example of how the selection of multiple drug resistance in pathogenic bacteria by indiscriminate use of antibiotics may result in unpardonable disaster.

Proponents of gene implantation technologies often suggest that it would be advantageous to find an easy, inexpensive source of insulin (10 p. 296). Most of the insulin which is administered to diabetic patients is a mixture of bovine and porcine insulin. The apparent shortage of insulin is supposed to be due to a shortage of pigs. At the very least, we could implement more efficient ways to collect and use the pancreata from the pigs that are already being slaughtered. This would be safer than running the risk of having *E. coli* churning out insulin in our guts. Bovine insulin differs from human insulin by two amino acid residues; porcine insulin differs by only one and is adequate for most patients (16,21,44). For those few who have allergic reactions even to porcine insulin, desensitization is almost always possible. For the very few who cannot be desensitized, we would recommend the development of a technology other than recombinant DNA for the production of human insulin in small quantities.

What about cancer? It has become clear that the immediate cause of most cancers is exposure to carcinogenic chemicals (32,37). Thus, the research which would enable us to deal with cancer would involve the identification of these carcinogens, and the elucidation of their mechanisms of action and of the complex relationships between the lengths of exposure and the effective doses of these substances.

If we really want to solve the cancer problem, we should spend more energy cleaning up the environment and changing our eating and smoking habits. At the NIH public hearings on recombinant DNA research, David Baltimore said he suspects that, since it is hard to change our habits, we should study oncogenic viruses to learn how to cure cancer as well as how to prevent it (10 p. 250). We would be very glad if there were a cure for cancer, but we think that we must review our priorities. He encourages us to maintain the status quo, even though one in six Americans may die from doing just that (19).[3] We certainly do not intend to accuse Dr. Baltimore and many others of malevolent intentions; however, we can hardly trust ourselves to be altruistic where our own work and self-interest are concerned. This is why it is so important to have people other than animal virologists and molecular geneticists on committees for the review of these

---

[3] A scientist of Dr. Baltimore's eminence is in a position to urge the public to change its habits: this kind of advice would be very powerful coming from an expert on cancer. Scientists must now start to take an initiative in public affairs, especially in pointing out the limitations and the hazards of science, as well as the possible applications. This would encourage people to take more responsibility for the environment and for their own health.

questions. It is much too easy to argue that one's own work is the most important kind.

So far my discussion of the applications of gene implantation research to the cancer problem has focused on the pursuit of a more productive avenue—that is, prevention rather than cure. There are also serious drawbacks to be considered. We would worry that new cancers might be produced in the effort to cure the known ones.

Recently a group from Belgium reported their work with *Agrobacterium tumefaciens,* which infects certain dicotyledonous plants thereby causing the neoplastic disease called crown gall. A few days after infection a tumor is seen, and the presence of the bacterium is no longer necessary to maintain the tumorous state (22,28,29). Montague and co-workers have shown that the cancer is due to an oncogenic plasmid which is transferred from the bacterium into the plant cell (10,25,28 p. 179). This is a remarkable discovery, and, although the detailed mechanism of action has not yet been elucidated, the parallel is striking: if in fact oncogenes are present in animal viral genomes and, therefore, in mammalian cells, there is obvious danger in combining mammalian DNA with coliform plasmids.

2. Possible treatments of genetic diseases. As regards curing or preventing genetic diseases, the cure appears worse than the malady. That a host of scientists are willing to plunge into this form of research without first carefully examining the political and social repercussions may be an example of the compartmentalization of our universities and of our minds. Even though one person can act in the various capacities of biologist, philosopher, parent, and artist, very rarely are these disciplines merged in the mind; rather, they draw us along on parallel paths. To begin to do genetic manipulations on human beings is to take a step toward the "Brave New World." Each step counts, and the society as a whole must ask: Is this the direction of choice?

A more obvious concern is that eager clinicians may jump in with imperfect cures before the basic research is done. Furthermore, consider the perhaps equal probability of spreading disease through laboratory accidents, which would result in trading one disease for another. Of course, this is a fear associated with many medical technologies. It is especially worrisome in this case because of the self-perpetuating nature of the projected afflictions.

3. Agriculture: the Green Revolution. At first, the agricultural applications seem not merely acceptable but exciting and wonderful. But let us take a step back in time and look at the "Green Revolution," which involved introducing a new genetic variety of rice to underdeveloped countries. Because this new rice required special fertilizer and a different growing season, and had stalks which were unsuitable for use as fodder or thatch, the rich farmers got richer and the poor farmers got poorer. Changing a crop may affect the economy and the ecological environment, but it cannot change the political conditions. Any increase in welfare becomes illusory.

Although there are people in the United States who go hungry, it is not because we do not know how to grow food. No matter how much food is produced, it will not keep people from starving until it is distributed to the people who need it, regardless of the likelihood of profit to producers and distributors.

Neither is it trivial to worry about the possibility of contamination by other DNA's, or the possibility of creating a nitrogen-fixing crabgrass which would confound farmers by its virile growth (10 p. 296).

4. Intellectual gains? We have been accused of wishing to restrict freedom of inquiry. In a sense, this is true. But we do not deny the right to ask questions and to seek the answers; we deny only the right to create hazards in the process. We question the style, methodology, and timing of the technology of gene implantation. This sort of restriction is not new. It is already commonly accepted that human experimentation which endangers the subject either physically or psychologically is abhorrent; remember the Tuskegee study (36). Gene implantation research is another example of a field of investigation in which there must be constraints because of the hazards to public health.

One of the most valuable lessons to be learned from the exercise of problem solving, and one of the most exciting things about the human mind, is that there is never only one way to solve an intellectual problem. Of course, there may be a most elegant way, an easiest way, or a best way, but never just one way. What the best way may be is always open to debate and must be judged by the circumstances in which the problem is found. Questions of safety and ethics are among the relevant circumstances. In the case of the problem of genetic controls, there are sure to be alternative techniques which may be more cumbersome but less dangerous and controversial than recombinant DNA. In fact, alternative approaches to all of these questions were being pursued before the potential of gene implantation was realized; there is no reason to ignore these possibilities now.

EVOLUTION

As yet, I have not discussed the ecological implications of the transfer of genetic material between widely disparate organisms (33). The creation of barriers to genetic contact between groups of organisms allowing them to diverge is central to the evolutionary process. Molecular biologists can now short-circuit these barriers in the laboratory. Containment cannot be absolute, particularly if research proceeds under the present Guidelines.[4]

[4] There is another interesting case of possible damage to the biosphere resulting from technological manipulations. Some years ago there was talk of melting chunks of the polar ice cap so as to bring cold, fresh water to Southern California. The idea was abandoned because it was predicted that the effects on global weather conditions might be disastrous. Just recently, the Saudi Arabians confirmed that they had commissioned a study on the feasibility of bringing icebergs from Antarctica to melt for irrigation and drinking water. Clearly, there should be some means for international discussion before such drastic actions are taken. It was puzzling that there was no mention of ecological dangers in the *New York Times* article (2).

It is impossible to predict what impact the escaped recombinant organisms might have on the biosphere because not enough is known about evolutionary biology or ecology. However, if the effect turned out to be significant, there is little doubt that it would be disastrous. We all accept that human experimentation must be restricted (23,24). Is the biosphere a more appropriate experimental system for unrestricted investigation (10 p. 237; 24,27,36)?[5]

Because of the possible impact of escaped recombinant organisms on the communities in the immediate vicinity of our research universities, we strongly recommend that this research be done only in a small number of laboratories in sparsely populated areas. Access to these laboratories should be available to qualified investigators from around the country, and the laboratories should provide maximal possible containment. Until we know more about the repercussions of this research, we would advise doing all work involving eukaryotic genes in a few such isolated maximal containment facilities.

HISTORY OF SCIENCE

Before closing I want to refer again to the history of the atom bomb. Here is an example of a scientific breakthrough which resulted in a product even more dangerous than its inventors expected it to be. It is well known that many scientists were reluctant to pursue this work and that, at the last minute, many urged the government to refrain from using it. They were too late (24; S. D. Warshaw, *personal communication*).

We think that there is a direct parallel between the story of this bomb and the problem of gene implantation. In both cases some scientists warned of the danger, although not as forcefully as they could have. In both cases there was pressure to continue the work. There are, however, some clear differences. The bomb was created during a war, and we are not under such pressure. Furthermore, there was only one atomic pile and one bomb, whereas research in recombinant DNA is taking place all over the world, thereby increasing the hazards to all of us. We think that the hazards of gene implantation may be equally great.

It is most important to realize that, although we can choose to stop using nuclear technology, should any of the organisms created by DNA recombination escape, they will propagate themselves. We will have no way to monitor them or to stop their proliferation.

In the history of science many great discoveries—that is, changes in the status quo—were greeted by opposition from traditionalists. Our specific

[5] For a discussion of risks to the public and problems of containment, see, for example, refs. 14, 31, 34, 35, 38, 41, and 42. [*R-1* (35) by Stickgold and Noble is a book that was conceived as science fiction and based on the recent history of the recombinant DNA issue. It has since turned out to be more fact than fiction.]

criticism should not be viewed as a support of a traditionalist–status quo criticism of science. We agree with many of the supporters of this technique in believing that it is qualitatively different from anything that has ever been investigated until now. We oppose not its newness *per se,* but rather the intangible and incalculable hazards inherent in the technique of gene implantation, and therefore the haste with which the technique is being pressed into service.

SUMMARY

I shall briefly summarize our point of view.

1. The decision whether and under what conditions this work may be continued should not be up to any one scientist, or even up to any group of scientists. Every citizen should have a vote.
2. The supposed benefits to be gained from the pursuit of work in gene implantation are not real benefits. This research will not solve the world's agricultural or medical problems.
3. If the projected hazards of this work become fact, these dangers will far outweigh the supposed benefits. The interpretation of data about hazards is dependent on the interpreter's interests and politics.
4. This is not a problem of freedom of inquiry but of protecting the public from a new hazard. We question the freedom to manufacture novel organisms.

If this work is to be pursued, there must be an active search for a host other than *E. coli,* preferably one that is not naturally promiscuous and whose habitat is limited.[6]

That there is at present no mechanism for including the public in a decision-making process is no excuse to proceed without public participation. It is high time that we work to create such a mechanism. We recommend that this research be delayed and that, during the delay, there be created a political institution for bringing public representation into the decision-making process.

CLOSING STATEMENT

The continuation and application of research in gene implantation are likely to have global consequences. Some of the applications may (although

---

[6] We do not accept *Bacillus subtilis* as an appropriate alternate host for recombinant DNA molecules. Although *B. subtilis* has no direct ecological contact with human beings, it is found in every hay field, and so has indirect ecological contact with every living thing in our immediate experience. Furthermore, *B. subtilis* is a spore-former, making wide dissemination of any possible recombinant genomes even more likely than in the case for *E. coli* (9,15,18,20).

not necessarily) be beneficial; should there be any accidents, they are likely to be irreversible and very damaging (10 p. 237).

We hope that more and more scientists will follow the example of R. L. Sinsheimer and E. Chargaff (8,32,40) who question the wisdom of continuing this research. Scientists have enormous power and must be responsible in handling it.

We urge the scientific community to take pause, to reflect. This technology will remain exciting; for now, we must approach it with great caution.

ACKNOWLEDGMENT

I wrote this paper with support and generous criticism from the Group on Genetics and Social Policy of the Boston Area Science for the People, of which I am a member. This chapter is meant to represent the view of this group. Three philosopher friends of mine, J. Long, W. Berkson, and S. Richmond, were good enough to remind me to keep in mind the History of Science. Special help came from R. Goldstein, C. Orrego, P. Ward, J. King, E. S. Allen, D. J. Kayman, and S. Kayman.

REFERENCES

1. Anonymous (1975): *About Health and Safety: Yours and Everyone's: A Questionnaire.* Ad hoc Biology Workers' Health and Safety Committee, Massachusetts Institute of Technology, Cambridge.
2. Anonymous (1976): Saudi Arabia commissions iceberg study. *New York Times,* CXXVI, November 3, 1976, p. 2.
3. Baltimore, D. (1976): Genetic research . . . safe or not? U.S. Guidelines give protection. *Boston Globe,* September 6, 1976, p. 41.
4. Barrett, F. F., Tabor, L. H., Morris, C. R., Stephenson, W. B., Clark, D. J., and Yow, M. D. (1972): A 12 year review of the antibiotic management of Haemophilus influenzae meningitis. *J. Pediatr.,* 81:370–377.
5. Beckwith, J. (1976): Social and political uses of genetics in the United States: past and present. *Ann. N.Y. Acad. Sci.,* 265:46–58.
6. Beckwith, J. et al. (Genetic Engineering Group of Science for the People) (Feb. 1973): Open letter to the Asilomar Conference on Hazards of Recombinant DNA.
7. Bergman, K. et al. [for the Genetics and Society Group (Boston) Science for the People] (1974): Proposals on Research Involving Gene Manipulation.
8. Chargaff, E. (Feb. 1976): Correspondence. In: *Recombinant DNA Research, Vol. 1* (Documents Relating to "NIH Guidelines for Research Involving Recombinant DNA Molecules," February 1975–June 1976), pp. 439–440. DHEW Publication No. (NIH) 76–1138. U.S. Government Printing Office, Washington, D.C.
9. Curtiss, R., III (Aug. 6, 1974): Memorandum to P. Berg, D. Baltimore, et al., on Potential Biohazards of Recombinant DNA Molecules.
10. DHEW Publication No. (NIH) 76–1138 (1976): Proceedings of a conference on NIH guidelines for research involving recombinant DNA molecules: Public hearings held at a meeting of the Advisory Committee to the Director, NIH. In: *Recombinant DNA Research, Vol. 1,* pp. 140–349. U.S. Government Printing Office, Washington, D.C.
11. DHEW Publication No. (NIH) 76–1138 (1976): Proposed guidelines for research involving recombinant DNA molecules. In: *Recombinant DNA Research, Vol. 1,* pp. 71–139. U.S. Government Printing Office, Washington, D.C.
12. DHEW Publication No. (NIH) 76–1138 (1976): *Recombinant DNA Research, Vol. 1* (Documents Relating to "NIH Guidelines for Research Involving Recombinant DNA

Molecules," February 1975–June 1976). U.S. Government Printing Office, Washington, D.C.
13. Duncan, M., Goldstein, R., Orrego, C., and Primakoff, P. (Nov. 24, 1975): Critique of the NIH Guidelines for Recombinant DNA Research by the Boston Area Recombinant DNA Group.
14. Echols, H. (March 9, 1976): Correspondence. In: *Recombinant DNA Research, Vol. 1* (Documents Relating to "NIH Guidelines for Research Involving Recombinant DNA Molecules," February 1975–June 1976), pp. 497–498. DHEW Publication No. (NIH) 76–1138. U.S. Government Printing Office, Washington, D.C.
15. Falkow, S. (undated): The Ecology of *Escherichia coli.* Memorandum to Members, Recombinant DNA Molecule Program Advisory Committee, NIH.
16. Goldman, R. A., Lewis, A. E., and Rose, L. I. (1976): Anaphylactoid reaction to single-component pork insulin. *J.A.M.A.*, 236(10): 1148–1149.
17. Goldstein, R. (Feb. 13, 1976): Correspondence. In: *Recombinant DNA Research, Vol. 1* (Documents Relating to "NIH Guidelines for Research Involving Recombinant DNA Molecules," February 1975–June 1976), p. 453. DHEW Publication No. (NIH) 76–1138. U.S. Government Printing Office, Washington, D.C.
18. Goldstein, R. et al. (Boston Area Recombinant DNA Group) (May 1976): Analysis and Critique of the Curtiss Report on the *Escherichia coli* Strain Intended for Biological Containment in DNA Implantation Research.
19. Gonzalez, N. (1976): Preventing cancer. *Family Health/Today's Health.* May 1976, pp. 30–33 and 70–74.
20. Hardy, K. G. (1975): Colicinogeny and related phenomena. *Bacteriol. Rev.*, 39(4):469–515, esp. p. 490.
21. Joslin Clinic and Diabetes Foundation, Boston (1976): Conversations with several clinicians about the use of bovine and porcine insulins.
22. van Larebeke, N., Engler, G., Holsters, M., van der Elsacker, S., Zaenen, I., Schilperoort, R. A., and Schell, J. (1974): Large plasmid in *Agrobacterium tumafaciens* essential for crown gall-inducing ability. *Nature,* 252:169–170.
23. Luria, S. E. (1965): Directed genetic change—perspectives from molecular biology, Chapter 1. In: *Control of Human Heredity and Evolution,* edited by Sonnebom. Macmillan, New York.
24. Luria, S. E. (1972): Slippery when wet. *P.A.P.S.*, 116(5):351–356; especially 355 and 356.
25. Montague, J. et al. (1976): (Abstract). In: *Abstracts of the 41st Cold Spring Harbor Symposium on DNA Insertions.* Cold Spring Harbor Symposium Lab.
26. Nightengale, E. O. (Feb. 27, 1976): Correspondence. In: *Recombinant DNA Research, Vol. 1* (Documents Relating to "NIH Guidelines for Research Involving Recombinant DNA Molecules," February 1975–June 1976), pp. 489–494. DHEW Publication No. (NIH) 76–1138. U.S. Government Printing Office, Washington, D.C.
27. Opinion Research Company (1974): Public attitudes toward science and technology. In: *Science Indicators.* The National Science Board, New York.
28. Schell, J. (1975): The role of plasmids in crown gall formation by *A. tumefaciens.* In: *Genetic Manipulations with Plant Material,* edited by L. Ledoux, pp. 163–181. Plenum Press, New York.
29. Schilperoort, R. A., and Bomhoff, G. H. (1975): Crown gall—a model for tumor research and genetic engineering. In: *Genetic Manipulations with Plant Material,* edited by L. Ledoux, pp. 141–162. Plenum Press, New York.
30. Schreier, H. A., and Berger, L. (1974): On medical imperialism. *Lancet,* 1:1161.
31. Sherris, J. C. (April 16, 1976): Correspondence. In: *Recombinant DNA Research, Vol. 1* (Documents Relating to "NIH Guidelines for Research Involving Recombinant DNA Molecules," February 1975–June 1976), p. 528. DHEW Publication No. (NIH) 76–1138. U.S. Government Printing Office, Washington, D.C.
32. Sinsheimer, R. L. (Feb. 5, 1976): Correspondence. In: *Recombinant DNA Research, Vol. 1* (Documents Relating to "NIH Guidelines for Research Involving Recombinant DNA Molecules," February 1975–June 1976), pp. 436–438. DHEW Publication No. (NIH) 76–1138. U.S. Government Printing Office, Washington, D.C.
33. Sinsheimer, R. L. (May 3, 1976): Correspondence. In: *Recombinant DNA Research, Vol. 1* (Documents Relating to "NIH Guidelines for Research Involving Recombinant

DNA Molecules," February 1975–June 1976), pp. 537–538. DHEW Publication No. (NIH) 76–1138. U.S. Government Printing Office, Washington, D.C.
34. Sinsheimer, R. L. (1976): Genetic research . . . safe or not? Extreme caution called for. *Boston Globe,* September 6, 1976, p. 41.
35. Stickgold, B., and Noble, M. (1976): *R-1, submitted for publication.*
36. Subcommittee on Health, Committee on Labor and Public Welfare, United States Senate (1973): *Quality of Health Care-Human Experimentation, 1973,* pp. 1108–1135, esp. pp. 1131–1134. U.S. Government Printing Office, Washington, D.C.
37. Train, R. E. (EPA Administrator) (April 1975): Address before the National Audubon Society. *Nature's Way,* 7(8):25. Syndicate Magazines, Inc.
38. Trumbull, R. (April 15, 1976): Correspondence. In: *Recombinant DNA Research, Vol. 1* (Documents Relating to "NIH Guidelines for Research Involving Recombinant DNA Molecules," February 1975–June 1976), p. 524. DHEW Publication No. (NIH) 76–1138. U.S. Government Printing Office, Washington, D.C.
39. Van, A. D., Bieth, J. G., and Bouanchand, M. D. H. (1975): Résistance plasmidique à la tétracycline chez Haemophilus influenzae. *C.R. Acad. Sci.* [*D*] (*Paris*), 280:1321–1323.
40. Wade, N. (1976): Recombinant DNA—a critic questions the right to free inquiry. *Science,* 194(4262):303–306.
41. Wedum, A. G. (Jan. 20, 1976): *The Detric Experience as a Guide to the Probable Efficacy of P4 Microbiological Containment Facilities for Studies on Microbial Recombinant DNA Molecules.* (Unpublished report based on research sponsored by the National Cancer Institute under Contract No. NO1-CO-25423, with Litton Bioretics, Inc.)
42. Weisenfeld, S. L. (Member of Western Recombinant DNA Committee) (March 19, 1976): Memorandum to D. Fredrickson, M.D., and Advisory Committee to the Director of NIH, entitled "Position on Recombinant DNA Research."
43. Wright, S. (March 24, 1976): Correspondence to D. Fredrickson and members of the Recombinant DNA Molecule Advisory Committee expressing "views on appropriate containment and monitoring of Recombinant DNA research."
44. Yue, D. K., and Turtle, J. R. (1975): Antigenicity of "monocomponent" pork insulin in diabetic subjects. *Diabetes,* 24:625–632.

Recombinant Molecules: Impact on Science and Society, edited by R. F. Beers, Jr. and E. G. Bassett. Raven Press, New York © 1977.

47. Discussion

Moderator: Kenneth Murray

Dr. S. D. Ehrlich: Dr. Teich, can you induce lysis by mixing DNA of one species with a cell culture of that species?

Dr. N. M. Teich: The transcription experiments have been done with pro-viral DNA synthesized *in vitro* in the same species if it is a susceptible host. As I mentioned, some of the animal cells we have been discussing have both the ecotropic viruses, those that can replicate in its host, and the xenotropic viruses; so it is a matter of overcoming the host strain problem. If you use an ecotropic virus with DNA of a particular virus, transfection may occur in that cell. If it is a xenotropic virus, the main problem seems to be at the penetration level. That experiment has not yet been done within one species.

Dr. Ehrlich: You can, in other words, repeat experiments that you reported, namely, extracting DNA from the cat, treating tissue cultures of human cells with it, and inducing human viruses.

Dr. N. M. Teich: I should say no one has done that with human cells that I know about, that is, looking for endogenous human virus, but there are indications that human cells do contain some material, DNA-related, to the C-type viruses.

Dr. B. Rosenberg: I would like to ask Dr. Jacobs to comment on the implications of his statement that the public should not be brought into the debate on problems of recombinant research and genetic engineering, and that decisions which should involve relatively objective and disinterested ethical considerations and will affect the entire population are best made by the few scientists directly involved in the research.

Dr. L. Jacobs: I do not remember saying that. Shall I read you my speech? Let me read the comments I made. I did abbreviate so there may not have been enough emphasis on it, but I think I had better make that point pretty clear.

I stated:

> Nobody can question the sincerity of many of the public interest groups that are seeking to preserve our environment, etc., and of their right of participation when the issues are transscientific. However, I would hope that they all behave in as responsible a manner as possible. In the case of research on DNA recombinant molecules, there is ample evidence that the expertise and the public responsibility reside in the scientists who raised the issues in the first place, and

who have labored hard to provide us with some guidelines on how work with recombinants of various types should be done. We have also asked for public participation in the work. We cannot hope that public representatives will be able to understand all of the science and make their own assessments of risk, but they can make sure that the scientists are engaging in the process of making rules in a fair way.

In another place I stated:

I guess I am revealing the discomfiture that Dr. Weinberg mentions, which scientists feel when they must engage in public debate of issues. I am in complete agreement with him that the public has a right to participate in the debate when questions arise that cannot be settled by scientific means. As he points out, we have a responsibility to define, in such debate, how much we know and what we cannot know, where science ends and trans-science begins. I am not worried about that debate; I am worried about the tactics which may confound honest debate, and I am stating my hope that they will not be used regarding DNA recombinant research.

Those are statements that I made.

Now, I have some trouble in defining trans-science in the particular, in relation to the particular problems we are discussing. I do not know if you remember Alvin Weinberg's definition of what are trans-scientific issues, but one is the impracticable experiment, say the experiment to define the lowest possible threshold of safety for radiation. The second is the situation in which the subject matter is so variable one cannot hope to obtain a definitive answer that cannot be challenged. The first relates to the esthetics of science, what science should be done, and how much support it should receive versus another area of science.

I do not know that DNA recombinant research, which is at this stage only talking about theoretical risks, is trans-scientific or is simply a whole series of scenarios that scientists can dream up and talk about with others and with the public; but there is really no good way that any of us have of measuring.

Maybe that is the definition of trans-science, and the public certainly has a right to participate, but I would hope that it can be done the way we have been doing it. I have few complaints about how guidelines have been developed up to now. One of the things I did not particularly like about Miss Warshaw's chapter, even though I complimented her on her excellent delivery, was the statement that there was no public participation in the Frederickson meetings, and so on.

Dr. S. Cohen: I have two questions to ask Professor Richmond. First of all, in order to assess the relevance of your data to some of our biological concerns, it is important for us to know more about the kinds of plasmids that were used in your experiments. As you know, there are both conjugative and nonconjugative plasmids. Of the conjugative, self-transmissable plasmids, the derepressed are transmitted at a much higher frequency than repressed plasmids. I do not know of anyone who is carrying out recombinant DNA experiments with conjugative plasmids, let alone with derepressed conjugative plasmids. My first question is: could you tell us something about the kinds of plasmids that were used in the transmission experiments, and what you see as the relevance of your data to recombinant DNA experiments carried out with nonconjugative plasmids?

Dr. M. H. Richmond: The feeding experiments were done with normal, what you would call repressed, conjugative plasmids. The plasmid which emerged in the study on the person who was not receiving antibiotics came out as a conjugative plasmid with a frequency of transfer of *E. coli* K-12 of about 10^{-5}.

So I accept the point I think is behind your question, namely, that the studies there reflect the behavior of conjugative plasmids in the human gut. As you point out or imply, a lot of the work which has been done by you and others involves nonconjugative plasmids, that is, plasmids that can not transmit themselves between bacterial species.

You say no work with conjugative plasmids is going on. At a meeting I attended a week ago in Europe, an American described experiments in which he had cloned bacterial DNA in conjugative plasmids. So it is going on and I think the experiments are relevant. That is one point.

I think I agree with you that if one is working with a nonconjugative system, the probabilities are reduced significantly.

Dr. S. Cohen: The second question: your study with tetracycline showed a large overgrowth of naturally occurring tetracycline-resistant bacteria in the gut, as many previous studies have shown when patients or animals are receiving antibiotics. I wonder if you or others have done studies to determine the relative ability of a cell carrying a recombinant plasmid to propagate itself in the gut under such circumstances. If a large number of naturally occurring tetracycline-resistant bacteria are already present in the gut, it would seem that any bacteria carrying tetracycline resistance genes on recombinant molecules would have to compete with the resident tetracycline-resistant microorganisms for survival. Do you have any data on whether a tetracycline-resistant recombinant molecule does or does not take over the gut from the resident resistant organisms in this competitive situation?

Dr. M. H. Richmond: You mean an artificial recombinant molecule?

Dr. S. Cohen: Yes.

Dr. M. H. Richmond: I think the information one has about the incidence of tetracycline-resistant organisms in people who are not receiving antibiotics is that tetracycline-resistant coliforms will be found in about the first million organisms examined in about half the people tested. In other words, practically everyone is carrying a few resistant organisms all the time.

When people are given tetracycline, often the result is an almost pure culture of one particular tetracycline-resistant organism. One never knows if it is the one that was present in low levels before the feeding. Occasionally two strains are obtained, sometimes three. The most we have ever seen is seven distinct "O" types with tetracycline-resistant plasmids of various kinds, which together make up the total coli flora. So I think that tetracycline treatment does select these things.

Now, what I think is going to decide which comes out on top is not the resistance to tetracycline but the colonization problems of the *E. coli*. If the χ1776 strain, the one carrying the recombinant element, is used, of course there is the health situation.

But my argument that antibiotics should not be fed to people who are working in the lab with resistant cloning vehicles, I think, is a general one. If antibiotic therapy is used for people working in that environment, it is probable that the chance is going to be increased. It is as simple as that.

Miss T. McLellan: I have two questions for Leon Jacobs. My profession is as an ecologist. I am concerned about what you have said about theoretical risks to the environment. The human species has a rather poor record in trying to manipulate its environment and in introducing species into environments where they are not native. What were often innocent intentions have backfired and been disastrous. The prospect of having an *E. coli* which carries enough genes from a fruit fly to produce insect hormones is something that goes beyond what has been discussed in the possibility of recombinant DNA influencing human diseases. To what extent is the NIH studying the environmental implications of recombinant DNA research? There is a law in this country about environmental impact statements that must be filed when anything is planned and which might possibly harm the environment.

My other question is about the decision-making process leading to the NIH guidelines: to what extent have you sought information from environmental scientists, and what are the implications of this research and the guidelines?

Dr. Jacobs: I will try to answer your question briefly. We have tried to engage some environmental scientists and ecologists in discussions on the guidelines. I do not have the roster of the committee with me now. But we did appoint *ad hoc* consultants. Dr. Peter Day, an agricultural scientist, joined the committee. Perhaps some of the other members of the committee can identify others who have expertise in this respect. I do not remember any now but we have been engaged with the Department of Agriculture and the Department of the Interior and asked them to identify people who could help us with respect to the development of guidelines which could relate to the environment.

All of us are familiar with various situations in which insects or mammals have been introduced inadvertently or purposefully into various new regions with some disastrous effects. I think we are all aware of that and we will continue to work on that in relation to what we might expect to find happening because of manipulation of DNA molecules and their cloning in *E. coli*.

I do not think we are unsympathetic to that approach at all. I am getting the feeling that people think I am completely unconscious or disdainful of DNA, of any dangers involved in this. I am not at all unconscious or disdainful of dangers. I have simply been trying to emphasize that I want the debate and the revelation of these individual concerns to be open and honest.

Dr. M. Pollard: I would like to direct a comment to Dr. Teich. Her chapter indicated that some viral vectors are important in DNA recombination. There are, however, other vectors in the laboratory which I am sure she is aware of, the most important one being the HeLa cell. How much of a hazard does the HeLa cell actually represent when it contaminates almost every tissue culture that we have in the laboratory and confuses many of the experiments in which we are involved?

Dr. N. M. Teich: I am really not too worried about the confusion that comes out because someone will eventually disprove the particular paper. One should recognize, however, that HeLa cells in Europe contain the Mason-Pfizer virus, a classic type B virus that has been associated with mammary carcinomas in primates. It seems that that happened somewhere in the mid-Atlantic because the U.S. HeLa cells do not seem to have this virus.

We have a big problem in what to do about these viruses that are ubiquitous. Can we cut out all mammalian cell tissue culture? In my laboratory, we have safety guidelines with a restriction in use of any of the viruses which have xenotropic properties for types of human cells; we contain these in what we call "P2½." I have tried to point out that in nonvirology labs ordinary procedures, such as cell fusions, are efficient in inducing these viruses. Unfortunately, nonvirologists do not think about these hazards. So perhaps we should extend our guidelines not only to this recombinant DNA as a genetic engineering problem, but to some of these other more common problems.

I do not work with HeLa cells because I think it is not worth the effort, for one thing, to try to contain the problem with the type B virus that is there. It is so much more difficult, if you are trying to study enzymes or synthetic pathways, to distinguish between what viral controls are influencing the host mechanism. I think there is a danger with working with HeLa cells. I see no reason for using them. There are other human cells which seem a lot cleaner.

Dr. M. Pollard: As a facetious remark, we who are involved with it are calling it the Gila monster.

Dr. T. Powledge: I am a little concerned that all our wrangling over the adequacy or inadequacy of the NIH guidelines is distracting us from the fact that a lot of this work will not be done under NIH jurisdiction and, therefore, is not subject to the guidelines except possibly voluntarily. Do any of the speakers or does anybody in the audience have any practical suggestions for what we should do about this situation?

Dr. K. Murray: The concern is about people doing experiments on the side, as it were, and not being affected by guidelines, and not being subject to control of guidelines. Does anyone have any comment on this? What sort of steps might one take to prevent this type of moonlighting?

Dr. J. King: In the first Science for the People position paper on this topic, we were worried about this same thing. It was not addressed at Asilomar. We proposed that these experiments be regulated from below, not from above, and that any place they were going on had to have a safety committee composed of people representing all the people who physically worked in the building, whatever work categories they were, and their okay had to be given independent of national policy on it. So if a group of people in South Carolina were going to do these experiments, they would have to have a local committee. That local committee would have to give its okay independent of any questions of NIH guidelines, making it a community health problem that has to get community okay.

Dr. M. H. Richmond: Just to comment on that one, it seems to me that one can suggest things you could not do. In my personal view, the thing not to do is legislate because I do think it really does no good at all. By "legislate" I mean really try and have enforceable laws because if people are going to break codes of practice or codes of honor in this area, they will break the laws, it seems to me.

So I tend to agree with the previous speaker that one wants a consensus view of the laboratory. One hopes to build maturity of view in the laboratories of the country. Peer group pressure, I am sure, is an important element of this.

Dr. K. Murray: I wonder if I might make one comment about the situation in Britain. We have something called the Health and Safety at Work Act—you have the equivalent here but I forget its name—which is generally believed to have enough teeth in it to require no further legislation.

The best safeguards against the situation that has just been raised is through just the sort of thing that Dr. King mentioned, local safety committees. I think that is a fair point.

Mr. B. M. Kacinski: Dr. Teich, you mentioned that xenotropic viruses evidently have a lysogenic state in some animals, following which they can infect others and produce infection. Is there any indication that they themselves transfer genes from one species to another? And if they do, what would be your estimate of the frequency of this?

Dr. N. M. Teich: We do not know that the xenotropic viruses cause any neoplastic disease. They do replicate in other cells. There is evidence that when they grow in heterologous cells and are brought back to their own host cell, they can pick up approximately 5 to 10% of cellular DNA, presumably from the site of incubation. In particular, it has been shown that with something like a Friend virus of mouse, which causes the erythroblastic leukemia I was talking about, a good proportion, perhaps 10^{-3}, of the particles carry the globin messenger that the cells are producing. We know they can either carry genes as noncovalently linked parts just gobbled up into the virus particle or bring in covalently linked genetic material from the cells in which they have been grown. This is reversible. If you put a chicken virus through duck cells, it picks up duck material. If you put it back in chicken cells it starts losing the duck material and takes up chicken.

Dr. K. Murray: Will Dr. Van Montagu speak next? One of his particular pieces of work was mentioned by Miss Warshaw, and I am sure he would like to comment on it.

Dr. M. C. Van Montagu: I want to stress that we did not take the oncogenic plasmid from *Agrobacterium* and transform it in *E. coli,* a thing that we could have done several years ago when we found the plasmid. We deliberately postponed this until the situation was more clear how this type of work would evolve. We have just seen that on transferring between *Agrobacterium* strains a recombinant is formed between a P-type plasmid and the oncogenic plasmid. We realized that this recombinant, which we had analyzed afterward, can now transfer with high efficiency to any Gram-negative bacterium. Indeed, we are checking the *E. coli* strains we have used in the lab but which have not been in contact with that *Agrobacterium* strain; we think there are some affected *E. coli* strains in the lab, but we surely did not deliberately permit it to occur.

I want to stress the way this plasmid was formed. We could call it a condition of *in vivo* genetic engineering. There are some arguments to the effect that this situation probably occurs readily in nature.

Dr. K. Murray: Thank you for clarifying that point.

Miss F. Warshaw: I am sorry, Dr. Van Montagu, if I misrepresented your observations. I was not at the Cold Spring Harbor meeting. I heard from various reports that that is what you had done. I am glad to hear that you did not do it.

On the other hand, if you do find that the plasmid is in *E. coli* strains in your laboratory, that would be interesting information for us to have. The fact that it can move around so easily, if it does, indicates that we need much more containment so that it stays inside the laboratory. Of course, you may object that that happens in nature. But the strains you are working with in the laboratory are different than the strains you work with in nature. So you really do not know what the effects are going to be.

Dr. F. R. Blattner: Dr. Teich, do you feel it is dangerous to eat raw meat of cattle?

Dr. Teich: That is an unfair question. I will say this: I prefer mine cooked. For the vegetarians or nonmeat eaters among you, avian cells contain these viruses. You will find that the standard eggs you buy in the supermarket contain these viruses. We do not know that they are oncogenic to humans.

Dr. F. R. Blattner: This is sort of what I was driving at. These things are ubiquitous. The experiment probably has been done.

Dr. N. M. Teich: I think you are wrong about that. You are talking about ingestion which I do not consider to be the sole criterion.

Dr. F. R. Blattner: How about butchering?

Dr. N. M. Teich: I do not know. Perhaps there is a higher incidence of leukemia among butchers. I have no data about the epidemiology of it. I think it is something that should be looked into. We cannot just consider that ingestion is the only means of introducing these things.

The other thing we do know from animal model studies is that generally the neonate and the immunosuppressed person, rather than the adult, normal human being, are at risk. It depends on a particular immune response system and on particular histocompatibility loci. These factors are now emerging as important ones in viral leukemogenesis, particularly with regard to the Epstein-Barr virus and Burkitt's lymphoma in nasopharyngeal carcinoma. I do not want to put any of these things down. It is certainly a multifactorial disease.

Dr. S. Lederberg: This type of question ignores the fact that few people in the audience disbelieve that heated DNA can renature.

Dr. N. M. Teich: May I make one more point on that? We know that the pathogenesis of many viruses changes. The one important thing I would like to bring out, since you are all talking about replicating units, is that we know that if some viruses are inactivated or made defective so that they cannot replicate, they are even more pathogenic. A case in point is herpes viruses, which will transform cells after they have been UV-irradiated. Additionally, the vesicular stomatitis virus I mentioned earlier will, in its nonreplicating form, cause a slow paralytic disease.

Dr. J. Beckwith: Dr. Jacobs, I also sensed from your chapter, even on the rereading of the passages, a downplaying of public involvement. And that raises in my mind many questions about the whole process that has been going on with NIH committees if this represents the attitude of NIH. In particular, from my conversations and from what I have heard from members of the so-called Public Advisory Committee, there was a good deal of feeling on that committee, which was communicated to the NIH people and Dr. Frederickson, that much stronger restriction should be put on the research. A large number of people in that group felt an organism other than *E. coli* should be used.

It is also my understanding that these suggestions by this committee essentially will in no way be incorporated into the final guidelines. I think that indicates the limits that the NIH seems to be putting on public participation.

Dr. L. Jacobs: First of all, Dr. Beckwith, I speak for myself. I am not here as a representative of NIH. I speak because I was asked to describe the difference between rules and guidelines, regulations and guidelines. Beyond that, I do not know how many times I have to reiterate that I am perfectly happy to have participated in the debate. I have told you that when I speak about the public, I am not talking about the scientific public because the scientific public has to engage in it all the time.

I am talking about public participation by people who are not trained as scientists. We have welcomed them into the meetings. I perceive their ability to contribute as that of policing to make sure that all of what goes on and is said is fair and that the debate is fair. I do not know what else they can do.

So far as whether the guidelines incorporate new conditions which now satisfy the arguments of the Boston DNA Recombinant Group, who spoke at the February meeting that Dr. Frederickson called, I think you will have to wait until the week of June 21 to find out what has been decided there. I am not privileged at this time to reveal that information.

Dr. G. Wald: I want to mention something that seems a little disturbing and that perhaps was covered insufficiently, if at all, in this meeting. Dr. Frederickson and the guidelines are under heavy pressure. A couple of weeks ago 20 representatives of the pharmaceutical and chemical industry met with him. They made three points: first, that the guidelines would be very hard on industry and should by no means be turned into regulations; second, that they cannot reveal what they are doing or what they intend to do because that would spoil their business; and third, that the guidelines are unnecessarily restrictive. For example, they want to begin large-volume research which is also part of the business.

Mr. P. Youderian: I would like to address a question to Dr. Jacobs. I would like to argue that the people making the decisions of what is safe are not the best people to decide. You will find on the tapes that Dr. Boyer stated that the antibiotic used in his lab was controlled under P1 and P2 conditions. He did not elaborate on that. If the P1 and P2 conditions are like those in everybody else's laboratory, as he said, those conditions are inadequate.

Dr. Curtiss argued that since there were some limited situations under which his strain might survive, we need not go on and test other environmental situations in which the strain might survive, such as sewers. Excuse me, Dr. Curtiss, this may be a slight misrepresentation.

The point is that the attitudes of the committees that have reviewed these programs so far have been almost totally ignorant of ecological considerations. Not only that, there is a conflict of interest among the members of the committee since many of them are involved in the same kind of research they are proposing guidelines for.

Estimates of frequencies of escape are nonsense. It took only one Japanese beetle—actually two Japanese beetles—to initiate the massive number of these insects in America. It took one snail, one hermaphroditic snail, to bring in the snail blight in Florida.

The public confoundment of debate on these issues is a misrepresentation of what the public should be doing and can be doing. It ignores the responsibility that we have to educate the public on these matters which concern their health so that they can have a directed, informed input into these decision-making processes.

Dr. Szybalski stated yesterday that mutants in gene *Q* and *S* could be recovered from a lambda vector in normal *E. coli* strains, with frequency of 10^{-10}. That is not correct. The frequency is 10^{-4} to 10^{-7}. If you read Ira Herschkovitz' thesis, these numbers appear there.

I conclude with the following question: why are the NIH guidelines, which seem to be in a final form, which are going to be printed on June 23 and are supposedly open to the public, not revealed to anyone right now?

Dr. L. Jacobs: It is merely a matter of logistics. We have to print them. They are in production right now. They will be out of the printing office some time during the third week in June, and then we will distribute them as rapidly as possible. The first printing will be approximately 2,000 copies. It will not satisfy everybody but we can not print 100,000 very fast. So we are printing 2,000 very fast.

Dr. K. Murray: Although I am sure many of you would like to continue, we are already 5 min beyond our extended period. I am terribly sorry but I think we should end the discussion here or continue informally, if you wish.

Appendix

Because the proposed NIH Guidelines were such an active issue of debate during the Symposium on which this volume is based, we have chosen to include herein their table of contents and introductory text. In addition, we have appended similar matter of the Draft Environmental Impact Statement that was issued subsequently. Copies of both documents in their entirety may be obtained from Dr. Rudolf G. Wanner, Bldg. 12A–4051, NIH, Bethesda, Maryland 20014.

The Editors

Guidelines for Research Involving Recombinant DNA Molecules

June, 1976

Appendices

INTRODUCTION

The purpose of these guidelines is to recommend safeguards for research on recombinant DNA molecules to the National Institutes of Health and to other institutions that support such research. In this context we define recombinant DNAs as molecules that consist of different segments of DNA which have been joined together in cell-free systems, and which have the capacity to infect and replicate in some host cell, either autonomously or as an integrated part of the host's genome.

This is the first attempt to provide a detailed set of guidelines for use by study sections as well as practicing scientists for evaluating research on recombinant DNA molecules. We cannot hope to anticipate all possible lines of imaginative research that are possible with this powerful new methodology. Nevertheless, a considerable volume of written and verbal contributions from scientists in a variety of disciplines has been received. In many instances the views presented to us were contradictory. At present, the hazards may be guessed at, speculated about, or voted upon, but they cannot be known absolutely in the absence of firm experimental data—and, unfortunately, the needed data were, more often than not, unavailable. Our problem then has been to construct guidelines that allow the promise of the methodology to be realized while advocating the considerable caution that is demanded by what we and others view as potential hazards.

In designing these guidelines we have adopted the following principles, which are consistent with the general conclusions that were formulated at the International Conference on Recombinant DNA Molecules held at Asilomar Conference Center, Pacific Grove, California, in February 1975: (*i*) There are certain experiments for which the assessed potential hazard is so serious that they are not to be attempted at the present time. (*ii*) The remainder can be undertaken at the present time provided that the experiment is justifiable on the basis that new knowledge or benefits to humankind will accrue that cannot readily be obtained by use of conventional methodology and that appropriate safeguards are incorporated into the design and execution of the experiment. In addition to an insistence on the practice of good microbiological techniques, these safeguards consist of providing both physical and biological barriers to the dissemination of the potentially hazardous agents. (*iii*) The level of containment provided by these barriers is to match the estimated potential hazard for each of the different classes of recombinants. For projects in a given class, this level is to be highest at initiation and modified subsequently only if there is a substantiated change in the assessed risk or in the applied methodology. (*iv*) The guidelines will be subjected to periodic review (at least annually) and modified to reflect improvements in our knowledge of the potential biohazards and of the available safeguards.

In constructing these guidelines it has been necessary to define boundary conditions for the different levels of physical and biological containment and for the classes of experiments to which they apply. We recognize that these definitions do not take into account existing and anticipated special procedures and information that will allow particular experiments to be carried out under different conditions than indicated here without sacrifice of safety. Indeed, we urge that individual investigators devise simple and more effective containment procedures and that study sections give consideration to such procedures which may allow change in the containment levels recommended here.

It is recommended that all publications dealing with recombinant DNA work include a description of the physical and biological containment procedures practices, to aid and forewarn others who might consider repeating the work.

Draft Environmental Impact Statement
(*August, 1976*)

Guidelines for Research Involving Recombinant DNA Molecules

Appendices

FOREWORD

Recent developments in molecular genetics, particularly in the last 4 years, open avenues to science that were previously inaccessible. In the "recombinant DNA" experiments considered here, genes—deoxyribonucleic acid (DNA) molecules—from virtually any living organism can be transferred to cells of certain completely unrelated organisms. For example, the genes from one species of bacteria have been transferred to bacteria of another species. And genes from toads and from fruit flies have been introduced into the bacterium *Escherichia coli.*

If the recipient bacterium is then allowed to multiply, it will propagate these newly acquired genes as part of its own genetic complement. It appears likely that any kind of gene from any kind of organism could be introduced into *E. coli* and certain other organisms.

This ability to join together genetic material from two different sources and to propagate these hybrid elements in bacterial and animal cells has resulted in a profound and qualitative change in the field of genetics. Now, for the first time, there is a methodology for crossing very large evolutionary boundaries, and for moving genes between organisms that are believed to have previously had little genetic contact.

The promise of recombinant DNA research for better understanding and improved treatment of human disease is great. There is also a possible risk that microorganisms with foreign genes might cause disease or alter the environment should they escape from the laboratory and infect human beings, animals, or plants. However, in the absence of further experimental data neither the benefits nor the risks can be precisely identified or assessed.

On June 23, 1976, the Director of the National Institutes of Health released Guidelines governing the conduct of NIH-supported research on recombinant DNA molecules. Promulgation of these Guidelines followed 2 years of intensive discussion and debate within the scientific community and NIH itself, with public participation, concerning the possible hazards of such research and the best means for averting them, although the possible hazards remain speculative. The Guidelines prohibit certain kinds of recombinant DNA experiments and, for those experiments that are permitted, they specify safety precautions and conditions designed to protect the health of laboratory workers, the general public, and the environment should the putative hazards prove real.

The issuance of Guidelines establishing conditions and precautions with respect to such experiments is viewed by NIH as a Federal action that may significantly affect the quality of the human environment, and NIH Director Dr. Donald S. Frederickson ordered the preparation of this statement pursuant to the National Environmental Policy Act.

Although NEPA assumed that such Federal actions will not be taken until the NEPA procedures are completed, the Director of NIH concluded that the public interest required immediate issuance of the Guidelines, rather than deferral for the months that would be required for completion of the NEPA process. This was because the escape of potentially hazardous organisms was more likely in the absence of NIH action. Further, prompt issuance of the Guidelines was believed necessary in order to promote their acceptance by scientists in the United States and abroad who do not come under the purview of NIH.

Issuance of and compliance with the Guidelines is, in itself, expected to decrease the chance of any detrimental environmental impact. However, since there has been little actual experience to date with recombinant DNA experiments, the indicated confidence in the Guidelines rests essentially upon the judgment of scientists. Their confidence is based on two premises. First, it is believed that the containment measures specified in the Guidelines make the escape of potentially harmful recombinant organisms into the environment highly improbable. Second, it is believed that, even if an experiment performed in accordance with the Guidelines does result in accidental release of recombinant organisms, adverse effects will either not occur or not be serious.

In the absence of an adequate base of data derived from either experiments or experience, it must be recognized that future events may not conform to these judgements. There is some statistical probability that recombinant organisms will find their way into the environment either from experiments under NIH auspices or from the activities of others. It is not difficult to construct scenarios in which injury could result. Although the possibility of significant environmental consequences is entirely speculative, the chance of an event that could cause severe injury, however low the probability, must be treated as an environmental impact.

The NIH Guidelines, in addition to ensuring the safety of NIH-supported researchers, the general public and the environment, are serving as a model for other laboratories throughout the world, thereby promoting environmental protection beyond that achievable through other actions available to the Federal Government. And the experiments themselves may be expected ultimately to lead to an increase of knowledge and the advancement of medicine and other sciences.

Although the action in question—that is, issuance of the Guidelines—has already been taken, the Director of NIH believes that the NEPA review will further enlighten the public and focus attention on the important issues involved, in the interest of gaining the understanding and views of the broadest possible segment of the American people. In issuing the Guidelines, the NIH Director pointed out that they will be subject to continuous review and modification in the light of changing circumstances. Constructive modification could result from information received during the NEPA process.

Epilogue

In the interval since the symposium, the concern and interest of the scientists and public grew to a crescendo of activity, culminating in the *Research With Recombinant DNA Forum* of the National Academy of Sciences on March 7 to 9, 1977.

Concurrent with this vigorous display of both protest and encouragement of recombinant DNA research has been the steady progress in formulating a national policy implemented with regulations about to be formalized into law by Congress. The fears generated by the scenarios ranging from the Andromeda strain to human genetic engineering have yet to be justified by the authenticity of these scenarios. Most important, perhaps, is the growing realization that even some of the hoped for benefits of recombinant DNA research may never be realized as the complexity of the practical application of recombinant DNA technology becomes more apparent with each new study.

Thus, both the proponents and opponents must temper their predictions somewhat as the knowledge about the possibilities of recombinant DNA research and technology becomes more complete and sobering in its implications for man and his environment.

Subject Index